# Basics of ECONOMIC GEOLOGY and MINING PRACTICES

***A Course Book for Universities and Polytechnics***

**(With Emphasis on Africa)**

**Moses. A. Olade**

Prof. Moses A. Olade, Ph.D., MBA
2301 Harbor Chase Dr.
Pearland, Texas 77584
United States of America
mosesolade48@gmail.com

ISBN 978-1-63877-849-3
Prescott Resource Publishers
https://www.prescottbooks.com

Library of Congress Control Number: 2021907039

## *Dedication*

*This book is especially dedicated to* ***the "Father of Nigerian Geology"***

***Professor M.O. OYAWOYE (1928- )***

*He was the first Nigerian to attain a Ph.D. in Geology and the first Nigerian Professor of Geology and Head of Department at the premier Nigerian University of Ibadan. He actively supported my aspirations to study Economic Geology and he became a mentor to my successful academic career at the University of Ibadan.*

# Preface

Mineral resources are Nature's endowment to humanity; they are to be discovered, extracted and used for our economic benefit. However, economic minerals are very 'scarce' and non-renewable resources that require the training and expertise of geoscientists to find, explore and exploit. The insatiable demand for mineral raw materials to meet the needs of a growing world population, and rising standard of living has created the urgent need to train more economic geologists who understand the complex processes responsible for the occurrence, distribution and origin of mineral deposits, and know how to apply the acquired knowledge in the search for new mineral deposits. In the teaching of geology and mining courses at higher institutions of learning in developing countries of Africa, it is noticeable that students do not have adequate access to textbooks that can broaden their knowledge due to the fact that textbooks are scarce, and even if available are too expensive to be affordable for students from mostly low-income families.

This book represents an invaluable and systematic review of the whole subject of economic geology integrated in a continuum with mineral exploration and mining geology. It covers the major ore deposit types including metallic minerals and industrial minerals and rocks including gemstones, with focus on examples from Africa - a continent with the world's largest mineral industry. Non-metallic deposits of industrial minerals and construction materials are fully covered to reflect their growing importance in the mineral industry of most developing countries.

I have written this book in a simple and concise language that is easily understood, and organized it into two parts with eight chapters: Part One is focused on the Principles of Economic Geology with five chapters: Chapter 1: Introduction to Economic Geology; Chapter 2: Nature of Mineral Deposits; Chapter 3: Ore Forming Processes and Associated Deposits; Chapter 4: Economic Geology of Metals; and Chapter 5: Industrial Minerals and Rocks, Part Two of the book is dedicated to Applied Economic Geology with three chapters: Chapter 6: Mineral Exploration; Chapter 7: Elements of Mining Geology; and Chapter 8: Environmental Geology and Mining Operations. A Glossary of Common Terms in Economic and Mining Geology is included in the book.

The book is intended for Geology and Mining students in Universities and Polytechnics, and also for practicing geoscientists and mining professionals who want to brush up their knowledge in economic geology. The format of the book is topical and well organized for teaching courses on Mineral Deposits and Mining Geology at undergraduate and postgraduate levels. It is an offshoot of my lecture notes used over the years to teach at Universities and Community Colleges, augmented by extensive field experience and literature research. The work was finally put together during the 2020-21 Coronavirus Pandemic when most of the world was under lockdown or similar restriction which provided me with the singular opportunity of devoting long hours of minimum distraction. This book has benefited from information produced by innumerable sources including the research work of many of my colleagues at Nigerian Universities and elsewhere.

Dr. Moses A. Olade
Pearland, Texas, USA
June, 2021

# Acknowledgement

In writing this book, I am very much indebted to my mentors and colleagues from whom I have benefited during my over forty years of academic and professional career. I am thankful to my mentors who inspired my interest in Economic Geology and supported my struggles to attain an excellent education in this discipline, particularly Prof. M.O. Oyawoye and the late K.C. (Kevin) Burke, (University of Ibadan); the late Prof. R.D. (Roger David) Morton (University of Alberta), and Profs. Emeriti W.K (William Kenneth) Fletcher and A.J. (Alastair) Sinclair, (University of British Columbia). .I appreciate the support of my former students at the University of Ibadan, particularly, the late Prof. A.A. (Azubuike) Elueze, Prof. A.O. (Akin) Oyinloye, (University of Ado-Ekiti), Prof. S.J. (Solomon) Ekwere (University of Calabar), Prof. M.I. (Minipuye) Odigi (University of Port Harcourt) and Dr. J. A Ogundipe (Afe Babalola University) who have carried on the "torch" of Economic Geology and spread the "gospel" to the current generation of University students in Nigeria and abroad. Many thanks to my former colleagues especially, (now retired) Prof. J.A. Adekoya, Prof. T.A. Badejoko and Prof T.R. Ajayi. I appreciate Messrs. A.O Adegbuyi, J.A. Dada and S.K. Ariyo (Achievers University) for sharing their useful geological field experiences.

Lastly, I want to extend my appreciation to my wife Mrs. Esther Seymour-Olade, and my adult children, particularly, Dr. Roger Babatunde, Mr. Lincoln Olalekan, Mr. Randall Ademola, Dr. (Mrs.) Tolu Olubori and Ms. Sheena Jardine for their support and encouragement while writing this book.

# Table Of Contents

# Chapter 1
# Introduction

## 1.1 What is Economic Geology?

**Economic geology** is the scientific study of *mineral resources* (or mineral deposits) and the application of the acquired knowledge in the exploration and mining of minerals. Economic geology is considered as a major branch of *Geosciences,* focused on the understanding of the distribution, characteristics and origin of mineral deposits, and the use of such knowledge in the discovery and evaluation of new mineral occurrences, as well as the continued investigation and profitable extraction of metallic ores at optimum grade. of recovery. *Mineral resources* is a general term used to describe the concentrations of valuable earth materials (minerals and rocks) that are of economic and/or industrial value, whereas *mineral deposits* are substantial mineral accumulations .that can be potentially extracted.

Mineral resources are of three types: *metallic ores* (e.g. gold, copper), *non-metallic minerals* and *rocks* (e.g. barite, limestone) and energy or mineral fuels *(*e.g. coal, petroleum). In economic geology, the mineral resources of academic interest are mostly restricted to the solid earth materials which form mineral deposits *(solid mineral resources),* and tend to exclude the *non-solid* fuel resources, such as oil and gas (petroleum) which are currently studied in the academic discipline of *petroleum geology.* Among all the known *natural resources,* minerals are very unique because they are of instant value and are non-renewable (exist in finite amounts). Consequently, as economic minerals are exploited, new deposits must be discovered to replace them. Economic geologists and associated geoscientists (geophysicists, geochemists and mining professionals) have an important roles to play in ensuring that new mineral deposits are discovered and extracted.

Mineral resources are virtually indispensable to human existence and their exploitation and utilization have become an important mainstay of the world economy. Since time immemorial, mineral resources have been exploited and used as raw materials for industrial production and to generate revenue for investors and national economies through payments of royalties and taxes, and for foreign earnings through export to other countries. However, the quantities of mineral resources are limited, and as the world's population grows exponentially and living standards increase globally, the demand for mineral raw materials continues to escalate widely. Nonetheless, economic geologists and other geoscientists continue to successfully discover new deposits and expand known ore reserves. Africa which has the world's largest mineral ndustries is a major supplier of metals and gemstones to world trade and economy.

As an academic discipline, the focus of economic geology is the study of various aspects of "*mineral deposits*". The terms '*mineral deposit*' and '*ore deposit*' are sometimes used interchangeably but they are not exactly the same. Although all ore deposits are mineral deposits, the reverse is not true because - not all mineral deposits are ore deposits. For a mineral deposit to become an ore deposit, it must satisfy not only the geological requirements. such as sufficient tonnage and ore grade, but also satisfy economic factors that ensure its viability for exploitation, such as metal prices, market conditions, transportation costs etc. Therefore, the difference between the two terms; mineral deposit and ore deposit does have economic connotation. The term "mineral deposit" can be considered as a *geologic* term while "ore deposit" is regarded as an *economic* term. In the study of mineral resources, economic geologists most often use the term *mineral deposit* without any economic connotation, whereas mining geologists will preferably use the term ore *deposit* because all minerals that are mined belong to that category. In this book, the study of mineral deposits will be undertaken without detailed emphasis on their economic viability but with more focus on their distribution and geological characteristics.

### 1.1.1 Definition of Common Terms

Several terms are commonly used in Economic geology to describe the nature and accumulation of useful minerals in the Earth's crust.

- ***Mineral*** is a naturally-occurring element or chemical compound that is normally crystalline and formed as a result of geological processes.
- ***Economic mineral*** is a valuable mineral that can be extracted and used for human benefit. Out of the over 4000 minerals that are known in nature, only about 400 are considered as of economic value.
- ***Mineral deposit*** is a substantial accumulation of useful minerals in a form that has economic potential.
- ***Ore deposit*** is a mineral deposit that is sufficiently rich to be extracted at a profit. An ore deposit usually contains one or more ore bodies.
- ***Ore body*** is an extensive or continuous mass of mineralization (ore) distinguishable by form and character from the enclosing host rocks.
- ***Ore*** is an aggregate of valuable mineral(s) and gangue from which one or more metals can be extracted at a profit.
- ***Ore mineral*** is an economic mineral from which one or more metallic elements can be extracted.
- ***Ore grade*** is a measure of the concentration of a metal(s) contained in an ore body
- ***Mineral occurrence*** is a concentration of a mineral that is considered valuable or of scientific or technical interest.
- ***Mineral showing*** is a surface indication of mineralization or a mineral occurrence that is very small or of low grade.
- ***Mineral prospect*** refers to any promising mineral occurrence that has been partially explored or developed.

## 1.2 Purpose and Scope of Economic Geology

The main aim of economic geology is to promote the understanding of the geological characteristics, mode of occurrence and formation of mineral deposits, and to apply such knowledge in the exploration and exploitation of ore deposits. Mineral deposits are products of complex geological processes that contributed to their localization and origin, and affect their shape, size and economic potential. Economic geology is both a field and laboratory geoscience. The field aspects involve geological investigations of exposures of ore bodies and associated minerals which may entail geological mapping, structural analysis, petrology and mineralogical studies. Laboratory investigations include petrographic and ore microscopic studies, and chemical analyses and isotope studies of host rocks, ore minerals and gangue. More advanced studies may involve the use of sophisticated equipment such as the X-Ray Diffractometer, X-Ray Fluorescence Spectrometer, Atomic Absorption Spectrometer, Scanning Electron

Microscope, Electron Microprobe Analyzer, Fluid Inclusion Geothermometer and Electron Mass Spectrometer.

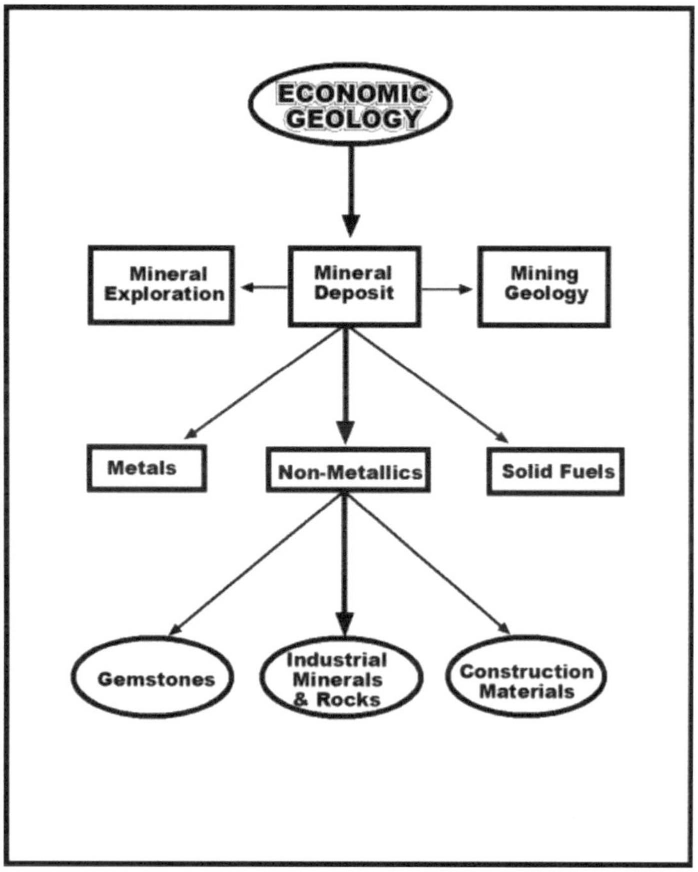


Figure 1.1: Scope of Economic Geology

The *scope* of economic geology has somewhat evolved over time, since its inception in the late 19th century. Its growth as a scientific discipline was initially influenced by the dominance of mining of metallic ores and mineral discoveries in the late 19th century. However, in the early part of the 20th century, economic geology emerged from being mainly descriptive of mineral occurrences in mines, to being more scientific, with the addition of more sophisticated laboratory studies of minerals to augment field observations. Waldemar Lindgren in 1933 described economic geology as the "applications of geology...." so initially, the scope of this emerging scientific discipline was very broad, and covered a wide range of activities where geology is beneficial to the study of earth resources, such as groundwater resources, metallic and non-metallic minerals, as well as solid and liquid fuels (coal, oil and gas). This broad scope has shrunk over time with the development of some major sub-disciplines. With the rapid advancement in the body of knowledge in economic geology, new sub-disciplines emerged, and developed into full-blown scientific areas of study, including *hydrogeology* which focused on groundwater resources, and *petroleum geology* on liquid and gaseous fuels.

In the first half of the 20th century, modern economic geology evolved both in field and laboratory studies that were focused mainly on metallic minerals because of the abundance of exposures in mines and growth in mineral exploration worldwide. However, towards the end of that century, interest in the study of non-metallic minerals, particularly *industrial minerals* and *gemstones* had grown considerably due to the increasing demand and use of these minerals in industrial processes and jewelry, with focus on their physical characteristics and industrial properties. In the past three decades, the world production of non-metallic minerals has outpaced that of metallic minerals.

The scope of *modern economic geology* continues to evolve with a broader definition of economic minerals that include numerous metals, non-metallic minerals and solid mineral fuels (Fig. 1.1) with focus on their physical characteristics, distribution, mode of occurrence, localization and genesis. In the past few decades, mineral deposits geology has seen the widespread use of laboratory techniques for oxygen, hydrogen, sulfur and carbon isotope geochemistry of ore and gangue minerals which has helped economic geologists determine the sources of ore-forming fluids and genesis of mineral deposits. Significant interest in the distribution of ore deposits and their relationship to major tectonic features, known as "metallogeny" has occupied a prominent place in the study of economic geology partly because of the potential economic advantage of using such knowledge in predicting the location of the next big mineable deposit. This has led to the development of *geodynamics* as a subdiscipline that promotes a better understanding of the formation of a number of poorly-understood mineral deposit types, such as sedimentary exhalative and volcanogenic deposits.

The practice of economic geology which is referred to as *applied economic geology* is distinguished by its main objective of utilizing the principles of economic geology in the search for and extraction of mineral resources. These applied aspects that relate to ore search (*mineral exploration* or *exploration geology*) and mineral extraction *(mining geology*) are growing into major sub-disciplines of economic geology. Petroleum geology which has developed into a major discipline of its own is defined as "the application of geology to the exploration and production of oil and natural gas"

***Mineral Exploration*** or exploration geology is the application of geological knowledge and related techniques in the search, discovery and evaluation of economic minerals. Mineral deposits are delineated during exploration by using a combination of geological, geochemical and geophysical techniques coupled with drilling and reserve estimation applying geostatistics and computer modelling to evaluate the quantity and quality of ore deposits.

***Mining Geology*** is the application of geological knowledge and expertise in the extraction (mining) of ore deposits. It is an applied science which combines the principles of economic geology and mining to the development of a defined mineral resource.

### 1.2.1 Types of Economic Minerals

In the descriptive aspect of economic geology, the economic minerals of interest belong to three main groups: *metallic minerals, non-metallic minerals* and rocks and solid *mineral fuels* (Fig. 1.2).

***Metallic minerals*** are those minerals containing one or more metals. About 80% of the known chemical elements possess some metallic properties, but the term metal, is reserved for those chemical elements that possess two or more of the characteristic physical properties of metals (opacity, ductility, malleability, fusibility) and are also good conductors of heat and electricity. With this definition, only about 40 chemical elements are metals that are extracted through the mining and smelting of the minerals in which they occur. There are over 300 metallic minerals, the majority being sulfides and sulfosalts (60%) followed by oxides (25%) and native elements (10%). Metals can be subdivided into precious, ferrous and non-ferrous, and examples include gold, silver, iron, aluminium, copper, lead, and zinc (Table 1.1).

***Non-metallic minerals*** are minerals that are not metals or fuels. They have recently attained some prominence in economic geology because of their increased value in industrial and/or economic applications. Non-metallics include a group of minerals that are used for industrial purposes known as *industrial minerals,* such as barite, feldspar and clays. Earth materials such as sand, gravel, crushed stone aggregates and ornamental stones that are used for building and construction are known as construction *minerals* while the rocks that are used in the manufacturing of products like cement, glass, lime, concrete etc., are described as *industrial rocks*. *Gemstone*s are non-metallic, precious minerals that are used in making jewelry and other adornments. While gemstones can be considered as a special class of industrial minerals because of their use as raw material in the jewelry industry, it is preferable to describe them as a separate group of minerals with glitter and color. They are the focus of study in the emerging sub-discipline of *gemology*.

***Mineral fuels*** constitute the third group of mineral resources that are covered in economic geology. They are the solid mineral fuels: coal, lignite, tar sand, and bituminous sand and shale. The liquid fuels oil and gas (petroleum) are covered in petroleum geology, although several principles used in economic geology are also applied in the description, exploration and extraction of petroleum deposits. However, some prominent economic geologists such as W. L. Pohl of Germany still treated oil and gas as part of economic geology.

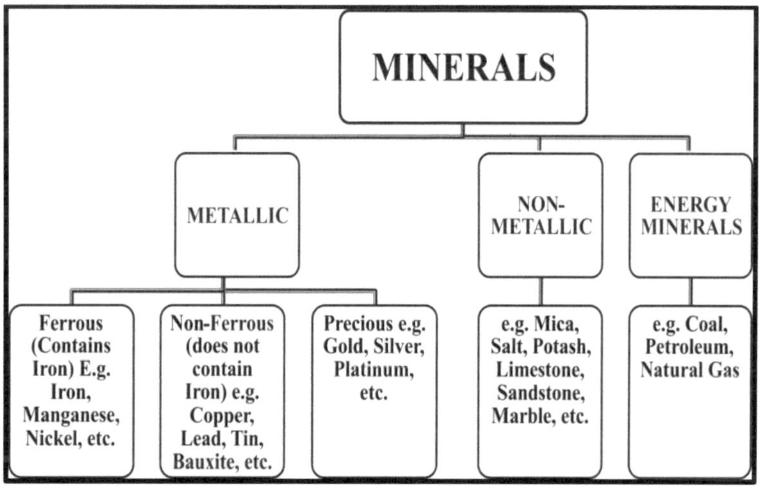

Fig. 1.2: Classification of economic minerals

# 1.3 Tools of Economic Geology

Economic geology can be considered as an interdisciplinary geoscience in which practitioners use several tools in the study of mineral deposits, including the knowledge of various aspects of geology, and application of the principles and methods of various other fields of the geosciences, including geophysics, geochemistry and geostatistics. The basic understanding of the characteristics and genesis of mineral deposits requires the ability to identify ore minerals and mineralized host rock types, and undertake detailed geological mapping which requires a good knowledge of mineralogy, petrology, and stratigraphy. The understanding of structural features that control ore localization necessitates an adequate knowledge of structural geology and remote sensing. Mineral deposits are made up of minerals, both ore and gangue which both occur in ore bodies and host rocks. The ability to identify metallic and non-metallic minerals is crucial to ore discovery, ore recovery and mapping of alteration halos that are often associated with mineralization. The sub-disciplines of geosciences that are relevant to the study of economic geology include petrology, mineralogy, stratigraphy, structural geology, remote sensing, geochemistry, geophysics and geostastics.

Table 1.1: List of Common Economic Minerals

| Metals | Non-Metallic Minerals and Rocks | Mineral Fuels |
|---|---|---|
| | | |
| Gold | Diamond | Coal |
| Platinum | Barite | Lignite |
| Silver | Kaolin | Tar Sand |
| Iron | Limestone | Bitumen |
| Lead | Marble | Oil Shale |
| Zinc | Muscovite | |
| Chromium | Feldspar | |
| Manganese | Fluorspar | |
| Tin | Gemstones | |
| Cobalt | Gypsum | |
| Tungsten | Bentonite | |
| Aluminium | Rock Salt | |
| Nickel | Talc | |

### 1.3.1 Petrology

Petrology is a branch of geology that is concerned with the composition, distribution and formation of rocks. Mineral deposits are usually found in a variety of rocks, and the relationship of such deposits to host rocks may be related to their origin. Consequently, a good understanding of the host rocks and their spatial relationship to mineralization is critical in evaluating the economic potential of mineral resources. During the extraction of ore deposits, the composition and characteristics of the host rocks can be useful in facilitating extraction and determining economic viability. If mineralization is epigenetic, such as in hydrothermal deposits, host rocks may show considerable mineralogical and chemical changes. On the other hand, if mineralization is formed at the same time as the host rocks (syngenetic), their characteristics may provide information on the processes of ore formation. Therefore, the geologist's knowledge of petrology, may promote a better understanding of the mode of occurrence and origin of associated mineral deposits.

### 1.3.2 Stratigraphy

Stratigraphy is the study of the layering and chronology of sedimentary, and some volcaniclastic rocks. Some base metal deposits and non-metallic and energy minerals, such as coal, bitumen and oil shale are contained in layered (sedimentary) rocks, sometimes within thick sedimentary basins. To understand the localization of sedimentary ore deposits requires a good knowledge and interpretation of the sedimentary sequence. Stratigraphy is therefore of paramount importance in understanding the placement of sedimentary ores and the exploration for mineral deposits in sedimentary basins. The scientific exploration for any such deposit requires that the geological events which determined its formation and location be well known. Also important is an understanding of the depositional history and post-depositional changes of the sedimentary host rocks, which are very relevant for the accumulation of mineral deposits.

### 1.3.3 Mineralogy

Mineralogy is concerned with the physical properties, chemical composition, occurrence, distribution and origin of minerals. Mineral deposits are composed of ore and gangue minerals and their composition and physical characteristics need to be identified visually and studied, sometimes using ore microscopes, and other advanced diagnostic techniques. A good knowledge of mineralogy can be very helpful in the field and laboratory identification, microscopic examination and related study of the minerals found in ore bodies. The equipment of relevance are: petrographic and ore microscopes, scanning electron microscope, the electron microprobe analyzer and X-ray Diffractometer. Results of such studies have contributed to the advancement of economic geology.

### 1.3.4 Structural Geology and Remote Sensing

Structural geology deals with rock deformation on both large and small scale. Geotectonics is a branch of structural geology concerned with the structure of the crust and deformation on a regional scale. Understanding these aspects are very important in economic geology because of their relationship and/or controls on mineralization. Mineral deposits that occur as epigenetic veins and lodes are often localized within structural features, such as faults, folds, fractures and joints. Folded and faulted rock strata often serve as traps for migrating mineralizing solutions which accumulate in high concentrations. Deformation of ore deposits are sometimes very complicated and may disrupt the continuity of ore bodies. Structural geology helps the economic geologist to understand structural relations and controls of mineralization. Remote sensing using aerial photographs and spectral imagery is a tool widely used in exploration to identify features, such as lineaments and altered zones that localize or indicate the occurrence of mineralization, sometimes without making physical contact.

### 1.3.5 Geochemistry

Geochemistry is a study of the chemical composition of rocks and minerals. Minerals in ore deposits and host rocks are composed of elements or metals of economic interest. The value of ore deposits is usually based on the ore grade which is obtained from the chemical analysis (assay) of the ore. A good knowledge of geochemistry helps the economic geologist study chemical constituents of ore deposits and evaluate the quality of the ores, as well as provide a better understanding of the nature of chemical processes involved in ore deposition. The interpretation of geochemical exploration data using the identification of primary or secondary dispersion halos of chemical elements requires a good understanding of geochemistry.

### 1.3.6 Geophysics

Geophysics is concerned with the physical processes and physical properties of the Earth and associated rocks. Geophysical techniques in mineral exploration measure the gravity, magnetism, electromagnetism, radioactivity and geo-electrical properties of rocks that indicate the presence of mineral deposits. Geophysics is a widely used tool in the exploration for mineral deposits, and in regional geological mapping of remote terrains. Economic geologists should be acquainted with geophysical techniques and their interpretation, as well as being capable of selecting what technique is appropriate in delineating mineralized zones and their extensions.

### 1.3.7 Geostatistics

Geostatistics is a branch of applied geology and statistics that offers a collection of tools for understanding and modeling variability in spatial data. The power of the geostatistical tools and models in evaluating data related to ore deposits and the estimation of mineral resources is well known to professionals in the mineral industry. Economic geologists are often involved in the important process of ore-deposit evaluation as it relates to estimating the physical parameters (size, shape, tonnage, ore grade) of a mineral deposit during the advanced stage of mineral exploration. While the services of a geostatistician may be required, it is often useful for the economic geologist to possess sufficient knowledge of statistics in evaluating data related to the quantity and quality (ore reserves) of economic minerals present in ore deposits. Knowledge of ore grades and ore reserves as well as error estimation of these values, is fundamental for mining geologists who are involved in mine development and ore-grade control.

## 1.4 The Practice of Economic Geology

Economic geology is an applied geoscience and considered as a "professional" discipline practiced by geologists who are trained and capable of applying their acquired knowledge of mineral deposits in mineral exploration, and exploitation. The practice of economic geology provides employment for a large number of geologists in the mineral sector. On a worldwide basis, probably more than one third of those persons employed in the geologic sciences are engaged in work that touches on the economic aspects of geology. These include economic geologists whose main interests lie in diverse fields of exploration and mining of solid minerals (metallic and non-metallic) and in related geosciences (geophysics, geochemistry). Apart from employment by private companies in the mineral industry, a very large number of economic geologists are hired by government agencies and parastatals such as Geological Surveys, Bureaus of Economic Geology etc., where they conduct geological mapping and evaluation of mineral resources, while others serve as consultants or are involved in academic teaching and research.

Careers in the mineral industry require exceptional knowledge in a wide range of geological subdiscipline, including mineralogy, petrology, geochemistry, geophysics, sedimentology and stratigraphy, structural geology, tectonics and volcanology. Relevant skills that are useful in professional practice include hand-sample, mineral and rock identification skills, excellent mapping techniques (at scales ranging from extremely detailed (e.g. 1:100 scale) to regional (1:1,000,000 scale), drill core logging techniques, knowledge of geographic information systems (GIS), excellent computer applications skills, a desire to keep current with the geological literature, the ability to travel and work effectively in remote areas. Exceptional people skills and ability for teamwork are absolute requirements to become an effective and successful economic geologist. The energy mineral subsector, including the petroleum industry, is collectively the largest employer of "economic" (petroleum) geologists who have expertise in areas such as petroleum geology, sedimentary petrology, structural geology and stratigraphy.

### 1.4.1 Scope of Professional Practice

Economic geologists are trained geoscientists who possess sufficient knowledge and training in mineral deposits geology, including the geological characteristics, mode of occurrence, mineralogy and origin of mineral deposits and are capable of applying such knowledge in the search for mineral discoveries and extraction of ore bodies. They are also involved in the development of new concepts of ore genesis that have facilitated the discovery of previously unknown deposits. The responsibilities of economic geologists involved in mineral exploration can transition or evolve into that of a mine geologist, who assists the mining engineer in mine planning and the exploitation of ore deposits. This role derives from the detailed knowledge of the mineralogical characteristics and mode of occurrence of mineral deposits. Economic geologists also play a role in environmental protection in mining operations, including environmental monitoring, implementation of environmental mitigation plans and the supervision of the remediation of contaminated soils and groundwater.

The major areas involved in the practice of economic geology are outlined as follows:

- Geological Surveying and Mapping
- Mineral Evaluation
- Mineral Prospecting
- Reconnaissance and Detailed Exploration
- Mineral Sampling and Ore Reserve Estimation
- Valuation of Mineral Deposits
- Geotechnical mapping
- Mine Geology
- Environmental monitoring
- Environmental Protection and Remediation

Mining geology is an integral part of economic geology. A mining geologist is a trained and/or knowledgeable applied scientist with considerable knowledge of economic geology and ability to work closely with mining engineers in the development of mining operations on both surface and underground mines. Drilling and blasting of rocks and ores in mining operations require a knowledge of the strength of the host rocks. So, a mine geologist provides professional advice on ore blasting, sampling and geotechnical issues in mine excavations. Mine geologists also prepare geological maps and routine reports on orebodies, and assist the mining engineers to determine the best way to extract ores and also assist operations to maximize ore recovery and implement ore grade control. Environmental protection of mining sites and surrounding areas is often the responsibility of the mine geologist.

### 1.4.2 Professional Role and Responsibilities

The primary role of an economic geologist is to use knowledge and experience to investigate mineral deposits. Responsibilities may vary depending on the institutions or agencies they work for, and the type and stage of activities involved in exploring and developing mineral resources. Some of these responsibilities are outlined as follows, and there are could be some overlaps:

- Review and interpret geological and structural information
- Develop conceptual framework for ore search
- Conduct aerial photo and remote sensing analysis of prospective geological terrains
- Conduct reconnaissance field investigations of mineral occurrences and host rocks
- Conduct geological mapping of potential host rocks and ore bodies
- Participate and interpret geochemical and geophysical Surveys
- Conduct petrographic and mineralogical studies of host rocks and f ore minerals
- Supervise exploratory drilling programs
- Conduct drill core logging, sampling and ore reserve estimates
- Participate in technical and economic evaluation of ore deposits
- Prepare geological and geotechnical maps and reports
- Supervise drill hole blasting, sampling and ore grade control
- Conduct environmental monitoring and mitigation

## 1.5 Brief History of Economic Geology

Although ancient peoples looked for and exploited minerals in Neolithic times, the "science" of mineral resources is a relatively recent field of study. Historical records are replete with stories of the discovery and mining of minerals by ancient people in North and Central America from as early as 7,500 years ago. The Mayan Indians searched for, and mined copper deposits. They followed copper veins down into the ground, sunk shafts or tunnels up to 60 feet deep in hard rock, and extracted the copper by primitive techniques using fire and water. Over 7000 years ago, Egypt sent out expeditions to parts of Africa and the Middle East in search for gold, turquoise, silver, tin, and lead. In Europe, mining had a very long history in which ancient Greeks and Romans mined several metals, fighting wars and sometimes using slave labor as well as hydraulic methods in mining. For all their ability to appreciate the worth of such resources, pre-modern peoples however possessed little in the way of scientific theories regarding the formation or the means of extracting minerals of a large scale.

The study of metallic minerals began in Germany in the 16th century. In 1556, *Georgius Agricola*, a German physician who became fascinated with minerals published a handbook on mining called *De Re Metallica*. In the book which focused mostly on mining practices, Agricola summarized his observations on the metallic sulfide ores in the region of Saxony, Germany, and proposed the first classification of mineral deposits. Regarded as the father of both mineralogy and economic geology, Agricola introduced several ideas that provided a scientific foundation for the study of the Earth and its products. He divided mineral deposits into two large groups - *alluvial* and *"in place"*. In addition, he also made two fundamental observations, that: ore channels are secondary features in rocks, and ores were deposited from solutions migrating through those channels.

After the pioneer work of Agricola, the next advance came in 1669 when *Nicolas Steno*, a Danish scientist published a book in which he noted that ore minerals may represent the condensation of vapors traveling through fissures, and that not all ores had to be deposited by fluids. He got this idea from looking at volcanoes and precipitates around hot springs. After Steno, the next major contribution to economic geology was that of Johann von Charpentier (1728-1805), a professor at the Mining Academy of Freiberg, Germany who was an expert in mineralogy, mining and economic geology in the late 1700s. He studied the lead-zinc mines around Freiberg, Germany and published several books on mining geology. From his observations he concluded that metals and minerals in ore veins were the result of reactions between the wall rock hosting the ore, and heated water passing through them. His main evidence for this was the gradation of some mineralized veins into the wall rock. These observations represent the start of the hydrothermal ore theory, and the beginning of what has been called the "lateral secretion theory" of ore deposits in which metals were considered to have been leached out of the wall rock by heated water and then deposited in openings in the same rock.

Towards the end of the 1700s, there developed two groups of thought in geology, known as the "Plutonists and Neptunists". The founder of the plutonist idea of ore formation was *James Hutton*, a Scottish geologist regarded as the "father of modern geology", who proposed, in 1788, that not only igneous rocks but also all ore deposits were derived from molten magmas, and were thus transported in a liquid state to where they are found. Opposed to Hutton and the Plutonists were the Neptunists led by Abraham Gottlieb Werner, a German professor at the Freiberg Mining Academy, who proposed that all rocks and ore deposits were formed as sediment in a primordial ocean. The argument between the two camps eventually led to ore deposits being subdivided into two separate groups: those formed by *igneous processes* (such as tin and copper) and those formed by *sedimentary processes* (such as bedded deposits of iron ore). In the mid 1800's there was a return to Carpentier's ideas with the writings of *Eli de Beaumont* (1798-1884), a French scholar and professor of geology at the École des Mines (School of Mines). He became the first geologist to organize (classify) hydrothermal (hot water) deposits related to igneous centers. He also recognized replacement type deposits (skarns) and magmatic segregations (chromite).

Modern economic geology, as it is known today, did not emerge until the end of the 19th century in Europe and the United States. In the early 1900's, geologists had concluded that ore deposits were formed from a variety of geological processes; and that, structures in the rocks were important in localizing ores and in the movement of fluids. The origin of modern economic geology revolved around two nuclei: the North American school, from which sprang the Economic Geology Publishing Company founded in 1905 (that started publishing the journal of Economic Geology), and the European school, from which the term 'metallogeny' was coined in 1913.by de Launay, a French geologist (1913). There were attempts to classify mineral deposits based on their origin, and the most practical and insightful classification system (partly in use today) was proposed in 1915 by Waldemar Lindgren (1860-1939), a German-trained, Swedish-American professor of geology at the Massachusetts Institute of Technology who in 1905 helped found the first journal of Economic Geology. Lindgren who is regarded as the "father of modern economic geology" classified mineral deposits into two groups based on whether they were products of (1) *chemical* or *mechanical* concentration and, (2) if *chemical*, whether the deposit formed from hydrothermal water, magma, or sedimentary processes. Lindgren's published writings run to nearly 200 titles, that circulated widely including his landmark book entitled, "Mineral Deposits", a widely used textbook with several editions published in 1913, 1919, 1928 and 1933.

During the second half of the 20$^{th}$ century, economic geology had become a well-established discipline which developed rapidly due to field research on mineral deposits and advances of scientific knowledge in the areas of fluid inclusion studies, isotope geochemistry, and underwater geothermal systems. It moved from a descriptive phase to a deeper understanding of ore genesis. Among the prominent economic geologists of that era was Brian J. Skinner (1928-2019), Alan Bateman (1889-1971) who published the book "Economic Mineral Deposits" in 1942, and Charles Park (1903-1990), who proposed new ideas about ore formation embodied in his widely used textbook "Ore Deposits", written with Roy McDiarmid as co-author. After three editions (1964, 1970, 1975), the book was extensively revised in collaboration with John Guilbert and retitled Geology Ore Deposits (1986). Their work led to a modified Lindgren classification scheme for mineral deposits, and supported the idea that most ore metals were deposited from hot solutions originating in or associated with intrusive igneous bodies as well as other sources. Their book has undergone several revisions, but still remains the main textbook used in the teaching of mineral deposits.

In the 1970s and 80s, the conceptual understanding of geodynamic processes through plate tectonics theory changed ideas about how mineral deposits form along plate margins and further enhanced the identification of metallogenic processes and provinces that have encouraged the search and discovery of new areas with considerable mineralization potentials. There has been a great deal of change in the exploration for ore deposits, and in our understanding of their origin and mode of occurrence. This has led to some new ore deposit classification schemes.

In the early 21$^{st}$ century, within a period of about 100 years, a significant body of knowledge had emerged, and economic geology has become an established discipline that is taught as a very important course in geology programs at universities worldwide because of the importance of mineral resources in the economy of most countries. Economic geology as a geoscience developed with the applications of geophysics, geochemistry and geostastics, and has seen the emergence of several sub-disciplines, including Mineral Exploration, Mining Geology, Coal Geology and Industrial Mineralogy. As an applied science, Economic Geology has been influenced by developments or discoveries in pure or theoretical science, including physics, chemistry, mathematics and statistics. Furthermore, there has been a significant increase in the sheer volume of material published on ore deposits. From 2000 to the present; more than 30,000 papers, field trip guides, special symposia volumes, reports, and special issues of journals have been added to the literature on Economic Geology. Advances in analytical techniques, isotopic dating methods and isotopic ratio determination (sulfur, oxygen, hydrogen, carbon isotopes) have contributed immensely to our understanding of ore element sources, ore fluid sources, age of mineralization, ore transport and depositional processes.

## 1.6 Economic Importance of Mineral Resources

Minerals have been extracted from the bosom of the Earth since ancient times, and the history of human civilization and industrial development is closely linked to the use of economic minerals. It is often said that *"human civilization would not occur without the benefit of mineral resources"*. Mineral resources provide the mineral commodities on which modern civilization is heavily dependent for domestic consumption or export to other countries. Consequently, there is a constant need to find new mineral resources as the known reserves are depleted. Mineral resources play a significant role in the economic and industrial development of nations by generating revenues as well as providing raw materials for the construction of infrastructures and the manufacture of several products that affect our human existence and quality of life. However, mineral resources are not evenly distributed around the world, and only a few nations are completely self-sufficient in all commodities. Adequate supplies of minerals are essential for manufacturing, construction, energy requirements and agriculture and thus for the sustainable development of a modern economy

As the world's population grows exponentially and living standards increase globally, the demand for mineral raw materials continues to escalate widely. To satisfy the world's hunger for mineral raw materials, new mineral deposits must be discovered on a regular basis which requires the trained expertise of economic geologists and related professionals. Minerals are basic and essential raw materials for a lot of human needs including the following:

a. manufacturing involving the production of a wide range of industrial and consumer goods;
b. construction to develop, maintain and enhance our built environment and transport infrastructure
c. generation of electricity for use in the home , commerce and industry
d. transportation in the operation of vehicles, trains, aircraft and ships and
e. agriculture to improve soil productivity and enhance food security

Most developed and developing countries of the world produce as well as consume minerals. Minerals from indigenous sources are very valuable assets to where they can be utilized locally or exported as a source of revenue for economic development. Currently, there is a mineral resource crisis in the world in which mineral consumption is growing even faster than the population. Most minerals are used in the more developed countries (MDC) while relatively small consumption occurs in less developed countries (LDC). For example, MDCs account for 15% of the world population but consume 70% of world aluminum, copper, and nickel, 60% of world oil, 50% gas and 40% coal. As the standard of living increases in LDCs, these nations will start to increase consumption of their minerals. This pattern is already apparent in countries such as India and Brazil. If global population increases as rapidly as consumption rates, the pressure to find and produce minerals will be enormous. Obviously, there are limits to the discovery, extraction and utilization of non-renewable mineral resources, and the need for substitution by other natural or synthetic products is growing.

The use of renewable energy sources, recycled materials and industrial by-products are becoming popular and should be used to meet part of the demand for mineral raw materials and to reduce waste for a sustainable world economy. The so-called "green revolution" has just started with the goal of conserving our mineral resources and protecting the environment from pollution and global warming.

Africa is very much blessed with abundant mineral resources, and the continent ranks either first or second in quantity of known reserves of several minerals including diamond, bauxite, phosphate, platinum, cobalt, uranium, chromium and gold. Africa is described as having the largest mineral industry in the world, and in many African countries, mineral exploration and production constitute significant parts of their economies and remain keys to economic growth. Africa's share of world production of platinum, gold, diamond, and phosphate reaches over 75% (Table 1.2). Revenues generated from mining activities and mineral exports often constitute a substantial component of the GDP which explains why some countries are so heavily (and dangerously) dependent on their mineral resources as the main source of revenue. For example, mineral fuels (coal, petroleum) account for more than 90% of the export earnings for Algeria, Equatorial Guinea, Libya, and Nigeria. Minerals account for 80% for Botswana (diamond, copper, nickel, and gold), Congo (Brazzaville) (petroleum), Congo (Kinshasa) (diamond, petroleum, cobalt and copper), Gabon (petroleum and manganese), Guinea (bauxite, alumina, gold, and diamond), Sierra Leone (diamond), and Sudan (petroleum and gold).

Minerals and mineral fuels accounted for more than 50% of the export earnings of Mali (gold), Sierra Leone (diamonds), Mauritania (iron ore), Namibia (diamond, uranium, gold, and zinc), and Zambia (copper and cobalt), and Mozambique (aluminium), However, mineral production in Africa is beset with several problems, including illegal mining, political instability, and a lack of security, systemic corruption and poor stewardship to the local communities affected by mining activities. Also, the informal and unregulated extraction of precious metals by thousands of untrained artisanal and small scale miners has produced large environmental impacts on landscape, air quality, surface water quality, groundwater quality, soils, vegetation, and ecosystem damage.

Table 1.2: Share of World Production of Minerals from Africa (2019)

| **Metal** | **% of World Production** | **Countries** |
|---|---|---|
| Platinum | 92% | South Africa 97% |
| Gold | 89% | South Africa 56%; Ghana, 13%; Tanzania, 10%; and Mali, 8% |
| Cobalt | 71% | Congo (DR) 95%,; Zambia 5% |
| Diamond | 75% | Botswana 35%; Congo (Kinshasa) 34%; South Africa 17%; Angola, 8% |
| Chromium | 60% | South Africa 98%; Zimbabwe 2% |
| Manganese | 49% | South Africa 60%; Gabon, 30%; Ghana 10% |
| Phosphate | 16% | Morocco and Western Sahara 98% |
| Uranium | 16% | Namibia 46%; Niger 44%; South Africa less than 10% |
| Bauxite | 17% | Guinea 95%; Ghana 5% |

Over the past two decades, Africa has experienced a surge in the exploration of mineral and energy resources, driven by demand for precious metals and industrial raw materials. In this modern era, with most mineral exploration for resources occurring in the developing countries of Africa, South America and Asia, international exploration and mining companies have recognized the strategic economic, social and political values of partnering with local communities to build capacity within the countries in which they operate. Apart from its infrastructural challenges, Africa poses particular challenges with regard to developing the necessary indigenous high-level skills required to find new resources and extract them. The training of economic and mining geologists should be in the forefront of educational and professional development at Universities and Polytechnics to which this book is devoted.

## 1.7 Mineral Resources and Mineral (Ore) Reserves

Mineral resources are not only of instant value, but also are finite in supply and non-renewable. The totality of Nature's mineral endowment consists of two major parts: *identified mineral resources* and *undiscovered mineral resources*. Identified resources are those resources whose location, quality, and quantity are known or can be estimated from specific geological information, whereas undiscovered resources are those whose existence is extrapolated on the basis of broad geological knowledge. Mineral resources and mineral reserves (ore reserves) are parameters that are used widely in economic geology to represent the value and availability of mineral treasures. The differences between "resources" and "reserves' should be noted. Resources represent a large universe of potentially useful minerals that exist in a designated area, compared to ore reserves which are a portion of resources that are ready to be mined at a profit. As shown in Fig. 1.4, reserves are usually a small fraction of known resources. This distinction is important particularly when comparisons are made between the values of various deposits, or when investment decisions are to be made based on estimated resources and reserves.

In the process of ore search, a mineral resource (mineral occurrence) is discovered and explored, and a *resource evaluation* is undertaken to provide information on the geological characteristics, and the quantity and quality of economic minerals present in the deposit. This is achieved primarily by a chain of activities involving detailed exploration and feasibility studies which may ultimately lead to mine development and exploitation. Figure 1.5 illustrates the dynamic relationships between the extent of known mineralization, and identifiable resource and ore reserve. Resource and reserve only makeup a part of the true extent of any mineralization that has been roughly defined by geological exploration, but cannot be reported as part of the identifiable resource because the geological confidence in the extent and economic prospects may not be sufficient to meet the needed criteria. The boundary of known mineralization is far more extensive than the identifiable mineral resource, while the ore reserve is a small portion of the mineral resource. The relationship continues to change as more data becomes available.

However, as the ore deposit is exploited, known reserves may diminish with time, but further exploration may identify more reserves from identified resources. This may be accompanied by the delineation of more resources from the surrounding poorly delineated zone of mineralisation. This can extend the initially stated life of a mine by several years, thereby making resources and reserves to potentially grow coincident with production. The dynamic relationship among these components should not be a surprise in mining projects as ore deposits are exploited. In addition to the internal factors, changes in external factors such as commodity prices or new mineral extraction technologies can prolong the life of a mine beyond earlier projection.

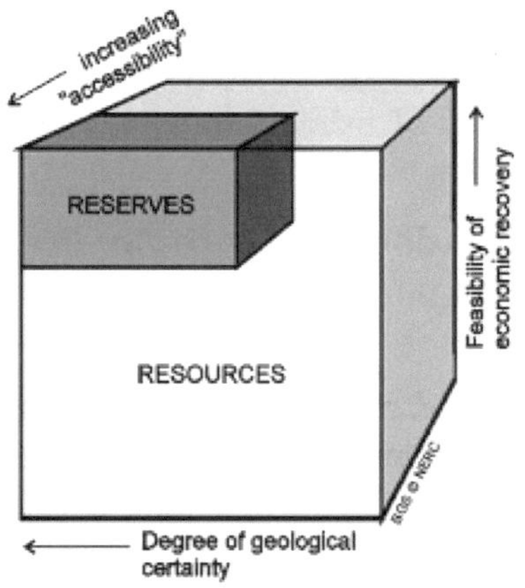

Fig. 1.4: Schematic diagram showing reserves and resources.

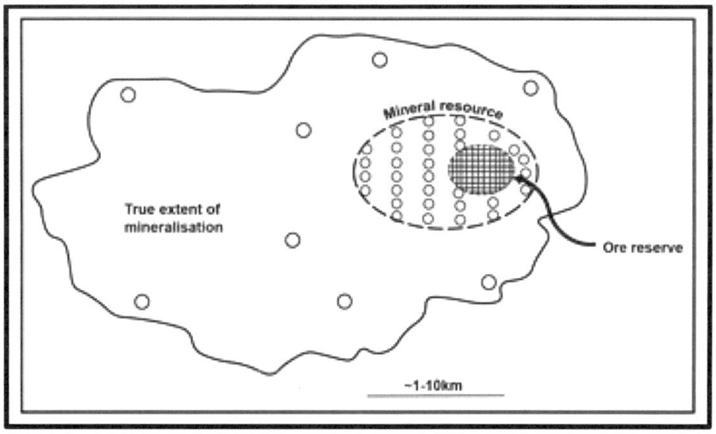

Fig.1.5: Diagram of relationship between extent of mineralization, mineral resource and ore reserve

## 1.7.1 Mineral Resource Classification

Mineral resources can be classified into three categories based on an increasing level of geological knowledge and confidence about economic value: infe*rred, indicated* and *measured* resources. Such classification can be used to organize information obtained on the economic value of a mineral commodity during exploration, particularly prior to making decisions regarding the viability and commencement of a mining project. For example, investors in a mineral project may like to know how much of the mineral resource is readily available for exploitation at a profit, and the confidence level of the estimates.

***Measured Mineral Resource*** is that part of a mineral resource for which quantity, grade or quality, shape, densities, and physical characteristics of the mineral deposit are estimated with a high degree of confidence based on geological information derived from very detailed and reliable exploration, sampling and testing, and is sufficient to confirm continuity of geological extent and ore grade between points of observation.

***Indicated Mineral Resource*** is that part of a mineral resource for which quantity, grade or quality, shape, densities and physical characteristics are estimated with sufficient confidence based on geological evidence derived from fairly detailed and reliable exploration, sampling and testing.

***Inferred Mineral Resource*** is the part of a mineral resource for which quantity, quality and mineral content are estimated at a low level of confidence based on limited geological information and sampling. It is reasonably expected that the majority of inferred mineral resources could be upgraded to 'indicated mineral resources' with further exploration.

***Resource Base*** of a mineral or commodity refers to the totality of the identifiable resources including *inferred, indicated* and *measured.* It is a projection of the totality of all the known quantities of a mineral resource (Fig. 1.6)

***Reserve Base*** is that part of an identified resource that meets specified minimum physical and chemical criteria related to current mining and production practices, including those for grade, quality, thickness, and depth. The reserve base is the *measured* plus *indicated* resource from which reserves are estimated (Fig. 1.6).

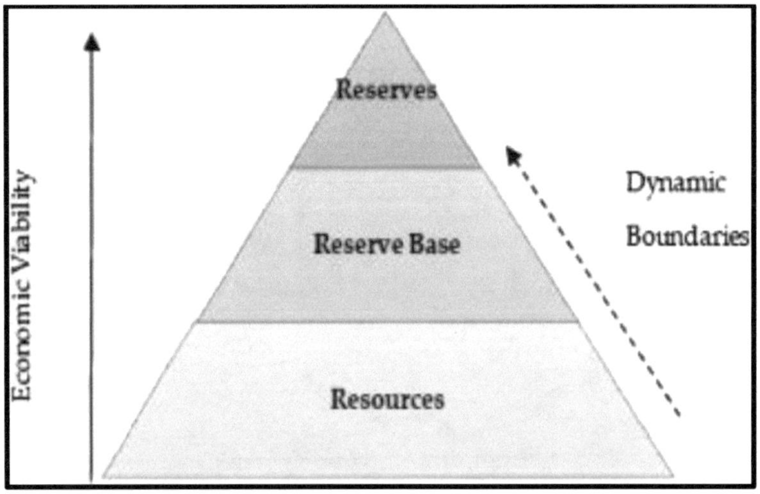


Fig. 1.6: Resource Base and Reserve Base

## 1.7.2 Mineral (Ore) Reserves Classification

A mineral or ore reserve is the economically mineable part of the reserve base (measured or indicated mineral resource) demonstrated by at least a preliminary feasibility study (Fig. 1.6). Compared to mineral resource evaluation, the delineation of ore reserves requires a more rigorous or intensive and technical evaluation aimed at statistically quantifying the ore grade continuity and the size of the orebody. This process may involve more closely-spaced diamond drill-core sampling to obtain more detailed structural information on weaknesses within the rock mass such as faults, foliations, joints and shea zones. Reserve definition also includes the preliminary feasibility assessment of several so-called '*Modifying Factors*' including (but not restricted to geotechnical, mining, processing, metallurgical, infrastructure, economic, marketing, legal, environmental, social and governmental factors. In the context of land-use planning, the term mineral reserve should be further restricted to those minerals with legal access and for which a valid permission for extraction also exists (i.e. permitted reserves). At the end of this process, a feasibility study is published, and the mineral deposit may be either deemed viable or not.

Ore reserves are estimated after statistical analysis of exploration data. Three are two categories: *probable* reserves or *proved* (proven) reserves, in accordance with the level of sampling and data confidence, feasibility of economic extraction and accessibility (Fig. 1.7). These guidelines on how to report resources and reserves are followed by exploration and mining companies when submitting annual reports.

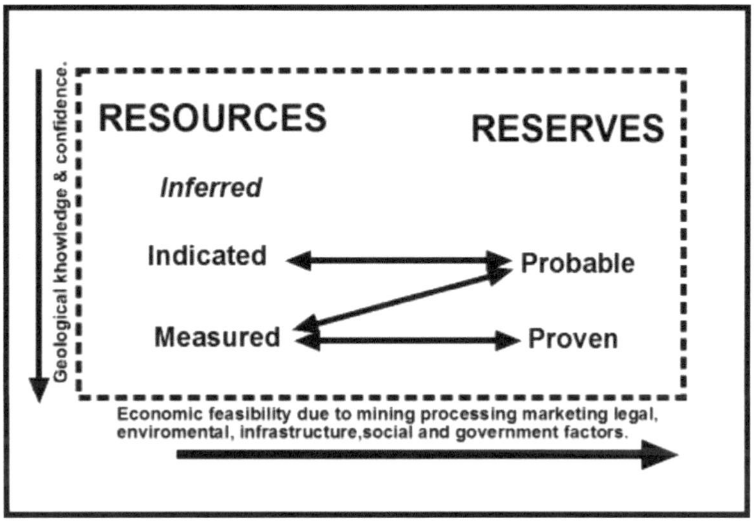


Fig. 1.7: Mineral resources and Mineral (Ore) Reserves

***Proven Ore Reserve*** is the economically mineable part of measured resources. A proved (or proven) ore reserve represents the highest confidence category of ore reserve estimates indicating the measured tonnages and quantified ore grades. The ore reserves now become an economic asset for the company and its investors.

***Probable Ore Reserve*** is the economically mineable part of an indicated, and in some circumstances, measured mineral resource (Fig. 1.7). A probable ore reserve has a lower level of confidence than a proved ore reserve, but is of sufficient quality to serve as the basis for decision on the development of deposit.

## 1.8 Life Cycle of Ore Deposits

Mineral deposits usually pass through a life cycle, in what is sometimes described as passing from "cradle to grave". A mineral deposit is "born" with the *discovery* of a "mineral occurrence "that progresses with further evaluation to a "mineral deposit" and if rich enough becomes an "ore deposit," which is exploited, and ends with the *closure* of mine workings after the ores are exhausted (Fig. 1.8). Below is an outline of the stages in the life cycle of an ore deposit.

- Prospecting and Discovery
- Exploration and Evaluation
- Feasibility Study
- Mine Development
- Mine Production
- Mineral Processing (beneficiation/milling)
- Marketing
- Closure and Reclamation

The life cycle of ore deposits begins with the discovery of the mineral occurrence through prospecting or exploration, except if it is an already known occurrence or prospect. This is followed by detailed exploration to define the extent and value of the mineral deposit which leads to ore reserves estimation that helps determine the size and grade of the deposit. The reserve estimates obtained are used to conduct a pre-feasibility study to establish the potential economic viability of the ore deposit. The next step is to conduct a full feasibility study to evaluate the operational costs, financial viability, the technical and financial risks, and the robustness of the project. At this point, the exploration or mining company makes the decision whether to develop the one deposit into a mine or to "walk away" from the project. Such detailed evaluation and decision to proceed with the mine project are usually followed by the mine design and planning for extracting the resource.

Major steps are taken to determine the economically recoverable portion of the deposit, the beneficiation process and ore recoverability, marketability and payability of the ore concentrates. The engineering concerns, milling and infrastructure costs, finance and equity requirements are also evaluated, and an analysis of the proposed mine from the initial excavation all the way through to reclamation are prepared. Once the analysis determines a given ore body is worth recovering, development begins are implemented to create full access to the ore body. To get direct access to extract ore, it is often necessary to remove waste material which is not of immediate interest to the mining operation. The total movement of ore and waste constitutes the mining process. Often, more waste than ore is mined during the life of a mine, depending on the nature and location of the ore body. Waste removal and placement is a major cost to the mining operation.

Fig. 1.8: Life Cycle of Ore Deposits

The mine buildings and processing plants are built, and any necessary equipment is obtained during mine development. The operation of the mine to recover the ore begins and continues as long as the company operating the mine finds it economical to do so. Once all the ores that the mine can produce profitably are recovered, the mining operation ceases and the mine is closed. Reclamation, which is the process of making the land previously used by the mine suitable for future alternative use and to be environmentally safe, the final process of soil remediation may be required. The life cycle of a mine from discovery to closure usually lasts for an average period of 30 years.

## 1.9 Useful Terms in Economic Geology

Several terms are commonly used in Economic Geology that require initial definition to assist the reader in the understanding of the subject matter.

**Mineral occurrence** - a concentration of a mineral that is considered valuable or that is of scientific or technical interest (or a locality or wherever a mineral of potential economic interest is found, without reference to quantity or quality).

**Mineral deposit** – a mineral occurrence of sufficient size and grade that it might, under the most favorable of circumstances, be considered to have economic potential.

**Mineral showing** – a surface indication of mineralization in which the extent and economic value are unknown: (or a mineral occurrence that is so small and only of geological interest but not necessarily of economic interest).

**Mineral prospect** - any promising mineral occurrence that has been partially developed by underground or by above-ground techniques, or by subsurface drilling to determine the extent of mineralization.

**Mineral processing** - the process of extraction and concentration of economic minerals contained in ore.

**Ore deposit** - a mineral deposit that has been tested and found to be of sufficient size, grade, and accessibility to be mined at a profit.

**Ore body** - an extensive or continuous mass of ore and gangue distinctly distinguishable by form and character from the enclosing host rocks. An ore deposit may contain one or more ore bodies.

**Ore** - the valuable material extracted from an ore deposit. It is an aggregate of minerals from which one or more metals can be extracted *profitably*.

**Orel reserves** – the portion of mineral resources that is economically feasible to extract.

**Ore mineral** - a mineral from which one or more metallic elements can be extracted.

**Ore grade** - a measure of the concentration of a mineral or metal contained in an ore body.

**Cut-off grade** - the minimum metal concentration or lowest permissible grade of ore that is considered minable at a profit which may change with time and changing economic considerations.

**Syngenetic** –ores and minerals that formed at the same time as their host.

**Diagenetic** – ores emplaced during the compaction and consolidation of sediments

**Epigenetic** – ores emplaced into pre-existing rocks of any origin (e.g. veins, metasomatic

**Hypogene** – primary ores that were formed by ascending solutions.

**Supergene** – ore formation by descending meteoric water during weathering.

**Endogenic** – concentration caused by processes in the Earth's interior (magmatism or metamorphism)

**Exogenic** – concentration caused by processes at the Earth's surface (e.g sedimentation, weathering).

**Disseminated ore** - the scattering of ore mineral uniformly throughout host rock.

**Massive ore** – mineralization composed of uniformly a large proportion (>60%) of ore minerals.

**Host rock** - the rock or rocks containing or surrounding a mineral deposit.

**Tenor** – the percentage of the ore mineral that is actual metal to be extracted. Often used as synonymous with ore grade.

**Tonnage** - the quantity of metal or useful mineral in a deposit. Usually expressed as the number of tons available at a specific average grade.

**Metallogeny** - the study of the origin (genesis) of ore deposits, particularly in relation to tectonic activity.

**Metallogenic province** - an area of the earth's crust characterized by a particular assemblage of mineral deposit types.

**Metallogenic epoch** - a period of geological time favorable for the deposition of mineral deposits characterized by specific deposit types.

**Stratiform** – Mineralization that is conformable with host rocks and laid down like strata.

**Stratabound** – Mineralization restricted to a particular stratigraphic unit or interval but may be discordant in places.

## 1.10 Review Questions

1. What is economic geology? Describe its relationship to mineral exploration and mining geology.
2. Discuss the life cycle of ore deposits. When does a mineral deposit become an ore deposit?
3. Differentiate the following pairs of terms:
   a. Mineral occurrence and Mineral prospect
   b. Mineral deposit and Ore deposit
   c. Metallic and Non-metallic minerals.
4. Write a short essay on the history of economic geology
5. List the three groups of economic minerals, and give examples
6. Discuss the responsibilities of an economic geologist. What are the tools of economic geology?
7. Describe the following terms and their relationship. Ore, Ore mineral, Ore body and Ore grade
8. Define and classify mineral resources. Why is it important to have a good knowledge of a country's mineral resources?
9. With the aid of a labeled diagram, show the relationship between ore reserves and mineral resources.
10. Describe the components of resource base and reserve base. Illustrate your answer with a triangular diagram.

## References


Bateman, AM. (1950). *Economic mineral deposits,* 2nd ed. New York: Wiley, 916p. Google Scholar

Guilbert, JM and Park, CF. Jr. (1986) *The geology of ore deposits*: W.H. Freeman and Co., 985 p.

Hughley, DE, Chapman, GR and Bound, KA (2004) The economic importance of minerals to the UK. Brit. Geol. Surv. Publication, 42p.

Jebrak, M (2006) Economic geology: Then and Now. Geoscience Canada, 33(2), 81-93.

Kesler, SE. (1994), *Mineral resources, economics, and the environment*: Macmillan College Publishing Company, 391 p.

Lindgren, W. (1933). *Mineral Deposits,* 4th ed. New York: McGraw-Hill, 930p.Google Scholar

Misra, KC. (2000) Understanding mineral deposits: Kluwer Academic Publishers, 845 p.

Pohl, WL (2011) Economic Geology: Principles and Practice. Wiley-Blackwell, 655p.

Skinner, BJ., *ed.* (1981), Economic Geology 75th anniversary volume 1905-1980: The Economic Geology Publishing Company, 964 p.

Skinner, BJ. (2005) The founders of Economic Geology: SEG Newsletter, v. 60, p. 1-4.

Skinner, BJ. (2015) Mineral deposits. Encyclopedia Britannica, 15p. https://www.britannica.com/science/mineral-deposit.

United States Geological Survey (1080) Principles of a Resource/Reserve Classification For Minerals. USGS Circular 831, 12p.

# Chapter 2
# Nature of Mineral Deposits

## 2.1 Introduction

Mineral deposits are the natural accumulations of useful minerals in sufficient size and grade to be considered as having economic potential. The focus of economic geology is the study of mineral deposits. Mineral deposits are known to occur both in rocks that form the continents (continental crust) and those that lie beneath the oceans (oceanic crust), although the only deposits that have been mined are found mainly in the continental crust where mineralization is associated with igneous, sedimentary and metamorphic environments. Mineral deposits in the oceanic basins are too deep to be studied or recovered at a profit. Geochemical distribution of elements in the Earth's crust has shown that all the common ore-forming elements or metals that make up mineral deposits are also found in ordinary rocks, although at relatively low levels: usually in the range of parts per million (ppm) or parts per billion (ppb). For a mineral deposit to form in rocks at a particular location (Fig. 2.1), requires the unique enrichment of one or more metals. Such localized accumulation of metals in ten to several thousand times the normal concentration in rocks, requires some extraordinary energy source and/or unique geological processes that can produce metal enrichment.

The level of concentration of an element in an ore body is of considerable importance in economic geology because it ultimately defines the economic value of a mineral deposit. *Ore grade* is an expression of the concentration of a metal contained in an ore body. It is usually expressed in percentage (%) or grams per ton (g/t) or ounce per ton (oz/t) depending on the quantity of concentration. For the more abundant metals, such as iron (Fe), it is expressed in percentage, whereas for precious metals that occur in small quantities such as gold, it is expressed in grams per ton of ore. The minimum grade required in order for a mineral or metal to be economically mined (or processed).is called *cut-off grade.*

Fig. 2.1 Outcrop of a chromite ore deposit overlying c host rock

## 2.2 Geochemical Concentration of Metals in Mineral Deposits

From the geochemical viewpoint, ore deposits represent anomalous accumulations of particular elements in the Earth's crust. Elements (metals) found within mineral deposits can be classified into two groups based on their relative abundance in the Earth's crust. The geochemically abundant metals, such as Al, Fe, and Mn that range in concentration from 0.5 - 10%, and the geochemically scarce metals, such as Cu, Pb, Ni, and Au that occur in low concentrations in the parts per million to parts per billion range. The *concentration factor* of an element is a measure of the level at which it is concentrated relative to its abundance in the Earth's crust. This is the concentration necessary in a mineral deposit for profitable extraction of a metal. In other words, it is a reflection of the enrichment of an ore body. Concentration factor can be calculated by dividing the average concentration of the metal or element in an ore deposit by its average crustal abundance. Table 2.1 lists the average crustal abundances and concentration factors for some of the important metals that are commonly extracted.

Concentration factors are highly variable and depend not only on the particular element but also on the type of deposit. They range from about 5–10 for geochemically abundant elements (e.g., aluminum and iron) to about 100–1,000 for less abundant elements (e.g., copper, zinc, and uranium). For example, as shown in Table 2.1, aluminum (Al), which has an average crustal abundance of 8%, requires a concentration factor of 3 to 4 to form a mineable deposit, which means, containing 24 - 32% Al to be considered economically viable. For some precious metals such as gold (Au) and silver (Ag), the level of concentration required to form a mineable deposit is usually in the thousands, and that may explain why gold and silver deposits are relatively rare.

The enrichment and concentration of ore elements in mineral deposits are controlled by certain geological processes in which each element has a specific geological association with host rock or ore-forming process. Such association ultimately leads to a heterogeneous and very uneven distribution of mineral deposits around the world because there is such a wide variety of geological environments and ore-forming processes. It is indeed remarkable that geologic processes which normally result in further dispersal of trace elements, have been able to produce mineable deposits of geochemically scarce elements, such as gold and mercury, involving concentration factors in the thousands. Therefore, the concentration factor is a good indicator of how difficult it is to enrich a metal to the level of a viable ore deposit.

Table 2.1: Concentration Factors of Metals in Mineral Deposits

| Substance (Element) | Average Crustal Abundance | Typical Grades of Ore Deposits | Concentration Factor |
|---|---|---|---|
| **Aluminum (Al)** | 8.0% | 24 - 32% | 3 to 4 |
| **Iron (Fe)** | 5.8% | 40% | 6 to7 |
| **Titanium (Ti)** | 0.86% | 25 – 45% | 25 to 100 |
| **Chromium (Cr)** | 0.0096% | 40% | 4,000 to 5,000 |
| **Zinc (Zn)** | 0.0082% | 4% | 300 |
| **Copper (Cu)** | 0.0058% | 0.5 – 4% | 100 to 200 |
| **Silver (Ag)** | 0.000008% | 0.01% | ~1000 |
| **Platinum (Pt)** | 0.0000005% | .0003 | 600 |
| **Gold (Au)** | 0.0000002% | 0.0001 – 0.001 | 4,000 to 5,000 |
| **Uranium (U)** | 0.00016% | 0.3 | 500 to 1000 |

## 2.3 Distribution and Size of Mineral Deposits

The *distribution* of mineral deposits worldwide varies both in space and time due to the geological processes that formed them and the timing of such events. Economic geologists have tried to unravel the reasons for the variation in the spatial and temporal distribution of mineral deposits. Research has however shown that, as a result of geological and/or tectonic processes, such as orogenic events or plate tectonic regimes, mineral deposits of a certain metal or group of metals, tend to be clustered in specific geographic areas or regions, known as *metallogenic provinces* or *mineral belts*. Some of these mineralized regions or districts are associated with specific metals that have been exploited for centuries, for example the base metal districts associated with the Kuperschiefer belt in north-central Europe.

The spatial groupings of mineral deposits occur because geological and tectonic processes that involve mineral deposition, such as the formation of sedimentary basins, emplacement of igneous bodies, and metamorphism in orogenic belts tend to be controlled by larger processes of crustal evolution. For example, the Cretaceous Benue Trough in Nigeria is characterized by the abundance of lead-zinc deposits within a linear sedimentary basin formed by rifting during the breakup of Gondwanaland and opening of the Atlantic Ocean. The distribution of those lead-zinc deposits is linked to the continental rift evolution of that intracratonic sedimentary basin. Another good example of a mineralized district is the Jos Plateau of north-central Nigeria where high-level, Jurassic anorogenic alkaline granites were emplaced during a period of regional uplift due to hotspot mantle plume activity are associated with tin and niobium-tantalum deposits. These examples cited above are indications that deep-seated mantle processes may have contributed significantly to the distribution of mineral deposits in the continental crust as a result of their effects on tectonic processes and anorogenic magmatism that produced the associated mineral deposits. While a lot is known about the spatial (horizontal) distribution of mineral deposits, the *vertical distribution* within the continents and their link to tectonic processes or orogenic cycle are still not well understood, although it is quite evident that the sources of ore metals can be found at several levels with the lithospheric crust and upper mantle, which are controlled by crustal architecture, in terms of heat transfer, permeability, redox potential and other parameters.

The *timing* of ore-forming processes also varies, and is known to spread across the geological time frame, from Precambrian to Recent. *Metallogenic epochs* are periods of geologic time during which conditions were particularly favorable for the formation of a specific type of mineral deposit. There are geological records that show certain periods when the formation of some deposit types was prevalent. These deposits have commonly formed through a combination of geological factors closely related in time. For example, the Banded Iron Formations (BIFs) which are the dominant source of iron ore worldwide were most widespread during the Archean to Early Proterozoic time frame. This period was a metallogenic epoch for iron ores, which has been attributed to the combination of certain factors, including an abundant supply of ferrous iron, and the absence of oxygen in the atmosphere during this period allowing iron to accumulate without oxidation. The recognition and delineation of metallogenic provinces and epochs are of fundamental importance to Economic Geology because of their relevance to the search for new mineral deposits. It is most probable that future discoveries of new "world-class" metallic deposits will be made within these mineralized regions and within established geological time frames.

The *size* of a mineral deposit is an important factor in its economic viability. It is a geometric reflection of the quantity and spatial disposition of an orebody. Mineral deposits vary in size from very small to super-large, usually depending on the ore-forming process and geological environment. Size is very relevant to whether a mineral deposit can be mined at a profit. Large low-grade deposits that occur near the surface can easily be mined by surface excavation whereas narrow vein deposits have to be extracted by expensive underground mining methods.

Size distribution of mineral deposits has been studied extensively by economic geologists, and two distribution patterns have emerged: and are referred to as fractal and exceptional. A *fractal* distribution is associated with those deposits in which their sizes are dependent on the rate of accumulation within a defined space, such as a magma chamber or a fault zone. Such constraint is observed for both magmatic and hydrothermal deposits, reflecting a limitation in size but continuity and similarity in the processes that created both small and large deposits. *Exceptional* size is observed in deposits that appear to be truly exceptional in size which may reflect extreme or unique events in the Earth, such as meteoritic impacts that create massive space for ore accumulation, or mantle plume events that create large igneous bodies such as the Bushveld Igneous Complex of South Africa , or major climatic factors that produced exceptional conditions for the widespread deposition of mineral deposits such as banded iron formations, or accumulation of thick lateritic bauxites. Such unique conditions provided the geological environment for exceptionally large but isolated mineral deposits such as the nickel deposits in the Sudbury basin of Ontario, Canada, or the chromite deposits in the Bushveld Igneous Complex, South Africa. Other examples include the large widespread bauxite deposits across West Africa that formed during a late Tertiary pediplanation that accumulated these deposits along an old erosion surfaces.

## 2.4 Mode of Occurrence and Morphology

Mineral deposits occur in various forms, shapes, and sizes, and are localized within a variety of host rocks of different composition. The *mode of occurrence* of a mineral deposit refers to its attitude or physical disposition. It is a descriptive term that connotes the general appearance and relationship of the mineral deposit to host rocks incorporating several characteristics that can be observed in the field. In general, magmatic deposits are usually massive, layered, lenticular, or pod-like in form, while sedimentary ore deposits are roughly tabular or stratiform. Hydrothermal deposits may occur as lodes, reefs, veins, veinlets, stringers, stockworks, and replacement bodies, while most other deposits may appear in a variety of modes including a blanket form for lateritic or residual deposits (Fig. 2.2).

From cross-cutting relationships and relative age, mineral deposits may be classified into two broad groups: *epigenetic* and *syngenetic*. Epigenetic ore body cuts across surrounding rocks and is believed to have been emplaced after the host rock was formed, whereas a syngenetic mineral deposit conforms to surrounding rocks and formed contemporaneously with host rocks. The relationships between mineral deposits and their host rocks are very important in understanding the processes by which mineral deposits were formed.

The *morphology* of mineral deposits depicts the shape and form of ore bodies as they appear in two or three dimensions. Morphology may be part of the mode of occurrence but both terms are not exactly synonymous. Table 2.2 shows some of the common terms used in describing morphology of mineral deposits.

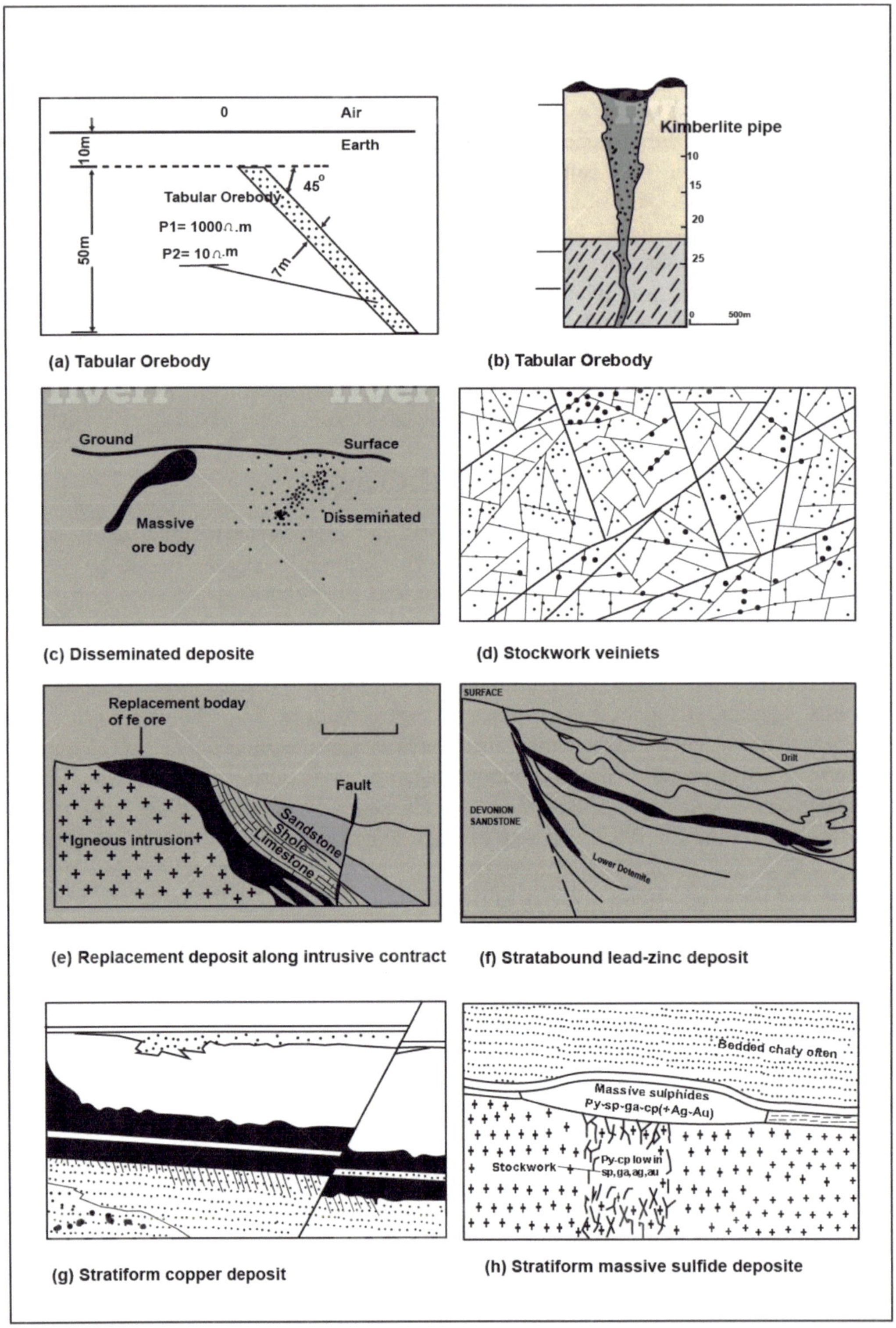


Fig. 2.2: Morphology of mineral deposits

*Discordant* mineral deposits are those that cut across the host rocks as, either regularly-shaped or irregularly shaped bodies. Three types of regularly-shaped discordant ore bodies can be distinguished by shape: *isometric, tabular,* and *tubular*. Isometric ore bodies are equidimensional or approximately equal in all dimensions. *Tabul*ar *or* flat ore bodies such as sheets, veins and lenses, have two long dimensions and one short dimension (Fig. 2.2a). The sheet, which is the most common shape in sedimentary deposits, is a tabular body separated from other rocks by bedding planes. *Veins* are ore bodies formed when mineral substance fills cavities or when there is a metasomatic substitution of mineral substances for rocks along cracks. The plane of contact between the vein and the enclosing rocks is called the hanging wall or footwall. A lens is a lenticular ore body that tapers out markedly in all directions; its thickness is smaller compared to its length. In terms of morphology, lenses and lenticular beds are sometimes transitional forms between isometric and flat ore bodies. *Tubular* ore bodies are pipes or pipe-like structures elongated in one direction, and oval in cross-section. A good example is the diamond-bearing kimberlite pipes of South Africa (Fig. 2.2b).

*Disseminated* ore occurs when economic minerals are scattered throughout the body of the host rock (Fig. 2.2c). Examples are the disseminated cassiterite (tin) found in granites of the Jos Plateau, northern Nigeria, and disseminated diamond in kimberlites. In other deposits, the disseminations may be mainly along closely-spaced fractures, such as in some porphyry copper deposits. *Stockworks* are characterized by long, closely-spaced, multidirectional veinlets cutting the host rock and forming an interlacing network. Examples are copper, molybdenum and tin deposits in granitic and granodiorite rocks (Fig. 1.2d).

*Replacement* textures occur when host rocks are replaced by mineralization products. Many ore deposits have formed by replacement of preexisting rocks, particularly at the contact between intrusive igneous rocks and carbonate-rich sediments at low to medium temperatures (Fig. 2.2e). On a smaller scale, host rocks surrounding mineralized quartz veins or lodes may show evidence of replacement by impregnation with fine-grained quartz or silicification. Figure 2.2 illustrates the morphology of several deposits.

*Concordant* ore bodies show conformity with the host rocks. They are usually *syngenetic* and show considerable development in two dimensions (parallel to bedding) and are limited in the third dimension. There are two types: stratabound and stratiform. *Stratabound* refers to any type of ore body that is restricted or confined to a particular part of the stratigraphic column or lies within a defined stratigraphic unit. They are sometimes remobilized in forms that are discordant to the bedding. *Stratiform* **is** a term used to describe a mineral deposit that is stratified or layered or occurs as a specific stratigraphic (or sedimentary) unit. Concordant ore bodies are found in a variety of sedimentary and igneous rocks. Magmatic ore deposits, such as chromite within mafic and ultramafic complexes occur as alternating layers with host rocks (Fig. 2.2g). Massive sulfide deposits also form concordant ore bodies within volcanic sequences within volcanic sequences as shown in Fig. 2.2(h).

Limestones are host to several types of concordant base metal sulfide deposits because of their increased permeability, such as the Mississippi Valley-type lead-zinc deposits (Fig. 2.2f). Shales, argillites, mudstones, slates are important rocks for concordant ore bodies e.g. Kupferschiefer of Germany which is a copper-bearing shale. The Zambia copper belt is also a good example where copper of diagenetic origin occurs as layers in shales and sandstone with local remobilization. Some deposits in sedimentary rocks may be both stratiform in some parts of the orebody and stratabound in other parts.

## 2.5 Mineralogy of Ore Deposits

Mineral deposits are of economic value because of the useful minerals they contain. The term '*ore*' is used to describe the assemblage of minerals found in mineral deposits: *ore minerals* and *gangue minerals*. The ore minerals are the useful minerals that are extracted whereas the gangue minerals are the valueless minerals demobilization with the ore minerals. Other useful non-metallic minerals that are not gangue minerals may also occur in mineral deposits, such as the industrial mineral barite. Both ore and gangue minerals are believed to be products of the same mineralizing process.

## 2.5.1 Ore Minerals

Ore minerals are metallic minerals from which one or more metals, such as iron and copper can be extracted and smelted easily for commercial use. Some important metallic minerals are magnetite (a source of iron), gibbsite (a source of aluminum), sphalerite (a source of zinc) and galena (a source of lead). Ore minerals contain one or more elements combined with anions or anion complexes, commonly oxygen, sulfur, sulfate and hydroxyl. Occasionally but rarely, these minerals occur as a native element (e.g. native gold or native copper). Ore minerals are generally classified into four groups: oxides, sulfides, native elements, and rarely silicates. The most common ore minerals and their chemical formula are listed in Table 2.2.

### 2.5.1.1 Native Metals

These are minerals or metals in native state that do not occur in combination with other elements. Only two metals, - *gold* and *platinum*, are found principally in their native state, although silver, nickel, copper, iron, osmium, and several other metals also occur in the native state, but only a few occurrences are large and rich enough to be ore deposits, e.g. the native Cu deposits in the Keweenaw basalts of Michigan, USA.

### 2.5.1.2 Sulfides

Sulfides are minerals composed of a combination of metal and sulfur. They constitute the largest group of ore minerals. The following metals copper, lead, zinc, nickel, molybdenum, silver, arsenic, antimony, bismuth, cobalt, and mercury; all form sulfide ore minerals in mineral deposits (Table 2.2). Because of the similarity in physical and chemical properties and mineral association, the sulfide class also includes the telluride, selenide, and arsenide minerals Gold and silver form tellurides under certain conditions, while platinum forms an important arsenide ore mineral.

### 2.5.1.3 Oxides and Hydroxides

These are minerals that combine metal and oxygen. The ore minerals of the most abundant metals such as aluminum, iron, manganese, and titanium belong to this group. Other less abundant metals, chromium, tin, tungsten, tantalum, niobium, and uranium also occur as oxides and hydroxides. Vanadium is found mainly as a substitute for Iron in magnetite, and uranium in uraninite.

### 2.5.1.4 Silicates

Silicate minerals are rarely found as ore minerals. Exceptions are garnierite $(Ni,Mg)_6Si_4O_{10}(OH)_8$, a clay-like mineral that is a major ore mineral in Ni laterites; and grunerite $(Fe_7Si_8O_{22}(OH)$, an iron-silicate mineral that occurs in lateritic iron ores. These silicate ore minerals are of secondary origin in residual deposits.

Table 2.2: List of Common Metallic Ore Minerals and Uses

| Metallic Minerals | Type of Ore | Uses |
|---|---|---|
| Argentite: $Ag_2S$ | For production of silver | Coinage, plate, and jewelry |
| Bornite: $Cu_5FeS_4$ | (Copper ore) | Copper-bearing products |
| Cassiterite: $SnO_2$ | Tin ore | Alloys and plating of steel |
| Chalcocite: $Cu_2S$ | for production of copper) | Copper-bearing products |
| Chalcopyrite: $CuFeS_2$ | Copper ore | Copper-bearing products |
| Chromite:$(Fe,Mg)Cr_2O_4$ | for production of chrome | Alloys, mainly with iron, nickel |
| Cinnabar: HgS | (production of mercury) | Scientific instruments |
| Columbite-Tantalite: $(Fe, Mn)(Nb, Ta)_2O_6$ | Ore of niobium-tantalum | Electronic components; capacitors |
| Galena: PbS | Ore of lead | Car batteries, pigments |
| Gold: Au | Native gold: | Jewelry |
| Hematite: $Fe_2O_3$ | Iron ore | Iron and steel |
| Ilmenite: $FeTiO_3$ | Titanium ore | Alloy in steel |
| Magnetite: $Fe_3O_4$ | Iron ore | Iron and steel |
| Molybdenite: $MoS_2$ | (Molybdenum ore) | special steel and alloy |
| Pentlandite:$(Fe, Ni)_9S_8$ | Nickel ore | Ferro-alloy, Special alloy |
| Pyrolusite: $MnO_2$ | (Manganese ore) | Alloy in iron and steel |
| Scheelite: $CaWO_4$ | Tungsten ore) | High-grade tool steel |
| Sphalerite: ZnS | Zinc ore) | Alloys and plating |
| Uraninite $UO_2$ | for uranium production | Nuclear energy |

## 2.5.2 Gangue Minerals

Gangue minerals are the non-valuable minerals associated with ore minerals in ore bodies. The most common gangue minerals are silicates and carbonates, including quartz, feldspar, mica, chlorite, clays, calcite or dolomite, tourmaline and garnets. They are found closely associated with certain ore minerals. For example, quartz is the gangue in most vein gold deposits. Minerals that are produced by hydrothermal alteration in epigenetic deposits may constitute part of a gangue assemblage. Examples of such minerals include sericite, biotite, kaolinite, chlorite and epidote in porphyry copper deposits. Contact metasomatic deposits formed in carbonate rocks may be associated with a unique assemblage of calc-silicate minerals called "skarn", including wollastonite, diopside, forsterite, grossularite and epidote.

## 2.5.3 Non-Metallic Minerals

Non-metallic minerals can form extensive mineral deposits that are exploited although they are not metals or fuels but valued for their physical and chemical properties. Non-metallic minerals have their own gangue minerals. For example, barite mineralization may contain barite as the economic mineral while the associated quartz is the gangue mineral. Non-metallic minerals include industrial minerals, and gemstones. Table 2.3 provides a list of industrial minerals and those used in construction. A list of the most important gemstones is shown in Table 2.4.

Table 2.3: List of Common Non-Metallic Minerals and Rocks

| Non-Metallic Minerals and Rocks | | |
|---|---|---|
| **Industrial Minerals** | | **Construction Materials & Rocks** |
| Barite | Halite | Granite |
| Bentonite | Zircon | Limestone |
| Beryl | Mica | Marble |
| Clays | Monazite | Phosphate |
| Diatomite | Spodumene | Gravel |
| Feldspar | Talc | Sand |
| Fluorite | Sillimanite | Slate |
| Gypsum | Zircon | Trona |

Table 2.4: List of Precious and Semi-precious Gemstones

| Precious Stones | Semi-precious Stones |
|---|---|
| Diamond | Garnets |
| Ruby | Amethyst |
| Sapphire | Aquamarine |
| Emerald | Beryl |
| Pearl (black) | Tanzanite |
| | Tourmalines |
| | Topaz |
| | Zircon |

## 2.6 Textures and Structures of Ore Minerals

The textures of ores refer to the appearance, forms, and interrelationships of the minerals. The identification and interpretation of mineral textures can be very helpful in understanding the processes involved in the formation of mineral deposits. The nature of ore textures is usually determined by many factors, including the composition of the mineralizing fluids, the physical conditions of ore formation, and the nature of available space within the host rocks. There are quite a variety of textures observed in ore minerals, but they can be simply classified into three types: (1) *magmatic precipitation textures*, (2) *open-space filling textures* that permits the precipitation of minerals, and (3) *replacement textures* that are derived from changing the pre-existing conditions in the host rocks.

### 2.6.1 Magmatic Precipitation Textures

Magmatic precipitation textures are uniquely characteristic of magmatic ore deposits in which oxides and sulfides crystallize contemporaneously in the magma. The growth of ore minerals in silicate melts during the early phase of crystallization generally results in the development of euhedral to subhedral crystals because there is a little obstruction to the growth of mineral faces. Thus, primary chromite, magnetite, and platinum mineral phases that crystallize early from mafic magmas are usually refractory enough to retain their original textures. However, in many cases, these early minerals may subsequently form phenocrysts that may be resorbed by later silicate phases. *Resorption textures* are therefore common in magmatic chromite and iron-titanium ores. (Fig. 2.3).

*Cumulus textures* are developed when early formed crystals of ore minerals accumulate or settle from liquid melt to the bottom of a crystallizing magma. This is a texture reflecting accumulations of the same mineral in solid form during magmatic crystallization (Fig. 2.4). Examples include the chromite crystals in magmatic segregation deposits in mafic/ultramafic rocks. *Intergranular textures* occur where the ore mineral appears as anhedral crystals lying between the grains of gangue minerals because the ore mineral crystallized later in the paragenetic sequence than the gangue minerals. Examples include numerous sulfides that in many cases, are crystallizing from liquids immiscible with, and of a lower melting point than the silicate magma. *Exsolution textures* occur when one mineral phase separates from another as a result of incomplete miscibility during cooling, with a tendency to concentrate along cleavage planes.

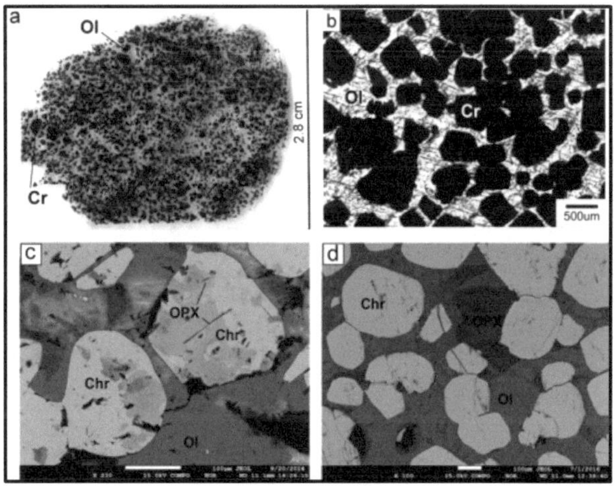

Fig. 2.3: Resorbed crystals of chromite in silicate melt

## 2.6.2 Open Space-Filling Textures

Ore forming solutions may precipitate along fissures or cavities created by fractures and faults. The initial deposition and growth of the ore and gangue minerals in many deposits occur in open spaces in vugs or fractures developed by structural deformation or mineral dissolution. Minerals may grow from the outer contacts of the wall to the center of the vein, and are often characterized by well-developed crystal faces or by crystals that exhibit growth zoning, e.g., cassiterite in greisen veins. Such textures are most common in vein deposits. The deposition of layers of different minerals in the walls of a cavity leads to the development of what is known as *crustification*. Successive deposition of layers of crystals by hydrothermal solutions which are developed within the cavities, give rise to what is called *comb* structure or *colloform banding* which is common in low temperature deposits (Fig. 2.5). The open spaces left after the filling up of the cavities with minerals are known as *vugs* or *druses*. If cavity filling deposits are composed of one mineral only, and devoid of crustification, they are sometimes described as *massive texture*.

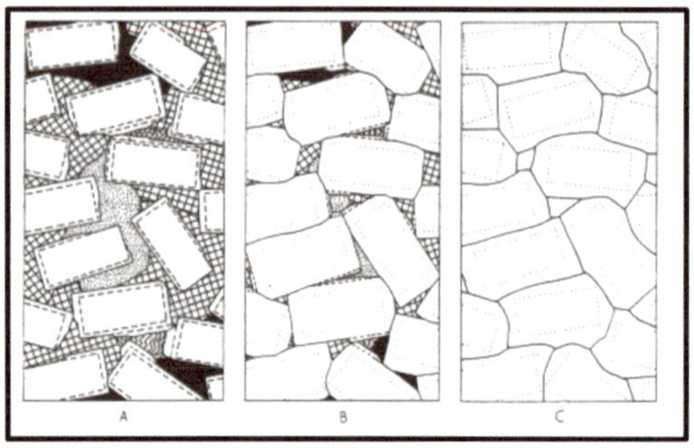


Fig. 2.4: (a) Cumulus textures in metallic ores

(b) Cumulus texture in Chromite

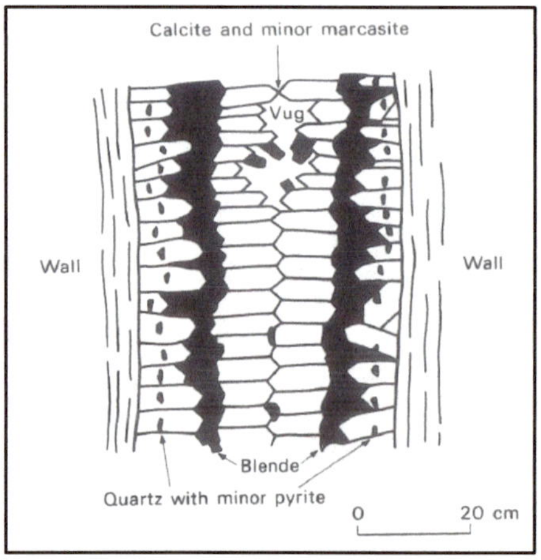


Fig. 2.5: (a) Open space filling

(b) Colloform banding of calcite-sulfide ore

### 2.6.3 Replacement Textures

Replacement textures are formed when one mineral replaces another mineral or the surrounding wall rock. Such processes may take place either at high temperatures during crystallization or at low temperatures during weathering. Replacement textures are more common at high T and P where open spaces are very limited or unavailable, and fluid flow is rather difficult. Replacement may result from one or more of the following processes: (I) dissolution and subsequent re-precipitation, (2) oxidation, and (3) solid state diffusion. The resulting boundary between the replaced and the replacing mineral is commonly either sharp and/or diffuse. It also depends largely on the chemical composition and reactivity of both the host rock and the incoming hydrothermal solution.

Based on the observation of how certain minerals preferentially replace other minerals, a set of "rules" has been proposed regarding the pattern of mineral replacement in ores:

a. Sulfides replace gangue or ore minerals
b. Gangue minerals replace host rock, but not the ore minerals
c. Oxides replace host rock and gangue, but rarely replace sulfides

Replacement of one ore mineral by another or by a secondary mineral formed during chemical weathering and supergene enrichment is common in many types of ores, such as lateritic iron and manganese ores and niobium deposits. However, the replacement textures are often difficult to recognize because of the low crystallinity of the secondary minerals.

## 2.7 Mechanisms of Ore Formation

Mineral deposits form in a great variety of ways, and one or more geological processes may be involved in ore deposition. The three basic geological requirements for the formation of mineral deposits (often referred to as the *"source-transport-trap"* components) are (1) *source(s)* of the ore components; (2) transport *mechanism* for the components to reach mineralization site, and (3) *depositional mechanism* (trap) that precipitates the ore components. The *preservation* involves the geological process or environment that allows the mineral deposit to be preserved (Fig. 2.6). These components constitute a system for which an external source of energy acts as the engine that drives the processes. The mechanism, duration, and relative importance of these factors vary among the different ore-forming processes, but they are all controlled essentially by a similar set of physical and/or chemical conditions in the depositional environment. The energy that drives the transport mechanism includes thermal, gravitational and tectonic. The source and transport mechanism may be different.

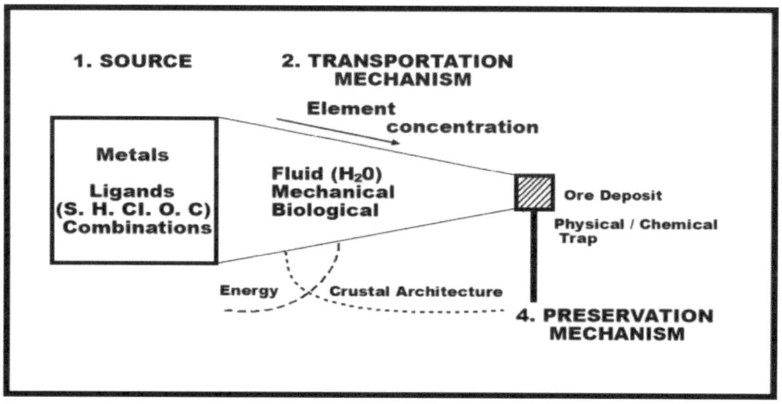


Fig. 2.6: Basic requirements for ore formation (After McQueen, 2005)

### 2.7.1 Sources of Ore Constituents

A source or sources of ore constituents in mineralizing solutions is necessary for the accumulation of mineral deposits. However, a unique source may not exist or be recognized; instead, multiple sources may be involved in providing the metals. The mantle and crust are believed to be ultimate or major sources of metals found in mineral deposits through the processes of partial melting and leaching. Magmas which are silicate melts containing several ore constituents are believed to originate from either the upper mantle or the crust, or by partial melting of subducted oceanic and lithospheric crust (plate tectonics). For example, magmas carrying heavy metals may also scavenge additional ore metals from the crust or surrounding rocks. Host rocks of mineral deposits are potential sources of ore constituents, as well as sea water from which metals precipitate by chemical and biochemical reactions or evaporation. The potential sources and extraction processes of ore elements are outlined as follows:

a. Magmatic fluids from deep-seated sources, in the mantle or crust.
b. Volcanic exhalative processes through volcanic eruptions beneath oceanic or sea water.
c. Lateral secretion or diffusion of metals from country rocks into faults and other structures.
d. Leaching of metals from sediments by heated meteoric or connate waters
e. Recrystallization of existing metals or leaching from metamorphic rocks.
f. Leaching of metals from pre-existing minerals by secondary processes including supergene enrichment.

Geochemical methods, particularly isotopic methods, have contributed immensely to our knowledge of the sources of ore elements. Age determinations using radiogenic isotopes such as Rb/Sr, Os/Rh have helped establish the age of mineralization relative to host rocks and point to a source and/or process. For example, gold and associated pyrite in Archean conglomerates of South Africa were found to be older than the host rock, thereby ruling out a hydrothermal origin and suggesting a sedimentary process and detrital source. The widespread use of oxygen, hydrogen, sulfur and carbon isotopes allows today's economic geologists to distinguish among surficial, oceanic, magmatic, and mantle sources of fluids that transport ore elements. For example, in the Benue lead-zinc deposits in Nigeria, isotopic ratios for sulfur and hydrogen indicate that the sulfur was from seawater sulfate in connate brines while the lead (Pb) isotopes indicate a crustal source for the Pb.

## 2.7.2 Transport of Mineralizing Fluids

The formation of mineral deposits requires the transport of the ore constituents from the source region to the site of deposition. Transport agents include silicate melts, hydrothermal fluids or gaseous medium which is very rare. Transport mechanism is mainly mechanical means, mass transfer, and fluid action. The movement of mineralizing fluids is assisted in places by environmental factors, such as temperature and pressure changes, the permeability of the host rock, and structural controls created by fractures and faults. Two main types of transport medium have been identified for ore elements; *silicate melts* (magmas) and *hydrothermal solutions* (epigenetic deposits). Transport of ore metals in a *gaseous* or *vapor phase* and subsequent precipitation by de-sublimation (changing from a gas to a solid) has been recognized in some high temperature hydrothermal and volcanogenic deposits, but is believed to be rare. Ligands have a major role to play as chelating agents in the transportation of metals.

Magmas which are silicate melts are a major medium of transporting metals such as chromium, titanium and platinum that are found in magmatic deposits. The ore metals transported by silicate melt from deep within the upper mantle or lower crust get concentrated in the magma chamber by processes such as gravitational settling and density separation, as well as liquid immiscibility. However, recent geochemical studies have shown that the mechanisms are much more complex than originally imagined a few decades ago. For example, the sulfur in magmatic sulfide deposits is partly of crustal origin and sulfur saturation may not be related to the cooling of an enriched magma but to the interaction between magma and crustal rocks to produce sulfide immiscibility.

The mechanisms that operate at the transition between magmatic and hydrothermal processes are also better understood. Residual volatile-rich fluids from fractional crystallization in magmas are believed to fractionate into a water-rich fraction by exsolution of an aqueous fluid to form a distinct phase in the silicate melt. The aqueous fluid may boil and gas bubbles from metals are transported at relatively lower temperatures often accompanied by sodium or potassium enrichment and metasomatism. Such hot solutions may be vapor-rich and carry a variety of elements, particularly those that cannot be incorporated in the rock during crystallization. Some rare-earth metal-rich pegmatites and hypothermal deposits are commonly associated with such processes.

Hydrothermal fluids are a major transport medium for many ore systems. They are hot saline waters with lesser and variable amounts of $CO_2$, $H_2S$, $SO_2$, $CH_4$, $N_2$, NaCl and other salts, as well as dissolved metal complexes that are capable of transporting ore elements as ionic complexes. There are various sources of hydrothermal solutions, including magmatic fluids, connate brines, sea water, and meteoric water. At upper levels in the crust, hydrothermal fluids are typically hotter than the rocks they traverse and characterized by variable pH and Eh values that may be charged with a range of metal complexing agents including fluorine, chlorine and hydrogen sulfide. Magmas contain water that can reach up to 10% of its volume. Magma-derived hydrothermal solutions accumulate towards the end of magmatic crystallization to form a distinct phase of the silicate melt that may be ultimately ejected out of the magma chamber. Connate waters are saline brines trapped during sedimentation. They are capable of transporting metals within sedimentary basins and migrate and deposit minerals in response to changes in fluid pressure, tectonic activity, or diagenesis. Sea water and meteoric water can be mobilized as transporting medium for ore elements by heat from submarine volcanism or shallow intrusive or near-surface magmatic activity.

Over the years, the understanding of ore transport mechanisms by hydrothermal fluids has improved due to studies of the contents of *fluid inclusions* which are the remnants of hydrothermal solutions trapped in mineral lattices. Moreover, our knowledge of fluid movement has been enhanced due to advances in the field of crustal permeability studies that shed light on depths at which fluids could be found in the crust. Apparently, hot and saline hydrothermal fluids can migrate vertically and laterally within rocks of the lithospheric crust and upper mantle.

### 2.7.3 Deposition of Ore Minerals

For mineral deposits to accumulate, there must exist a "trapping mechanism" in the form of a geological environment conducive for the localization and deposition of the ore minerals and associated gangue. If the ore-forming solutions are not trapped, there will be no mineral deposit (Figure 2.7). *Localization* is used to describe the process of creating open spaces in rocks and channeling ore solutions through them, while *deposition* is the accompanying process of precipitation of ore minerals in open spaces due to physical and/or chemical reactions. The environment of ore deposition is varied and complex; while some deposits are formed at great depths, some are deposited at shallow depths, near the earth's surface or on the seafloor. Ore deposition will occur within a mineralized zone when metal-carrying fluids become chemically unstable during transportation, due to changes to physicochemical parameters, including temperature, pressure, pH, redox state, and the total concentration of ligands. These changes are also associated with such processes as the addition of components by contamination, phase separation, cooling across a temperature gradient, pressure decrease, fluid mixing, and reaction with host rocks. These factors that could lead to ore deposition are outlined for more clarity as follows:

1. Changes in the thermo-dynamics of magmatic crystallization and crystal fractionation processes
2. Falling temperature and/or loss of pressure in mineralizing fluids which renders the metal-bearing complex unstable or metal insoluble;
3. Physical and/or chemical reactions within the ore-forming fluid;
4. Changes in the pH (acidity) of ore-forming solutions
5. Exchange reaction of fluids with chemically reactive wall rocks such as carbonate rocks or other rocks of reduced oxidation state, including iron-bearing sediments, and mafic or ultramafic rocks;
6. Mixing of two fluids that can lead to cooling and modification of the prevailing ore fluid properties, and destabilization of existing metal complexes
7. Degassing of hydrothermal fluids into a gas and water phase, or what is called "boiling", which alters the metal carrying capacity of the solution and may even destroy metal-carrying chemical complexes;
8. Interaction with physical barriers created by structural disposition that suddenly forces ore-bearing solutions to deposit their ore load.

Ali the above factors or processes can act independently or in combination, under varying scenarios to facilitate the deposition of ore minerals in various geological environments.

## 2.8 Classification of Mineral Deposits

The classification of mineral deposits is a process of grouping deposits into similar categories based on common or unifying features. Classification is a widely used procedure in geosciences to help simplify relationships and promote a better understanding of the substances being studied. The main purpose of classifying mineral deposits is to facilitate description and permit generalizations regarding environments of formation and genesis of mineral deposits.

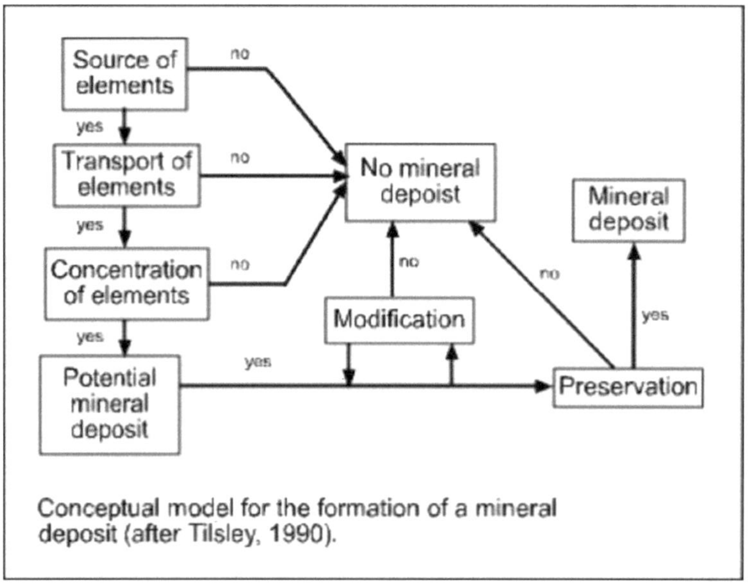


Fig. 2.7: Conceptual model for the formation of a mineral deposit

### 2.8.1 Criteria for Classification

Several criteria have been used in classifying mineral deposits since the days of Agricola, the *founder* of the science of ore deposits who wrote the book *De Metallurgica* in 1556. Successive authors on the subject have attempted various classifications, none of which has received unanimous endorsement of geologists, miners, and mining engineers who are all interested in ore deposits. Many economic geologists have proposed classification schemes that are mainly of academic interest. Mining engineers and miners naturally have little regard for a classification that does not help them find ore. Mining geologists may share this feeling and deplore the overemphasis that may be placed on classification by their academic peers. However, the use of certain criteria have been accepted and evolved with time as more scientific data became available and increased our body of knowledge in such areas as fluid inclusion studies and isotope geochemistry.

The criteria that have been accepted and used for classification of mineral deposits include; (1) metal content (commodity); (2) shape or form of deposits (morphology); (3) host rock association; (4) ore forming processes, (5) conditions of deposition (temperature, pressure), (6) tectonic setting, (7) origin (genesis). Each of these criteria may have its own merits or demerits, but classifying mineral deposits can be controversial because no two deposits are the same. Moreover, the formation of a deposit may be due to multiple processes superposed on each other, thereby creating binary uncertainty. To be most useful, particularly for students, any classification of substances as complex and variable as mineral deposits should be based on criteria that can be easily identified in the field, as well as being as simple as possible.

Thus, if a classification is too complicated, it may not allow for generalization that can be useful in understanding ore-forming processes.

## 2.8.2 Metal Content Classification

Economic minerals are often classified according to their metal content or mineral extracted from them. For example, minerals containing copper include several sulfides and oxides in the same way, mineral deposits can be classified according to their metal content. The metals can be classified into four groups, namely; ferrous metals, precious metals, base metals, and non-metallic minerals. Ferrous metals refer to those metals associated with iron and steel production, such as Fe, Cr and Mn. The base metals, also known as 'industrial metals' refer to a group of common metals that are widely used in manufacturing. The precious metals are the very valuable metals, while non-metallic minerals do not contain metals. Not all metals fit into this classification, particularly the so-called rare metals. The benefit of this classification is the emphasis on the economic value and the relationship of the metals. Ores of the same metal from different geological environments are grouped together. A classification based on the type of metals may not have geological connotation, and may be of lesser value to geologists, although it can be used for the purpose of economic analysis and organizing the general description of metallic mineral deposits, as used in Chapter 4 of this book. Table 2.5 is an example of such classification of Nigeria's mineral deposits.

Table 2. 5: Metal Content Classification of Nigeria's Deposits (Olade, 2020)

| Class | Metal | Deposit Type | Major Minerals |
|---|---|---|---|
| Ferrous Metals | Iron | Ferruginous Quartzite<br>Magmatic Iron Ore<br>Sedimentary Iron ore | Hematite, Magnetite<br>Magnetite, Ilmenite<br>Goethite, Magnetite |
| | Manganese | Metamorphic Gonditic | Manganite, Pyrolusite |
| | Chromium | Disseminated Magmatic | Chromite |
| | Nickel | Residual/Disseminated | Native Nickel |
| | Molybdenum | Hydrothermal veins | Molybdenite |
| | Tungsten | Hydrothermal veins | Wolframite |
| Base Metals | Copper | Hydrothermal veins | Chalcopyrite, Azurite |
| | Lead | Hydrothermal veins | Galena |
| | Zinc | Hydrothermal veins | Sphalerite |
| | Tin | Hydrothermal Veins<br>Pegmatitic & Placer | Cassiterite<br>Cassiterite |
| Precious Metals | Gold | Hydrothermal Veins<br>Placer | Gold<br>Gold |
| | Silver | Hydrothermal Vein | Argentite, Ag-Galena |
| | Platinum | Disseminated | Platinum-Palladium |
| High Technology Metals | Niobium | Disseminated<br>Placer | Columbite<br>Columbite |
| | Tantalum | Pegmatitic<br>Alluvial | Tantalite, Coltan<br>Tantalite, Coltan |
| | Titanium | Magmatic and Alluvial | Ilmenite, Rutile |
| | Bismuth | Accessory | Bismuthinite |
| | Zirconium | Accessory/Alluvial | Zircon |
| Radioactive Metals | Uranium | Hydrothermal Veins<br>Sedimentary Uranium<br>Peralkaline Granite | Uraninite, Autunite<br>Betafite<br>Uranoan-Pyrochlore |
| | Thorium | Alluvial | Monazite |

## 2.8.3 Morphological Classification

This classification is based on the shape, size or attitude of mineral deposits, Vein-like deposits are distinguished such as layered deposits; and massive ores are distinguished from disseminated ores, etc. Emphasis is based on the form of the deposit in relation to host rocks (Table 2.6) This classification has found more relevance in mining, where the shape and attitude of ore bodies are relevant in mine design and development. Classification based on the *morphology* of ore bodies could be quite useful, but two ore bodies of similar shape may be very different in the type of ore, texture, origin, temperature of formation etc. which are factors of greater geological interest.

Table 2.6: Simplified Classification Based On Morphology of Mineral Deposits

| Form | Mode of Occurrence | Typical Examples |
|---|---|---|
| *Disseminated* | Ore minerals dispersed through the host rock | Diamond in kimberlites |
| *Stockwork* | Intersecting network of small, narrow, closely spaced, ore-bearing veinlets in host rock | Footwall alteration zone of volcanogenic massive sulfides |
| *Massive* | Mineralization comprising >50% of host rock | Volcanogenic massive sulfides |
| *Tabular* | Mineralized zone that is extensive in two directions but has limited development in the third dimension | Sandstone uranium deposits |
| *Tubular* | Mineralization that has a pipe-like, vertical disposition , narrow in two dimensions and extensive vertically | Diamond in kimberlite pipes |
| *Vein-type* | Mineralization in veins, commonly discordant to the host rock | Base metal and precious metal veins |
| *Stratiform* | Mineralization confined to a specific bed, and broadly conformable to the host rock layering | Kuperschiefer –type stratiform copper deposits |
| *Stratabound* | Mineralization confined within a stratigraphic unit, but may be discordant in places | Mississippi Valley type deposits; Unconformity uranium deposits |

## 2.8.4 Host-Rock Classification

This classification uses the type of host rock or rock association as the criteria for classification, for example, ores in mafic-ultramafic rocks, ores of felsic association or ores in sedimentary rocks etc. This classification emphasizes the relationship between ore and host rock, which may also be partly genetic. The lithological association of ore with host rock is in most instances a genetic relationship that has significant application in the exploration for mineral deposits. Host rock criteria can also be combined with other criteria, such as genesis for a detailed sub-classification of mineral deposits as used in Chapter 3 of this book. Table 2.7 is an example of a genetic classification proposed by geologists at the University Of Quebec at Montreal, Canada (UQAM). Examples of types of deposits are shown in the column next to the lithologic association.

Table 2.7: Host Rock Classification (Jebrak, 2005)

| Association | Type |
|---|---|
| **Mafic and ultramafic plutonism** | Chromium, copper-nickel layered complexes<br>Nickel komatiites<br>Chromium and PGE ophiolites<br>Platinum in ultramafic intrusions<br>Titanium anorthosites |
| **Alkaline volcano-plutonism** | Carbonatites<br>Diamondiferous kimberlites<br>Differentiated alkaline magmatism<br>Cu-U-Au-REE iron oxides (IOGC) |
| **Felsic plutonism** | Granitic pegmatites<br>Tin-tungsten cupolas Uranium episyenites<br>Copper porphyries<br>Molybdenum and tin porphyries<br>Metasomatic Cu-Pb-Zn-W deposits |
| **Aerial felsic volcanism** | High-sulfidation epithermal Cu/Au deposits<br>Low-sulfidation epithermal Au/Ag deposits<br>Gold-bearing alkaline maars and diatremes<br>Replacement gold deposits<br>Replacement base metal deposits (mantos) |
| **Vein deposits in the mid- and lower crust** | Gold shear zones<br>Pb-Zn-F-Ba veins<br>Co, Ag veins<br>Sb veins |
| **Submarine volcanism** | Cu-Zn ophiolites<br>Bimodal Pb-Zn-Cu volcanism<br>Sedex deposits |
| **Sedimentary deposits** | Iron formations (BIF)<br>Oolitic iron<br>Sedimentary phosphates Manganese deposits<br>Barite in black shales |
| **Diagenetic deposits** | Copper in pelites<br>Uranium sandstones Discordant uranium<br>Pb-Zn-deposits in sedimentary cover rocks<br>Extensional veins |
| **Alteration deposits** | Nickel laterite deposits<br>Oxidized copper<br>Supergene gold<br>Bauxite |
| **Placers and paleoplacers** | Fluvial placers<br>Gold-bearing deltaic paleoplacers<br>Black sand marine placers |

## 2.8.5 Classification by Environment of Deposition

This classification scheme uses known or inferred physico-chemical conditions of ore deposition, such as temperature, (hypothermal, mesothermal, epithermal), depth of deposition (deep, intermediate and shallow) or estimated pressure, fluid salinity etc. Lindgren proposed this classification in 1933 as shown in

Table 2.8 according to the estimated temperature of deposition based on mineral association and texture. The scheme is still valid and used today. With advances in research on ore deposits and more precise analytical techniques, it has become feasible to measure the physical and/or chemical conditions of mineral deposition using fluid inclusion and isotope geochemistry. Whenever possible, physical parameters such as temperature of formation, can be used to subdivide hydrothermal deposits.

Table 2.8 Classification of Ore Deposits by Temperature of Deposition (Evans, 1933)

| | Depth | Temperature (°C) | Occurrence | Metals |
|---|---|---|---|---|
| **Telethermal** | Near surface | ±100 | In sedimentary rocks or lava flows; open fractures, cavities, joints. No replacement phenomena | Pb, Zn, Cd |
| **Epithermal** | Near surface to 1.5 km | 50–200 | In sedimentary or igneous rocks; often in fault systems; simple veins or pipes and stockworks; little replacement phenomena | Pb, Zn, Au, Ag, Hg, Sb, Cu, Se, Bi, U |
| **Mesothermal** | 1.2–4.5 km | 200–300 | Generally in or near intrusive igneous rocks; associated with regional faults; extensive replacement deposits or fracture fillings; tabular bodies, stockworks, pipes | Au, Ag, Cu, As, Pb, Zn, Ni, Co, W, Mo, U etc. |
| **Hypothermal** | 3–15 km | 300–600 | In or near deep-seated felsic plutonic rocks in deeply eroded areas. Fracture-filling and replacement bodies; tabular or irregular shapes | Au, Sn, Mo, W, Cu, Pb, Zn, As |

## 2.8.6 Tectonic Classification

Mineral deposits can be classified according to their tectonic setting, either in relation to orogenic cycle or the plate tectonics theory. Using the stages of orogeny, mineral deposits are classified as pre-orogenic, synorogenic such as orogenic gold, and late orogenic or post-orogenic deposits (e.g pebble conglomerate gold deposits). Such classification schemes are no longer common except for deposits of Precambrian age. The 20th century was associated with the development of the theory of plate tectonics which spawned a swarm of classifications based on tectonic setting (Table 2.9). Mineral deposits are classified in relation to plate boundaries: convergent, divergent and collision plate margins. Deposits in ocean basins are distinguished from those in intracratonic basins or back arc basins, and so on and so forth. Tectonic classification is still used, particularly when discussing the large-scale or global distribution of mineral deposits, and may be useful in the study of metallogenic belts. However, this type of classification is faced with complications that arise from considering a large number of variables (lithological, structural, geochemical and tectonic etc.), which may make the classification very complex.

Table 2.9: Tectonic Classification of Mineral Deposits

| | | | |
|---|---|---|---|
| 1 | ***Deposits in divergent plate margins*** | Volcanogenic massive sulfides | (Cu, Zn) |
| | | Sedimentary exhalative deposits | (Zn, Cu, Pb, Au and Ag |
| | | Manganese nodules | Mn |
| | | Stratiform igneous complexes | Cr, PGE |
| 2 | ***Deposits in convergent plate margins*** | Porphyry deposits | Cu, Mo |
| | | Veins and contact metasomatic | Pb–Zn–Ag |
| | | Precious metals | Pt, Au, Ag |
| | | Other metals | Sn, W, Hg, Sb |
| 3 | ***Deposits in cratonic rift systems*** | Mineralization in alkaline granites | Sn, fluorite, barite |
| | | Evaporites in rift basins | Na, K< B |
| | | Carbonatites containing | Nb, P, REE, U |
| | | Hydrothermal veins in basin sediments | Pb-Zn, Cu, U, Ba, F |
| 4 | ***Deposits in intracontinental settings*** | Stratiform deposits in layered intrusions | Ni, Platinum |
| | | Mineralization in anorthosites | Ti-Fe |
| | | Iron-oxide Cu-Au deposits | Fe-Au-Cu |
| | | Pb–Zn–Ag deposits in limestones | Pb-Zn-Ag |
| | | Sedimentary Cu deposits | Cu |
| | | Diamonds in kimberlites | Diamonds |

## 2.8.7 Genetic Classification

The genetic classification is broadly based on the supposed origin of the mineral deposits. The genetic types include deposits associated with magmatic, sedimentary, metamorphic and surficial processes (Table 2.10). Geologists generally prefer integrated genetic classification schemes that also incorporate elements of composition, ore forming processes, form and rock association. From these, it is possible to construct predictive models that can be used to search for geological environments in which appropriate ore-forming processes have probably operated. A genetic classification scheme that includes all deposit types and integrated ore-forming processes is presented in Table 2.10. Further subdivision on the basis of morphology, rock association and environment of deposition can be applied to some deposit types as used in Chapter 3 of this book. Such an integrated genetic classification scheme can be easily understood while more details can be added depending on the level of knowledge. For example, in this scheme the deposit types are magmatic, hydrothermal, sedimentary, metamorphic and surficial. Each deposit type is subdivided on the basis of ore-forming processes. The subtypes can be further classified according to other parameters such as host rock, mineral composition, form and depositional environment (e.g. temperature and pressure conditions).

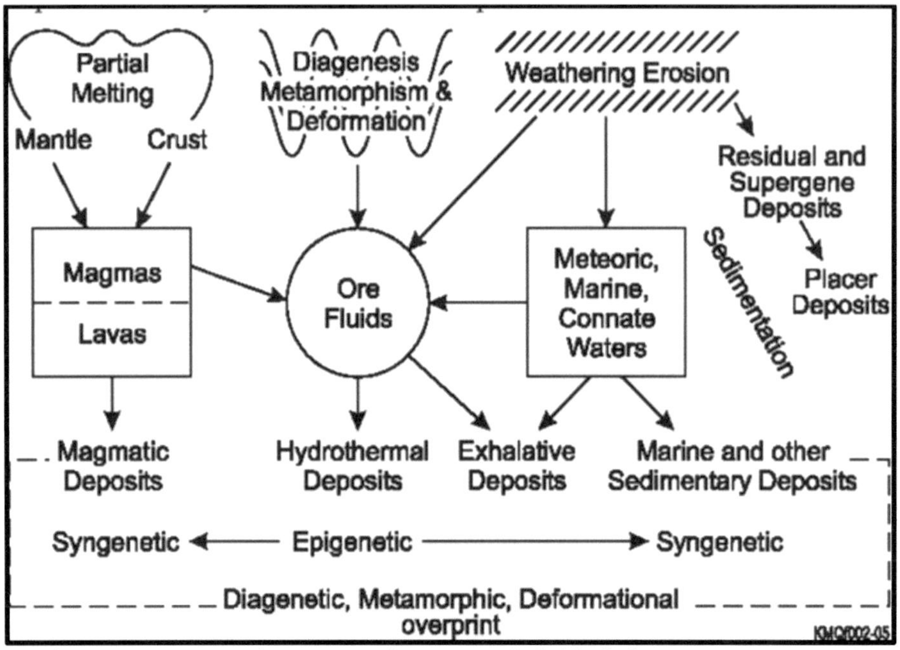

Fig. 2.8: Ore Deposit Types and Sources of Fluids (After McQueen, 2005)

Genetic classification is very easy to understand because the mineral deposits are classified in the same way as major rock groups: igneous, sedimentary, metamorphic and weathering products. Using the *genesis* (origin) of mineral deposits has always been the most fascinating and widely used criteria for the classification as it provides a simple and broad classification scheme that can be related to host rocks observed in the field. It has also proven to be the most consistent criteria that can be best adapted for further subdivision according to other fundamental factors, such as ore-forming processes, lithology or morphology. Both metallic and non-metallic minerals can be classified under such scheme.

## 2.8.9 Ore Deposit Types

A simple and widely accepted genetic classification will be used in this book to classify and describe mineral deposit types, with further subdivision by ore-forming processes, and/or morphology or temperature of deposition (Table 2.10). Figure 2.8 is a schematic diagram showing the major deposit types, and the sources of ore-forming fluids and processes responsible for ore formation.

Table 2.10: Genetic Types of Mineral Deposits and Ore-Forming Processes

| | Genetic Type | Ore forming Processes | Deposit Type | Typical Examples |
|---|---|---|---|---|
| 1 | **Magmatic** | Dissemination | Disseminated Deposits | Diamond in kimberlites, Nb, REE in carbonatites & alkaline granites |
| | | Gravitational Settling | Stratiform Layered Deposits | Chromite in layered igneous complexes, (Bushveld) Fe-Ti in gabbros (Allard Lake) |
| | | Immiscible liquid injection | Massive sulfide /oxide deposits | Nickel-Cu sulfides in komatite and gabbros (Sudbury) |
| | | Fractional Crystallization | Pegmatite deposits | Pegmatites (Tanzania) Rossing mine Namibia |
| 2 | **Hydrothermal** | Magmatic fluids | Magmatic-hydrothermal deposits | Sn-W deposits in granites (Nigeria) |
| | | Volcanic-related fluids | Volcanogenic massive sulfides | Cu-Pb-Zn deposits (Kuroko, Japan) |
| | | Connate/ water | Mesothermal/Epitherm al deposits | Pb-Zn deposits (Nigeria); |
| | | Metamorphic fluids | Metamorphic-hydrothermal | Orogenic gold deposits (Ghana) |
| | | Volcanic exhalation | Sedimentary exhalative deposits | Polymetallic Pb-Zn-Ag deposits (Broken Hill) |
| | | Heated Meteoric Water Geothermal hot springs | Geothermal-epithermal deposits | Gold-Ag-Hg deposits (New Zealand) |
| 3 | **Sedimentary** | Chemical precipitation | Chemical precipitate deposits | Banded Fe formations; Sedimentary Fe & Mn deposits. Copper deposits (Zambia) |
| | | Evaporation | Evaporite deposits | Gypsum & Halite deposits |
| | | Clastic Sedimentation | Placer deposits | Alluvial Au, Pt &Sn deposits; Paleo-placer uranium and Au deposits |
| | | Biogenesis | Biogenic deposits | Coal, Lignite, Tar Sand Diatomite and phosphate |
| 4 | **Metamorphic** | Metamorphic crystallization | Metamorphic deposits | Graphite, Talc & garnet deposits |
| | | Metamorphic recrystallization | Metamorphosed deposits | Iron ores, |
| | | Dynamic metamorphism | Lateral secretion deposits | Gold deposits (Yellowknife, Canada) |
| | | Contact Metamorphism | Skarn deposits | Copper, Fe, W deposits |
| 5 | **Surficial** | Weathering | Residual deposits | Lateritic, bauxite, Lateritic Fe, Ni & Au deposits |
| | | Supergene processes | Supergene deposits | Gold and Ag bonanzas,; Blankets of porphyry Cu; Supergene Fe and Mn |

#### 2.8.9.1 Magmatic deposits

Magmatic deposits are those valuable minerals concentrated in igneous rocks during magmatic crystallization. Processes involved in ore formation include crystal settling, dissemination, liquid immiscibility and fractional crystallization. They occur mainly as disseminations, layered bodies of massive ores of oxides and sulfides in plutonic and volcanic rocks and pegmatites. Examples include disseminated Nb in alkaline granites, chromite and Fe-Ti oxides, massive Ni sulfides and pegmatite minerals. Further subdivision of the magmatic deposits can be achieved by using ore-forming processes, host rock association and morphology.

#### 2.8.9.2 Hydrothermal Deposits

Hydrothermal deposits are those deposits formed by hot mineralizing fluids from various sources, including magma-derived solutions, heated connate water, seawater, meteoric water and metamorphic fluid. The deposits in this group are epigenetic or diagenetic and are usually emplaced in open spaces within pre-existing rocks. Hydrothermal deposits include veins, porphyry type deposits, volcanogenic massive sulfides, sedimentary exhalative deposits, Mississippi Valley-type deposits, and epithermal gold deposits. They can be further subdivided on the basis of temperature of formation and morphology (see Table 2.9).

#### 2.8.9.3 Sedimentary Deposits

Sedimentary deposits are hosted by sedimentary rocks and formed within sedimentary basins as products of precipitation, evaporation, clastic sedimentation, biogenesis and diagenesis. The deposits can be stratiform or stratabound and predominantly syngenetic or diagenetic. Sedimentary mineral deposits include chemical precipitate deposits, evaporite deposits, placer deposits, diagenetic deposits and biogenic deposits. These deposits are subdivided by ore-forming processes and morphology (e.g. stratiform, stratabound)

#### 2.8.9.4 Metamorphic Deposits

Metamorphic deposits are products of crystallization of new minerals or recrystallization of existing minerals due to the effects of increased temperature and/or pressure during regional or contact metamorphism and/or metasomatism. These deposits are either *metamorphic* (formed by metamorphism) or metamorphosed. *Regional metamorphic deposits* include new economic minerals formed by metamorphism such as graphite deposits, and deposits produced by recrystallization and metamorphic changes or enrichment of pre-existing rocks. Skarn deposits are formed along the contact of igneous intrusions and sedimentary host rocks.

#### 2.8.9.5 Surficial Deposits

Surficial deposits are accumulations of valuable minerals near the Earth's surface as a result of intense chemical weathering, leaching and supergene enrichment. Mineral deposits include *residual* deposits concentrated by the removal of unwanted materials and concentration of valuable minerals, and *supergene* deposits formed by oxidation, leaching and concentration of enriched secondary minerals near the water table.

## 2.9 Review Questions

1. What is the concentration factor of metals, and how is it calculated?
2. Define metallogenic province and metallogenic epoch. How can these concepts be applied in mineral exploration?

3. Differentiate the following pair of terms: epigenetic and syngenetic; Veins and lodes; tabular and tubular orebodies, concordant and discordant, stratiform and stratabound.
4. List the three main types of textures of ore mineral. Describe how these textures are produced by the mode of emplacement.
5. Define and classify ore minerals. What are the most common gangue minerals in ores?
6. Describe open-space fillings and replacement processes in hydrothermal deposits.
7. List the four factors that control the formation of mineral deposits.
8. Briefly describe the three main media in which ore-forming solutions are transported. Explain how such inferences can be determined.
9. What is the purpose of classifying mineral deposits, and list several criteria used to classify mineral deposits.
10. Briefly describe the five main types of mineral deposits based on genetic classification.

## References

Arndt, N, Kesler, S and Ganino, C (2012) Metals and Society: An introduction to Economic Geology, 232p.

Carlson, CA (1991) spatial distribution of ore deposits: Geology, v. 19, p. 111-114. Craig, JR. (1991): Textures of the ore minerals. In Advanced Microscopic Studies of Ore Minerals (JL Jambor & DJ. Vaughan, eds.). Mineral. Assoc. Can., Short-Course Notes 17, 213-261.

Condie, KC (2001) Mantle plumes and their record in Earth history: Cambridge University Press, 320 p.

Courtillot, V. Davaille, A., Besse, J., and Stock, J. (2003) Three distinct types of hotspots in the Earth's mantle: Earth and Planetary Science Letters, v. 205, p. 295-208.

Cox, DP and Singer, DA., *eds.*,(1986) Mineral deposit models: U.S. Geological Survey, Bulletin 1693, 379 p.

Craig, JR and Vaughan, DJ. (1994) Ore Microscopy and Ore Petrography (second ed.). Wiley Interscience, New York, N.Y.

Evans, MA (2006) The mineralogy of economic deposits. In: Introduction to mineral exploration, p. 19-32. Mason, Whateley and Evans (eds), Blackwell Publishers.

Evans, MA and Moon, CJ (2006) Mineral deposits geology and models. In: Introduction to mineral exploration, p. 33-52. Mason, Whateley and Evans (eds), Blackwell Publishers.

Guilbert, JM., and Park, CF., Jr. (1986) The geology of ore deposits: WH. Freeman & Co, 985 p.

Hedenquist, J.W., and Lowenstern, J.B. (1994) The role of magmas in the formation of hydrothermal ore deposits: Nature, v. 370, p. 519-526.

Hutchinson, CS., (1982) Economic deposits and their tectonic setting: John Wiley 365 p.

Jebrak, M (2006) Economic geology: Then and Now. Geoscience Canada, 33(2), 81-93.

Kesler, SE (1994) *Mineral resources, economics, and the environment*: Macmillan College Publishing Company, 391 p.

Kirk, J. Ruiz, J, Chesley, J, Walshe, J. and England, G. (2002) A major Archean gold- and crust-forming event in the Kaapvaal Craton, South Africa: Science, v. 297, p. 1856-1858.

Kirkham, RV., Sinclair, WD., Thorpe, R., and Duke, JM., *eds.*,(1995) Mineral deposit modeling: Geological Association of Canada, Special Volume 40, 798p.

Kyser, K. (2000) Fluids and basin evolution: Mineralogical Association of Canada, Short Course 28, 262 p.

Lowell, JD and Guilbert, JM. (1970) Lateral and vertical alteration-mineralization zoning in porphyry ore deposits: Economic Geology, v. 65, p. 373-408

McQueen, KG (2005) Ore deposit types and their primary expressions. In: Regolith Expression of Australian Ore Systems (p.1-14) Publisher: CRC LEME, Bentley, WA (Editors): C.R.M. Butt, I.D.M. Robertson, K.M. Scott, M. Cornelius (Accessed at Researchgate)

Misra, KC (2000) Understanding mineral deposits: Kluwer Academic Publishers, 845 p.

Mitchell, AHG and Garson, MS. (1981) Mineral deposits and global tectonic settings: Academic Press, Geology Series, London, 405 p,

Olade, MA. (2020) Mineral deposits and exploration potential of Nigeria. Prescott Publishers, 402 p.

Pirajno, F. (2000) Ore deposits and mantle plumes: Kluwer Academic Publishers, Dordrecht/Boston, 556 p.

Taylor, R. (2010) Ore textures: recognition and interpretation. Springer Verlag. 250p

# Chapter 3
# Ore-Forming Processes and Associated Mineral Deposits

## 3.1 Ore-Forming Processes

The formation of mineral deposits involves complex processes in which ore-bearing fluids of various types are able to transport and deposit metals as ore minerals within favorable geological environments. There are a great variety of mineral deposits and several processes that lead to their formation, but what is most important for economic geologists is to understand these processes and apply such knowledge in the search for new mineral deposits. In describing mineral deposits in this chapter, the main genetic types will be used, and subdivided in accordance with other applicable criteria, such as ore-forming processes, morphology etc. Mineral deposits have been classified according to their origin which is based on the major geological processes.

Ore-forming processes are distinct offshoots of the main geological processes that form or modify rocks in the Earth's crust and land surface. There are four main geological processes responsible for the formation of rocks: (1) *magmatic*, (2) *metamorphic*, (3) *sedimentary,* and (4) *surface weathering* (surficial). The first two processes are described as *exogenous* processes because they occur deep within the earth, while the last two are *endogenous because* they occur near the surface. Thus, some mineral deposits are formed by magmatic and metamorphic processes, while others are produced by sedimentation or surface weathering. Ore-forming processes are closely connected with the geological processes that form or modify rocks. For example, a sulfide-rich magma can separate as a small immiscible liquid to form sulfide deposits while the 'main' magma is in the process of forming ultramafic/mafic igneous rocks as it cools. Multiple ore-forming processes may operate within a crystallizing mafic magma to form various magmatic deposits.

Apart from the four major geological processes identified above, there is a large group of deposits known as "hydrothermal" deposits that are formed by hot solutions derived from various sources related to the four geological processes, viz-a-viz magmatic processes (magmatic-hydrothermal solutions), sedimentary processes (connate water), surficial processes (meteoric water) and metamorphic processes (metamorphic fluids). The list of ore-forming processes obviously exceeds the number of known geological processes. However, ore-forming processes can be classified into five major categories as analyzed above:

1. *Magmatic* ore-forming processes: - dissemination, gravitational settling, liquid immiscibility and fractional crystallization
2. *Hydrothermal* ore-forming processes - magmatic-hydrothermal, metamorphic fluids, sedimentary -connate waters, and surficial-meteoric water.

3. *Sedimentary* ore-forming processes - chemical/biochemical precipitation, evaporation, clastic deposition, biogenesis and diagenesis.
4. *Metamorphic* ore-forming processes: - regional metamorphism and contact metasomatism.
5. *Surficial* ore-forming processes:- weathering and supergene enrichment

## 3.2 Magmatic Ore-Forming Processes

Magmas are hot molten rock materials that are liquid at high temperatures but crystallize to form solid minerals as they cool and temperatures fall. Magmas are silicate melts produced by melting of the mantle and/or the crust in various tectonic settings, including mid-oceanic ridges, continental rift zones, subduction zones, and hotspots. They migrate upwards from deep-seated sources because they are more buoyant than the surrounding solid rocks. Magmas move through the crust, and solidify underground to form intrusive bodies known as *plutons* or reach the surface to feed volcanoes, and are sometimes injected into surrounding rocks as dykes or sills. They are characterized by variable temperature, viscosity and chemical composition, and contain a variety of elements (including metals).

Based on their chemical composition, magmas are often classified into five different types: *ultramafic* (komatite), *mafic* (basaltic), *intermediate* (andesitic), *felsic* (rhyolitic) and *alkaline* (carbonatite). The physical and chemical characteristics of the magma types are summarized in Table 3.1. If a magma is rich in metals, the economic minerals may originate within the magma as the magma crystallizes and may form discrete crystals or accumulate as layers of massive ore. The type of minerals that precipitate from magmas depend on their chemical composition and the geological processes that take place.

Magmatic processes in a magma chamber can concentrate ore minerals by taking elements that were once widely dispersed at low levels in the magma and concentrating them in minerals that mostly accumulate in the body of the igneous rock or injected into surrounding rocks (Fig. 3.1). Sometimes, parts of the magma that are rich in ore minerals, such as oxides or sulfides can concentrate largely as an "ore magma" which can separate and be injected into surrounding rocks by a process called "filter pressing" or magmatic injection. During magmatic differentiation, some dense ore minerals may sink to the bottom of the magma chamber due to gravitational settling, others may float to the top if they are lightweight minerals associated with volatiles, to be subsequently injected as low-temperature pegmatite deposits. The dynamics of magmatic crystallization may sometimes involve movement of crystals by flow from one part of the magma chamber to the other. The magmatic processes that produce mineral deposits in metal-rich magmas are; *dissemination, gravitational (crystal) settling, liquid immiscibility* and *fractional crystallization.*

A large variety of mineral deposits are formed by various processes during the formation, evolution, emplacement, and crystallization of magmas in the upper mantle and in the Earth's crust. These magmatic deposits may be produced as a result of (1) solid phase crystallization as the magma cools and evolves, (2) the formation of an immiscible liquid fraction from the silicate melt, (3) minerals crystallizing from the enriched residual fluids formed by differentiation as magma cools and crystallizes, and (4) where a magma transports xenolithic or xenocrystic phases that it has picked up during movement through the Earth's crust.

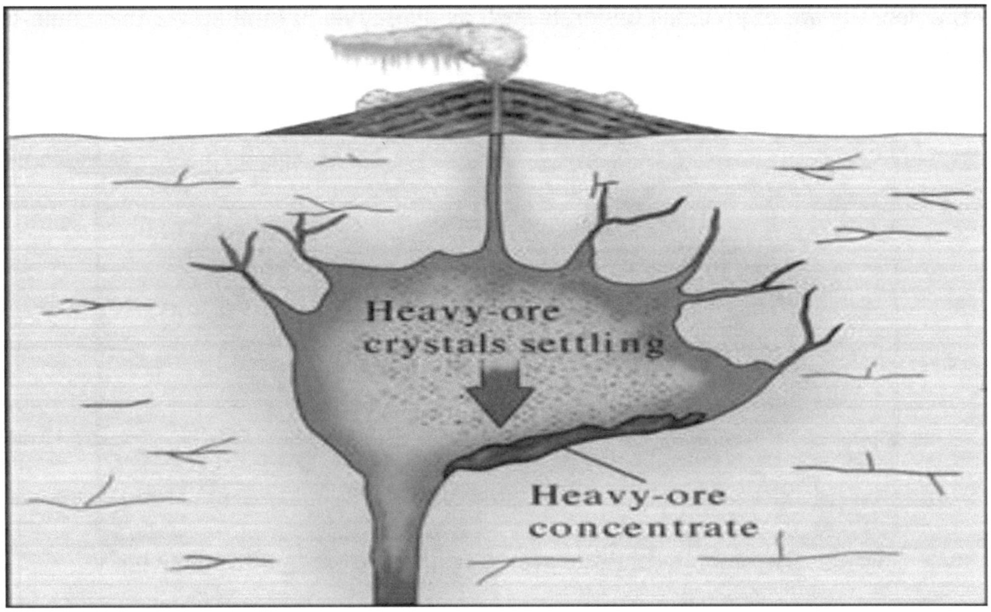

Fig. 3.1: Magma chamber and crystallization of silicate melt

Table 3.1 Physical and Chemical Characteristics of Magmas

| Magma Type | Solidified Rock | Chemical Composition | Temperature | Viscosity | Gas Content |
|---|---|---|---|---|---|
| Ultramafic | Komatite Peridotite | < 45% $SiO_2$, very high in Fe, Mg, Ca | 1300 - 1500 °C | Very Low | Very Low |
| Mafic (Basaltic) | Basalt Gabbro | 45-55 $SiO_2$ %, high in Fe, Mg, Ca, low in K, Na | 1000 - 1200 °C | Low | Low |
| Intermediate (Andesitic) | Andesite Diorite | 55-65 $SiO_2$ %, intermediate in Fe, Mg, Ca, Na, K | 800 - 1000 °C | Intermediate | Intermediate |
| Felsic (Rhyolitic) | Rhyolite | 65-75 $SiO_2$ %, low in Fe, Mg, Ca, high in K, Na. | 650 - 800 °C | High | High |
| Alkaline | Carbonatite Phonolite | >50% $CaCO_3$, high Na | 500 - 600 °C | Very High | Very High |

### 3.2.1 Dissemination

Dissemination is a simple process of magmatic crystallization involving the homogeneous scattering of economic minerals as accessory constituents in igneous rocks. This process may occur either during the early phase or late-stage crystallization of the magma. Examples include disseminated chromite or nickel sulfides in mafic-ultramafic bodies (Fig. 3.2), and disseminated diamond in kimberlites. Dissemination of columbite and cassiterite is also common in some felsic and alkaline granites. This process may not produce deposits that are rich enough to be mineable at a profit, except if they occur over a large area and at shallow depths or if the deposits are of precious minerals such as diamonds in kimberlites that could be mined at greater depth from surface.

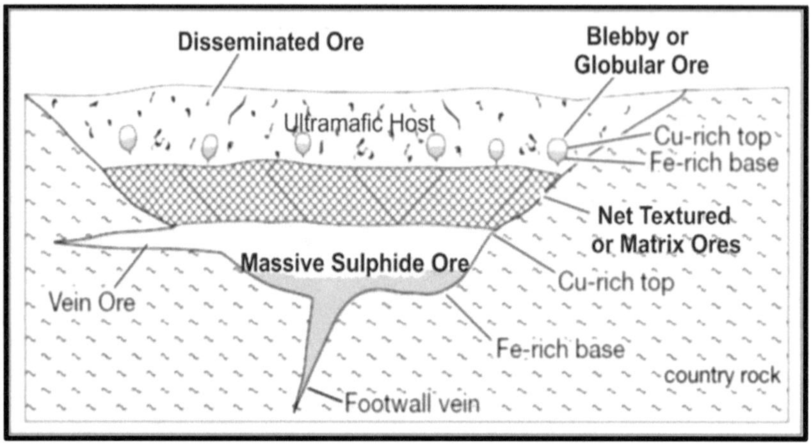

Fig. 3.2: Disseminated sulfide overlying massive ore in ultramafic body

### 3.2.2 Gravitational Settling

Gravitational settling is a process of magmatic segregation in which early-formed crystals of economic minerals of high density sink to the bottom of a magma chamber (Fig. 3.3). This occurs most commonly in ultramafic and mafic magmas of low viscosity due to low silica content that allows early formed minerals to float, sink, or be moved very rapidly by flowing magma that segregation can occur before crystallization is complete. Both the metallic ores and host intrusions often show rhythmic layering and funnel-shaped, with the neck forming a feeder system (Fig. 3.2). The best examples are layered chromite and titanium deposits. Chromite, the ore of chromium is a very high temperature mineral of high density which is sometimes alternating with layers of the host rock.

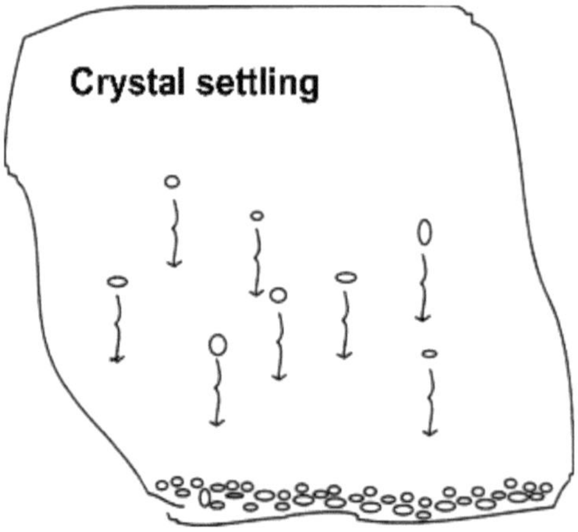

Fig. 3.3: Crystal settling of early formed crystals in a magma

### 3.2.3 Liquid Immiscibility

Liquid immiscibility is a process of magmatic segregation in which a portion of a magma separates physically into two or more magmas of different composition. Two liquids that will not mix are called *immiscible*, e.g. oil and water, but when immiscibility happens in magmas, there are usually two liquids, one silicate rich and the other, usually much smaller in volume, can be rich in metal *oxides* and *sulfides* (ore magma) or *carbonates* (carbonatite magma). The ore magma may separate and sink below the silicate-rich part of the intrusion, or be injected into the surrounding rocks by a process known as "filter pressing" to form massive ores. Fig. 3.4 shows how a diagram of an evolved magma rich in Fe-Ti oxides produces an immiscible melt by gravitational settling which forms a separate massive ore body. Such mineral deposits are found in mafic and ultramafic rocks, Examples of deposits originating from liquid immiscibility are the massive nickel sulfide and platinum deposits associated with layered intrusive rocks of Sudbury, Ontario, Canada; and the titano-magnetite deposits in anorthosite at Allard Lake, Quebec, Canada. Three types of immiscible melts have been recognized in magmas:

- *Oxide melts* rich in iron ($Fe_2O_3$, hematite) and titanium ($FeTiO_3$, ilmenite).
- *Sulfide melts* rich in Ni, Cu, and the platinum group of elements (PGEs), in addition to iron sulfide ore bodies (FeS, pyrrhotite).
- *Carbonate melts* (form carbonatites) separated from alkaline magmas rich in niobium, tantalum, rare earths, copper, thorium, and phosphorus.

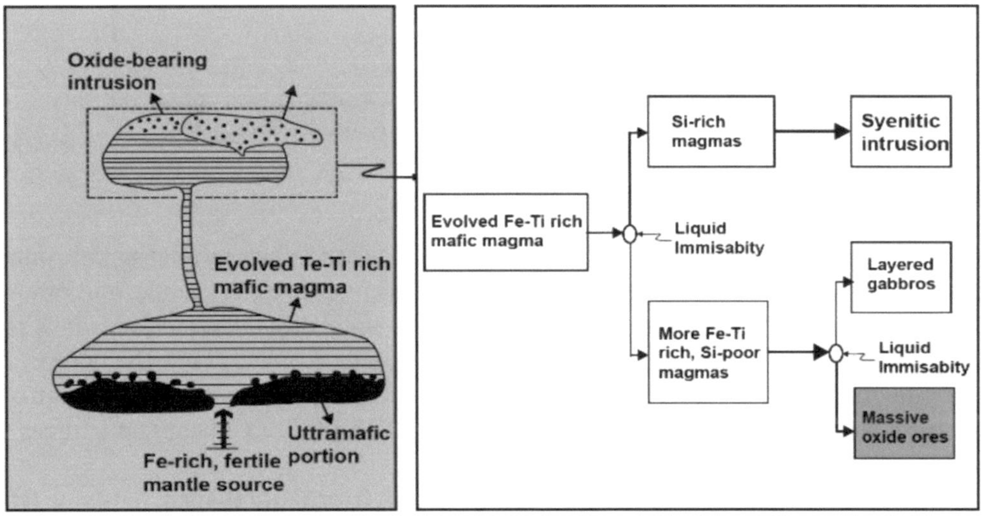


Fig. 3.4: Schematic diagram of 2-stage immiscible Fe-Ti melts

### 3.2.4 Fractional Crystallization

Fractional crystallization is one of the main processes of magmatic differentiation. This is the removal and segregation from a silicate melt of early formed crystals which changes the composition of the magma and results in the enrichment of selected elements in the residual fluids. These enriched elements are those that did not fit well into the structure of common rock forming minerals during normal crystallization. Instead, they tend to accumulate in the final volatile-rich residual liquids that float to the top of the magma chamber, and may crystallize within the igneous body or injected into surrounding rocks to form pegmatites which are very coarse-grained rocks that form at the very last stage of crystallization (Fig. 3.5). The residual fluids are usually rich in volatiles and water vapor enriched in the rare and highly-charged elements such as niobium, tantalum, tin, cesium, beryllium and uranium. Because of the rarity of some of these elements, they can form unique and valuable minerals that accumulate in pegmatites and granites. Examples include the rare-earth and tin-niobium mineralized pegmatites associated with the Older Granites of Nigeria.

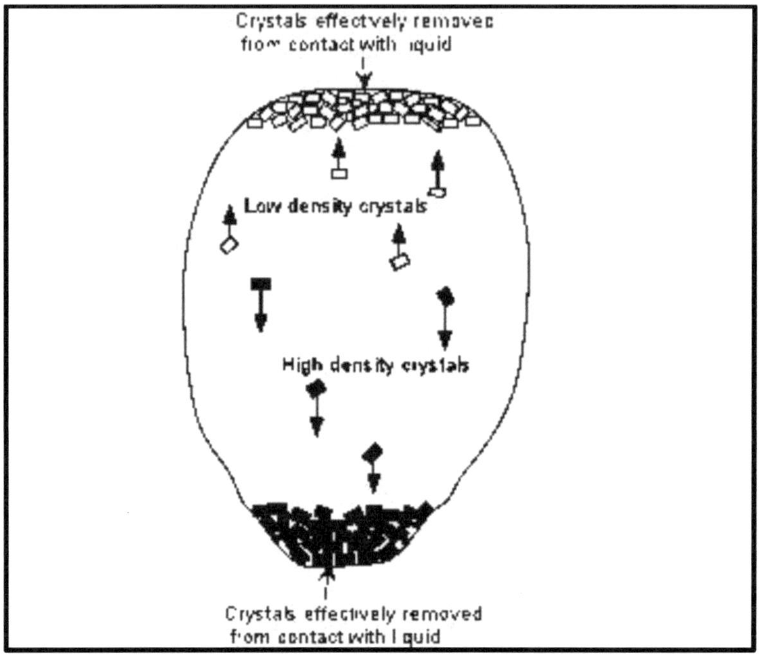

Fig. 3.5: Fractional crystallization in a magma chamber

## 3. 3 Magmatic Mineral Deposits

Magmatic mineral deposits (or orthomagmatic deposits) are the accumulations of economic minerals within igneous rocks. They are usually formed deep within the Earth's crust or upper mantle, and rarely as volcanic rocks formed at the surface. There are some deposits that are indirectly linked with igneous rocks but are not considered as magmatic deposits. Examples are the deposits formed along the contact between an igneous body and surrounding rocks, known as contact metamorphic or contact metasomatic deposits (skarn), and hydrothermal deposits formed by hot fluids derived from magmatic sources (magmatic-hydrothermal).

Magmatic mineral deposits vary in shape and size, from irregularly shaped, flat-lying (stratiform) to lenticular and pod-like form (or podiform), and from tabular to tubular The host rocks range from ultramafic to ultra-felsic in composition, and the deposits can form massive ores, layers and lenses in some cases, and disseminations of rare minerals in others. The ore and host rocks range in texture from fine-grained to coarse-grained.

Magmatic mineral deposits are the biggest and best-known sources of economic minerals including iron, diamond, copper, platinum, titanium and chromium. Examples of magmatic deposits are: diamond in kimberlites, chromite, nickel-copper sulfides and iron-titanium oxides, and platinum group metals (PGM) in mafic/ultramafic rocks, tin, niobium, uranium and copper in felsic rocks, and niobium and rare earths in carbonatites.

The close association between mineral deposits and igneous host rocks is well known and described in economic geology. The strong affinity of some metals and rare elements with a particular type of magma or igneous host rock is known as the “ore-rock association”. This relationship is believed to be genetic, and attributable to the fact that certain metals tend to originate from a particular type of magma (ultramafic, mafic, felsic, ultra-felsic and peralkaline). This relationship has been found to be a useful criteria in the classification of mineral deposits, and a guide in reconnaissance exploration for metallic ores.

The following are the well-known associations between mineral deposits and igneous host rocks:

- Deposits related to *ultramafic and mafic* intrusive rocks. (Cr, Pt, Ni-Cu and Ti-Fe).
- Deposits related to mafic and ultramafic volcanic rocks (Cu, Ni)
- Deposits related to *felsic granitic* rocks (Cu, Sn, W)
- Deposits related to *ultra-felsic* pegmatites (Sn, Ta, Nb, U)
- Deposits related to per*alkaline* granites *and carbonatites* (REE, Nb, U)

## 3.3.1 Classification of Magmatic Mineral Deposits

Magmatic mineral deposits can be classified into four main groups based on ore-forming process, morphology, and type of host rock (Fig. 2.5).

1. Disseminated Mineral Deposits ( Diamond, Ni, Cr, Sn, Nb, REE)
2. Stratiform (Layered) Oxide Deposits in Igneous Rocks (Cr, PGE, Fe, Ti, V)
    a. Associated with Layered Intrusive Complexes (Cr, Fe, Fe-Ti-V)
    b. Associated with Volcanic Rocks (Fe, Fe-Apatite)
3. Massive Sulfide Deposits
    a. Associated with Plutonic rocks (Ni, Cu, PGE);
    b. Associated with Volcanic rocks (komatite) (Ni, Cu, and PGE)
4. Pegmatite Deposits (Sn, Nb, Ta, Li, U)

## 3.3.2 Disseminated Deposits

Disseminated deposits are characterized by the random or uniform scattering of ore minerals (or valuable accessory minerals) in igneous host rocks. The disseminated ore usually contains small particles of valuable minerals, spread quite uniformly or randomly throughout the host rock, and could be a product of either early or late-stage crystallization. These deposits are found both in rocks derived from mafic and felsic magmas. In ultramafic-mafic rocks, the disseminated minerals are commonly sulfides of Ni-Cu or oxides of Cr-Fe-Ti. They can be scattered within the host rocks or situated above massive ores (Fig. 3.6)... In alkaline and felsic rocks, the disseminated ores are usually widely distributed in the rock body or confined in zones or sheets of disseminated ore lying within the roof zones or cupolas of the host granitic rocks. For many of the disseminated deposits, ore grades are usually low, but tonnage can be very large and extend over tens of square kilometers. Despite the low grades, the concentrations may be valuable enough to make the mineral deposit minable, particularly if the ore minerals are of high value, such as diamonds, or dispersed over a large area (large tonnage), or enriched by secondary processes such as lateritic weathering and supergene enrichment. For example, the lateritic disseminated pyrochlore in the carbonatites of Brazil are the largest sources of niobium in the world. The following are the various types of disseminated magmatic deposits (Fig. 3.6).

- Diamond in Kimberlite Pipes
- Disseminated Nickel Sulfide and Chromite/ Iron-Titanium Oxide Deposits
- Niobium and REE Deposits in Carbonatites
- Disseminated Tin-Niobium and Copper in Granitic Rocks

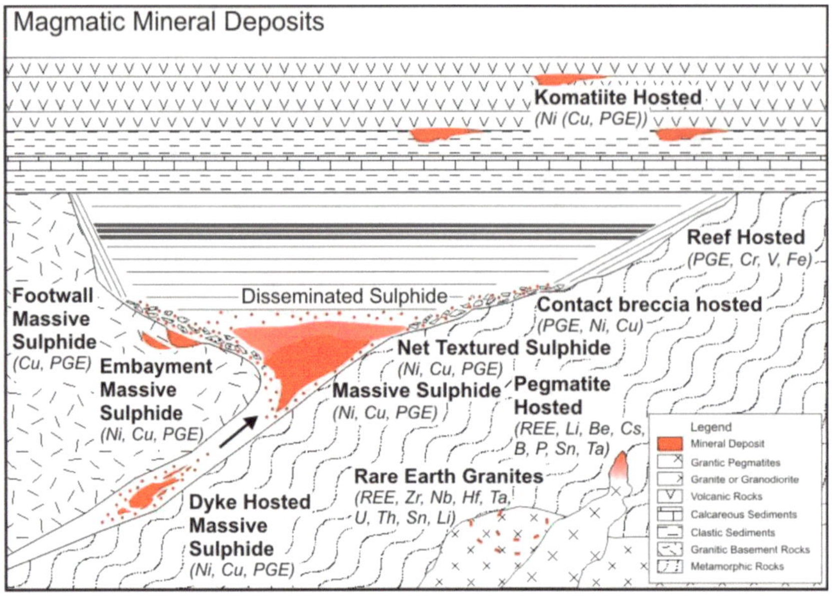

Fig. 3.6: Schematic diagram of magmatic mineral deposits

### 3.3.2.1 Diamond in Kimberlites

Kimberlites are unusual potassium-rich, low-silica volcanic rocks of ultramafic origin that commonly contain *diamonds* formed at very high pressure. They occur as pipes or tubular bodies and dikes of igneous rock that originate from about 150-200 km deep in the mantle, and were explosively transported as magma with diamonds from great depths. Kimberlites are associated with a suite of unique minerals such as chromium diopside, chromium spinels, magnesian ilmenite, and chromium-rich pyrope garnet that support their crystallization under conditions of high pressure and temperature within the mantle. The rapid and explosive rise of the kimberlite magma through the crust punched a circular hole and created a ring of tuff breccia usually underlain by massive kimberlite. Diamond, a high-pressure polymorph of carbon occurs as discrete crystals or fragments that are scattered randomly within the kimberlite that rarely shows any discernible alteration (Fig3.7).

Diamond-bearing kimberlites are known to occur within ancient cratonic shields of the world. They are found in several African countries including South Africa, Botswana, Angola, Democratic Republic of Congo, Tanzania, Lesotho, Ghana, Sierra Leone, Guinea and Liberia. Other world leading producers are Russia, Canada and Australia. In South Africa, once the world's leading producer of diamonds, the classic type area where kimberlite diamond was first discovered was in Kimberley near Johannesburg, South Africa (from where the word kimberlite was derived). Following the discovery at Kimberley, more valuable deposits were later found in other parts of the Kaapvaal craton in South Africa, Botswana and Lesotho, Swaziland and Zimbabwe. (Fig. 3.8). Diamond concentrations in kimberlites are highly variable, and several pipes are known to be barren. In fact, only about 1% of kimberlites contain diamonds in economic concentration,

Fig. 3.7: Schematic diagram of Kimberlite pipe and kimberlite breccia with diamond

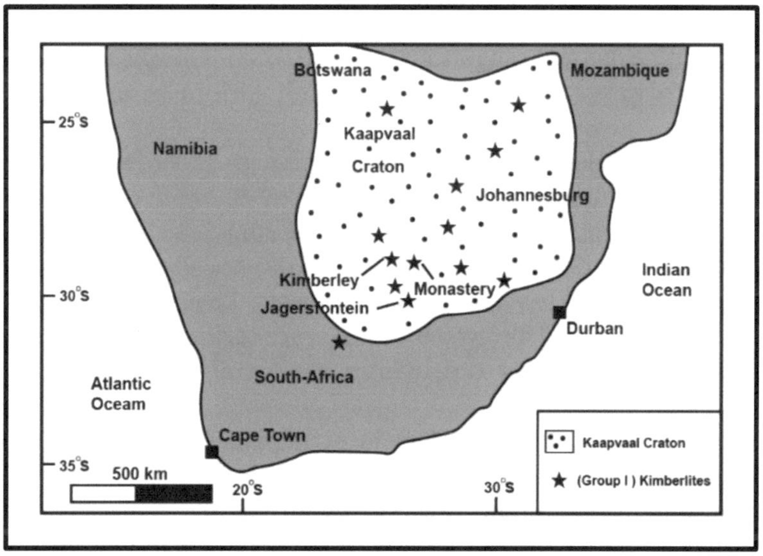

Fig. 3.8: Major diamondiferous kimberlite occurrences in Southern Africa

### 3.3.2.2 Disseminated Nickel Sulfide and Chromite Deposits

Disseminated Ni and Cu sulfides (pentlandite (Ni), chalcopyrite (Cu) and oxides, of Cr (chromite) and Fe-Ti (titaniferous-magnetite) are associated mafic and ultramafic intrusive rocks where they constitute about 2 - 10% of the rock. The disseminated ores sometimes occur above and beside the higher-grade massive sulfide or oxide layers, but because of their low grade may not be considered as valuable at current metal prices. However, some open pit mines have developed, particularly in Asia to extract Ni and Ti from these low-grade deposits. Examples include the disseminated nickel-copper ores in the Duluth Gabbro in Minnesota, USA and several deposits in Western Australia and Quebec, Canada. The Duluth Ni, Cu and PGE deposits are disseminated and considered as low grade, but contain over 140 billion tons of Cu (and N) that are still yet to be developed (Fig. 3.9a).

Apart from the layered intrusive complexes, disseminated chromite, usually of low grade, occurs within podiform orebodies in ultramafic rocks and serpentinites within ophiolite complexes where they may be associated with stratiform ores. The platinum group metals are sometimes found as inclusions in disseminated chromite bodies.

### 3.3.2.3 Niobium and REE in Carbonatites

Carbonatites are alkaline igneous rocks containing more than 50% carbonate minerals (calcite and dolomite) of magmatic origin. They are very rare and exotic rocks emplaced as small pipe-like intrusive bodies usually within alkaline igneous complexes containing nephelinites, ijolites and syenites associated with intracontinental rift systems and mantle hotspots. Carbonatites are the world's dominant sources of Nb and REE of which nearly 90% of global supply comes from carbonatites in Brazil. While primary mineralization is dispersed (disseminated) and low grade, it has been considerably enriched by lateritic weathering and supergene processes.

Primary niobium in carbonatites are characterized by the presence of pyrochlore and sometimes, columbite that are disseminated over wide areas, as large tonnage but low grade deposits (typically <1% $Nb_2O_5$) (Fig. 3.10). Pyrochlore, a family of complex Nb–Ta–Ti oxides and related primary minerals, may contain elevated values of rare earth elements (REE). Other economic minerals found in carbonatites include apatite, monazite, zircon, thorite, urano-pyrochlore, Ti-magnetite and copper sulfides. Carbonatites have been found on all continents; and range widely in age, from deposits in the East African Rift Valley that were formed during the present geologic age to South African occurrences dating from the early Proterozoic.

The best-known primary deposit is the open-pit mine at Araxá (Minas Gerais, Brazil) where the carbonatite complex consists of intrusions of dolomite carbonatite, some calcite carbonatite, and intensely fenitized micaceous rocks. The ore zone consists of crystal cumulates of magnetite, mica (phlogopite), pyrochlore, ilmenite, and apatite with average concentration of 1.5% $Nb_2O_5$. Reserves are estimated at 460 Mt sufficient to cover current world needs for about 500 years. Phosphate is also produced from mining operations. Carbonatites are also mined for pyrochlore and rare earths at Catalão (Brazil) and Saint-Honoré (Quebec, Canada). In Africa, mineralized carbonatites are found in South Africa, Angola, Kenya and Tanzania.

The origin of Nb and REE mineralization in carbonatites is attributed to both magmatic and hydrothermal processes. There is evidence of primary crystallization of Nb-bearing and REE minerals (e.g. pyrochlore, bastnasite (Ce) from the carbonate melt, followed in places by crystal fractionation, and redistribution by density currents. Late stage metasomatic-and hydrothermal activity are characterized by remobilization and precipitation of fluorocarbonates, fluorite and barite.

Fig. 3.9: (a): Disseminated sulfide mineralization in gabbro, Russia (b) Disseminated cassiterite in the Bushveld Granite

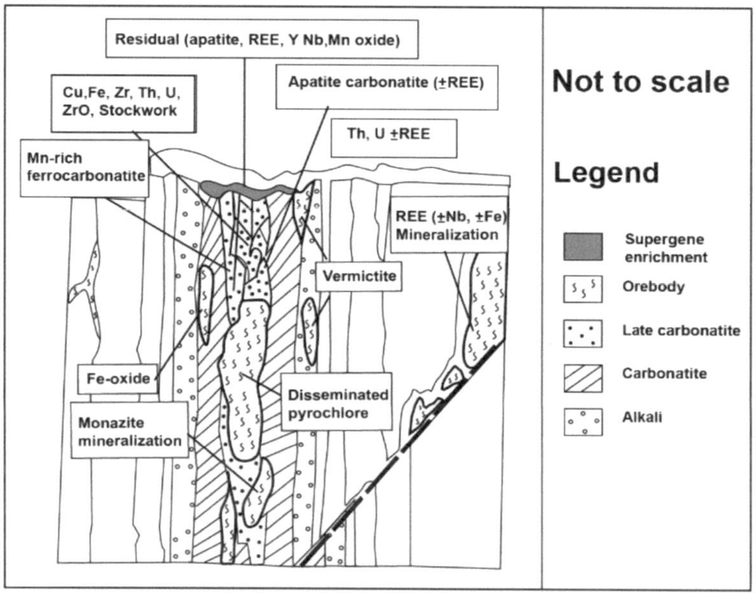


Fig. 3.10: Schematic diagram of a hypothetical carbonatite mineralizing system

### 3.3.2.4 Disseminated Tin-Columbium in Alkaline Granites

Minerals of tin (cassiterite) and columbium (columbite) are found together as widely dispersed crystals within alkaline granitic rocks. These economic minerals are believed to have formed primarily during the late stage crystallization of the magmas often accompanied by sodic and potassic metasomatism and hydrothermal alteration.

In alkaline to felsic intrusive rocks, cassiterite and columbite are dispersed in patches over large areas or in sheet-like bodies within the roof zones or cupolas of the granites. In several cases, mineralization is accompanied by late-stage albitization. Where these mineralized cupolas have been stripped by erosion, alluvial deposits are usually well developed, but only a relic of the disseminated mineralization is now preserved. Examples of disseminated Sn-Nb deposits include the Penouta orebody in NW Spain, the cassiterite and columbite (tantalite) deposits associated with the alkaline ring dyke complexes (Younger Granites) in northern Nigeria, and disseminated cassiterite in the Bushveld granites of South Africa.

In the Bobbejaankop Granite of the Bushveld Complex, mineralization is particularly well developed in the granite roof cupola at the Zaaiplaats tin mine where disseminated deposits form the main producing ores. Cassiterite which is commonly associated with scheelite and sulfides, is interstitial to the major silicate minerals in the disseminated ores (Fig. 3:9b). The ores are confined to a favorable zone that forms a tabular body of varying thickness (55-66 m) with an upper contact that is parallel to, and some 66 to 83 m below, the pluton's roof contact.

## 3.3.3 Stratiform (Layered) Magmatic Deposits

Stratiform magmatic deposits are those deposits formed by the accumulation of metallic minerals into layered ore bodies by gravitational settling or liquid immiscibility during the crystallization of mafic and ultramafic rocks (Fig. 3.11). They occur mainly as layers, bands, and lenticular bodies that usually alternate with layers of the host rocks that may extend for several kilometers. Ores that are found in these stratiform deposits are mainly chromium, platinum group metals, nickel-copper sulfides and titanium-iron oxides.

They produce most of the world's supply of chromium and the platinum-group metals. The host rocks are usually layered intrusive complexes that range in composition from ultramafic to mafic, with dominant lithologies of dunite, peridotite, pyroxenite, komatite, gabbro and anorthosite.

### 3.3.3.1 Chromite Deposits

There are two basic types of chromite deposits; *stratiform* and *podiform* chromite. Stratiform chromite deposits occur as massive layers of *chromitite* (greater than (>90% chromite) bodies and seams of disseminated chromite (>60 percent chromite). The host rocks are usually unmetamorphosed, repetitively layered intrusions of ultramafic and mafic rocks including dunite, peridotite, pyroxenite and gabbro, and less commonly of peridotite masses in deformed ophiolites. In metamorphic environments, the host rocks may be changed to serpentinite.

The layers of chromitite usually alternate with rock layers that are repetitive and laterally contiguous throughout the entire intrusive body (Fig. 3.11). The ore mineral, chromite (Fe, Mg) $(Cr,Al)_2O_4$ is the only commercial source of chromium. Notable examples of stratiform chromite deposits are found in layered intrusive complexes worldwide including the Bushveld Igneous Complex (BIC) of South Africa, the Great Dyke of Zimbabwe and Stillwater Complex in Montana, USA.

The origin and formation of stratiform chromite deposits have been extensively debated. It is believed that the various primary processes of magma mixing resulting from emplacement of new magma batches into an expanding chamber gave rise to economically important chromitite layers produced by gravitational settling, while fractionation combined with influx of magma caused the formation of base metal sulfides enriched in platinum group elements. Contamination of the parent magma by localized assimilation of felsic country rock at the roof of the magma chamber has also been proposed as a mechanism of ore formation. Others suggest that changes in pressure or oxygen fugacity may be responsible for the occurrence of massive chromitite seams in layered mafic-ultramafic intrusions.

Fig. 3.11 Chromite layers in mafic host rocks, South Africa

## (a) *Bushveld Igneous Complex, South Africa*

The Bushveld Igneous Complex (BIC) located in the Transvaal Province of South Africa is the world's largest layered mafic-ultramafic intrusive complex containing some of the world's richest ore deposits with extensive reserves of chromium and platinum group of metals (PGM) along with vast quantities of titanium and vanadium. The Complex which is approximately 2 billion years old covers an area over 66,000 km$^2$ (25,000 sq. mi). This large igneous province comprises the three main igneous suites are exposed as layered sequences of sheet like intrusions: the Rustenburg Layered Suite (8 km-thick layered mafic-ultramafic cumulate sequence), the Lebowa Granite Suite (large A-type granitic intrusions), and the Rashoop Granophyre Suite (subvolcanic granite porphyry) that are overlain by the Proterozoic Karoo sediments (Fig.3.12). The dominant lithologies within the Layered Ultramafic Unit vary from ultramafic peridotite, harzburgite, and bronzitite in the lower sections of the unit, to mafic norite, anorthosite, and gabbro towards the top.

The chromite ores in the BIC are located primarily in massive chromitite layers or seams which are referred to as *reefs* and, less abundantly, in disseminated chromite-bearing layers, both of which occur in the ultramafic section of the BIC. The host rocks of the chromitites are mainly pyroxenites and anorthosites. The chromitite seams that average about one meter thickness (range from 0.25-20 m), are cyclic in nature, as well as laterally contiguous throughout the entire intrusion. Gangue minerals include olivine and pyroxenes. Apart from chromium, the BIC is host to a variety of other ore deposits, including platinum group of metals (PGMs, primarily platinum and palladium), vanadium, iron (generally from magnetite), uranium, and tin from the Lebowa granites.

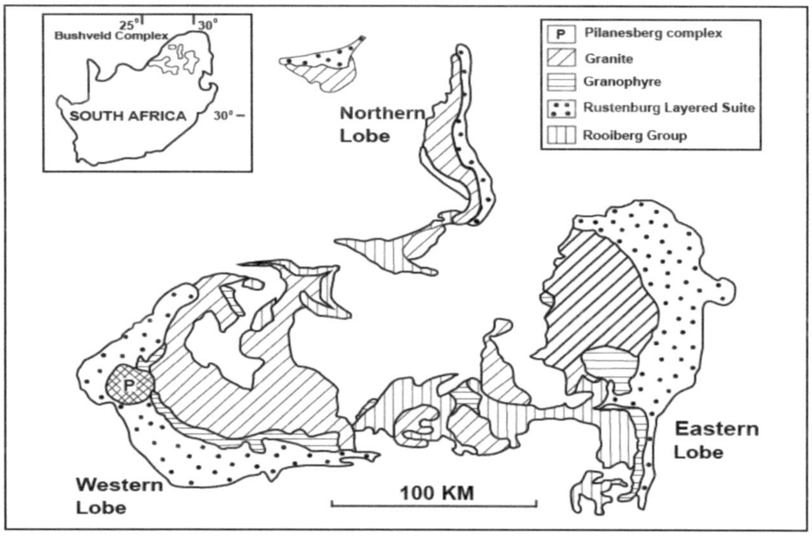


Fig. 3.12: Simplified geological map of the Bushveld Igneous Complex

## (b) *Great Dyke of Zimbabwe*

The Great Dyke of Zimbabwe, 2.5 billion years old is a remarkable layered mafic-ultramafic intrusion that extends north-south for over 500 kilometers and about 10 meters wide cutting across the Archean Shield of Zimbabwe (Fig. 3.13). Within this layered mafic-ultrabasic complex are several seams of chromite that are strikingly persistent and together constitute one of the world's largest resources of chromium. Structurally, the Great Dyke is a Y-shaped lopolith (synclinal basin) composed of several ultramafic layers overlain by a mafic gabbro layer with an underlying feeder dike (Fig. 3.13), The lithologic units within the lower ultramafic sequence are varieties of rocks rich in mafic minerals including dunites, peridotites, olivine bronzitites and pyroxenites together with narrow layers of chromitite that constitute the bases of cyclic units overlain by an upper mafic sequence consisting mainly of norites and olivine gabbro. The chromite seams are closely associated with peridotites and dunites.

The Great Dyke is host to significant mineralization of chromium, PGM and sulfides of copper, nickel and zinc which are mined extensively. Chromite occurs towards the base of the ultramafic sequence and below the ultramafic-mafic contact, and in the uppermost pyroxenite units are economic concentrations of nickel, copper, cobalt, gold, and PGM. The base metals occur as disseminated inter-cumulus Fe-Ni-Cu sulfides within an interval of layer referred to as the base metal subzone, below which is a sublayer enriched in platinum group metals called the PGE subzone.

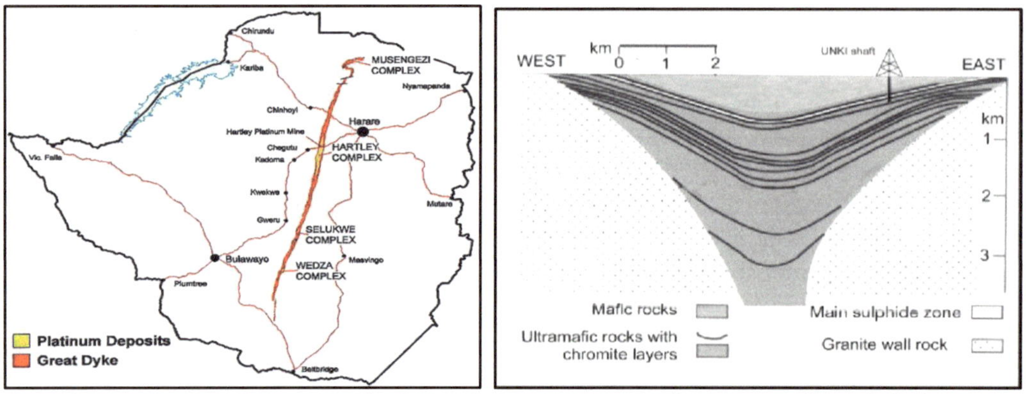

Fig. 3.13: Location map and schematic section across the Great Dyke, Zimbabwe

## (c) *Stillwater Igneous Complex, Montana*

The Stillwater Igneous Complex located in southern Montana, USA is a large layered mafic intrusion with many similarities to the Bushveld Igneous Complex of South Africa. The complex was intruded into existing gneisses during the Archean at about 2.7 billion years. Host rocks on the northern side are Archean sedimentary rocks while the southern boundary includes Precambrian rocks of granite, granite gneiss and schist. The intrusion forms a linear body stretching some 30 miles (48 km) and striking roughly N60°W and dipping from 50° to near 90° to the northeast. The exposed thickness is around 18,000 feet (5500 m) with an additional estimated 5,000 to 15,000 feet (1524 to 4572 m) having been removed from the top by Precambrian erosion. The cumulate stratigraphy comprises three distinct zones:

- The Basal series consists of a chilled fine grained gabbro overlain by gabbro, norite and feldspar pyroxenites. Thickness is up to 700 feet (210 m).
- The Ultramafic series is composed of a lower peridotite member (Peridotite zone) consisting of alternating dunite, chromitite, harzburgite and bronzite pyroxenite. The upper third is massive, monotonous bronzitite (Bronzitite zone). The Ultramafic series averages around 3500 feet (1100 m) in thickness.[6]
- The Banded series is composed of alternating norite, gabbro and anorthosite. The Banded series has a maximum thickness of 14,000 feet (4300 m).

The chromium orebodies are restricted to the Peridotite Zone of the Ultramafic Series. The platinum group orebody is located in the lower part of the Banded Series within a horizon referred to as the J-M Reef. The J-M Reef is similar to the Merensky Reef of the Bushveld Complex of South Africa. It is a continuous layer near the base of the banded zone consisting of one to three meter thick pegmatitic peridotite and troctolite with disseminated sulfide minerals including pyrrhotite, pentlandite (containing up to 5% Pd), and chalcopyrite along with lesser amounts of moncheite ($(Pt,Pd)(Te,Bi)_2$, cooperite ($(Pt,Pd,Ni)S$) and platinum-iron alloys. Overall the reef contains an average of 20-25 ppm platinum plus palladium in an average two meter thickness. The complex has extensive reserves of chromium ore and has a history of being mined for chromium. More recent mining activity has produced palladium and other platinum group elements.

### 2.3.3.2 Platinum Deposits

Platinum is one of the rarest and most expensive metals in the world. It belongs to the group of elements called the platinum group of elements (PGE) or platinum-group of metals (PGM) (platinum, palladium, rhodium, iridium, osmium and ruthenium). PGMs occur in similar environments and host rocks as chromite and nickel-copper deposits. PGM may be detected in the chromite layers but not in economic quantity. The economic concentrations are localized in stratiform PGE-enriched mineralized zones in mafic to ultramafic layered intrusions. These zones are called *"reefs"* which is a term used in Australian and South African literature to refer to this style of mineralization in which a rock layer is mineralized with precious metal with distinctive texture or mineralogy.

***Bushveld Igneous Complex***

In the Bushveld Igneous Complex (BIC) of South Africa, which is famous for its platinum mineralization, the platinum-bearing horizons are the *Merensky Reef* and *UG2 Reef* (Fig. 3.14a). The *Merensky Reef* is a layer of gabbro with extensive chromitite and Ni-Cu sulfide zones containing the platinum ore. It varies in thickness from 30 to 90 cm (1-3ft) with average concentration of 10 ppm PGEs or 5grams per ton. The PGE reefs of the BIC combine to produce about 80% of the platinum and 20% of the palladium mined worldwide each year, and contain approximately 90% of the world's known reserves of the PGEs. Other areas in the world where platinum is found are in the Great Dyke of Zimbabwe, Stillwater complex in Montana, USA and Sudbury basin in Ontario, Canada.

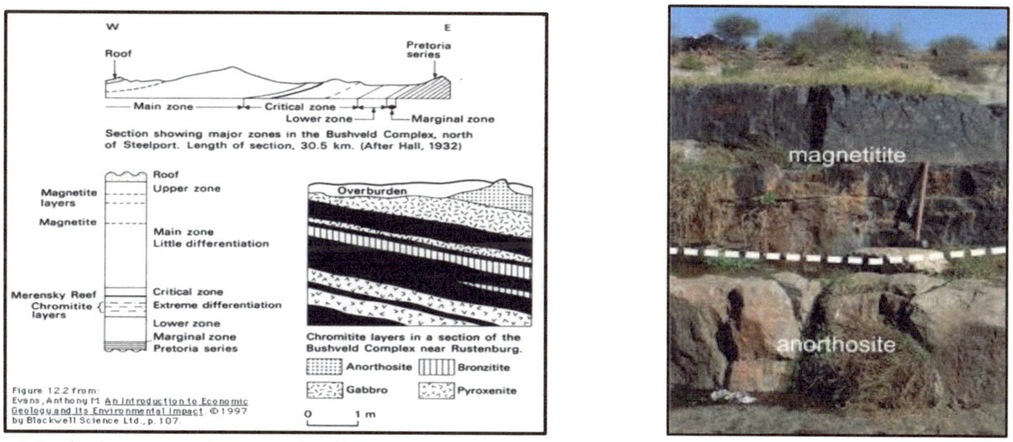


Fig. 3.14: (a) Platinum reefs in Bushveld Igneous Complex. (b) Titanium -magnetite layers

### 3.3.3.3 Titanium Oxide Deposits

Stratiform iron oxide (magnetite) deposits, some with high titanium (Ti) and vanadium (V) are known to be associated with mafic layered intrusions of anorthosite, norite and gabbro or their metamorphosed equivalents (amphibolites). Mineralization occurs as layers of mostly magnetite that are several meters thick alternating with mafic host rock. The magnetites may be rich in Ti and V and, in some instances, can be exploited as a polymetallic Fe–Ti–V deposit. In some deposits, the Ti mineral (ilmenite) is dominant over Fe. Other minerals present in these deposits include varieties of the principal minerals: Fe-bearing ilmenite, Ti- and V-bearing magnetite, and titano-hematite. Accessory minerals may include rutile, sphene, spinels, apatite, sulfides, and garnet. The Fe-Ti deposits are stratiform and conformable to primary layering of the host rocks. The V concentration is in the range of 0.2 - 1% $V_2O_5$, most of which is concentrated in the magnetite. These deposits have become the world's principal source of vanadium.

The origin of these stratiform magnetite deposits is connected to the evolution of mafic magmas in which Fe-Ti magnetite crystallized directly as magmatic cumulates by crystal settling or immiscible liquid separation. Examples include magnetite deposits in the Bushveld Complex, South Africa and the Ti-rich Allard Lake ilmenite deposits in Quebec, Canada. In the Bushveld igneous complex, Ti-bearing magnetite layers rich in vanadium occur throughout the entire Upper Zone in close association with anorthosite (Fig. 3.14b).

### *Allard Lake Ti-Fe Deposit, Canada*

The Allard Lake ilmenite deposits in Quebec, Canada are presently the world's largest titanium deposits with ore grade that averages 32% $TiO_2$ and 36% Fe and ore reserves of 300 million tons. Mineralization occurs in the northeastern part of one of the large Precambrian anorthosite masses. The ore bodies consist of a series of narrow dikes, irregular lenses, or sill-like bodies that appear to be genetically related to the norite-anorthosite rocks in which they occur. The main ore body is a funnel-shaped ilmenite-rich norite measuring about 1km long and 100–300 m-thick (Fig. 3.15a). Other small orebodies are separated by faults and anorthosite. The Allard Lake deposits originated by gravitational settling and/or residual liquid segregation and injection.

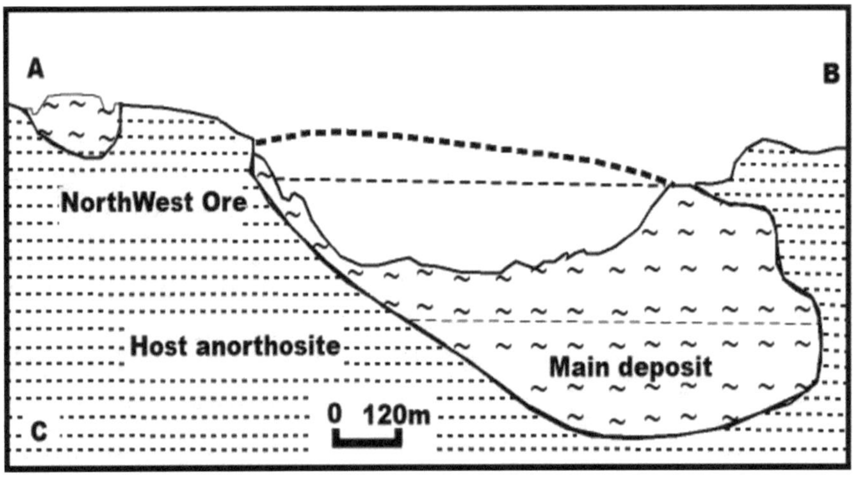


Fig. 3.15a: Schematic cross section of ilmenite deposit in anorthosite, Allard Lake

### 3.3.3.4 Magnetite-Apatite Deposits

Magnetite deposits have been found to be associated with volcanic rocks of mostly intermediate to felsic composition. These deposits occur as lenses or pods of magnetite intercalated within volcanic and sometimes pyroclastic rocks of diorite composition. There are sometimes effects of hydrothermal activity (alteration veining) superposed on the earlier products of magmatic crystallization. The dominant ore mineral is magnetite with varying amounts of apatite. Examples of these deposits include the El Laco in Chile, South America, Kiruna in Sweden, and several examples in China's Ningwu and Luzong basins in the Yangtze River Valley where Jurassic intermediate-mafic volcanic rocks host the iron deposits.

The origin of these volcanogenic magnetite orebodies has been controversial and attributed to various reasons including; (a) product of immiscible oxide magma, (b) magnetite lava formed through the melting of an assimilated previously formed sedimentary iron deposit, and (c) by hydrothermal solutions derived from the volcanic magma. However, the preponderance of evidence suggests the ores are of magmatic origin connected with volcanic activity.

### (a) *El Laco Iron Deposit, Chile*

In northern Chile, magnetite occurs within lava flows and pyroclastics that have been described as "magnetite lava flow" because the largest orebodies strikingly resemble lava flows in morphology. The Tertiary El Laco volcanic complex (ELVC) in the central Andes of Chile hosts world-class magnetite ($Fe_3O_4$) deposits with remarkable volcanic and subvolcanic features. The iron deposit consists of six massive magnetite ore bodies that are interbedded between andesitic to dacitic lava flows and pyroclastic rocks. Magnetite is fine-grained with small amounts of apatite and an iron phosphate mineral. Total estimated resources are 735 million tons (Mt) at an average ore grade of 49% Fe. The iron ores are volcanogenic, and recent evidence has identified the presence of "magnetite bombs" containing pyroclastic fragments of magnetite suggesting that magnetite crystallized earlier from an oxide magma produced by liquid immiscibility prior to emplacement by explosive volcanism.

### (b) *Kiruna Iron Deposit, Sweden*

The well-known Kiruna iron deposits in Sweden consists of huge apatite-magnetite deposits associated with alkali rhyolite, trachyte and trachy-andesite lava and ash flows, with subvolcanic intrusive rocks of quartz and syenite porphyries. (Fig. 3.15b). They represent one of the world's greatest magnetite-apatite deposits with reserves of about 2 billion metric tons of magnetite. The orebodies are tabular and concordant with the host volcanic rocks, forming mostly massive bodies that appear as irregular bulbous masses, often described as "ore breccias".

The Kiruna ore is dominantly magnetite, although hematite has been indicated by drilling. The ore contains about 30% apatite (mainly fluorapatite), together with accessory amounts of silicate gangue minerals. There are obvious effects of hydrothermal alteration of the host rocks. The origin of the Kiruna iron ores is controversial, ranging from magmatic to hydrothermal. However, the current understanding is that the Kiruna iron ores were formed dominantly through magmatic iron-oxide precipitation within a larger volcanic superstructure, while local hydrothermal activity resulting from low-temperature fluid circulation in the shallower parts of the volcanic system caused some modifications.

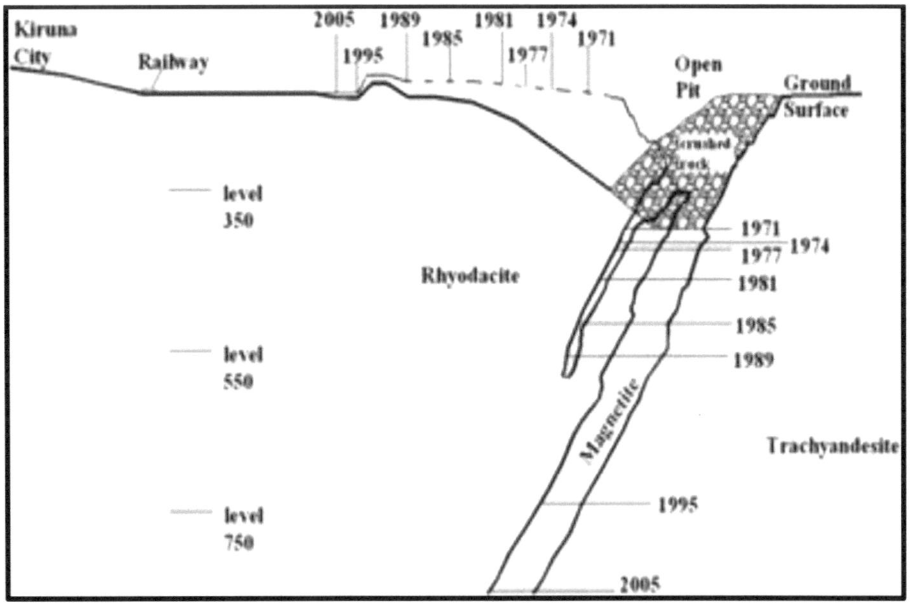


Fig. 3.15b: Geological cross-section of Kiruna magnetite deposit and host rocks

## 3.3.4 Massive Sulfide Deposits

This group of massive Ni-Cu sulfide deposits occur predominantly within mafic/ultramafic intrusions, and with rarer occurrence in ancient (Archean) basaltic rocks known as komatiites. They may be closely associated with PGEs and chromite layers. The massive sulfides can therefore be classified into two groups based on r host rock association; mafic intrusions and komatiite lavas.

### 3.3.4.1 Nickel-Copper Deposits (Mafic Intrusions)

Most of the very large Ni-Cu sulfide deposits around the world are found mostly in small mafic intrusions where they accumulate at or near the base of the intrusives. The deposits are stratiform, blanket-shaped bodies consisting of massive to stringer and disseminated sulfides, usually located in the basal portions of the layered intrusions that are commonly gabbro-norite or anorthosite in composition. Occasionally, this type of deposit may be found locally in ultramafic rocks (pyroxenite, peridotite) associated with the mafic intrusion. The sulfide ore minerals are pentlandite (Ni), chalcopyrite (Cu), and pyrrhotite/pyrite (Fe), and may be associated with platinum (PGE) and minor chromite. The ore grade is often low, but adequate for economic extraction of Ni and Cu, sometimes as supplementary metals to Cr and platinum. These deposits are formed by magmatic processes; as a mafic magma chamber begins to crystallize, nickel (+/-copper and PGE) become concentrated at the margins of the magma chamber, sometimes forming immiscible liquids that subsequently crystallize as massive sulfide ores. Examples of the Ni-Cu deposits are found in the gabbro complex of the Sudbury basin, Ontario and the Volsey's Bay deposit, Labrador, Canada. Others are the Stillwater complex in Montana and the Duluth gabbro in Minnesota, USA where disseminated sulfides overlie zones of massive sulfides.

***Sudbury Ni-Cu Deposits, Canada***

The massive Ni-Cu deposits that occur in the Sudbury Igneous Complex in Ontario, Canada are a classic example of this type of mineralization. The complex is a large, oval 65 by 35 km basin-shaped mafic intrusion about 1.85 billion years old, and believed to be of meteorite impact origin (Fig. 3.16). The massive lopolith dips inward throughout at around 30 degrees, and is composed of a sequence of norite, gabbros and granophyres. There are two main zones, a lower 500 to 2500m thick Lower Zone of norite and gabbro-norite, grading upwards into quartz- and oxide-rich gabbro. The Upper Zone varies in thickness from 1 to 2500m thick and is mainly a granophyre.

Overlying the Upper Zone at the center of the "basin" are the sedimentary rocks of the Whitewater Formation. Sudbury is the largest nickel-mining district in the world in which most of the Ni-Cu-PGE sulfide ores are located at or close to the lower contact of the intrusion. The individual ore bodies generally are moderately dipping, lenticular, or sheet-like bodies of disseminated, massive, or breccia ore or sulfide stringer zones. In some areas, the ore bodies coincide with steep faults along the norite contact. The origin of the deposits is attributed to the segregation of the massive sulfides from a mafic magma.

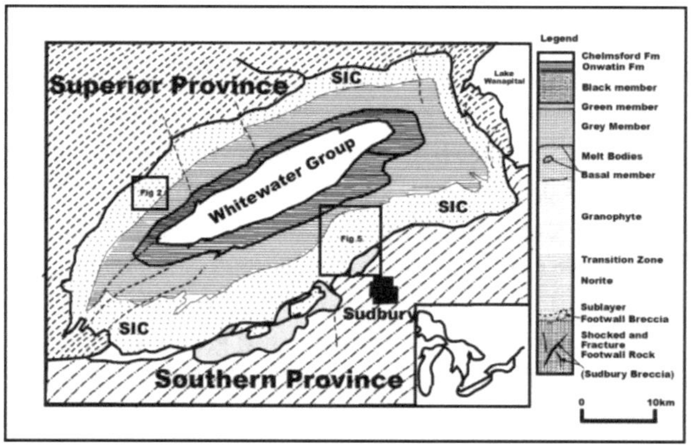


Fig3.16: General geology of Sudbury Igneous Complex, Ontario, Canada

### 3.3.4.2 Nickel-Copper Deposits in Komatiite Lavas

This is a group of massive sulfide deposits that are hosted in Precambrian basaltic to ultramafic lavas called "*komatiite*", named after a river in South Africa where the rocks were first described. Komatiites are unique ultramafic volcanic rocks containing low silica (40-45%) and potassium, but extremely high magnesium content (>18%). They are very rare and restricted mainly to the Precambrian. The ancient lavas composed mainly of olivine and orthopyroxene seemed to have cooled rapidly from temperatures as high as 1600°C, and formed a "spinifex" texture consisting of long acicular phenocrysts of olivine within interstitial plagioclase which give the rock a bladed appearance.

Mineralization of Ni and Cu ores occur within the lavas mainly as massive sulfides that are concordant with the host rocks. The mineralized zone typically consists from the base upwards, of (i) a zone of massive sulfides, (ii) matrix (net-textured) sulfides, (iii) disseminated sulfides, and (iv) cloud sulfides. The massive sulfides are usually composed of >90% Ni-Cu-Fe sulfides while the matrix sulfides are composed of 30-50% sulfides interstitial to olivine. The disseminated sulfides are the most widespread and contain 5-30% sulfides. The *cloud* sulfides are very fine disseminated sulfides that occur near the top (Fig. 3.17a).

The dominant ore minerals are the sulfides of Ni, and Cu (pentlandite and chalcopyrite) with pyrite and occasional PGEs. Komatiitic-type ore deposits are believed to have formed from immiscible Ni-Cu-Fe sulfide melt in which deposition occurred within a lava flow channel or depressions usually near the base or floor of the lava flows. Examples of this group of deposits are the Kambalda nickel deposits in Western Australia, the Raglan deposit in Quebec and the Alexo orebody in Ontario, Canada.

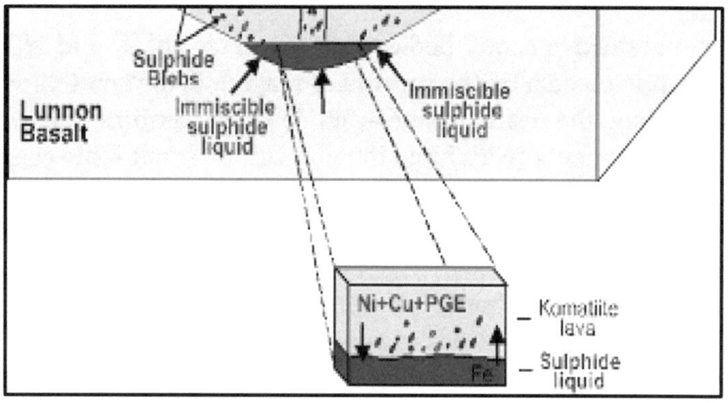


Fig. 3.17: (a) Schematic diagram of massive sulfide deposits in Komatiite lava

***Kambalda Ni-Cu Deposits, Western Australia***

The Kambalda Fe-Ni-Cu deposits occur within the Archean Yilgarn Block of Western Australia. The stratiform ore bodies are found within lava flows (channels) that have been metamorphosed (greenschist facies) and deformed. Ore minerals consist of Ni-Cu-Fe sulfides (pentlandite, chalcopyrite and pyrrhotite) and some PGE that are mostly restricted to the base of the lowermost flows of the ultramafic sequence and known as the "contact orebodies", although occasionally ores are found in the higher flows known as the "hanging-wall orebodies". The contact ore bodies which account for 80% of the ore reserves occur as elongate, lensoid and tabular bodies up to 3 km long and 300 m wide and usually <5 m thick, containing <0.5 to 10 Mt ore lenses. The individual orebodies are zoned, and grade upwards from about 2 m thick basal *massive* (>90%) sulfides to around 2 m of *matrix* (40 to 80%) sulfides, to *disseminated* sulfides and *blebby* sulfides. The massive sulfides are not always present in some deposits. Disseminated sulfides are the most common ore type and are composed of 5-30% Fe-Ni-Cu sulfides and grade upwards into subeconomic and barren olivine cumulate rocks. The origin of the Kambalda ores is attributed to crystallization from an immiscible sulfide liquid that had become segregated by liquid immiscibility from a magnesium-komatiite lava (Fig. 3.17b).

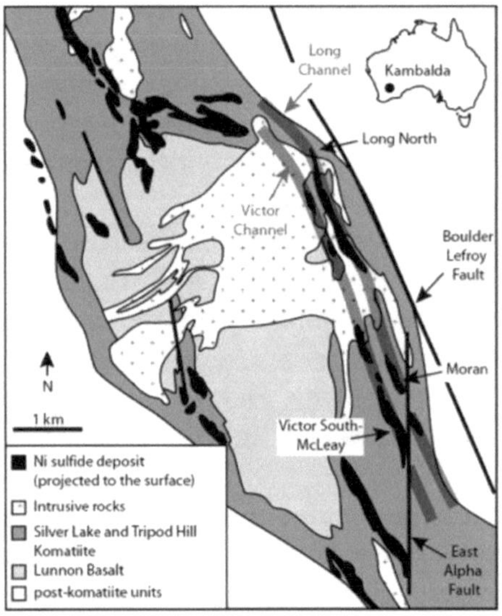


Fig. 3.17 (b): Simplified geology of Kambalda Ni-Cu Deposit, Australia

## 3.3.5 Pegmatite Deposits

Pegmatites are very coarse-grained igneous bodies that occur as dikes and sills formed from residual volatile-rich fractions of granitic magma by the process of magmatic differentiation. As a granitic intrusion reaches the final stages of cooling, the residual water-rich, low-temperature fraction concentrates at the top along with volatiles and other elements (with large ionic radius or small ionic charge) that could not enter the silicate structure of minerals. This residue may be injected into fractures within or surrounding the igneous intrusion to form pegmatites that often contain ore minerals**.** Zoning in complex pegmatites is common and it involves different mineral assemblages within each zone.

Pegmatite deposits are usually found as concentrations in pegmatite bodies and dikes that sometimes extend in length for several hundreds of meters. They may occur as swarms along the borders or inside the granitic intrusive rocks from which they emanate (Fig. 3.18(a). K-feldspar, quartz and micas (muscovite, biotite, and phlogopite) are the major minerals in pegmatites. Pegmatites have also been classified into two families based on the different enrichment trends in certain elements. The two groups are: (a) lithium–cesium–tantalum (LCT), and (b) niobium–yttrium–fluorine (NYF) pegmatites. The first family includes lithium, tantalum, and many gem-bearing pegmatites while the second family includes REE pegmatites and some gem-bearing pegmatites. The LCT and NYF families follow different enrichment trends as a result of different parent granites and their tectonic setting The following elements enriched in pegmatites are of economic and strategic importance, including lithium (Li), cesium (Cs), tantalum (Ta), niobium (Nb), boron (B), beryllium (Be), tin (Sn), uranium (U) and the rare earth elements (REE). Metasomatic alteration may result in the introduction of albite (albitization), tourmaline (tourmalinization) and fluorite which represent the effects of late-stage residual fluids rich in volatiles.

The origin of pegmatite deposits is attributed to the low-temperature crystallization of residual melts produced by fractional crystallization of volatile-rich granitic magmas. Several unusual elements and rare earths may accumulate under conditions of extreme fractionation resulting from extended crystallization of quartz and feldspars in granitic magma. Pegmatites are usually found within or close to the granitic intrusions from which they originated, and isotopic age dating has confirmed the genetic and temporal relationships. Fluorine, boron, and phosphorus are elements enriched in the late-stage residual melt of granite intrusions that are believed to play a special role in the formation of pegmatites because they form ligands (transporting agents) and fluxes (lowering the temperature of melts).

Fig. 3.18: (a) Schematic diagram of pegmatite dikes. (b) Massive pegmatites in Tanzania

Significant mineral deposits found in pegmatites are niobium, tin, beryllium, lithium, cesium, uranium and valuable semi-precious and precious gemstones including: aquamarine, tourmaline, topaz, fluorite, apatite and corundum. Examples of mineralized pegmatites are widespread, but important or classical deposits include the Bancroft Pegmatite in Ontario, Canada which is a major producer of uranium, with significant amounts of REE, Y, Ti, Zr and Be minerals and the Rossing pegmatites of Namibia which are sources of uranium. Much of the world's lithium (Li) resources is derived from pegmatites, from the minerals spodumene and lepidolite; such as those at King's Mountain, North Carolina, and Bikita in Zimbabwe. The great Tanco pegmatite in Manitoba, Canada, produces much of the world's cesium (Cs) from the mineral pollucite, and non-gem quality beryllium with reserves of over 900,000 tons of beryllium ore. Tantalum, tin, niobium, rare-earth elements are sourced from pegmatites such as Nigeria's Older Granite pegmatites. Feldspars used in ceramics, and gemstones, such as beryl, aquamarine, tourmaline, ruby, and emerald, are very rare and are found in pegmatites. Uranium has been exploited in pegmatites at various localities where it occurs as pitchblende.

***Rossing Pegmatite Uranium Deposit, Namibia***

The Rossing uranium deposit is located 70 km north-east of Swakopmund in Namibia, Africa. It is the fifth largest producer of uranium in the world and accounts for about 8% of the world's uranium oxide production. Mineralization is associated with pegmatites and alaskites (leucogranites) that occur as lenses, dykes and irregular intrusive bodies showing concordant, discordant and gradational relationships with the surrounding high-grade metamorphosed Precambrian biotite gneisses and migmatites (Fig. 3.19). The Rossing uranium deposit in Namibia is found within pegmatites and alaskites.

Uraninite (UO2), the dominant primary mineral, occurs as grains ranging in size from a few microns to 0.3 mm, with the majority in the 0.05 to 0.1 mm fraction. It is widely disseminated in the pegmatite and occurs as inclusions in quartz, feldspar and biotite. Betafite (Nb-Ta hydroxide) and monazite are widespread in some samples of ore and are commonly associated with uraninite. Secondary uranium minerals such as uranophane (Ca-U silicate hydrate) and carnotite (K-U-V hydroxide) are found near the surface as a result of oxidation/hydration of primary ores. The origin of the Rossing uranium deposit is related to the formation of uranium-bearing pegmatites, presumably from partial melts of uranium-rich country rocks during high grade metamorphism.

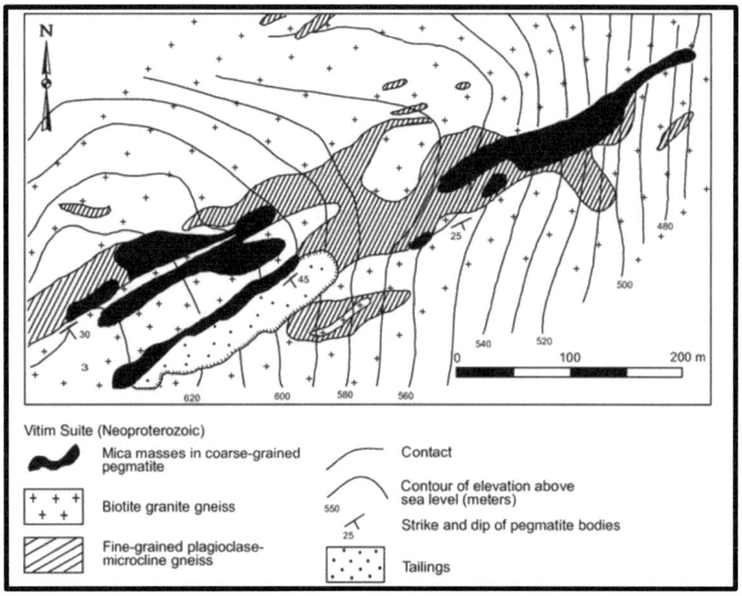


Fig. 3.I9: Geological map of mineralized pegmatite Rossing, Namibia

# 3.4 Hydrothermal Ore-Forming Processes

Hydrothermal processes take place within the earth's crust as a result of the activities of hydrothermal solutions and their reaction with surrounding rocks. Mineral deposits that are formed by these processes are known as hydrothermal deposits. They are usually epigenetic and commonly show sharp contacts with host rocks.

## 3.4.1 Nature of Hydrothermal Solutions

Hydrothermal solutions are the hot aqueous fluids (hot water) that serve as a very effective medium for the transportation and enrichment of certain elements which are subsequently precipitated from solution as ore minerals. The temperature of hydrothermal solutions is not that of "normal" hot water that boils at 100°C , but can be as hot as 500°C (range 50°C to 650°C), Hydrothermal solutions are "brines" composed mainly of water (H2O), carbon dioxide (CO2) and salt (NaCl) and are capable of dissolving and transporting heavy metals. The water in hydrothermal solutions originates from several possible sources including magmatic water from magmas, water expelled during metamorphism, deeply circulating water, shallow meteoric or groundwater, sea water and connate water held in sediments.

Hydrothermal fluids generally travel along temperature or pressure gradients, from hot areas to cool areas or from high pressure to low pressure. They migrate until they reach a suitable site for metal deposition, usually by filling open spaces in rocks or by replacement of host rocks. Hydrothermal solutions contain hot gases, metals and volatiles that are capable of penetrating surrounding rocks. The major factors controlling the movement of hydrothermal fluids are temperature and/or pressure gradients, permeability and the presence of structural openings (cavities) within folds, faults, fractures and joints. The driving forces for the migration of hydrothermal solutions include tectonic, gravity, thermal and hydrostatic pressures or gradients.

#### 3.4.1.1 Sources of Hydrothermal Solutions

Hydrothermal solutions can originate from the following five sources: magmatic water, connate water, sea water, meteoric/groundwater and metamorphic water. Figure 3.20 illustrates the various sources of hydrothermal fluids.

***Magmatic*** water is derived from magmas that contain dissolved water vapor and other volatiles that may reach up to 10% of their volume. At the late stages of crystallization, a granitic magma becomes water-saturated resulting in the separation or exsolution of an aqueous fluid to form a chemically distinct phase in the silicate melt. These volatile-rich fluids carrying dissolved salts, gases and metals may escape or be ejected from the magma and seep along fractures and pores in surrounding rocks. As these hot fluids travel through rocks, they often 'leach' additional metals from surrounding rocks to enrich their metal contents. In some cases, these mineralizing solutions may be derived from deeply circulating waters heated by a magmatic source. Most high temperature (hypothermal) vein deposits of tin and tungsten are formed by magmatic water.

***Meteoric*** water refers to rainwater that has percolated downward into subsurface rocks (as groundwater) capable of circulating by infiltration through fractured and permeable rocks and reacting and leaching metals from oxidized zones and concentrating in reduced zones. Meteoric water is responsible for the formation of many different types of hydrothermal mineral deposits at low temperatures and shallow depths such as sandstone-hosted (roll front) and surficial uranium deposits. Heated shallow subsurface waters near geothermal zones or volcanic areas are capable of circulating and leaching metals that are subsequently deposited as economic minerals. Examples include some epithermal mercury and gold deposits related to surface hot spring activity near geothermal zones.

***Sea water*** also acts as a solvent of ions from rocks beneath the sea floor. Wherever submarine volcanism occurs beneath the sea or ocean floor, sea water is most likely to penetrate into the volcanic rocks and fault zones get heated by a subjacent magma chamber and emerge from exhalative vents as hot brines capable of leaching scarce metals from enclosed rocks and depositing them on the seafloor as the solutions cool. Mineral deposits formed by this process include the volcanogenic massive sulfide (VMS) deposits and sedimentary exhalative (SEDEX) deposits which are described in more detail in a subsequent section of this book. Submarine volcanism can also contribute metals to the sea floor sediments through submarine exhalation or submarine hot-spring activity. Such activities are observed today in what is called "black smokers" from the Mid-Oceanic Ridge in the Atlantic Ocean. (A black smoker is a geothermal vent on the sea bed which ejects heated water containing black iron sulfide minerals).

***Connate*** water is the water trapped in the pores of marine sedimentary rocks. Originally, this water is from meteoric or sea water but has undergone considerable changes during the burial and compaction of marine sediments. Connate fluids are rich in NaCl (salt), and the salinity and density may increase with depth. Because connate water exists in thick sedimentary basins and contains salt brines, they can be heated and mobilized during diagenesis to leach metals from the sedimentary rocks to form ore deposits. Such deposits may range in temperature from about 50 to 100° C. The deposits that form from connate brines are usually confined within strata but may also show crosscutting relationships with the host rocks due to remobilization. Examples are the Mississippi Valley-type deposits that are sources of lead and zinc ores in carbonate rocks.

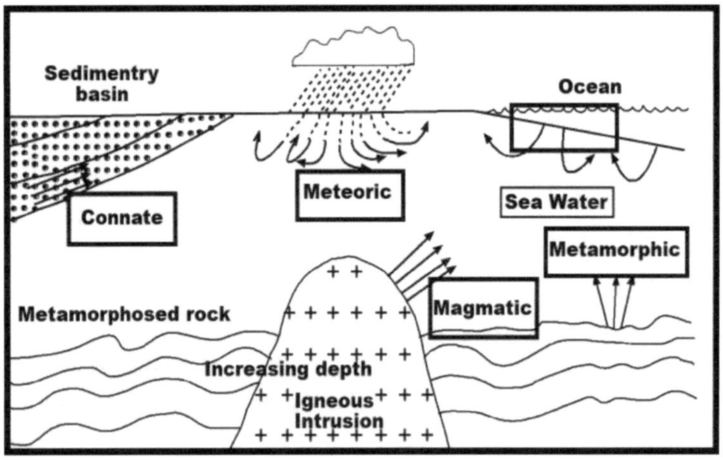


Fig 3.20: Hydrothermal solutions from various sources

***Metamorphic*** fluid is a solution derived from metamorphic processes. Metamorphism is induced in rocks by external heat or pressure or by a combination of both. Heat may be provided by the deep burial of a rock mass through time or alternatively by the intrusion of a magma body nearby. During regional metamorphism of rocks, heated interstitial water and the fluid released from water-containing minerals as they recrystallize can become potent "metamorphic fluids" that leach metals from rocks to form mineral deposits. Some "orogenic" gold deposits within Precambrian terranes are attributed to metamorphic fluids.

### 3.4.1.2 Composition of Hydrothermal Solutions

The composition of hydrothermal solutions can vary considerably, depending on their source and their interactions with rocks on the pathways of their migration. However, the main component of hydrothermal solutions is water (H2O). Pure water cannot dissolve metals and therefore, hydrothermal solutions are always brines containing dissolved salts such as NaCl, KCl, $CaSO_4$ and $CaCl_2$. The range in salinity varies from that of seawater (around 3.5 wt. %) to about ten times the salinity of seawater (35% NaCl). Such brines are capable of dissolving small amounts of metals such as Au, Ag, Cu, Pb and Zn. High temperatures increase the effectiveness of the brines to dissolve metals. Studies of fluid inclusions which are considered as remnants of hydrothermal fluids trapped in minerals, have confirmed the importance of water as the main component of hydrothermal solutions with varying salinity (trapped NaCl crystals) and sometimes CO2 contents. The CO2 is believed to have originated from the devolatilization of organic carbon from sedimentary sources during low grade metamorphism.

Metals and other ore constituents in hydrothermal solutions can be derived from several sources, including magmas from the mantle, subduction of oceanic and continental crust, leaching of metals from crustal rocks and oceanic volcanics by circulating solutions; dissolution of low-temperature minerals such as carbonates and mineral salts, and the breakdown of minerals due to elevated temperatures during metamorphism. Other minor sources include seawater and connate water.

### 3.4.1.3 Transport by Hydrothermal Solutions

Transport of ore constituents by hydrothermal solutions requires a salt (chlorides) or other soluble species which can form a metal-bearing complex. These metal-bearing complexes facilitate transport of metals within aqueous solutions, generally as hydroxides, but also by processes similar to chelation in which a metal ion is attached to a complex molecule. Also, the majority of metallic ore deposits formed by hydrothermal processes include sulfide minerals, which suggests sulfur may be an important metal-carrying complex. Therefore, it is app there may be several ligands that assist in the transport of metals in hydrothermal solutions.

Four kinds of aqueous complexes have been identified as likely transporting agents for many ore metals: (a) chloride complexes; (b) sulfide complexes; (c) organometallic complexes and (d) fluoride complexes. However, it is believed that organic complexes only play a minor role although not much evidence is available because they break down at temperatures above 300°C, and there are only low concentrations of organic acids in low-temperature hydrothermal solutions. The high concentrations of Cl in fluid inclusions supports the role of chloride complexes in the transport of some metals such as Pb, Zn, Cu and Ag. On the other hand, sulfide complexes are believed to be important in metal transport during formation of volcanogenic sulfide deposits of Ni, Cu, Zn and some epithermal gold deposits. Fluoride complexes are probably important ligands in the transport of some metals, including Sn, W, Nb and U.

The passage of hydrothermal solutions through enclosing rocks is commonly accompanied by wall rock alteration. This is characterized by the development of an assemblage of secondary minerals that reflect the composition of the affected rocks and the properties of the invading hydrothermal fluids.

### 3.4.1.4 Localization of Hydrothermal Deposits

Hydrothermal deposits are localized at the point of deposition within fractures, fissures and other available openings in host rocks, usually by precipitation of minerals from hot solutions due to changes in temperature, pressure or reaction with wall rock. There are two main processes of deposition; *cavity filling* and/or *replacement* of host rocks. The two processes may occur simultaneously, - the filling of an opening by precipitation may accompany the replacement of the walls of the opening.

When a hydrothermal solution moves slowly upward, as with heated groundwater percolating through an aquifer, the solution cools very slowly. If dissolved minerals were precipitated from such a slow-moving solution, they would be spread over a large volume of rock and may not be sufficiently concentrated to form an ore deposit. But when a solution flows rapidly, as in an open fracture, or through a mass of shattered rocks, or through a layer of porous rock where flow is less restricted, cooling can be sudden and can occur over short distances. This usually results in rapid precipitation and a concentrated mineral deposit is formed. The particular minerals precipitated in these deposits vary widely, in accordance with the composition of the hydrothermal fluids and the associated country rock. Both sources are believed to contribute to the content of hydrothermal deposits.

Many metals are found in hydrothermal ores, such as copper, lead, zinc, gold, mercury, tin, tungsten, molybdenum, silver and uranium. Since sulfur is common in hydrothermal fluids derived from igneous and sedimentary sources, many of the minerals created are sulfides, such as galena (PbS) and sphalerite (ZnS). Where the ore-forming solutions originate from connate or meteoric waters, carbonates, sulfates (which can be oxidized to sulfur) and other anions are combined with metals. Direct evidence for the presence of hydrothermal fluids in the Earth's crust are surface manifestations such as hot springs and fumaroles.

### 3.4.1.5 Conditions for Ore Precipitation

Hydrothermal deposits are epigenetic, which means they were emplaced after the host rocks were formed. The required open spaces for ore localization must be available in the host rocks through prior structural, stratigraphic or explosive volcanic processes. As heated hydrothermal fluids carrying metals migrate upwards within the crust, temperature falls and the dissolved minerals precipitate out of solution. To be effective in generating sufficient mineralization to form ore bodies, the mineralizing process must continue over a long period of time. If the upward movement is slow, mineral formation may be dispersed over a large area, but rapid movement into porous layers or open fractures may lead to sudden cooling and rapid precipitation of minerals over a limited region. The following conditions may be conducive for the formation of hydrothermal mineral deposits:

1. presence of hot solution to dissolve, enrich and transport minerals;
2. presence of fissures or interconnected openings in the rock to allow the solutions to migrate;
3. availability of sites or space for the deposits to accumulate, and,
4. chemical reaction that will result in ore deposition.

The majority of mineral deposits formed by hydrothermal processes contain sulfide minerals which suggests that sulfur from sulfate or sulfide is an important metal-carrying complex. The main reason why metal-bearing hydrothermal solutions precipitate economic minerals is due to the loss of physical and/or chemical stability attributable to one or more of the following processes;

a. Boiling of hydrothermal solutions due to reduced hydrostatic pressure
b. Falling temperature
c. Loss of pressure
d. Chemical reaction with wall rocks,
e. Mixing of different fluids, and
f. Degassing of the hydrothermal solution (separation into gas and water phases) which alters the metal carrying capacity of the fluid.

## 3.5 Hydrothermal Mineral Deposits

Hydrothermal mineral deposits are concentrations of economic minerals formed by precipitation from hot mineralizing solutions into open spaces or fissures in rocks. Hydrothermal deposits are formed at a wide range of depths, from near the surface to up to 10km in depth. The temperature of formation ranges from about 500°C to as low as 50°-25° C. These varying conditions of temperature and pressure can affect the conditions of ore formation as well as the stable mineral assemblage. At higher temperatures, certain ore minerals are more stable, and the same situation applies at low temperatures. Hydrothermal deposits are epigenetic and formed in a variety of host rocks, but sometimes there are obvious preferences for particular types of rock. For example, certain lead-zinc deposits have preferences for carbonate host rocks while hydrothermal copper deposits may occur preferentially in fine clastic sediments such as shales and siltstones. Structural control both at regional or local levels is a major factor in the localization of hydrothermal deposits which are the leading sources of metallic ores worldwide. They are of enormous significance in the extraction of many important metals such as copper, lead, zinc, gold, silver, and radioactive metals that occur mainly as veins and stratabound deposits.

## 5.3.1 Classification of Hydrothermal Deposits

Hydrothermal deposits are varied but can be classified according to their mode of occurrence which is largely determined by the structural and stratigraphic features that control their localization (e.g. veins, stratiform, stratabound). Other factors such as temperature of formation may be used, to subdivide hydrothermal deposits into smaller groups: (1) *hypothermal*, formed at great depths and temperatures ranging from 300° to 500°C; (2) *mesothermal,* formed at intermediate depths and temperatures ranging from 200° to 300°C, and; 3) *epithermal,* formed at shallow depths under relatively low temperatures ranging from 50° to 200°C.

Based on the mode of occurrence and temperature of formation the following types of hydrothermal deposits are recognized as shown in Fig. 3.21. The origin of the Mississippi Valley-Type (MVT) deposits and orogenic gold deposits are controversial but they are classified as hydrothermal deposits. The MVT deposits are predominantly epigenetic and formed by connate brines during diagenesis. Orogenic gold deposits are distinctly epigenetic and formed by metamorphic fluids.

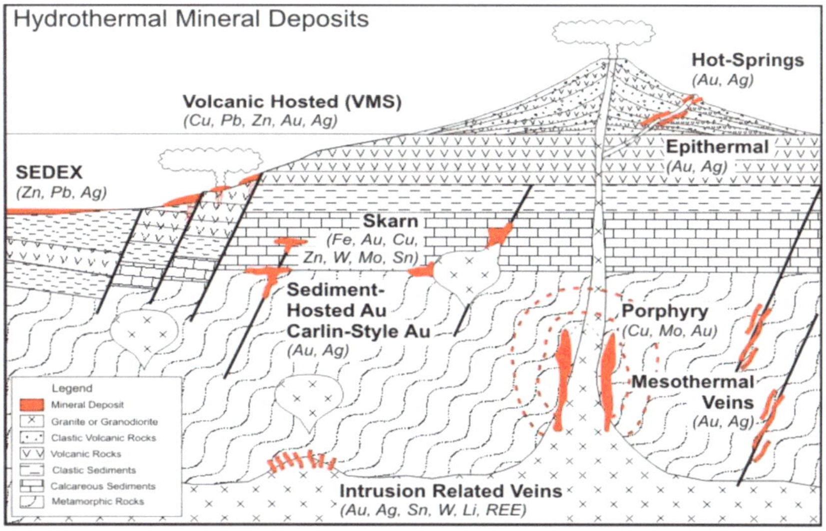

Fig. 3.21: Schematic diagram of hydrothermal deposits

1. Hydrothermal Vein Deposits
    a. Hypothermal (Intrusion-Related Veins)
    b. Mesothermal Veins
    c. Epithermal Veins
2. Porphyry Deposits (Cu, Mo, Au, Sn)
3. Volcanogenic Massive Sulfide (VMS) Deposits
4. Sedimentary Exhalative (SEDEX) Deposits
5. Mississippi Valley-Type (MVT) Deposits
6. 6 Sediment Hosted Carlin-Type Gold Deposits
7. Geothermal-Epithermal Deposits (Au, Ag)
8. Iron Oxide Gold-Copper Deposits
9. Orogenic Gold Deposits

### 3.5.2 Hydrothermal Vein Deposits

Veins are the most widespread form of hydrothermal mineral deposits, and are formed when hydrothermal solutions accumulate economic minerals in narrow open fissures in rocks, including faults, fractures, and joints, shear zones with mylonite and breccia, and other permeable zones in deformed rocks. Veins are defined mostly by their attitude and cross-cutting relationship with host rocks. They occur as tabular, elongated bodies that occupy fairly well-defined zones of mineralization that may be vertical or inclined and discordant, but are typically narrow when compared to their length and depth. Veins can be very small occupying tiny hairline fractures (veinlets) or extensive, reaching lengths of several hundred meters or even kilometers and width of a few centimeters to tens of meters (Fig. 3.22). A great many veins occur in faults or fractures or in shear zones usually close to igneous intrusions from which the mineralizing solutions originated. They are epigenetic, and in many instances, cut across a variety of host rocks.

Veins may occur singly, or as a group or system (vein system) composed of several *en echelon* (closely spaced) veins that are conjugate and parallel or subparallel. Small veins or veinlets may occur as *stockwork*, which is a small group is of closely-spaced and structurally controlled or randomly oriented veins and veinlets (also known as stringers) that form an interlacing network, Stockworks are common in many hydrothermal ores but are most prominent in greisens of tin deposits and vein systems associated with porphyry copper and molybdenum deposits in felsic igneous rocks. A vein that is rich in ore minerals is sometimes referred to as a (metalliferous) *lode*, most commonly applied to rich gold veins.

Vein deposits are notable sources of a variety of valuable ore minerals, such as native gold and silver, cassiterite (Sn), wolframite (W), pitchblende (U) or metal sulfides of copper, molybdenum, lead, zinc and mercury, which are deposited along with gangue minerals, mainly quartz and calcite. Most of the world's copper and molybdenum, and substantial production of gold, tin, silver, mercury, and uranium. Are derived from vein deposits. Hydrothermal wall-rock alteration is commonly associated with veins in which the intensity decreases from the center of the vein outwards. Typical hydrothermal alteration involves formation of secondary minerals including clays (argillic), sericite (phyllic) or sericite and chlorite (greisen), and chlorite (propylitic). The origin of vein deposits is well known: they are formed by precipitation of minerals from hydrothermal solutions due to several reasons including decreasing temperature, boiling or degassing of the fluids, and by chemical reactions with host rocks.

Some well-known examples of vein deposits are the Sn-Cu-Pb-Zn veins of Cornwall, England, the gold-quartz veins of Kalgoorlie, Western Australia, and Kirkland Lake, Ontario, Canada, and the silver-nickel-uranium veins of the Erzgebirge, Germany. In Nigeria, hydrothermal vein deposits of gold occur at various locations within the Precambrian schist belts: including Iperindo, near Ilesha in southwestern Nigeria, and in many other locations in northwestern (Maru, Anka) and north-central (Birnin Gwari, Okolom) Nigeria. In the Benue trough, lead-zinc veins with barite and copper are found within Cretaceous sediments near Abakaliki, Zurak, Azara, Wase and other localities in the Benue Trough. Within the Jurassic Younger Granite province of north-central Nigeria, cassiterite-bearing lodes containing wolframite and sphalerite have been mined in the Jos Plateau for over a century, and one of the prominent locations is the Ririwai lode in Kano State, Molybdenum and lead-silver veins are also found within these intrusive rocks or surrounding basement complex. Hydrothermal veins can be subdivided according to their temperature of formation into three groups: *hypothermal, mesothermal* and *epithermal*

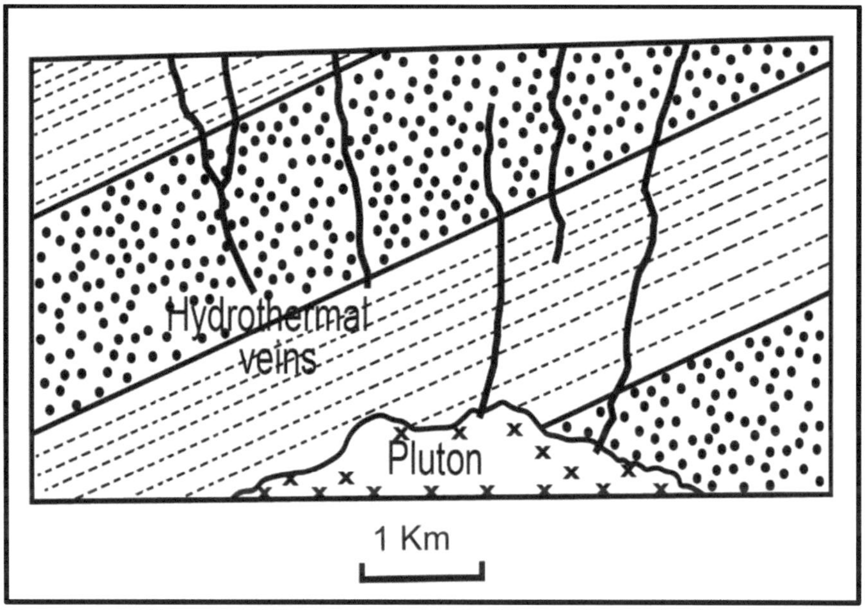


Fig. 3.22: Schematic diagram of hydrothermal veins

### 3.5.2.1 Hypothermal Deposits (Sn, W, Mo, Nb)

These are deposits formed at high temperatures and pressures, usually at great depths within or proximate to granitic or related intrusions. They are described as "intrusion related", and formed at temperatures that range from 300° to 500°C. Mineral associations include cassiterite-quartz veins; quartz-wolframite-cassiterite veins; and cassiterite-columbite-tourmaline veins; molybdenite-quartz veins, gold-quartz veins; and quartz-copper veins. Metals that may be extracted from hypogene deposits are tin (Sn), tungsten (W), copper (Cu), molybdenum (Mo), gold (Au), silver (Ag) and lead (Pb).

Principal ore minerals include cassiterite ($SnO_2$), wolframite ($(Fe,Mn)WO_4$), scheelite ($CaWO_4$), native gold (Au); with sulfides galena (PbS), chalcopyrite ($CuFeS_2$), pyrite ($FeS_2$), molybdenite ($MoS_2$), bismuthinite ($Bi_2S_3$), and arsenopyrite (FeAsS). Gangue minerals include quartz, fluorite, calcite, tourmaline, and topaz. The Sn-W deposits are often associated with "greisen" which is a product of hydrothermal alteration (greisenization) in which feldspar and muscovite are converted by acidic solutions to an aggregate of quartz, sericite ± chlorite, topaz, tourmaline, and fluorite. Ore fluids are derived from predominantly magmatic sources. Examples of hypothermal deposits are the Sn-W deposits in granitic intrusives of the Jos Plateau, Nigeria and Cornwall, England. Fluid inclusion studies have yielded fluid inclusion temperatures ranging from 400 to 500°C. Some porphyry Cu-Mo and Au deposits may have formed at high temperatures.

### 3.5.2.2 Mesothermal Deposits (Au, Ag, Sb, Cu, Pb-Zn)

Mesothermal deposits are formed at temperatures ranging from 200° to 300°C. Mesothermal veins are characterized by their relatively large size (several meters wide) and continuity at depth. Ore minerals found in these deposits include native gold (Au) and the sulfides; galena (PbS), sphalerite (ZnS), chalcopyrite ($CuFeS_2$), pyrite ($FeS_2$), bornite ($Cu_5FeS_4$), arsenopyrite (FeAsS), and tetrahedrite ($(Cu,Ag)_{12}Sb_4S_{13}$). Gangue minerals are usually quartz, carbonates and pyrite. Ore fluids are derived mainly from heated connate water and metamorphic fluids associated with orogenic activities. Classic examples of mesothermal vein deposits include the Archean lode gold deposits of Ontario-Quebec, Canada, Coeur d'Alene District, Idaho; South

Africa and Kalgoorlie, Australia. In West Africa, the "orogenic gold" veins in Ghana and Burkina Faso, and the gold-quartz veins in the Schist Belts of western Nigeria are classified as mesothermal deposits, based on fluid inclusion studies.

### 3.5.2.3 Epithermal Deposits (Au, Ag, Pb-Zn, Sb, Ba)

Epithermal deposits form at shallow depths under relatively low temperatures and pressures. Temperatures during formation may range from 50° to 200°C. Epithermal veins are usually short and narrow; and confined to shallow depths. Colloform and replacement textures are sometimes observed, but the majority of epithermal vein deposits are characterized by open-space filling textures (crustification, comb structures, symmetrical banding) - all indicators of low temperatures of crystallization. Ore minerals are native gold (Au); and the sulfides; realgar (AsS), cinnabar (HgS), acanthite ($Ag_2S$), pyrite ($FeS_2$), orpiment ($As_2S_3$), stibnite ($Sb_2S_3$),), and pyrargyrite ($Ag_3SbS_3$), galena (PbS), sphalerite (ZnS) and barite ($BaSO_4$). Gangue minerals found include quartz, opal, and chalcedony ($SiO_2$); calcite ($CaCO_3$), aragonite ($CaCO_3$), dolomite ($Ca,Mg(CO_3)_2$); and fluorite ($CaF_2$). Metals that are extracted from these deposits include lead, silver, gold and mercury. Ore fluids are derived predominantly from connate or heated meteoric water, particularly near centers of nascent volcanic activity such as hot springs and geysers. Examples of epithermal deposits are the gold and silver deposits in western United States, (Nevada, Idaho and Colorado), and the lead-zinc (-barite) veins in the Benue Trough of Nigeria.

## 3.5.3 Porphyry Deposits

One of the most unique groups of hydrothermal deposits is the "porphyry type" deposits which derived its name from the close association with *"porphyry"* or porphyritic intrusions, sometimes sub-volcanic and typically emplaced at shallow levels in the crust. A porphyry is defined as an igneous **rock** consisting of large-grained crystals such as feldspar or quartz dispersed in a fine-grained silicate-rich, generally aphanitic matrix or groundmass. Porphyry deposits are the world's most important source of Cu and Mo, and are major sources of Au, Ag, and Sn. They account for about 50 to 60% of world Cu production and more than 95% of world Mo production.

Porphyry-type deposits are large relatively low-grade epigenetic, intrusion-related, disseminated (stockwork), mineral deposits mainly of copper, but include molybdenum (porphyry molybdenum), gold (porphyry gold) and tin (porphyry tin) deposits. Some of the most common characteristics of the porphyry deposits include association with high-level (subvolcanic) porphyritic intrusions, large hydrothermal systems dominated by magmatic and meteoric fluids, pervasive hydrothermal alteration of host rocks, fracture-controlled mineralization, and association with breccia pipes and dikes (Fig. 3.23).

Porphyry-type mineralization occurs as pervasive disseminations and stockworks of ore minerals, mostly sulfides accompanied by narrow, closely-spaced veins and veinlets within hydrothermally altered rocks. Ore concentrations depend on the type of porphyry deposit. In porphyry copper deposits, Cu grades range from 0.2% to more than 1% Cu; while in porphyry molybdenum deposits, Mo grades range from 0.07% to nearly 0.3% Mo. In porphyry Au and Cu-Au deposits, Au grades range from 0.2 to 2 g/t Au. Molybdenum, gold, and silver are recovered as byproducts from a number of porphyry copper deposits which enhance the value of such deposits.

Hydrothermal alteration in porphyry type deposits is very distinctive, widespread and extensive, sometimes up to several kilometers wide, and is typically zoned on a deposit scale, as well as around individual veins and fractures. The classic alteration zoning pattern includes a potassium feldspar-biotite rich core, (potassic) surrounded by a muscovite/illite sericitic (phyllic) alteration zone, which is surrounded by a clay-rich zone (argillic) and finally by a chlorite-epidote rich zone (propylitic) (Fig. 3.24).

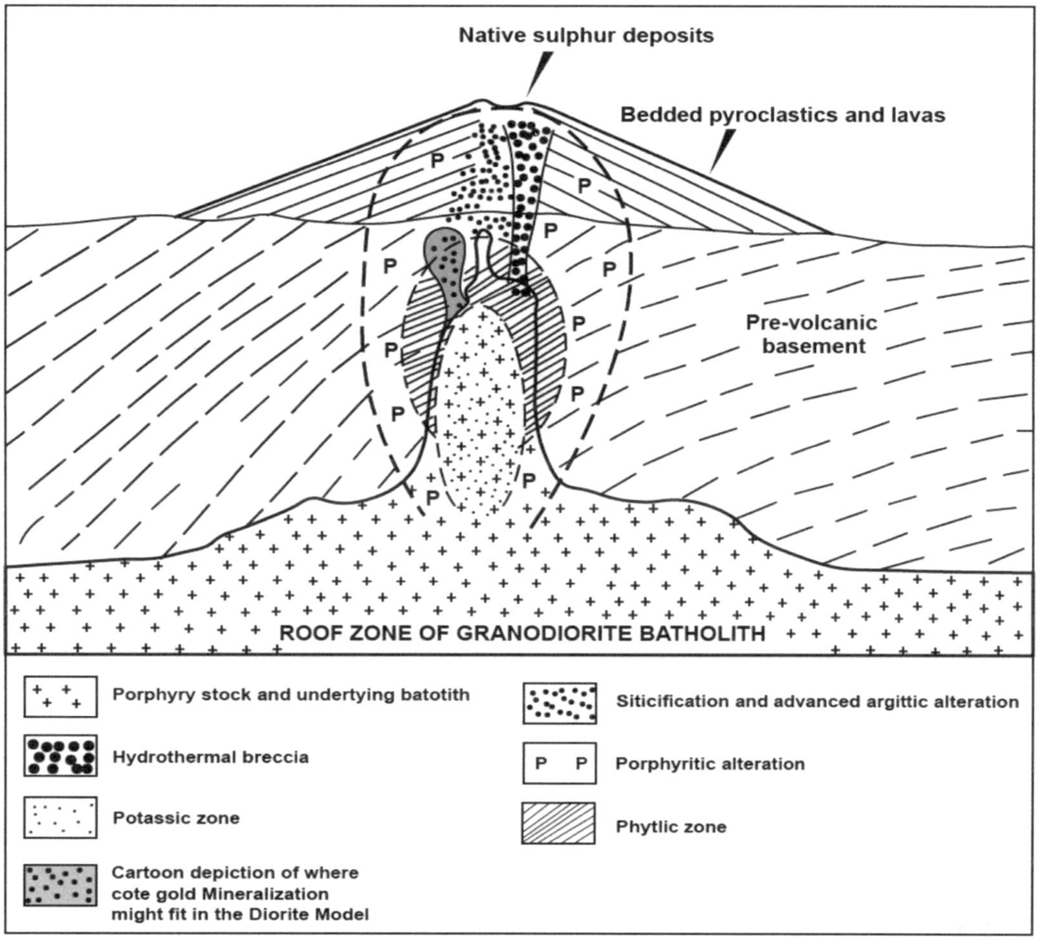


Fig. 3.23: Schematic diagram of porphyry-type system

### 3.5.3.1 Porphyry Copper

Porphyry coppers are the most common of the porphyry-type deposits. The major ores are copper with minor gold or molybdenum. Porphyry coppers are the world's leading source of copper and account for about 50% of world production. They are large tonnage and low grade deposits containing many billions of tons of copper ore with grades averaging between 0.5 and 1.5%Cu. Chalcopyrite is the dominant mineral, often with minor bornite, chalcocite and covellite. Secondary minerals that develop in the zone of oxidation or supergene enrichment are malachite and azurite with iron oxide-rich cap zone. Hydrothermal alliteration of the host rocks is distinctively zoned from the core outwards in the sequence; potassic, phyllic, argillic and propylitic (Fig. 3.24). The main gangue is quartz and pyrite, with minor magnetite. Zoning of ore minerals involving pyrite, chalcopyrite and bornite (Fig. 3.25) have been detected in some porphyry coppers including those of the well known Highland Valley in British Columbia where the author carried out his doctoral research work in the early 1970s.

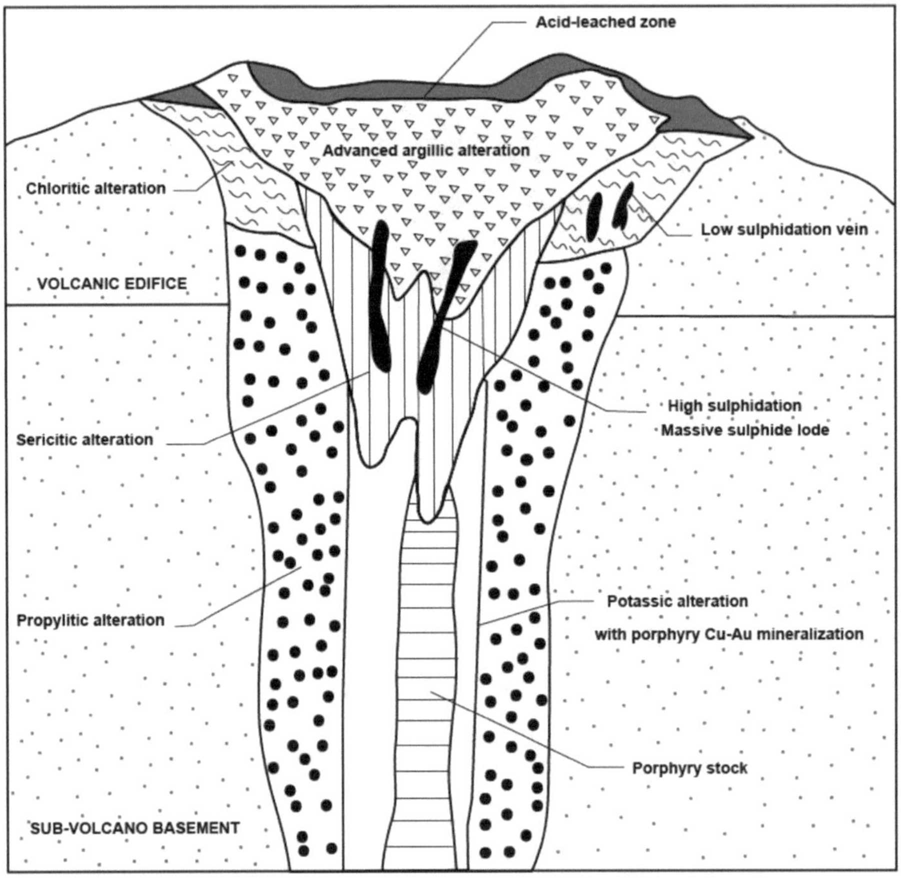


Fig. 3.24: Schematics of hydrothermal alteration zoning in porphyry type deposits

Porphyry copper deposits are among the largest of all hydrothermal deposits, with some giant-sized deposits. Wherever sulfide mineralization occurs on the surface, weathering and oxidation and supergene enrichment often produces rusty-stained bleached zones rich in Fe hydroxides called gossans, from which the metals have been leached, while zones enriched in copper accumulate below the water table. The supergene enrichment zones can be either oxide- or sulfide-dominated depending on the prevailing oxidation state at the site of formation, the depth of the water table, and climatic condition. There are several well-known porphyry copper deposits along the western continental margins of North and South America, particularly in the Tertiary Cordillera region of Mexico, western United States (Arizona, Montana, and Utah), and western Canada (British Columbia).

Porphyry copper deposits are believed to originate from hydrothermal fluids generated near the top of a cooling magma body at depths between 1 and 5 km. The close spatial and temporal association between the ore bodies and high-level porphyritic granitoid intrusions suggest that the calc-alkaline magmas are directly linked to the ore-forming process. Fluid inclusion studies suggest that the hydrothermal fluids were dense and highly saline and derived from waters of both magmatic and meteoric sources due to mixing. The mineralizing fluids seem to have boiled at temperatures of approximately 300 C to 450 C. The proportion of magmatic to meteoric waters varied considerably depending on several environmental factors. In terms of geotectonic setting, porphyry copper deposits are thought to have formed in island or continental volcanic arcs associated with subduction zones.

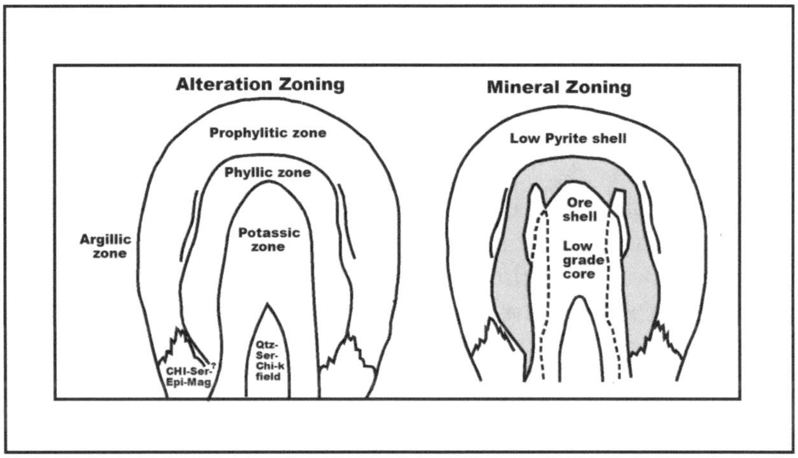


Fig. 3.25: Schematic sketch of alteration and mineral zoning in porphyry deposits

### 3.5.3.2 Porphyry Molybdenum

Porphyry molybdenum deposits also contain disseminated mineralization but with a lot more stockworks than porphyry copper deposits. The greatest concentration of porphyry molybdenum occurs in the western Cordillera of North America that extends from British Columbia (Canada) and adjacent parts of southeastern Alaska southward through the western Cordillera (Montana, Nevada, Utah, Colorado) to Mexico. There are two types of these deposits based on host rock types (a) quartz monzonite type, similar to porphyry coppers, and (b) alkali-granite-rhyolite type also known as the "Climax type" porphyry molybdenum (named after the Climax molybdenum mine in Colorado, USA).

The Climax type deposits are more important and contribute over 50% of world's molybdenum production. These deposits are not common, and only a few dozen deposits are known worldwide, including examples in Colorado (Climax, Henderson), Nevada (Mount Hope), Utah (Pine Grove), New Mexico (Questa) and Montana (Big Ben). Mineralization occurs mainly as stockwork consisting of thin veins and veinlets of quartz-molybdenite-fluorite assemblage and in discontinuous streaks within quartz veins, or as disseminations throughout the vein quartz, and rarely within the bedrock. The hydrothermal alteration zones are very similar to those of porphyry copper deposits. Many molybdenum deposits also contain some silver, tin and tungsten minerals which are recovered from the ores as byproduct.

### 3.5.3.3 Porphyry Gold

Porphyry gold deposits are disseminations of gold (only) in high-level intrusions of porphyries and related rocks of calc-alkaline to alkaline affinities. These Au porphyries are emplaced at higher levels in the crust than Au-bearing porphyry coppers, which are themselves at higher levels than the Cu-Mo porphyritic intrusives. Gold is disseminated in quartz-rich stockworks, veinlet swarms, and quartz veins within subvolcanic diorite, tonalite, and dacite porphyries. Gold ore (native Au and electrum) is accompanied by pervasive potassic (biotite) or phyllic (sericite) alteration zones that may be locally overprinted from above by advanced argillic (alunite-kaolinite) alteration, Gold tonnages are large, but grades are low, usually in the range of 1.5 g/ton. Examples include the Fort Knox deposit in Alaska and Racey, Oregon, USA, Marie and Lobo in Chile and the Dublin Gulch deposit in Canada. Indications are that the porphyry gold deposits formed at relatively higher temperatures at about 500 to 600$^{o}$ C.

### 3.5.3.4 Porphyry Tin

Porphyry tin deposits occur as disseminations, veinlets and breccia-fillings of cassiterite-quartz in subvolcanic intermediate to felsic quartz porphyry stocks (quartz-latite, dacite and rhyodacite) with associated pyroclastics and lavas (Fig. 3.26). The main ore mineral is cassiterite accompanied by sulfide minerals mostly pyrite but including stannite and arsenopyrite. Late veins may contain complex sulfo-stannates and silver minerals. Pervasive alteration in concentric zones grades from a central quartz-tourmaline core, outward to sericite-tourmaline, sericite and propylitic alteration with rare argillic alteration in the upper parts of some deposits. Porphyry tin deposits differ from many porphyry coppers in not being susceptible to supergene enrichment; and not possessing known potassic alteration zones. Examples have been described from some localities in Bolivia (e.g. Chorolque) extending into Peru and western Argentina, South America, These tin deposits are low grade with large volume of mineralized rock that contain grades of 0.2 to 0.3% Sn. Silver is derived from late stage tin-silver veins that postdate the hydrothermal alteration. The origin of the porphyry tin deposits is magmatic-hydrothermal in similarity to other porphyry deposits.

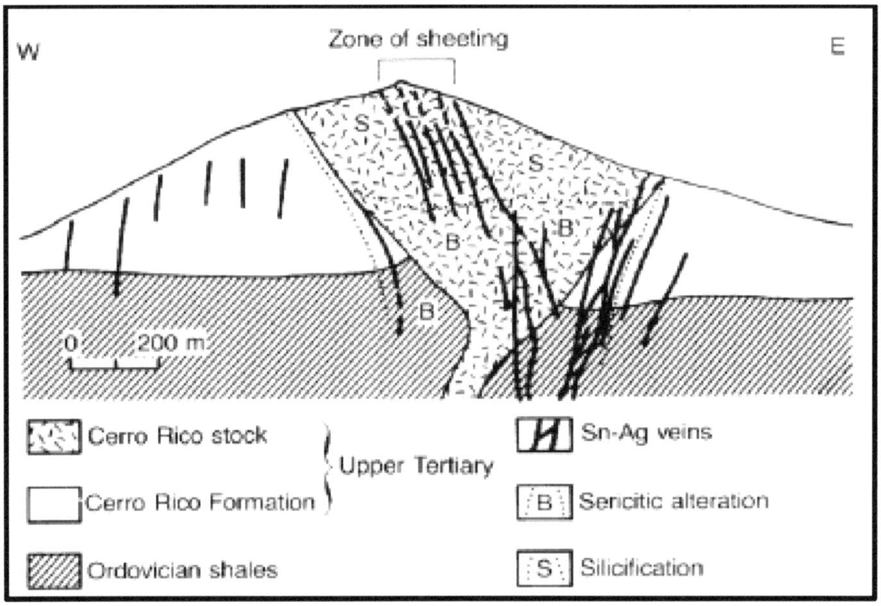


Fig. 3.26: Geological cross section of porphyry tin deposits, Bolivia

## 3.5.4 Volcanogenic Massive Sulfide (VMS) Deposits

Volcanogenic massive sulfide (VMS) deposits are metallic sulfide-rich (>60% sulfides) deposits hosted or associated with submarine mafic to felsic volcanic rocks. They are deposits created by volcanic-related hydrothermal events in submarine volcanic environments. The main reason why these intra-volcanic deposits are considered hydrothermal is because mineralization is epigenetic and deposited in volcanic rocks by hot fluids, mostly seawater that was mobilized by post-volcanic activity. VMS deposits have the following unique characteristics: (1) they are closely associated with volcanic rocks; (2) the ores are "massive" (with over 60% sulfides) and stratiform (or nearly so); and (3) they have a pipe-like zone of stockwork veins localized along the conduit zone beneath the massive ores (Fig. 3.27). Deposits range in age from Early Archean (3.55 by) to the present, and significant occurrences of VMS mineralization are found in greenstone belts of almost all Precambrian shield areas.

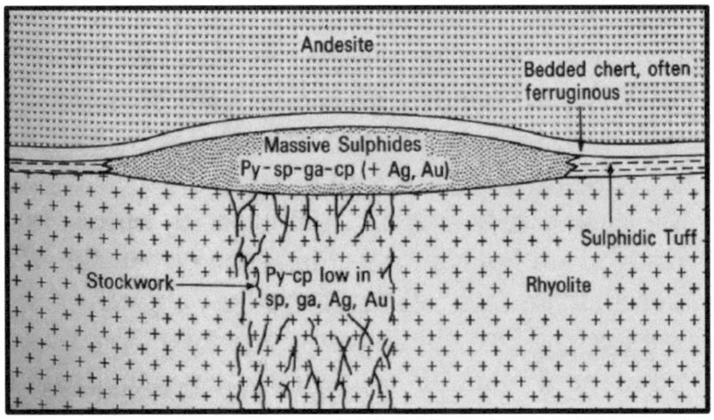


Fig. 3.27: Volcanogenic massive sulfides showing stockwork feeder zone

Mineralization is usually polymetallic and composed of the base metals; copper, lead and zinc (Cu-Pb-Zn) with minor gold and silver. These deposits are believed to have been created by volcanic-related, hydrothermal events in submarine environments on the seafloor. Characteristic host rocks are of three types: (1) felsic lavas (Kuroko type), (2) mafic lavas (Cyprus type) and, (3) mixed volcanic/pelitic sediments (Besshi type). VMS mineralization occurs mainly as lens-shaped bodies of massive ore underlain by a stringer ore zone consisting of stockwork-like veins, as well as pod-like replacements of the host rock. The stringer zone is roughly perpendicular to the massive ore and thus transects the volcanic host rocks. It narrows with depth and grades upwards into the massive ore.

The VMS deposits are important sources of the world's copper, zinc, lead, gold, and silver (Cu, Zn, Pb, Au, and Ag) ores. The different ore associations based on type of host rock in VMS deposits are: 1) felsic volcanic-hosted Kuroko type- Cu-Pb-Zn-Ag-Au; 2) mafic-volcanic hosted Cyprus type - Cu (-Zn, -Au); and 3) mixed volcanic/sedimentary Besshi type - Cu-Zn (-Au). The ore minerals are abundant iron sulfides (pyrite or pyrrhotite) with variable but subordinate amounts of chalcopyrite and sphalerite; bornite, tetrahedrite, and galena. Gangue minerals typically consist of quartz, barite, anhydrite, iron oxides, chlorite, sericite, talc, and their metamorphosed equivalents. Vertical zoning of the ore minerals within the volcanic pile from Fe-Cu at the base to Cu-Fe in the main ore zone and Zn-Pb-Fe at the top (Fig, 3.28). The size of the deposit ranges from small pods of less than a ton to supergiant accumulations such as the Rio Tinto deposit in Spain with over 1.5 billion metric tons (Bt) of ore.

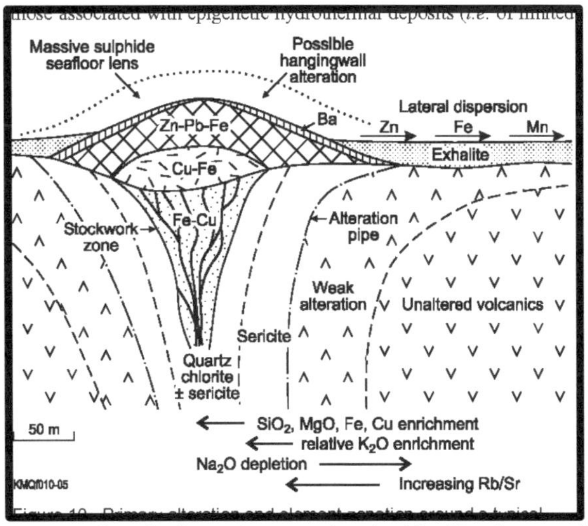


Fig. 3.28: Metal and alteration zoning in idealized VMS deposit.

Zones of hydrothermally altered rock are associated with all types of VMS deposits. These alteration zones characterized by silicate minerals (chlorite, Illite, sericite) are very distinctive, and cut across the volcanic stratigraphy. As shown in Fig. 3.28, the conduit pipe zone with stockwork mineralization beneath the massive sulfide orebody is usually intensely altered. The alteration was produced by the interaction of a hot (metal-rich) fluid with the volcanic or volcaniclastic wall rocks.

In terms of origin, the close association of VMS deposits with volcanic rocks suggests that ore deposits were formed at or near the sea floor during submarine volcanism. The circulation and mixing of two types of solutions, seawater and "juvenile" water (from the subjacent magma chamber) is thought to play a major role in ore formation and distribution of ore metals. A sub-seafloor hydrothermal system driven by heat from the emplacement of subjacent mafic intrusions mobilized heated connate seawater that rose and leached metals from the volcanic pile and precipitated the ores as the rising solutions interacted with cold seawater. The presence of pervasive alteration zones characterized by secondary quartz and phyllosilicate minerals are indicative of hydrothermal circulation through footwall volcanic rocks. Fluid inclusion studies indicate high salinity brines and temperatures ranging from 300 - 400 degrees centigrade. Commonly, there is an apparent zonation of ore metals from a central zone of copper-rich mineralization (yellow ore) to an outside zone of lead and zinc-rich mineralization (black ore) (Fig. 3.29). Barite ore forms the most distal portions of the zonation. The metals in the deposits were derived from leaching of the volcanic rocks and related magmatic sources.

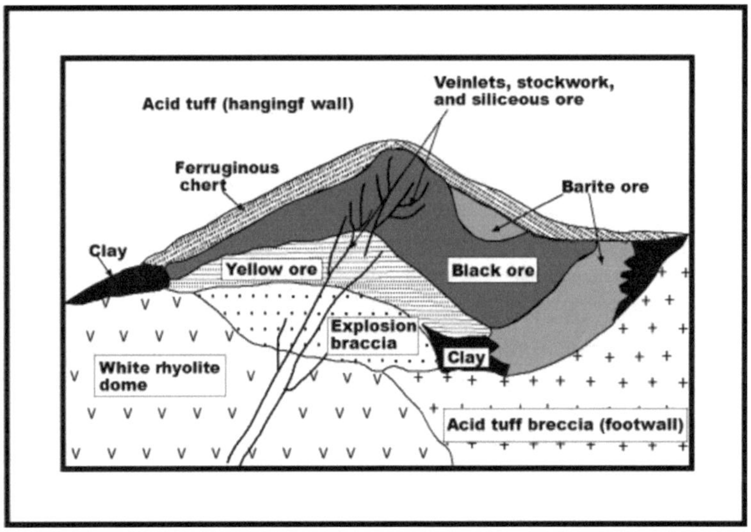


Fig. 3.29: Geology and ore distribution at Kuroko-type VMS deposit

### 3.5.4.1 Kuroko Type

The classic VMS deposits are the "Kuroko-type", named after the type locality in northeast Japan. In these deposits, the host rocks are unmetamorphosed felsic lavas and breccia which appear as dome-shaped intrusion of rhyolite and associated acid tuff breccia produced by explosive volcanism, and overlain by bedded ferruginous chert (Fig. 3.29). Most of the ore bodies are located within and above lava domes or lava flows within rhyolitic volcanic rocks or in adjacent depressions. Mineralization occurs in the form of massive sulfides of chalcopyrite and pyrite known as "yellow ore" hosted in brecciated tuffaceous rocks and overlain by thick deposit of "black ore" containing concentrations of finer-grained sphalerite and galena together with pyrite, chalcopyrite and tetrahedrite. The mineralized sequence is capped by ferruginous chert with large beds of barite, anhydrite or gypsum. Beneath the deposits are extensive, hydrothermally altered, pipe-like, conduit zones characterized by stockworks of sulfides, mainly pyrite and chalcopyrite. Individual mines commonly consist of a number of closely clustered ore deposits. The size of the orebodies range from less than 1 million ton of ore to over 60 million tons of ore at Hanaoka mine. In addition to Japan, notable examples of these deposits occur in the Iberian Pyrite Belt in Spain, the Bathurst district of New Brunswick, Canada and the unusually high grade deposits of the Buchans district, Newfoundland, Canada, and the Yebilemore deposits in the Urals, Russia.

### 3.5.4.2 Cyprus Type

Cyprus type massive sulfide deposits are usually smaller deposits consisting of pyrite, chalcopyrite and sphalerite mineralization in pillow basalts and mafic volcanic rocks associated with ophiolite complexes. The basaltic volcanics were formed in mid-ocean or back-arc spreading centers and much less frequently in intra-plate settings. In, the type area for this mineralization, the Troodos ophiolite consists of remnants of oceanic crust that was thrust unto the edge of the continental plate (obduction) during a Cretaceous subduction event.

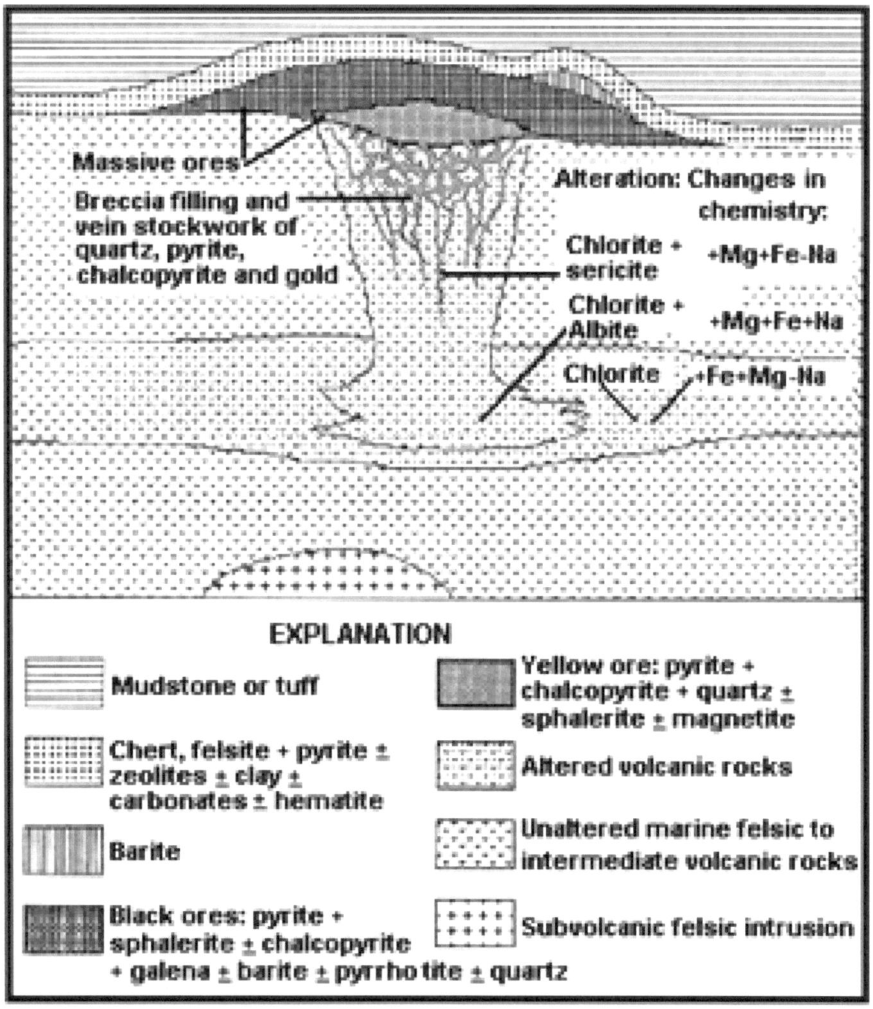


Fig. 3.30: Schematic cross section of Kuroko-type deposit

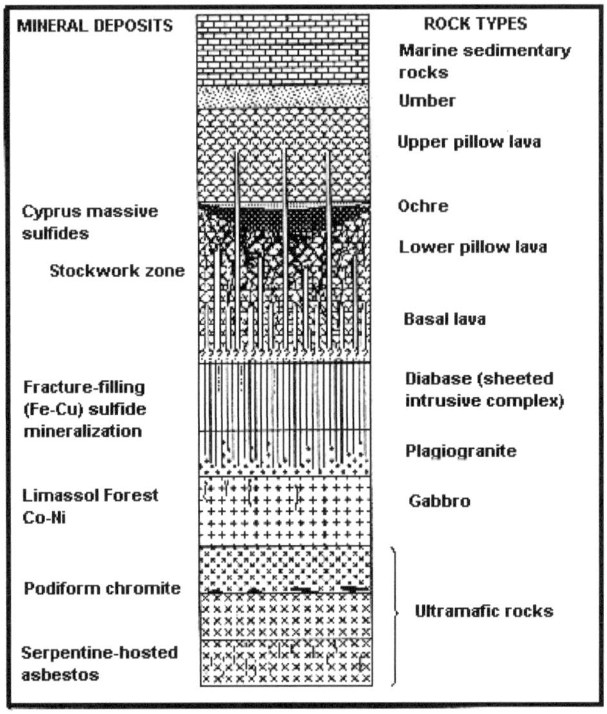


Fig. 3.31: Descriptive model of Cyprus-type VMS Deposit

All deposits are characterized by copper or copper-zinc ores that are enriched in Ni, Co, and in places Mn and As, but are very poor in Pb and demonstrate a low to moderate content of Ag and Au. Copper values range from 1–4% with similar Zn values. Ore bodies occur as a massive tabular cap overlain by sandy-textured and brecciated ore, in which the pyrite-rich massive core is cemented by chalcopyrite and sphalerite, and an underlying stockwork formed of a mixture of quartz and pyrite, with minor base metals (Fig. 3.31). Layers of iron and manganese are found in the distal pelagic sediments. The largest deposits in Cyprus are Mavrovouni (15 million tons) and Skouriotissa (6 million tons). In addition to Cyprus, examples have been described from ophiolites in Oman and Turkey.

### 3.5.4.3 Besshi Type

The Beshi-typ*e* Cu-Zn deposits are known to occur in the Sanbagawa Schist Belt of Japan where more than 100 deposits of similar characteristics are hosted by clastic metasedimentary rocks (metapelite and metagreywacke) with variable amounts of associated tholeitic metabasalt (greenstone, amphibolite). Most Besshi-type deposits are medium in size containing 5-30 million tons of ore with grades of 1-3% Cu, with Zn contents ranging from 0.3% to as high as 6% and no Pb. The deposits generally lack a well-defined footwall feeder zone, but the wall rocks may include beds or lenses of quartzite or chert, iron-formation and quartz-spessartine rock. In addition to Japan, examples of the Besshi-type VMS are described from Norway, Western Australia, the Southern Appalachians, USA, and British Columbia, Canada.

## 3.5.5 Sedimentary Exhalative Deposits (SEDEX)

Sedimentary exhalative deposits (SEDEX deposits) are sediment-hosted polymetallic ores of Pb-Zn-Ag formed in fine-grained sedimentary rocks by precipitation of metals into seawater sediments through the release of ore-bearing hydrothermal fluids along fault zones. The term "*exhalative*" is defined as the process of "venting hydrothermal solutions into submarine environments" and concentrating the ores in sediments. The geologic environments for Sedex deposits are mostly within fault-bounded intracratonic (rift) basins where thick sediments of mainly shales with minor carbonates are deposited, and volcanic rocks are rare. (Fig. 3.32) The deposits are believed to be a result of hydrothermal activity in areas of continental rifting.

SedEx deposits are the most important sources of lead and zinc worldwide, and account for more than 50 percent of the world's zinc (Zn) and lead (Pb) reserves, and furnish more than 25 percent of the world's production of these two metals. Ore minerals are sphalerite and galena (argentiferous), with little or no chalcopyrite. Moreover, other forms of SedEx mineralization are known; such as, the vast majority of the world's barite deposits are considered to have been formed by SedEx mineralization processes. Some of the characteristics of SedEx deposits particularly in comparison to VMS deposits are as follows:

- Form in fault-bounded sedimentary basins on continental or oceanic crust
- Host rocks are usually shales, with very rare volcanics
- Deposited almost at the same time as the enclosing host rocks
- Ores are fine grained and often banded
- Deposits form in clusters or stacked lenses
- Metals transported as chloride complexes in brines
- Metals derived mostly from the host sediments or weathered basement

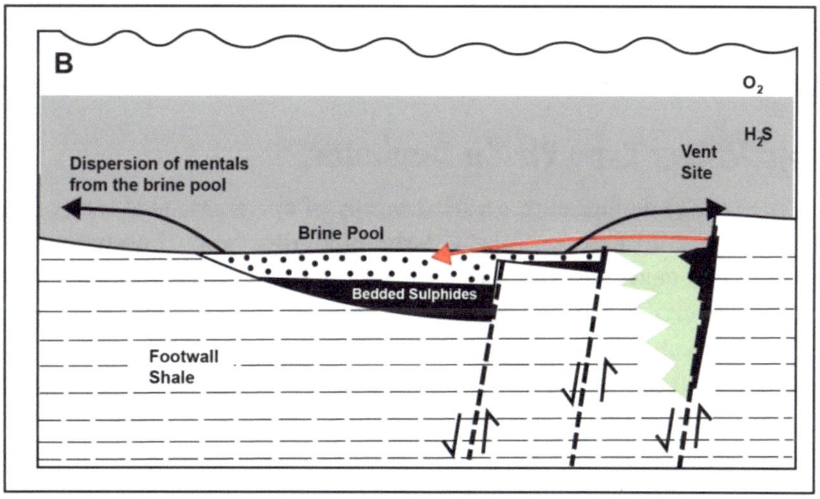


Fig 3.32: Schematic diagram of SedEx deposits in continental rift basin

The mode of occurrence is characterized by banded or layered, semi-concordant sulfide-rich bodies of Pb-Zn-Ag that are tens of meters thick, thinly bedded or laminated in carbonaceous shales (black shales and mudstones) or other fine-grained clastic rocks. Less commonly, the deposits are hosted in limestone units forming shelf-like layers near the edges of the sedimentary basins. The limestone-hosted deposits tend to occur more as veins, and may be associated with breccias. Sphalerite, the primary ore mineral occurs as dense accumulations of fine-grained sulfides in parallel layers or bands (Fig. 3.33). This banding is generally parallel to layering in the overall stratigraphy. Some of the faults that served as conduit for the passage of hydrothermal fluids may contain mineral deposits (see Fig. 3.32). Many SedEx deposits have been subjected to deformation and metamorphism which has produced further enrichment in some deposits. Examples of SedEx deposits are those at Broken Hill and Mt. Isa in Australia; Howards Pass, Yukon, Canada, and Sullivan mine in British Columbia, Canada.

The possible source(s) of the ore-forming fluids has been controversial. Suggestions include derivation from magma chambers beneath the seafloor or hydrothermal fluids generated by the heat of a magma chamber intruding into saturated sediments. Some geologists believe that seawater convection cells could play a big part in SedEx fluid sources. These sea-water convection cells could leach metals from the sediments or the underlying basement rocks. Fluid inclusion studies suggest the ore-forming fluids are saline brines (>15 wt% NaCl) at temperatures ranging from 100 - $300^{o}$.C. The source of sulfur in sulfides may be wholly seawater sulfate or a combination of seawater sulfate and magmatic sulfur.

Fig, 3.33: (a) SedEx Deposit, British Columbia (b) Typical Banded SedEx Ore

### 3.5.6 Mississippi Valley Type Pb-Zn Deposits

Mississippi Valley-Type (MVT) deposits are a varied family of epigenetic lead-zinc ore deposits that occur predominantly in carbonate host rocks, particularly dolomites and limestones of mostly Paleozoic (Cambrian) to Mesozoic (Cretaceous) age, They derive their name from the Mississippi River in south-eastern United States where classic deposits occur. Mineralization is epigenetic and commonly stratabound, but may be locally stratiform, and enclosed or parallel with surrounding sedimentary rock strata (Fig. 3.34). The most important ore controls are open spaces created by faults and fractures, dissolution collapse breccias, and lithological transitions. The deposits include tectonic or karst cavity breccias that were cemented by ore and gangue minerals; replacement bodies and cross-cutting veins. Texturally, the ores may be massive, banded or brecciated (Fig. 3.35). The principal ore minerals are galena and sphalerite with minor silver, Gangue minerals may include carbonates with minor barite and some fluorite. Silicification of the carbonate host rocks may accompany mineralization.

Some of the unique characteristics of MVT deposits are the localization of ores in open spaces and permeable zones within platform carbonate sequences along flanks or edges of sedimentary basins (Fig. 3.35). Sometimes the carbonate host rocks constitute a part of a reef, and the ores are localized relative to a particular reef facies. In other cases, the ores may occur in association with solution collapse breccias. Many of the larger orebodies occur as stratiform masses (i.e., parallel, or nearly so, with the bedding), and some as beds that extend from a few centimeters to several tens of meters with ore banding on a very fine scale. Other deposits occur as vein infillings or replacements along fissures (commonly faults or joints) that may cross the bedding. These epigenetic features have created controversy about the origin of these deposits that are still considered by some authors as syngenetic-diagenetic.

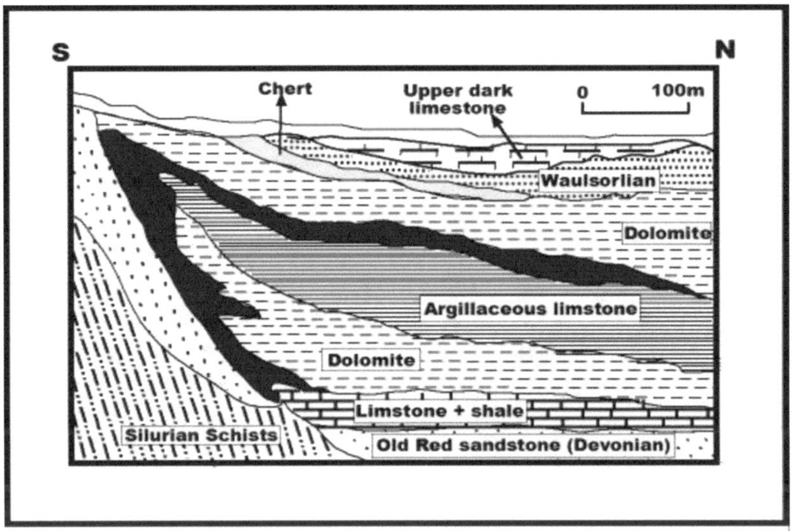


Fi. 3.34: Cross section of MVT Pb-Zn Deposits in Carbonates

MVT deposits are found throughout the world but are most abundant in Europe and North America. The United State examples are clustered around the central and southeast in the drainage basin of the Mississippi River in the states of Wisconsin, Missouri, Kentucky and Kansas. Other areas where these deposits are found include the Pine Point deposit in northern Canada, and those in Ireland, Spain and Morocco. The origin of the MVT deposits has been controversial, but the current consensus is that the deposits are epigenetic, and formed by highly saline basinal brines (10–30% NaCl) at low temperatures of 75-200°C. The source of the metals and sulfur are crustal sources and seawater sulfate respectively, but there is evidence of mixing with a separate H2S-rich fluid in the limestone environment that may have caused sulfide precipitation (Fig. 3.35b). Mobile hydrocarbons in the limestones and sulfate-reducing bacteria have been cited as important players in the conversion of dissolved sulfate to sulfide and the spatial link of Mississippi Valley deposits to oilfield-like brines. In the absence of any igneous activity, the metals are believed to have been leached from sediments and transported as chloride complexes. Migration of fluids took place in inter-stratal solutions and finally precipitated as ores in a diagenetic environment.

## 3.5.7 Sediment-Hosted Disseminated Gold Deposits (SHDG)

Carlin-type gold deposits are a small group of sediment-hosted, disseminated gold deposits that are very unique in their mode of occurrence, and should be of interest to economic geologists. This class of hydrothermal deposits called "Carlin gold" was first identified and named after the Carlin mine in Nevada, USA became a mass producer of gold in the 1960s. It was then recognized that other deposits of this type should exist elsewhere. These gold deposits are epigenetic and characterized by the following features: (1) gold is disseminated in fine-grained carbonaceous sediments; (2) mineralization occurs along favorable fault structures from where it permeates into favorable sedimentary rocks; (3) gold is extremely fine-grained and considered "invisible" because it is typically microscopic (micron-gold) and/or dissolved gold in pyrite and arsenopyrite or hematite when oxidized; (4) the host rocks are fine-grained carbonaceous or so-called "dirty" carbonate rocks, but ore concentrations are sometimes present in siliceous and silicified rocks as well as intrusive rocks; (5) hydrothermal alteration consists of decarbonization, silicification (jasperoid formation), and argillization; and (6) commonly associated with arsenide sulfides (orpiment and realgar).

Most of the deposits are known mainly in northern Nevada and northwestern Utah, USA where they are arranged in clusters and belts, although in recent years similar deposits have also been found in China, Iran, and North Macedonia.

Fig. 3.35: (a) Brecciated Ore/colloform banding (b) MVT deposits at basin flanks

The Carlin–type deposits represent some of the largest hydrothermal gold deposits in the world. The invisible nature of the gold in the deposit makes it difficult to find deposits of that kind easily. The deposits commonly exhibit significant structural (faults) and stratigraphic controls. They seem to follow old fault lines from where they are selectively deposited in silty carbonates which implied that the hydrothermal solutions migrated through major fault systems and preferentially deposited ores in carbonate rocks with sufficient porosity to form the "Christmas tree" pattern of mineralization (fig. 3.36). The ore mineral is dissolved gold with minor disseminated sulfides (pyrite), arsenic, antimony, mercury, and thallium. Base-metal sulfide minerals such as Cu, Zn and Pb are rare or absent in most deposits.

The origin of the Carlin-type gold deposits has been difficult to determine, although all evidence indicates they are epigenetic and hydrothermal, but the ore-forming processes are not crystal clear. Because of the fine-grained nature of the ores, they have in the past been considered as low-temperature epithermal deposits. However, fluid inclusion studies indicate relatively higher temperatures in the range of 120 - 255°C. Stable isotope data from pyrite and alteration minerals suggest that the ore fluids were dominated by deep-seated waters of magmatic or metamorphic origin, some of which had clearly exchanged oxygen with wall rocks during their passage through the crust. The alteration and ore mineral assemblage suggest the ore fluids were probably of near neutral pH, and gold was likely carried as a bisulfide complex. The source of metals is believed to be the sediments. The hydrothermal fluids may be of mixed origin (magmatic-hydrothermal mixed with connate water) and apparently 'invaded 'the host rocks preferentially within argillic host rocks and precipitated fine gold due to chemical reactions.

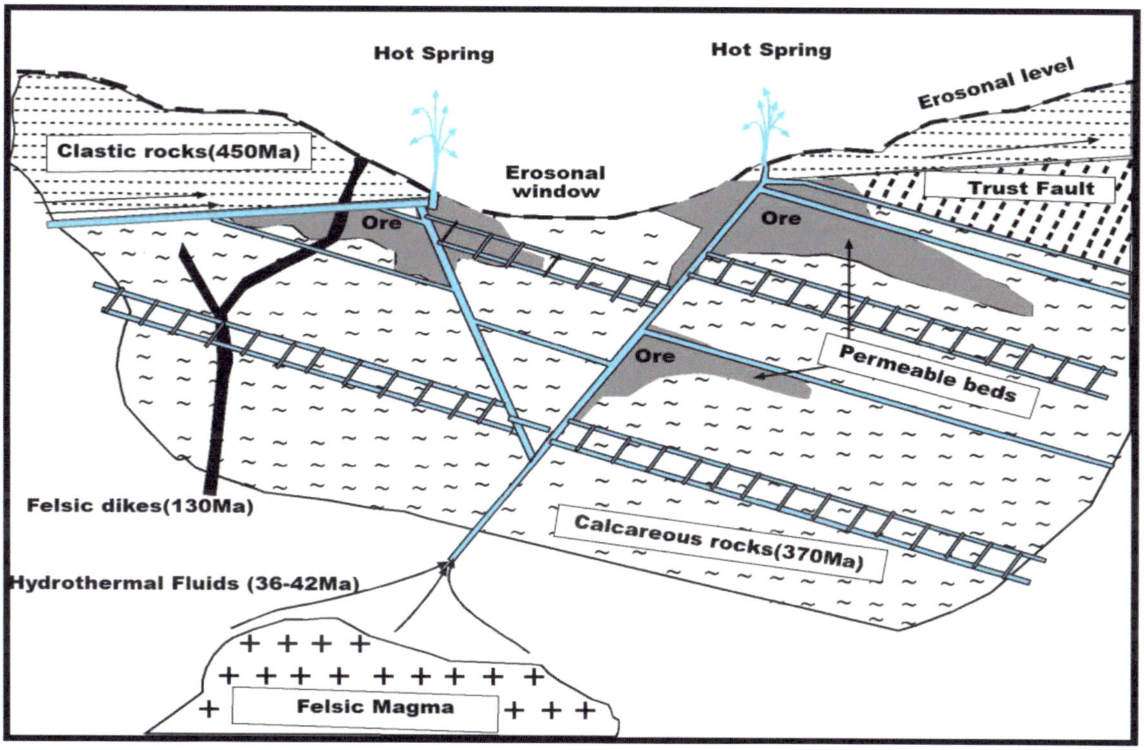


Fig. 3.36: Schematic diagram of Carlin-type gold deposits

### 3.5.8 Geothermal-Epithermal Au-Ag Deposits

Geothermal-Epithermal (geo-epithermal) deposits are a small group of hydrothermal deposits of Au-Ag closely associated with volcanic geothermal fields and formed at shallow depths and at low pressure and temperature. They are attributed to the action of heated groundwater (hot springs) associated with geothermal sources. One of their distinctive features is they are *low-sulfide* deposits composed mostly of the precious metals gold and silver, with some pyrite, antimony and mercury. Most geo-epithermal deposits occur as small veins, irregular branching fissures, stockworks or disseminations in volcanic rocks and breccia pipes, that are characterized by open-space filling textures (crustification, comb structures) and symmetrical banding with occasional colloform texture (Fig. 3.37). Host rocks are mostly volcanic rocks with subaerial pyroclastics and sub-volcanic intrusions. Hot springs and fumarole deposits (e.g. sulfur) may be present in locations that have not been deeply eroded.

Gold and silver are the main economic minerals in geo-epithermal deposits, with lesser amounts of mercury, arsenic and antimony. Ore minerals are native gold, argentite (Ag), tennantite (As), stibnite (Sb), and cinnabar (Hg). Gangue minerals are mainly quartz and calcite with lesser amounts of fluorite, barite, rhodochrosite, dolomite and chalcedony (Fig. 3.37b). Geo-epithermal gold deposits are found in New Zealand, Tasmania, and in the western Cordillera of the United States in Arizona, near Sulfur in Nevada and other localities in Colorado, and Montana. New Zealand is unique in that it has several commercial geo-epithermal gold deposits, but is also a country where the creation of observed today in the geothermal zone of hot spring and geysers in North Island, New Zealand.

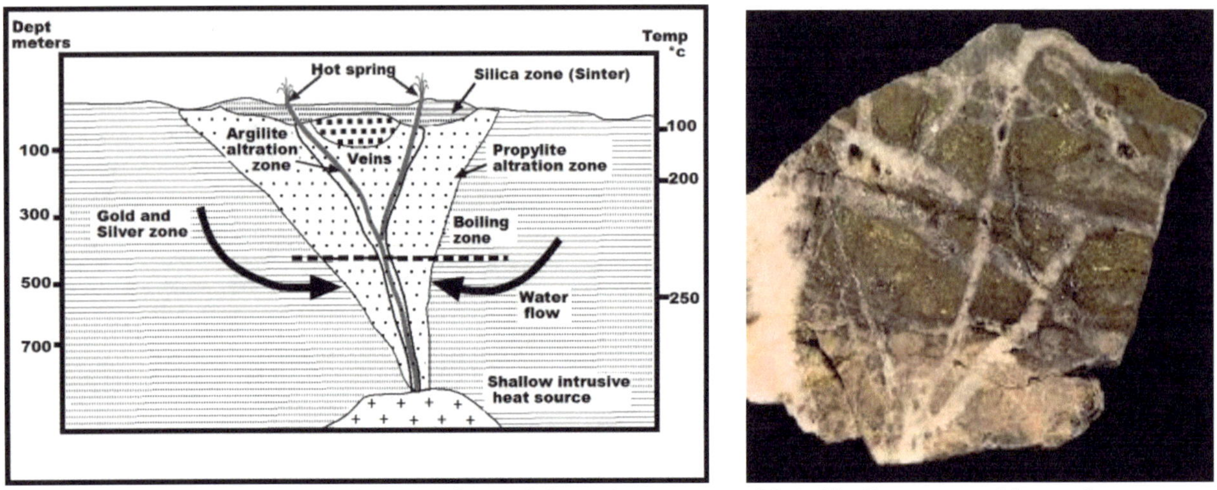


Fig. 3.37: (a) Diagram of geo-epithermal deposits (b) Gold-silver geo-epithermal ore

Geo-epithermal deposits have been studied and are believed to originate from low-temperature, geothermal-hydrothermal systems somewhat related to mobilization of heated groundwater due to sub-volcanic activity. Mineral deposition seems to have occurred as a result of Au-bearing hydrothermal solution of mixed magmatic vapor and meteoric water rises and "boils", thereby releasing $H_2S$ and destabilizing the $Au(HS)_2$ complex, which results in the precipitation of Au as a native metal. These hydrothermal systems, while active, discharge to the surface as hot springs or fumaroles (Fig. 3.36). Temperature of formation is in the range of 50 to 250°C based on geothermometry studies. Evidence from fluid inclusion chemistry, ore mineralogy and related texture suggests that these deposits were formed by dilute ((<5wt % NaCl)) solutions that underwent boiling or effervescent degassing at shallow depths (<2km). Apparently, hydrothermal deposits can be formed by heated groundwater due to geothermal activity near shallow subvolcanic environments as demonstrated by such activity in recent geothermal systems in New Zealand.

## 3.5.9 Iron-Oxide-Copper-Gold (IOCG) Deposits

The iron-oxide-copper-gold (IOCG) is a recently identified type of hydrothermal polymetallic, (Cu-Au-U) breccia-hosted deposit in which ore is spatially and temporally associated with granite and with iron oxide alteration. The deposit type was first recognized by the discovery in 1975 and study of the supergiant Olympic Dam mine, South Australia - the world's largest mine of copper-gold-uranium deposit (Fig. 3.37). IOCG deposits are classified as separate to other large intrusive related copper deposits such as porphyry copper deposits and other porphyry metal deposits primarily by their substantial accumulations of iron oxide minerals, association with felsic-intermediate type intrusives, and lack of the complex zoning of alteration mineral assemblages commonly associated with porphyry deposits. (Fig. 3.38).

The Olympic Dam deposit is located within a funnel-shaped, hematite-rich hydro-thermal breccia that formed close to the surface through progressive, polyphase fracturing and alteration of the upper part of a granitic intrusion. The ore minerals are diverse and complicated, comprising some 30 varieties of Cu, U, Au, Ag, Ni, Co sulfides, sulfosalts, oxides, carbonates, and native metals. They occur as veins, disseminations, irregular patches and breccia fillings that occur together with zones of intense calcic-sodic, iron and potassic alteration. Ore reserves are estimated as 8,330 million tons of ore at 0.8% Cu, 280 ppm $U_3O_8$, 0.76g/t Au, 3.95 g/t Ag and 151 Mt at 1.0 g/t Au. Similar but smaller deposits have been found in other parts of South Australia and South America.

The origin of the IOCG deposits is still not well known subject to continual revision because of its relatively recent discovery. However, most workers attribute mineralization to large-scale magmatic events that drive large-scale flow of oxidized probably magmatic hydrothermal fluids into mid to upper crustal levels along fault zones. Mixing of these fluids with near surface meteoritic fluids or brines is commonly invoked as the ore-forming process.

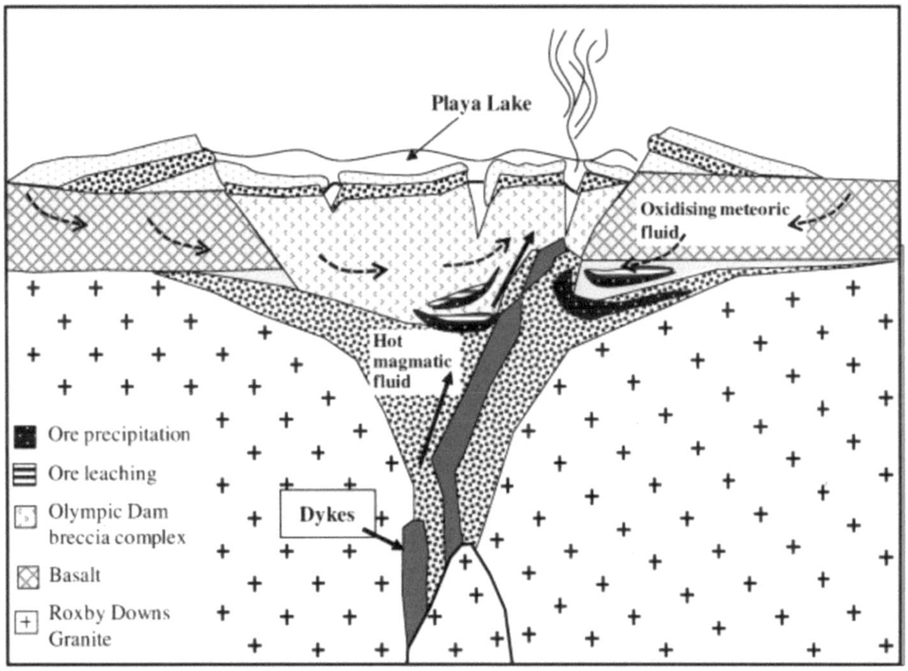


Fig. 3.38:Cross-section geology of Olympic Dam IOCG deposit, Australia

### 3.5.10 Orogenic Gold Deposits

Orogenic gold deposits are characterized by "mesothermal" quartz or quartz-carbonate veins in deformed and metamorphosed terranes of Precambrian greenstone belts and schist belts. They are described as "orogenic" because of their association with the deformation of orogenic belts or convergent plate margin. The metamorphic grade of the host rocks is greenschist facies to upper amphibolite facies or rarely to granulite facies. The gold-bearing veins are localized within major structural features spatially associated with deep-seated crustal deformation zones. Hydrothermal alteration surrounding the veins contains the same minerals as in the gangue – quartz, carbonates and sulfides with additional low-temperature silicate minerals such as sericite, albite, biotite, and chlorite. Well-known deposits of orogenic gold are found in the Archean greenstone belts of Australia and Canada, and the Proterozoic Birimian greenstone belts of the West African Craton in Ghana, Guinea, Mali and Burkina Faso. Orogenic gold constitutes a very substantial source of gold worldwide. The origin of the orogenic gold deposits is attributed to metamorphic fluids derived from intra-basin sources or underlying continental crust. Ores were deposited from low-salinity, near-neutral, and $CO_2$ rich fluids driven through the crustal sequence of metavolcanic and metasedimentary rocks, where they leach out gold transported as chloride or sulfide complexes. The fluids are channeled along major structural discontinuities where reaction with wall rocks resulted in the deposition of gold.

## 3.5.11 Wall Rock Alteration in Hydrothermal Deposits

Wall-rock alteration is one of the unique features of hydrothermal deposits. All types of hydrothermal deposits are associated with one form or the other of wall rock alteration which is attributed to interaction of host rocks with mineralizing hydrothermal solutions. Notable deposits with widespread alteration zones are the porphyry, volcanogenic massive sulfide and geo-epithermal deposits. Wall-rock alteration occurs over a wide range of temperatures, and is often characterized by changes in the mineralogy of the country rock as a result of a chemical reaction with hydrothermal solutions. The changes in mineralogy are most marked adjacent to the vein or center of deposit, and become less distinct further away. It may take place in advance of emplacement of the ore minerals, or during the final waning stages of hydrothermal activity. An alteration zone defines rocks that have been altered to a specific group or assemblage of secondary or alteration minerals, usually around the perimeter of a mineral deposit. Hydrothermal alteration has been recognized as a valuable tool in mineral exploration, because the alteration zones around many deposits are wider and easier to locate than the ore bodies themselves.

### 3.5.11.1 Types of Hydrothermal Alteration

Types of hydrothermal alteration are identified by the dominant mineral or mineral assemblage present in the alteration zone. Hydrothermal alteration can be described as “major” or “minor” depending on the areal extent of the alteration. The most common alteration types particularly associated with high-temperature solutions of magmatic origin are: (1) Potassic (2) Phyllic (3) Argillic and (4) Propylitic. Other minor forms include: chloritization, silicification, feldspathization, albitization, carbonatization, and tourmalinization. Zoning of wall rock alteration is common around vein deposits and porphyry copper deposits.

***Potassic Alteration***

Potassic alteration is characterized by the presence of new K-Feldspar and/or biotite, sometimes with minor sericite and quartz. It is distinguished by the pink to red coloration of the altered rock. Potassic alteration is associated with potassium metasomatism, and represents the highest temperature effects that are encountered in hydrothermal alteration processes, usually in the range of 450-600°C. It is exclusively associated with porphyry deposits and some tin deposits, forming the core of the alteration system and lying within the granitic intrusion.

***Phyllic Alteration***

Phyllic alteration is the most common alteration in a variety of hydrothermal deposits, and forms over a wide temperature range of 200-400°C. It is characterized by the assemblage of sericite (a fine-grained muscovite) and quartz. It is one of the most common forms of alteration around porphyry type base metal deposits, but also found in mesothermal precious metal deposits and VMS ores in felsic rocks (Kuroko type). Sericite is also found as an alteration mineral in tin greisens.

***Argillic Alteration***

Argillic alteration is characterized by the presence of clay minerals (hydrated aluminum silicates) and quartz. There are two types based on mineral assemblage: *intermediate argillic* and *advanced argillic*. In *advanced argillic alteration*, dickite, kaolinite and pyrophyllite (all hydrated aluminum silicates) and quartz are the alteration assemblage. Alteration involves the extreme leaching of cations and the concentration of H+ metasomatism. This type of alteration is characteristic of many epithermal precious metal deposits. *Intermediate argillic alteration* is dominated by kaolinite and montmorillonite. It is somewhat similar to advanced argillic alteration, but with a lesser degree of leaching of cations. Also, unlike advanced argillic alteration which is associated with vein type deposits, intermediate argillic alteration is more common in disseminated deposits such as porphyry and disseminated gold deposits. Kaolin is also produced in the pervasive hydrothermal alteration of granitic and felsic volcanic rocks.

***Propylitic Alteration***

Propylitic alteration is probably the most widespread type of hydrothermal alteration. It is made up of the assemblage chlorite + epidote + calcite and albite. Due to the presence of the green minerals -chlorite and epidote, this zone is usually easily recognizable by its green color. Unlike other alteration zones that are formed by the leaching of cations, this type of alteration seems to represent the addition of cations of calcium and magnesium. Propylitic alteration is common in mafic volcanic host rocks of vein and porphyry type deposits, as well as beneath volcanogenic massive sulfide (VMS) deposits (Cyprus type). *Chloritization* is a subcategory of propylitic alteration in which chlorite dominates but with minor epidote, whereas *carbonatization* is characterized by the dominance of calcite or dolomite.

***Minor Types of Alteration***

**Greisenization** is the hydrothermal alteration of granitic rocks into an assemblage of quartz + sericite + chlorite + tourmaline + topaz + (fluorite). It is believed that this type of alteration is aided by the action of water vapor containing fluorine. The associated metallic minerals are cassiterite and various tungsten-bearing minerals. This is a common form of alteration in association with tin deposits in granites.

**Tourmalinization** is the introduction of tourmaline crystals and quartz-sericite-tourmaline veins into altered granitic rocks and pegmatites at medium to high temperature. This type of alteration is most commonly associated with mineralization in pegmatites and granitic rocks containing tin and tantalum deposits. Tourmaline also occurs with auriferous quartz veins (orogenic gold) in metamorphic crystalline rocks.

**Silicification** is the addition of new quartz or amorphous silica minerals (such as chalcedony and jasperoid) to a rock during alteration. This type of alteration is common around hydrothermal quartz veins, such as in gold deposits where silicification accompanies replacement of adjoining host rocks.

**Hematitization** is characterized by the introduction of hematite as a result of oxidizing hydrothermal fluids. In porphyry copper deposits, disseminated hematite sometimes occurs with potassic alteration creating a reddened feldspar. In disseminated gold deposits, hematite is produced by the oxidation of pyrite. The iron oxide in IOCG deposits was introduced during hydrothermal mineralization and alteration.

**Albitization** is a form of alteration in which albite is the dominant product in replacement of existing feldspars in granitic rocks. This alteration is common in high temperature metasomatism of alkaline granites in association with tin and niobium deposits.

#### 3.5.11.2 Chemical Changes in Hydrothermal Alteration

Hydrothermal alteration of wall rocks is usually accompanied by mineralogical as well as chemical changes (Table 3.2). Potassic (K-feldspar) alteration and albitization occur at relatively high temperatures and correspond to gains in K and Na respectively. The added Na and K generally combine with the Al of pre-alteration feldspars to form alteration feldspars, and these replacement reactions are not always obvious or easy to recognize. They may induce K or Na gain accompanied by Ca-, Na-, and/or K-losses. Additional losses of Fe, Mg, and possibly Si may be necessary to form albitites (albite-rich) rocks. Phyllic alteration is characterized by the widespread development of fine-grained mica (sericite) and quartz (Table 3.2).This is an acidic alteration that destabilizes feldspar and may induce K loss. Depending on the precursor, by-products such as albite and quartz may be produced. Sericitization generally results in Ca- and Na-losses, accompanied by a K-gain or loss Argillic alteration that produces mostly clay minerals is a product of acidic solutions that result in the loss of K, Na and Ca. Propylitic alteration is a common weak alteration observed at the periphery of porphyry systems. This process produces calcite, chlorite, epidote, and albite (Table 3.2), which are greenschist facies minerals observed in unaltered rocks (calcite excepted). Propylitic alteration corresponds to carbonatization combined with hydration occurring at moderate temperatures.

Table 3.2. Alteration types and associated chemical changes and alteration minerals.

| | Type of Alteration | Mass Changes | Examples of Assemblages |
|---|---|---|---|
| | Potassic (K-Feldspar) Alteration | +K, Na | K-feldspar + biotite + quartz |
| | Albitization | +Na, -K | Albite + hornblende ± biotite-quartz |
| | Phyllic (sericite) Alteration | +K or -K, -Na, -Ca, +H | White mica + quartz + pyrite |
| | Argillic clay Alteration | -Na, K, -Ca | kaolinite, illite or montmorillonite |
| | Propylitic | +H, +C | Epidote + chlorite + albite ± carbonate |
| | Chloritization | +Fe, +Mg, +H<br>-Na, Ca, K | Chlorite + pyrite + white mica ± quartz |
| | Silicification | +Si | Any minerals + quartz |
| | Carbonatization | +C, +Ca | Carbonates, quartz-white mica-chlorite<br>Talc + chlorite + car |

## 3.5.12 Fluid Inclusions in Hydrothermal Deposits

Hydrothermal minerals commonly contain fluid inclusions that are used in scientific studies. These fluid inclusions are microscopic bubbles of liquid and gas that are trapped within crystals. They are believed to be remnants of the ore forming solutions (Fig. 3.39a). As minerals are formed from a liquid or aqueous medium, tiny *blebs* of that liquid can become trapped within the crystal structure or in healed fractures within a crystal. These fluid inclusions may be liquid or vapor, and may include aqueous solutions and volatiles such as $CO_2$, Nitrogen, and solid crystals of salt minerals. They often range in size from 0.1 to 1 mm and are usually only visible in detail under a microscope with high magnification. The trapped fluid in an inclusion provides useful information about ore-forming processes such as temperature, pressure, salinity, density and depth of trapping; and thereby providing direct information on the physical conditions and chemical composition of hydrothermal fluids, and constraints on pressure conditions at the time the minerals were deposited.

Fluid inclusions often contain one, two or more phases - liquid, gaseous and solids. If a vapor bubble is present in the inclusion along with a liquid phase, simple heating of the inclusion to the point of resorption of the vapor bubble gives a likely temperature of the original fluid. If minute crystals are present in the inclusion, such as halite, sylvite, hematite, or sulfides, they provide direct clues as to the composition of the original fluid. Fluid inclusions are classified as primary, secondary or pseudo-secondary depending on the timing of entrapment of liquid in the crystals. The inclusions occur either as $CO_2$, $CH_2$, $H_2S$, Cl, Br, F, I, $N_2$, S, Na, K, Ca, Mg and Fe.f trapped fluid varies greatly; commonly detected constituents include $H_2O$,

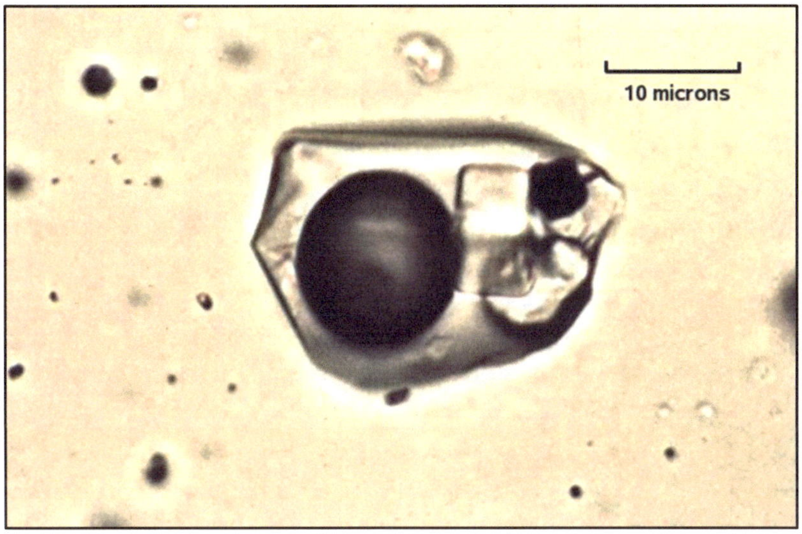

Fig. 3.39a: Fluid inclusion containing gas bubble and crystals of halite

### 3.5.12.1 Types of Fluid inclusions

Fluid inclusions can be classified on the basis of their mode of occurrence and composition of the inclusions. There are three types based on the timing of entrapment: primary, secondary and pseudo-secondary.

***Fluid Inclusions Based on Timing of Entrapment***

- *Primary inclusions* are those inclusions, which are trapped in the host mineral during the initial growth from the solution or during total recrystallization. These inclusions characteristically isolated inclusions.
- *Secondary inclusions* are those that are trapped along the fracture plane of the host mineral after the growth was completed. Consequently, they are definitely epigenetic in origin and may not be a reliable indicator of the conditions of mineral deposition.
- *Pseudo-Secondary Inclusions* are similar to secondary inclusions in their mode of occurrence but the entrapment seemed to have been followed by additional crystal growth by the mineral.

***Fluid Inclusions Based on Composition and Phase***

- Type 1: *Moderate salinity inclusions*, are the most common containing two phases, liquid mainly water and a small bubble of water vapor which forms *10-40 %* of the inclusion. The volume of vapor is usually proportional to temperature (Fig. 3.39).
- Type 2: *Gas-rich inclusions*, generally contain more than 60% vapor. They are dominantly aqueous but $Co_2$ may be present in small amounts. They appear to represent trapped steam. The simultaneous presence of gas-rich and gas-poor aqueous inclusions is good evidence that the fluids were boiling at the time of trapping.
- Type 3: *Halite-bearing inclusions*, have salinities ranging, up to more than 50%. They contain well formed, cubic halite crystals and generally several other daughter minerals, particularly sylvite and anhydrite.
- Type 4: *CO2-rich inclusions* have $CO_2$-$H_2O$ ratios ranging from 3 to over 30 mol. %. They grade into Type 2 fluid inclusions.

### 3.5.12.2 Characteristics of Ore-Forming Fluids

The following characteristics of ore-forming fluids can be deduced from fluid inclusion studies: temperature, composition and salinity, density and pressure and depth of ore formation.

***Temperature of the Fluid (Geothermometry)***

This is the most common information available from fluid inclusion studies. The theory behind fluid inclusion thermometry assumes that the partly filled vacuoles were completely filled with a single fluid phase when the mineral formed. The temperature at which the subordinate phase disappears - the homogenization temperature - marks a lower limit for the temperature of mineral formation. When a specimen is heated, a liquid inclusion will expand until the liquid occupies the entire vacuole, and a gaseous inclusion will simply lose its liquid fraction. Further heating should cause the vacuole to burst.

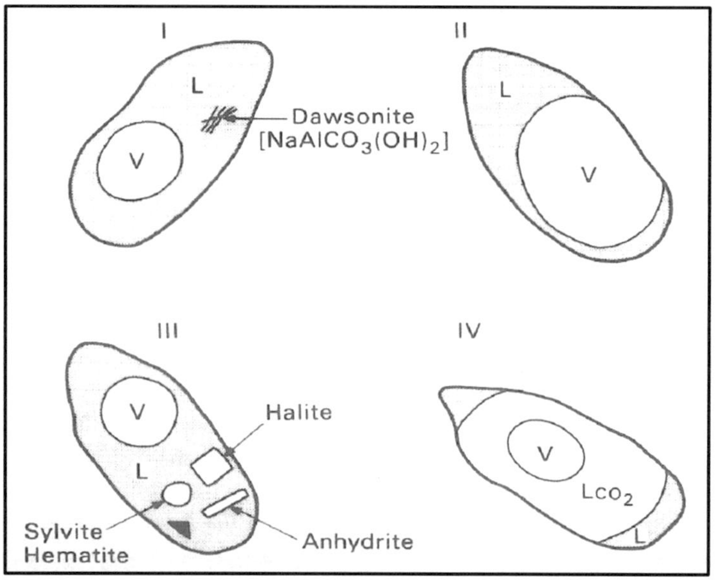


Fig. 3.39b: Types of Fluid Inclusions. *L= aqueous fluid, V=vapor, $LCO_2$= liquid*

*$CO_2$*. ***Composition and Salinity of Fluid***

Fluid inclusions can provide information about the composition and salinity of ore-forming fluids. If the inclusion is more than half liquid at room temperature, the ore fluid was hydrothermal, if more than half gas, the ore fluid was perhaps gaseous or *pneumatolytic* (high temperature). The composition of the gases, such as the presence of $CO_2$, Methane ($CH_4$), Nitrogen and inert gases (argon, helium) may provide a means for discriminating between fluids from magmatic, sedimentary and deep- or shallow-circulated meteoric sources. The presence of solid crystals of salt minerals suggests that the ore-forming liquid in this inclusion contained so much dissolved solids with a high salinity of NaCl of more than 23%. Those without salt crystals are of moderate salinity (10 – 15% NaCl) or low salinity (<10%). The presence of hematite and anhydrite crystals is indicative of the oxygen fugacity in the ore-forming fluid.

***Density of the Fluid***

Density of hydrothermal fluids is usually determined by the measurements of phase proportions in fluid inclusions. Fluid density is important in ore transport, and high salinity brines are usually of higher density. However, most hydrothermal solutions have a density similar to groundwater, close to one gram per cubic centimeter (1gm/cm3).

***Lithostatic Pressure Conditions***

The study of fluid inclusions provides quantitative data on fluid pressure in the mineralizing system, and the formation depth of hydrothermal deposits can be roughly estimated from data on pressure variations in fluid inclusions. Although the pressure estimates are not exact, they do set constraints on the depth of ore deposition, usually 2- 4 km or less.

## 3.6 Sedimentary Processes

Sedimentary processes are the low-temperature natural processes that are responsible for the accumulation of sediments and formation of economic mineral deposits at or near the Earth's surface. Geological processes in sedimentary environments involving *weathering, erosion, transportation, deposition* and lithification are somewhat complex because they commonly entail the combination of physical, chemical and biological factors that contribute to the formation of a variety of sediments and sedimentary rocks. Under favorable conditions, sediments and sedimentary rocks become selectively enriched in some elements of potential economic value.

Mineral deposits are formed in sedimentary environments by processes that can be grouped into three main types: (a) *physical* accumulation during sediment entrainment, transport and deposition; (b) *chemical* precipitation from surface water bodies; and (c) *biogenic* accumulation of the skeletal remains of organisms and plants. Sedimentary environments range from continental (fluvial), deltaic, lacustrine, and shallow and deep marine. Terrigenous (land-derived) materials in sedimentary deposits usually originate from the chemical and physical weathering of source rocks and the transportation of the fragments and particles as gravels, sands and clays by fluvial systems ultimately into standing water bodies. Mechanical enrichment of heavy minerals by the fluvial processes are called placers.

All surface waters in lakes, seas and oceans contain substantial concentrations of dissolved elements such as Fe, Mn, Cu that are chemically precipitated to form metallic ore deposits. Biogenic substances such as carbonates, phosphates, diatomite and evaporites (halite, gypsum and potash) are also formed by precipitation. Salts of Na, Ca, K, and Mg are often dissolved in seawater, and may be precipitated due to changes in solubility. Average seawater is known to contain about 3.5 wt% salts, comprising weight per cent levels of Na and Cl, and tens to hundreds of parts per million of a number of elements of economic interest. Plant remains may be buried in swamps to form coal and oil deposits Mineral deposits associated with sedimentary processes include diverse metallic and non-metallic minerals and mineral fuels.

### 3.6.1 Types of Sedimentary Processes

Some of the sedimentary processes that can lead to the accumulation of economic mineral deposits are outlined as follows:

- Concentration of heavy and resistant minerals by physical processes in fluvial and coastal environments (e.g., placer gold and beach heavy mineral sands)
- Evaporation of lake or sea water to form salts and sulfates (e.g., halite, gypsum, borax)
- Chemical and biochemical precipitation from solutions where minerals containing valuable substances are deposited directly out of lake or sea water (iron, copper, uranium).
- Submarine outflow of minerals through the discharge of metal-bearing fluids into the sea floor (e.g., sedimentary exhalative base metal deposits).
- Diagenetic changes resulting in the remobilization and deposition of metals in sediments (e.g., copper, lead-zinc and gold deposits).

- Biogenic processes involving the accumulation of valuable organic remains (e.g., phosphate, diatomite, coal).

#### 3.6.1.1 Clastic Deposition

The *physical* processes of erosion, transportation by rivers and streams and deposition of clastic materials including sand, silt, clay and resistant (heavy) minerals can lead to accumulation of valuable minerals. Streams and coastal currents are important agents for separating and sorting pre-existing precious or heavy minerals. Alluvial mineral deposits are formed in rivers and streams by hydraulic concentration of heavy, durable minerals such as gold, tin, platinum, and diamonds. Along the coastline, coastal sea-currents can concentrate dense minerals along beaches. Examples are magnetite (black sand), ilmenite, rutile and zircon found along beaches, and mined extensively in several countries including the Philippines, South Africa and Indonesia.

#### 3.6.1.2 Evaporation

Evaporation is a *physical* process in which water or aqueous solutions vaporize into the atmosphere due to the effect of heat which may lead to the concentration and precipitation of useful minerals. In the surface environment, a large body of water trapped in a shallow sea or lake may dry out during warm periods to produce extensive deposits of evaporites. There are two types of evaporite deposits: *marine* evaporites formed from ocean water, and *non-marine* evaporites formed from smaller standing bodies of water, such as lakes. Large mineral deposits of halite (NaCl), gypsum ($CaSO_4$), borax, and potash (KCl) are formed by this process.

#### 3.6.1.3 Chemical/Biochemical Precipitation

Chemical and/or biochemical processes can lead to the precipitation of metals and other valuable minerals held in solution in standing bodies of water due to reactions, changes in physical conditions or when their concentrations exceed solubility ranges. These metals or minerals are chemical precipitates that accumulate as sedimentary mineral deposits on the Lake Bottom or seafloor. The most noteworthy examples are the iron formations, and sedimentary copper deposits.

#### 3.6.1.4 Biogenic Processes

Biogenic processes can lead to the accumulation of the remains of plants and animal skeletons in near-shore to deep marine environments. Rocks that form from such processes may be of economic value. Examples include rocks formed from calcareous and siliceous skeletal remains of marine organisms such as limestones, diatomite and phosphate, and coal and lignite from plant remains.

#### 3.6.1.5 Diagenesis

Diagenesis is the totality of all the chemical, physical and biological changes that occur after sediments are initially deposited, compacted and consolidated (excluding metamorphism). As sediments and their contents are subjected to burial and compaction, temperature and pressure increase, and formation waters usually in saline and low redox state, are capable of alteration, remobilization and leaching of metals and concentrating them in open spaces within the sediments, thereby accumulating previously dispersed metals. Such deposits are called “diagenetic mineral deposits". They range from stratiform to stratabound, and syngenetic to epigenetic but are still contemporaneous with the enclosing rocks. Examples include copper deposits in shales (e.g. Kuperschiefer) and uranium in sandstones (e.g. Niger deposits).

The transformation of organic matter by burial during diagnosis can lead to the formation of various grades of coal, ranging from lignite, through sub-bituminous to bituminous coals.

## 3.7 Sedimentary Mineral Deposits

Sedimentary mineral deposits are mineral concentrations produced by sedimentary processes in sedimentary environments. The essential conditions for the formation of sedimentary mineral deposits include the following: (1) adequate source of ore materials: (2) gathering of the materials by dissolution or other processes: (3) transportation of the materials to the site of accumulation, (4) deposition of materials in the sedimentary basins; and (5) compaction and diagenesis. Sedimentary deposits are major sources of some important metals including, iron, manganese, copper, lead, zinc, uranium, gold, diamond, coal and several industrial mineral deposits, such as carbonates, phosphates, potash, celestite, barite and bentonite. Sedimentary mineral deposits can be classified into five main groups based on the processes of ore formation and environment of deposition.

### 3.7.1 Classification of Sedimentary Mineral Deposits

Sedimentary mineral deposits can be classified according to ore-forming processes into five major groups.

1. Placer deposits (alluvial and beach) (e.g., gold, cassiterite, diamond)
    - Alluvial Placers
    - Beach Placers
    - Paleo placers
2. Evaporite deposits (Halite, Gypsum, Borates)
    - Non-marine evaporites
    - Marine evaporites
3. Chemical (stratiform) sedimentary deposits
    - Iron Formations
    - Oolitic Ironstones
    - Manganese deposits
4. Diagenetic (stratabound) Sedimentary Deposits
    - Copper deposits
    - Uranium deposits
5. Biogenic deposits
    - Phosphate deposits
    - Carbonates and Diatomite
    - Coal deposits
    - Tar Sand and Oil Shale

### 3.7.1 Placer Deposits

The term "placer" is used to describe deposits formed by mechanical accumulation of economically valuable detrital and resistant minerals released by weathering from appropriate source rocks. Minerals that are commonly concentrated in modern placer deposits are heavy and stable minerals that may remain in place or slowly washed down slope into streams where they are further segregated. Most placers are found in alluvial or beach gravels and sands that are geologically young in age, although old placers, known as "paleo-placers" have been reported in indurated sediments. Notable examples of placer minerals are diamond, gemstones, some native metals (gold and platinum-group elements), and oxides (uraninite, chromite, magnetite, ilmenite, rutile, cassiterite, wolframite, columbite-tantalite),

Based on the geological environment of accumulation, placers may be classified into various types namely: *eluvial* placer (material in weathering profiles), *colluvial* placer (transported along slope by gravity), *alluvial* placer (transported by a river), *beach* placer (deposited along the edge of water bodies) and *paleo*-placer (ancient placers that are consolidated or metamorphosed). In arid regions where there is no flowing water, *aeolian* placers are created by the winnowing action of wind (Fig. 3.39).

There are certain preconditions for the concentration of placer minerals, including: (1) mechanical and chemical durability (2) resistance during weathering, erosion and transport, and (3) higher density compared to other minerals. Placers are important sources of diamond, gold, platinum, tin, titanium (rutile, ilmenite), zircon, rare earth elements (monazite) and gemstones.

### 3.7.1.1 Eluvial and Alluvial Placer Deposits

Eluvial placers are the mineral concentrations created *in situ* within the weathered regolith and disintegrated material. They are commonly found in tropical environments where the thickness of the weathering profile may exceed several tens of meters. Since eluvial placers have not been affected by stream action, they often need to be washed to separate the heavy minerals. Examples include gold, cassiterite and gemstone deposits in Nigeria, Thailand and Australia.

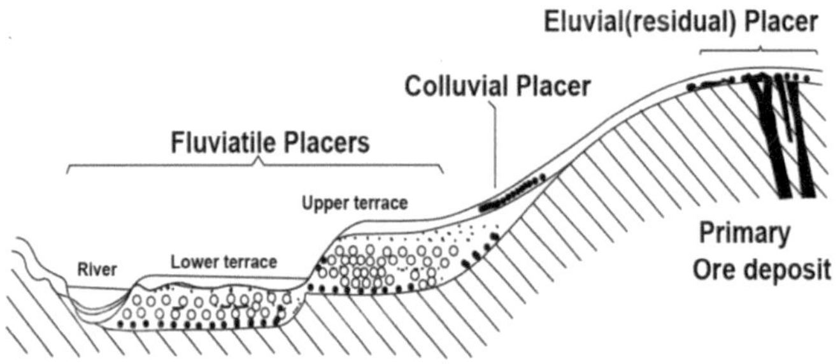


Fig. 3.40: Types of Placer Deposits

Alluvial or stream placers are heavy minerals transported and concentrated by the action of swiftly-flowing rivers and streams. They typically occur as heavy minerals within gravels and sands in sand bars or bottom sands, or in the bank materials that are deposited during episodes of high-water flow. Most transport of heavy minerals occurs during flood periods when the higher velocities can scour and dislodge grains from their previous sites. Obstructions, changes in channel shapes, or decrease in water flow results in the selective sorting of minerals according to size, shape, and specific gravity. Typical locations for alluvial placer deposits are on the inside bends (meander loops) of rivers and creeks; in natural hollows; at the break of slope on a stream; the base of an escarpment, waterfall or other barrier.

Unconsolidated placers, as old as Tertiary are found on river terraces, where they form blankets of sands and gravels adjacent to rivers and streams that have become more deeply incised (see Fig. 3.40). The most common heavy minerals in ordinary stream placers are magnetite and ilmenite. Economic alluvial placer deposits are almost restricted to gold, uranium, diamond, platinum, gemstones and tin. Examples are numerous, including the tin (cassiterite) deposits in Nigeria's Jos Plateau, Malaysia, Thailand, and Indonesia, and the diamond placers in Congo (Kinshasa), Angola, and Orange Free State, South Africa (Fig. 3.41). Several gold placers also occur in West Africa including Nigeria, Ghana, Sierra Leone and Burkina Faso (Fig. 3.42a).

### 3.7.1.2 Beach Placer Deposits

Beach placers are concentrations of heavy minerals along modern coastal beaches or shore areas. Beach terrace deposits are found where sea level has fallen relative to the land surface. They usually occur as thin, irregular, and discontinuous layers and lenses of dark minerals, called "black sands" which are composed of dark, heavy and resistant minerals such as magnetite, ilmenite, rutile and tourmaline. Beach placers are formed by the winnowing action of sea shore waves that act on pre-existing or newly formed alluvial deposits. They are a major source of what is called "mineral sands" - ilmenite, rutile, monazite, and zircon (Fig. 3.42b) that supply most of the world's zirconium and thorium resources. Placer gold and tin may also occur in beach-type deposits similar to those that contain major "black sand" (titanium) deposits. Examples are found in Western Australia and Queensland, in Australia; and Trail Ridge and Green Cove Springs, Florida, USA.

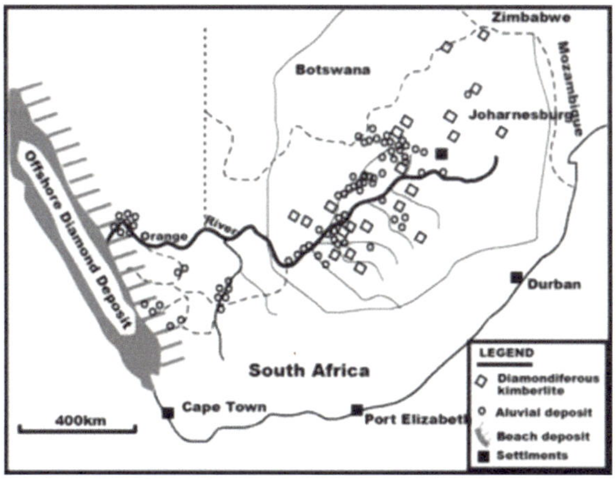

Fig. 3.41: Alluvial diamond placers in South Africa

Fig. 3.42: (a) Alluvial placer deposit, South Africa (b) Beach placer mineral sand, India

### 3.7.1.3 Paleo-Placer Deposits

Paleo-placer or "fossil placer" deposits are placers preserved in old indurated sediments. Ancient conglomerates containing gold and uraninite ores have been reported in the geologic literature. The gold and uranium ores are disseminated in beds and lenses of coarse pebble conglomerates which are parts of clastic sedimentary sequences. Generally, the conglomerates are of quartz pebbles and appear to be of fluviatile or shallow-water deltaic origin. The major examples of paleo-placers are Precambrian in age, including the deposits of gold and uranium in the well-known pebble conglomerates of the Witwatersrand Supergroup, South Africa (Fig. 3.43); Elliot Lake (Blind River), Ontario, Canada (major uranium deposit); and in Jacobina, Brazil, (gold).

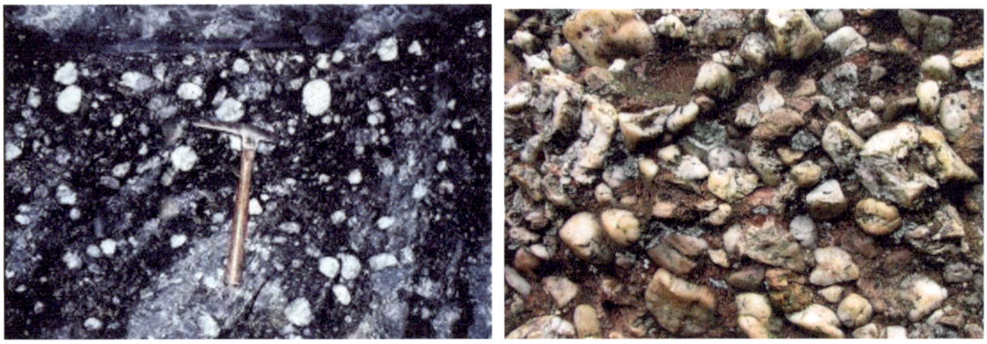

Fig. 3.43: Gold-bearing paleo-placer in pebble conglomerate, South Africa

The origin of placers in ancient conglomerates is believed to be similar to that of modern placers. The ore minerals were deposited as placers in alluvial conglomerates in a high energy depositional environment that were lithified and subjected to low-grade metamorphism which may have produced some post-depositional redistribution of the ores. The reason why these paleo-placers are confined only to the Precambrian is not well understood.

## 3.7.2 Evaporite Deposits

Evaporite deposits are those deposits formed by evaporation of lakes or sea water which results in the accumulation of dissolved substances that precipitate from the water. They consist mainly of sulfates such as gypsum and anhydrite and of chlorides such as halite (NaCl, common salt). Evaporite deposits usually occur as thick beds or thin layers of salts that accumulate in thick sequences as sedimentary rocks. The most common evaporite deposits are salts (most commonly sylvite and halite), but also include gypsum, borates, potash and nitrates (Fig. 3.44). There are two types of evaporite deposits; 1) *non-marine* (lakes), and; 2) *marine* (sea or ocean).

### 3.7.2.1 Non-Marine Evaporites

Non-marine salts that precipitate from lake water of suitable composition include sodium carbonate, sodium sulfate and borates. Huge evaporite deposits of sodium carbonate occur in the Green River basin of Wyoming, USA. Borax and other boron-containing minerals are mined from evaporite lake deposits in Death Valley and Searled and Borax Lakes, all in California; and in Argentina, Bolivia, Turkey, and China. Lake Chad in northeastern Nigeria is associated with evaporites including trona and potash salts.

Borate deposits are a good example of non-marine evaporites that are currently accumulating in playa lakes in the desert region of the western United States, and are well developed in the Tertiary sequence in Turkey. The formation of borate deposits in Turkey consisting of a sodium– and calcium–borate hydrates group is associated with playa-lake sediments and explosive volcanic activity. Some conditions are essential for the formation of economically viable borate deposits. They include playa-lake volcano-sedimentary sequences: formation of playa-lake environment; concentration of boron in the playa lake, sourced from andesitic to rhyolitic volcanics, direct ash fall into the basin, or hydrothermal solutions along faults; thermal springs near areas of volcanic activity; arid to semi-arid climatic conditions; and lake water with a pH of between 8.5 and 11. A large number of minerals contain boric oxide, but the three that are most important from a worldwide commercial standpoint are borax, ulexite, and colemanite, which are produced in a limited number of countries. Turkey has the largest borax, ulexite and colemanite reserves in the world and all the world's countries are dependent upon the colemanite and ulexite reserves of Turkey (Fig. 3.44a).

### 3.7.2.2 Marine Evaporites

Much more common and important than non-marine evaporites are the marine evaporites formed by evaporation of seawater. The most important salts that precipitate from seawater are: Gypsum ($CaSO4.2H2O$). Halite ($NaCl$) and Carnallite ($KCl.MgCl2.6H2O$) (Fig. 3.44b). Low-grade metamorphism of marine evaporite deposits causes another important mineral, sylvite ($KCl$), to form from carnallite. Marine evaporite deposits are widespread: in North America, for example, strata of marine evaporites underlie as much as 30 percent of the land area. Marine evaporites produce most of the salt that we use, the gypsum used for plaster and the potassium used in plant fertilizers. Rock salt (halite) is the main source of table salt and also used in livestock breeding, while potash is a main source of fertilizer. They are found in many parts of the world in Canada, USA, Siberia, India and the Middle East. Evaporites are found in the Middle East in the platform sediments of the Arabian Peninsula and offshore in the Persian Gulf.

Fig. 3.44: (a) Borate deposits, Turkey (b) Gypsum beds thinly bedded

## 3.7.3 Chemical Sedimentary Deposits

Chemical sedimentary deposits (also known as chemical precipitate deposits) are those deposits in which the ore minerals are precipitated from solution as a result of chemical or biochemical reactions in lake or sea water to form stratiform mineral deposits. The principal metals that occur in this type of deposit are mainly iron and manganese, in which the metals or mineral constituents are transported in solution from outside the basin and then precipitated. The common characteristics of these deposits include their stratiform and syngenetic nature, the relatively fine grained and uniform texture of the ore minerals. The stratiform nature, the presence of sedimentary features, and a close association with the surrounding host rocks indicate these deposits are syngenetic.

### 3.7.3.1 Sedimentary Iron Deposits

Sedimentary iron deposits are iron-rich sedimentary rocks that contain >15% iron. They are classified into three types, in decreasing order of importance: (a) banded iron formation (BIF), (b) sedimentary ironstones and (c) bog iron ores. From both the technical and economic points of view, iron is the most important metal on earth, and sedimentary iron deposits are the main source of iron ores, and are therefore of considerable value.

### *Banded Iron Formations (BIFs)*

Banded Iron Formations (BIFs) are unique rocks of Precambrian age composed mainly of alternating layers of iron-rich and silica-rich (chert) components from which the term "banded" is derived. BIFs and their metamorphosed equivalents constitute over 90% of iron ore produced worldwide and represent over 60% of global reserves of iron ore. The BIFs are usually well-bedded, elongated bodies which can extend for hundreds of kilometers. They are characterized by alternating layers of iron oxides and silica layers in the form of chert (Fig. 3.45). The layers vary from 0.5–3 cm thick while the laminations within the layers may be millimeters thick. Hematite ($Fe_2O_3$), and magnetite ($Fe_3O_4$) are the major ore minerals, although lesser amounts of iron carbonate (siderite), silicates (greenalite, minnesotaite, stilpnomelane), and sulfides (pyrite and very minor pyrrhotite) are also found in these deposits. BIFs usually contain about 15 to 40% Fe, but can be subjected to lateritic weathering and supergene enrichment which may create secondary iron minerals such as goethite and goethite-hematite, and increase ore grade to as high as 65% Fe. For BIFs to be economically viable they should attain thicknesses of at least 30 meters, strike lengths of several kilometers and grades of above 30% Fe.

There are three facies of iron formations: *oxide, carbonate* and *sulfide* facies. Among the various facies of BIFs, the oxide facies is the most common, and can be subdivided according to the dominant iron oxide into the hematite and magnetite subfacies. The ore grade of the oxide facies BIF typically averages 30 - 35%Fe which is viable if it can be mined and beneficiated to at least 63%Fe. The carbonate facies commonly consist of interbedded chert and siderite in roughly equal proportions. It may grade through magnetite-siderite-quartz rock into the oxide facies, or, by the addition of pyrite, may grade into the sulfide facies. The grade of the carbonate facies averages 15 - 20% Fe. Three types of oxide facies BIFs have been recognized: Algoma type, Superior type and Rapitan type.

BIFs are found mainly in Precambrian cratonic shields worldwide, notably in Canada, Venezuela, Brazil, Russia, India, West Africa, South Africa, and Australia. Notable examples are the extensive deposits in the Lake Superior district of Minnesota-Wisconsin-Michigan in the USA, and Ontario and the Labrador-Quebec belts in adjoining areas of Canada. BIFs are also prominent in the Hamersley Basin of Western Australia, and several deposits in Brazil, India and South Africa where they have been subjected to low-grade metamorphism, and known as "itabirites". In many parts of West Africa, similar BIFs can be found in Sierra Leone, Guinea, Liberia and Cameroon. In Guinea, the Simandou iron ore deposit located in the Simandou mountain range in south-eastern Guinea is believed to be the world's biggest high-grade iron ore deposit. It comprises BIFs within Early Proterozoic metasedimentary sequences in the southern portion of the West African Craton. As a result of lateritic weathering, the deposit contains friable hematite and goethite-hematite as ore minerals. Ore grade averages 65% Fe and is a direct shipping ore that requires no beneficial ion.

The origin of BIFs has been widely debated. They are unique rocks that are restricted to the Precambrian. They appear to be the product of *chemical precipitation* from sea water in shallow basins where they were commonly interlayered with other sedimentary rocks, such as carbonates and shales. Most BIFs were formed during the Archean/Early Proterozoic, between 2.7 and 1.8 billion years ago. Prior to this period, the Earth's primitive atmosphere and oceans had little or no free oxygen to react with iron or silica, resulting in high iron and silica concentrations in seawater. The cause of precipitation remains uncertain; and suggestions include the existence of photosynthetic organisms that began producing oxygen, the oxygen reacted with the iron dissolved in the seawater and caused it to precipitate on the seafloor. The Rapitan-type BIF is found in only a few locations in Canada and Australia. The close association with glaciogenic sediments may indicate a genetic link.

Fig. 3.45: Banded Iron Formation Lake Superior Type

***Lake Superior-type* BIF is** named after the important deposits around Lake Superior, Michigan, USA which formed in shallow marine shelf sediments, and is the most common and important type of BIF. The deposits are usually free of clastic material, and distinguished by distinct rhythmic banding of iron-rich and iron-poor chert layers, which normally range in thickness from as thin as 0.5 mm (0.02 inch) to as thick as 2.5 cm (1 inch), and the alternation of the layers is remarkably invariable. The bands can be traced for several kilometers. Compared to Algoma type BIF, the Lake Superior-type BIFs are large in dimensions; usually >100 m in thickness and over 100 km in lateral extent, and the bands are well differentiated and consistent (Fig. 3.44). The main iron-bearing mineral is hematite with occasional magnetite. They are believed to have for5ed in marine shelf environments in the absence of volcanic activity. The Lake Superior type deposits are only found in rocks ranging in age from 2.7 to 1.8 by. Examples include the Superior deposits in Michigan and Minnesota, the Labrador Trough deposits of Canada, Serra dos Carajás in Brazil, the Transvaal Basin deposits of South Africa, and the Hamersley Basin of Australia.

***Algoma-type*** **BIF** occurs also as finely-layered intercalations of silica and an iron mineral, generally hematite or magnetite, but the individual layers lack the lateral continuity of Lake Superior-type BIFs (Fig. 3.46a). They are found within Archean (.2.75 by) greenstone belts in association with rock sequences containing a significant proportion of submarine volcanic rocks and turbidites, and are believed to have formed in submarine volcanic settings. Algoma-type BIFs are relatively small with ore beds that are usually less than 50 m thick and display a transition from oxide through carbonate and silicate to sulfide facies. Oxide facies iron occurs mainly as magnetite, while hematite is rare. Large deposits of Algoma type BIF of middle Archean age occur in the Abitibi greenstone belt in Canada and the Guyanan and Liberian shields (West Africa).

***Rapitan-type BIF*** is named after the occurrences in the Rapitan Group of northwestern Canada. This BIF type is generally much younger (0.8 -0.6 by), and characterized by its association with glaciogenic marine sediments. Rapitan-type BIF is found as small lenses in the Late Precambrian rocks and is the least important in terms of volume of occurrence. The Rapitan BIF consists mainly of hematite and chert (Fig. 3.46a), and is distinctly different from that of most major banded iron-formation types of Archean and Early Proterozoic age. Examples occur in Rapitan and Yukon, Canada, and Bremer iron formation in South Australia. The origin is related to glaciation. Rapitan-type BIF was deposited immediately during glaciations when land and oceans were widely covered with ice causing a nearly global anoxia and the presence of dissolved ferrous iron in seawater. Inter and post- glacial melting of the ice caps resulted in formation of glacial sediments, re-oxidation of the oceans, and precipitation of ferric iron and manganese oxides.

## *Sedimentary Ironstones*

Sedimentary ironstones are low-grade, iron-impregnated sandstones, usually but not always characterized by oolitic and sometimes pisolitic textures. Sedimentary ironstones are not banded and contain no primary colloidal silica (chert). They occur as reddish-brown, earthy sandstone beds ranging in thickness from a few centimeters to tens of meters (Fig. 3.47). In tropical environments, the iron-rich beds may be subject to lateritic weathering and supergene enrichment, and develop a lateritic cap several meters thick. Oolitic ironstones are all Phanerozoic in age (compared to BIFs that are Precambrian). The ore minerals are mostly hematite, goethite, chamosite, siderite and occasional magnetite. Texturally, the ironstones appear as oolitic, pisolitic or pelletal grains that may have been entirely or partially replaced by interlayered limonite and hematite. The oolitic texture (or oolite) consists of concentric rings of Fe-bearing minerals, namely hematite siderite and chamosite assemblage or goethite around sand grains or fragmental fossil or mineral material (Fig. 3.47).

Fig. 3.46: (a) Algoma type BIF (b) Rapitan type BIF

The economic importance of oolitic iron ores in the world economy has been defined by their relatively low iron content (<35% Fe) and also their high content of undesirable elements such as phosphorus, aluminum, and silica. The sedimentary ironstones once formed the backbone of the iron and steel industries in Western Europe and North America but have been largely replaced in importance by BIFs, There are two types of sedimentary ironstones that can be distinguished simply on the basis of mineralogical assemblage and depositional environment: the *Clinton type* (hematitic) and the *Minette type* (limonitic).

***Minette-type*** (named after Lorraine, France) ironstone is the most common and widespread, and is composed of oolites of siderite ($Fe_2CO_3$), *chamosite*, a siliceous iron mineral (($Fe_4Si_4O_{16}(OH)_8$) and goethite. The deposits were formed in shallow, near-shore marine environments and are found extensively in Europe, particularly in England, the Lorraine area of France, Belgium, and Luxembourg.

***Clinton-type*** (named after an occurrence in New York, USA) deposits contain oolites of *hematite*, siderite, and chamosite, and are found mostly in the eastern Atlantic seaboard of the United States. The Cretaceous sedimentary ironstones in Nigeria (Agbaja and Enugu) belong to the Minette type, and are made up predominantly of goethite, magnetite and kaolinite but no chamosite.

The Phanerozoic sedimentary ironstones differ from the Precambrian BIFs in their age, absence of chert, presence of oolitic textures and containing more alumina and phosphorus. They are much smaller in size in both thickness and lateral extent. The iron concentrations in the oolitic ironstones are generally lower than in BIFs, ranging between 24% and 35%, while the phosphorus content can be as high as 1% $P_2O_5$ which is deleterious to iron and steel manufacture. The Minette-type deposits of Jurassic age occur extensively in England, the Lorraine area of France, Belgium, and Luxembourg, while the Clinton-type deposits of Silurian age are found in the eastern United States (Appalachians) from Newfoundland to Alabama.

The origin of sedimentary ironstones is well known. They were deposited in fluvial to shallow-water marine environments where they are associated with a variety of sedimentary rocks, including limestones, siltstones, shales and sandstones. The ultimate source of iron may have been by continental erosion and precipitation due to changes in pH-eH conditions. However, many observations favour a synsedimentary supply of iron followed by chemical and biochemical precipitation. The unlithified sediments that form the oolitic iron ore composed of ooids, pisoids and peloids that originated by rolling on the marine floor. The matrix between these larger grains consists of detrital minerals, fossils, organic material and fine-grained iron-bearing kaolinitic clay. Subsequent diagenesis, mostly under reducing conditions, caused by bacterial decomposition of organic matter, has resulted in the transformation of kaolinite into chamosite (an iron-rich chlorite).

Fig. 3.47: (a) Sedimentary Ironstones, Nigeria (b) Oolitic-Pisolitic texture of Ironstone

### 3.7.3.2 Sedimentary Manganese Deposits

Manganese is very similar to iron in chemistry and in the way it is distributed and concentrated in rocks. Sedimentary manganese deposits are formed by the precipitation of Mn from seawater. They are the main source of manganese worldwide, and are quite similar to sedimentary iron deposits in their mode of occurrence and spatial and chemical affinities. Manganese deposits, can be classified into three types; (a) Proterozoic (BIF-related) manganese deposits; (b) Cenozoic manganiferous (oolitic) deposits, and (c) Manganese nodules (equivalent to bog iron ores).

#### *Proterozoic BIF-Hosted Manganese Deposits*

These deposits of Proterozoic age are similar to, and commonly interlayered with Superior-type BIFs in sedimentary (or metasedimentary) sequences. They are usually small deposits that occur as beds or bands several meters thick (Fig. 3.48a). The deposits are difficult to mine except when they are enriched by lateritic or supergene processes. Occasionally, some high grade manganese ores do occur as mineralized bands that exceed 40%Mn. The primary ore minerals include braunite, and hausmannite with pyrolusite ($MnO_2$) and manganite ($Mn_2O_3.H_2O$) which may be supergene. Gangue minerals include rhodonite, manganous garnets, quartz, hematite, and magnetite. Examples of these deposits are found in South Africa (Fig. 3.40), Western Australia, Brazil, China, Mexico, Gabon, and Ghana.

The BIF-hosted manganese deposits of the Kalahari Manganese Field (KMF) in South Africa are by far the world's largest deposits holding some 4,200 Mt of manganese metal that represents about 80% of the world's known land-based Mn resources. The main ore bands are about 5-45 m thick, with grades ranging from 25-50% Mn. The manganese ores have been affected by supergene enrichment.

***Oolitic Manganese Deposits***

These deposits occur as laterally-extensive Oligocene sedimentary beds enriched in manganese in association with marine sediments, such as dolomite, limestone, chalk and black shales. They are analogous to the Minette-type sedimentary ironstones. The manganese beds are often more than 50 meters thick and extend laterally for over 50 kilometers. Textural features characteristic of these deposits include the presence of oolites and pisolites, rhythmic laminations, slumped bedding and abundant fossils. The dominant ore minerals are housemanite ($Mn_3O_4$), manganite ($Mn_2O_3.H_2O$), psilomelane, pyrolusite, rhodochrosite and other manganese minerals.

The most important deposits are located in the region just north of the Black Sea stretching from Turkey through Bulgaria, the Ukraine, Russia and Georgia and finally into Kazakhstan where they represent a substantial portion of the world's known resources of high-grade manganese. These deposits formed in shallow marine, near-shore depositional environments, commonly in sheltered sites around islands along some areas of continental shelf and the interior basins. Manganese precipitation is believed to have taken place in stratified water masses at the interface between anoxic seawater and near surface oxygenated waters (Fig. 3.48b).

***Manganese Nodules***

Manganese nodules of marine origin are a potential source of manganese that occur on the deep seafloor as concretions formed by concentric layering of iron and manganese hydroxides around a core which may be rock or fossil fragments. They grow at a very slow rate, on the order of millimeters per million years (Fig. 3.49b). The nodules are formed by slow precipitation of metals from seawater. Nodules are most abundant at depths between 4,000 m and 6,000 m in areas of low sedimentation rate. Nodules contain 29% Mn, 5% iron, 1.2% copper, 1.37% nickel, 1.2% cobalt and 15% $SiO_2$. Mn-nodules have a botryoidal or smooth surface and consist of concentric shells of amorphous and crystalline manganese and iron oxyhydroxides. The ability to exploit manganese nodules from the bottom of the ocean at reasonable cost is being investigated worldwide, but there is considerable doubt if these vast manganese resources will ever be exploitable at a profit.

Fig. 3.48: (a) BIF-hosted Manganese, (b) Oolitic Sedimentary Manganese, Australia

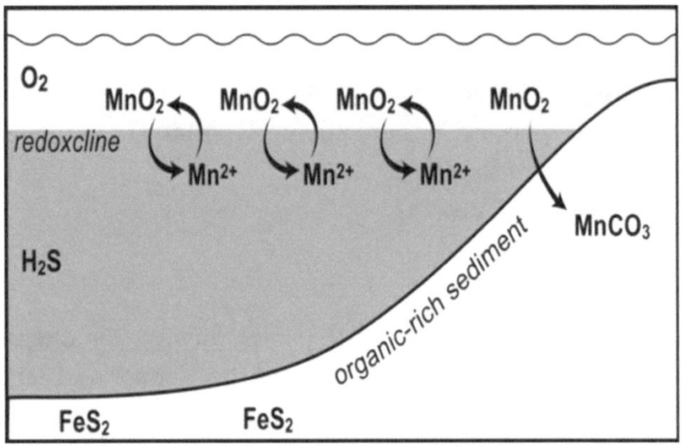

Fig. 3.49: (a)) Manganese origin in basin

(b) Manganese Nodules

***Origin of Manganese Ores***

The origin of manganese deposits is similar to that of sedimentary iron ore. These deposits formed in shallow marine depositional environments (15-300 m deep) in lakes and seas, commonly in sheltered sites around islands, along some areas of continental shelf and the interior basins. There is evidence of Mn deposition in deep-sea sediments (in association with black shales) where it is widely distributed as nodules, as coloring matter, and as coatings on fossils. Most deposits overlie oxidized sediments, but change basinward to carbonate deposits that may be in reducing environments. Manganese precipitation is believed to take place in stratified water masses at the interface between anoxic seawater and near surface oxygenated waters (Fig. 3.49a). Extreme Fe fractionation is caused by a low solubility of iron in low Eh environments where Fe precipitates as iron sulfide. A subsequent increase in Eh and/or pH of Mn-water may produce Mn-rich, Fe-depleted chemical sediments. The sources of Mn include submarine volcanic activity and chemical weathering and transportation in continental environments.

## 3.7.4 Diagenetic (Stratabound) Sedimentary Mineral Deposits

Diagenesis is "the sum of all chemical, physical and biologic changes undergone by sediment after its initial deposition, and during and after its lithification, exclusive of metamorphism. Diagenetic mineral deposits are a group of stratabound base metal and uranium deposits that have been concentrated or remobilized during diagenesis. The metals were probably precipitated initially with the sediments (syngenetic) but subsequently concentrated during diagenesis. They are enclosed by, and parallel to the sedimentary strata in which they occur, but do not necessarily follow sedimentary bedding (compared to stratiform deposits). They are described as "stratabound" because, while the deposits are concordant and restricted to particular stratigraphic units, they may sometimes be discordant due to post-depositional changes. This observation has led to the ongoing controversy on whether these sediment-hosted deposits are sedimentary (syngenetic) or hydrothermal (epigenetic). Most research workers believe these deposits are mostly syngenetic and diagenetic; and any traits of epigenesis are due to post-depositional remobilization.

Diagenetic deposits are commonly restricted to a narrow range of layers within a sedimentary sequence and characterized by base metal mineralization of copper, cobalt, lead-zinc and uranium-vanadium. The host rocks are a variety of *clastic* sedimentary rocks (argillaceous and arenaceous), mostly shales, grey beds of siltstone and sandstone, and very rarely in carbonate rock layers.

Base metal mineralization occurs as finely disseminated sulfides or as thinly or massively-bedded layers in argillites and arenites in which the ore minerals are precipitated in the pore spaces between sedimentary grains. The main ore minerals are chalcopyrite, bornite, chalcocite, cobaltite, galena, sphalerite and uraninite or pitchblende). Some of the world's most important ores of copper, lead-zinc and uranium occur as sediment-hosted mineral deposits.

Examples of diagenetic deposits that will be described are as follows:

a. Sedimentary copper-cobalt deposits (e.g. Zambia copper and Kuperschiefer, Germany)
b. Sandstone-type uranium-vanadium deposits (e.g. Colorado, USA).
c. Unconformity-related uranium deposits (e.g. Athabasca, Canada)

### 3.7.4.1 Sedimentary Copper Deposits

Sedimentary copper (and cobalt) deposits are a large, diverse group of deposits formed by chemical precipitation in clastic rocks and modified during diagenesis. They include some of the richest and largest copper deposits in the world. Twenty to twenty five percent of world copper production comes from this type of deposit which is also an important source of silver and cobalt. The argillaceous host rocks are shales, mudstones, argillites, and siltstones that are often continuous and extensive. Other host rocks are sandstones or quartzites and minor carbonates (dolomite and limestone). Copper mineralization occurs as finely disseminated, stratabound, sulfides in fine-grained to coarse sediments of which chalcocite and bornite are the dominant copper minerals with lesser amount of native Cu, chalcopyrite, galena, hematite and pyrite. Mineralization is usually continuous within beds or may follow old channels and in some places the mineralization may cut across beds due to remobilization. Disseminated sulfides are often seen along bedding planes in stratiform pattern.

Bleaching of the sediments due to reducing conditions is well developed in many areas because mineralization is usually localized at the reduction front between oxidizing and reducing conditions. The pH-Eh conditions changed from the edge to the far deeper parts of the depositional basin which may result in the zonation of sulfide minerals. (Fig. 3.50). Lateritic weathering may produce reddish crusts while secondary copper minerals (azurite and malachite) form gossans due to supergene enrichment. Classic examples of these deposits are the European (Kupferschiefer) copper shale in Germany and Poland, Central Africa Copper belt in Zambia and the Democratic Republic of Congo.

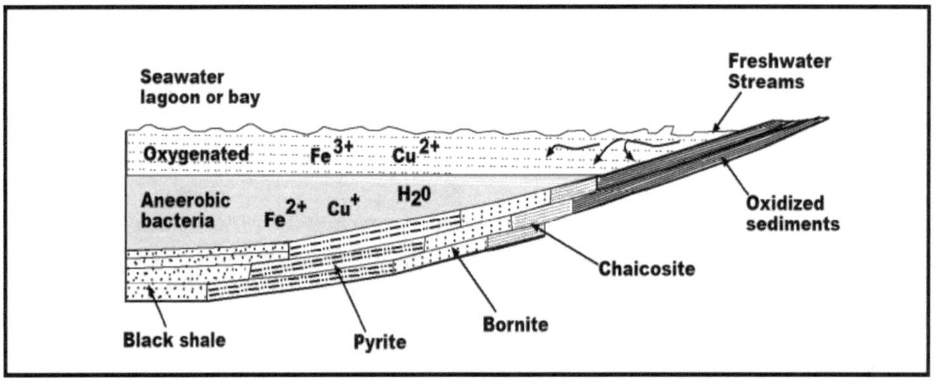


Fig. 3.50: Idealized facies of stratabound copper deposits

## *European (Kupferschiefer) Copper Shale*

The Kupferschiefer copper deposits of northern central Europe occur in the Upper Permian Shale sequence that extends for hundreds of kilometers in Germany into Poland as a regional stratigraphic unit with several copper deposits that were mined for centuries and still actively mined in Poland. These deposits are not only one of the largest sediment-hosted accumulations of copper ores worldwide but are also one of the longest continuously documented mining histories. The host rocks are predominantly shales with minor grey sandstone, carbonate rocks and evaporites (Fig. 3.51a).

Ore bodies range in thickness from 0.3m to 50m and contained largely within the black shale rich in organic matter. The copper mineralization is dominated by chalcocite, covellite, chalcopyrite, and bornite. Galena, sphalerite and pyrite are found in minor to substantial concentrations. Ore textures include disseminated ores, disseminated replacement of diagenetic and framboidal pyrite, crosscutting and bedding parallel veinlets, impregnation and replacement ore of carbonate and anhydrite cements, replacement of fossil shells, and even replacement of detrital feldspar and feldspar in lithic clasts. The Kupferschiefer copper mineralization is concordant with the stratification, although epigenetic veins were later formed, but constitute only a small fraction of the total ore concentration.

The origin of the Kupferschiefer mineralization has been controversial, and genetic models range from syngenetic through diagenetic to epigenetic. The diagenetic model which is the consensus opinion, entails the upward and lateral flow of hot, oxidizing connate fluids from deeply buried red bed sediments and volcanics which percolated and oxidized originally pyritiferous organic matter-rich sediments and precipitated base metals derived from red beds into the reduced sediments.as a result of changes in temperature, pressure and Eh-pH conditions (Fig. 3.51b).

Fig. 3.51: European Copper Shale

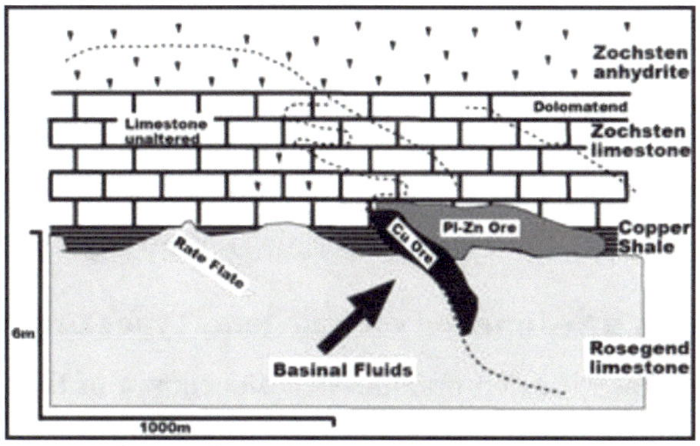


(b) Geological Diagenetic model

***Central Africa Copper Belt***

The Central Africa Copper Belt comprises several stratabound copper and cobalt deposits in Zambia and extending into the Democratic Republic of the Congo. The belt is more than 100 km wide and 700 km long and stretches from Katanga, the southern province of the DR Congo, almost to the southern border of Zambia. (Fig. 3.52). The host rocks are shales, sandstones and dolomite of the Neoproterozoic Katangan Supergroup, which has a maximum estimated thickness of 5-10 km. These rocks were deposited within a series of linked Proterozoic intracratonic rift basins.

Most of the copper deposits are confined to a stratigraphic interval of about 200 m thick sediments known as the Copperbelt Orebody Member which comprises argillite-hosted (~70% of ore) and arenite-hosted (~30%) ore types. Two main ore horizons have been identified with a combined thickness of 55m, each extending from hundreds of meters to kilometers. The distribution, geometry, and size of these deposits are fundamentally controlled by early sub-basin fault architecture and the availability of in situ and mobile reductant for the metals. Most deposits contain both disseminated and vein- or nodule-controlled sulfides of copper and cobalt, mostly chalcopyrite and bornite and cobaltite. The cobalt concentrations are high particularly in the Congo deposits which has made this area one of the world's leading sources of cobalt.

The origin of these copper deposits has been widely debated. The ore distributions indicate several generations of sulfides marking syngenetic, early diagenetic and late diagenetic processes with some evidence of epigenetic remobilization. Sulfur isotopic data on sulfides suggest the derivation of sulfur essentially from the bacterial reduction of seawater sulfates. The syngenetic metals that were initially precipitated in the sediments were affected by mineralizing brines generated from sea water in hypersaline lagoons during the deposition of the host rocks. Changes of Eh–pH and salinity probably were critical for concentrating copper–cobalt mineralization.

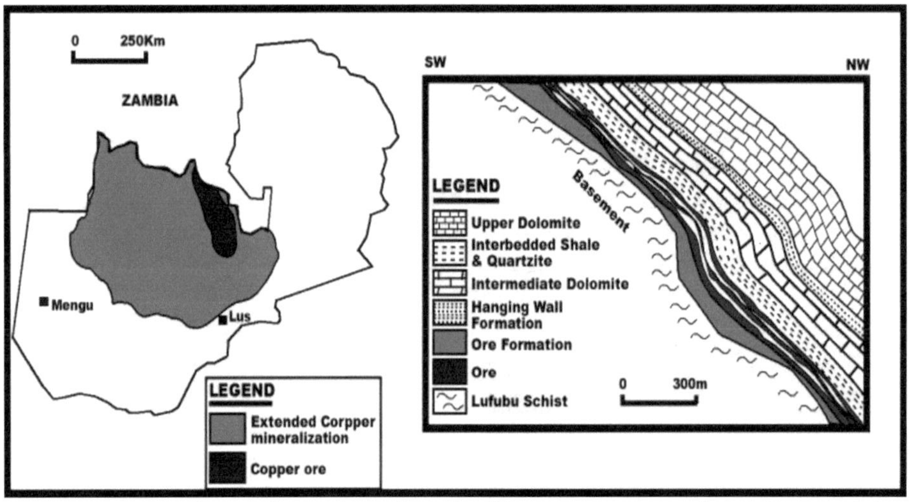


Fig. 3.52: Location and Stratigraphy of sedimentary copper deposits, Zambia

### 3.7.4.2 Sedimentary (Sandstone-Type) Uranium Deposits

Sediment-hosted uranium deposits consist of the minerals of uranium and sometimes vanadium that are found within clastic sedimentary rocks, mostly sandstones and some conglomerates that were deposited in a fluvial or marginal marine environment. Impermeable shale/mudstone units often occur immediately above and below the mineralized sandstone beds. These sandstone-hosted deposits are described as "sandstone-type" uranium deposits. They constitute about 28% of world uranium proven reserves and 40% of inferred resources, and are of major economic importance in Kazakhstan, Uzbekistan, USA and Niger Republic who are among world's leading producers of uranium. Orebodies of this type are commonly low to medium grade (0.05-0.35%U) and individual orebodies are small to medium in size, with a maximum range of 50,000 tons.

Mineralization occurs as lenses, pods, layers, or concave "rolls" and irregular masses of ore that are grossly conformable with enclosing sediments (stratiform), but may sometimes show cross-cutting relationship in detailed observations (stratabound),. There are three subtypes based on mode of occurrence: (a) tabular deposits, (b) channel fill deposits, and © roll front deposits. Most of the mineralization is concentrated within the intergranular spaces in the clastic host rocks, and may be associated with organic debris, or replacement of fossil logs. The dominant minerals in these deposits are uraninite (or pitchblende), and secondary uranium hydroxides such as coffinite [U(Sio4kiOH)4x], and vanadium mineral montroseite [VO(OH)]. Some sulfide minerals including pyrite, sphalerite and galena are sometimes found within the deposits. The amounts of uranium, (and vanadium) can vary considerably, both within and between deposits. Examples of sedimentary uranium deposits occur in Niger, Gabon (Franceville Basin), and in the eastern part of Africa, in the Karoo Formation (Malawi, Tanzania, Zambia and South Africa), and in the United States in the Colorado Plateau area of Colorado, Navajo county in Arizona, Utah, New Mexico, the Powder Basin of Wyoming, the Gulf Coast Plain in south Texas and several deposits in Chi (Fig. 3.53). In the Niger Republic, uranium deposits occur within Carboniferous to Cretaceous sandstones as tabular deposits (with rare roll fronts) that are irregular, elongate or lenticular and parallel to the depositional trend of the host arenites of fluviatile to deltaic in origin. Mineralization mostly within fluvial paleo-channels and unconformities. The main ore minerals are pitchblende and coffinite. Niger accounts for about 6% of world uranium production.

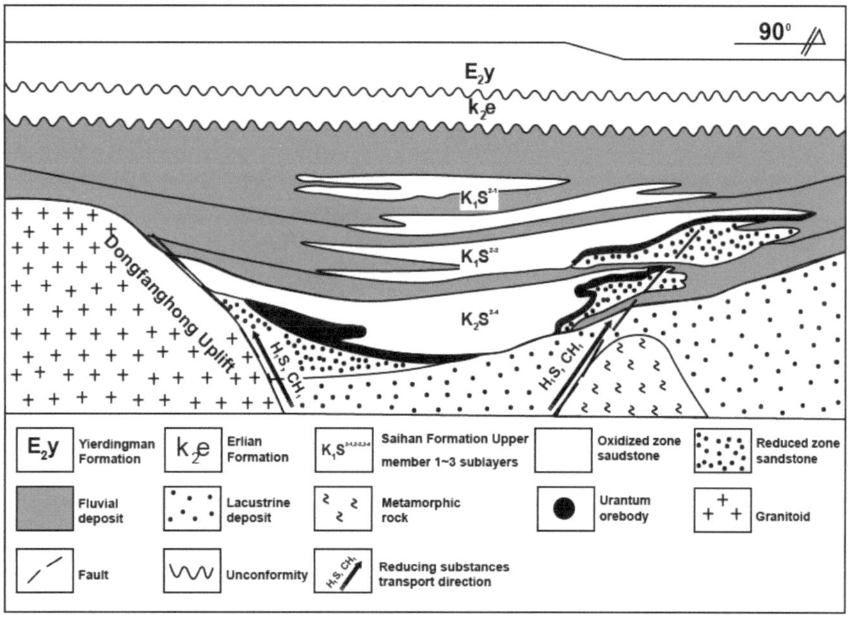


Fig. 3.53: Tabular uranium deposits in basal sandstones, China.

The origin of sedimentary uranium deposits is controversial, but overwhelming evidence suggests that the ore deposition involved a combination of processes that range from syngenetic to diagenetic. Most of the tabular deposits were formed from uranium leached from weathered uranium-bearing source rocks by oxidizing solutions and precipitated on encountering reducing environments within permeable sandstones. Uranium exists in nature as tetravalent (U4+) or hexavalent (U6+) forms. The latter is soluble and can be easily carried by surface or subsurface water until it reaches a reducing environment where it deposits again in tetravalent form. The reductants varied at different deposits and included organic matter such as woody fragments, fossils, coalified wood, structureless organic matter, oil and natural gas. Most of the uranium was introduced during the deposition of the host sediments and concentrated during diagenesis by formation waters. Most of the deposits are tabular (stratiform) or channel fills. (Fig. 3.52). Fluid inclusion studies of associated sphalerite in the Niger uranium deposits indicate the temperature of deposition was as high as 160 °C, and the sulfur was derived from seawater sulfate.

The roll front that developed in some deposits can be attributed to fairly recent groundwater remobilization or infiltration. These roll front deposits are definitely epigenetic and formed by movement of oxidizing groundwater solutions passing through porous sandstones leaching metal ions from associated strata at low temperature along an oxidizing surface that follows behind a reducing front where precipitation of U-V minerals occurs (Fig. 3.54).

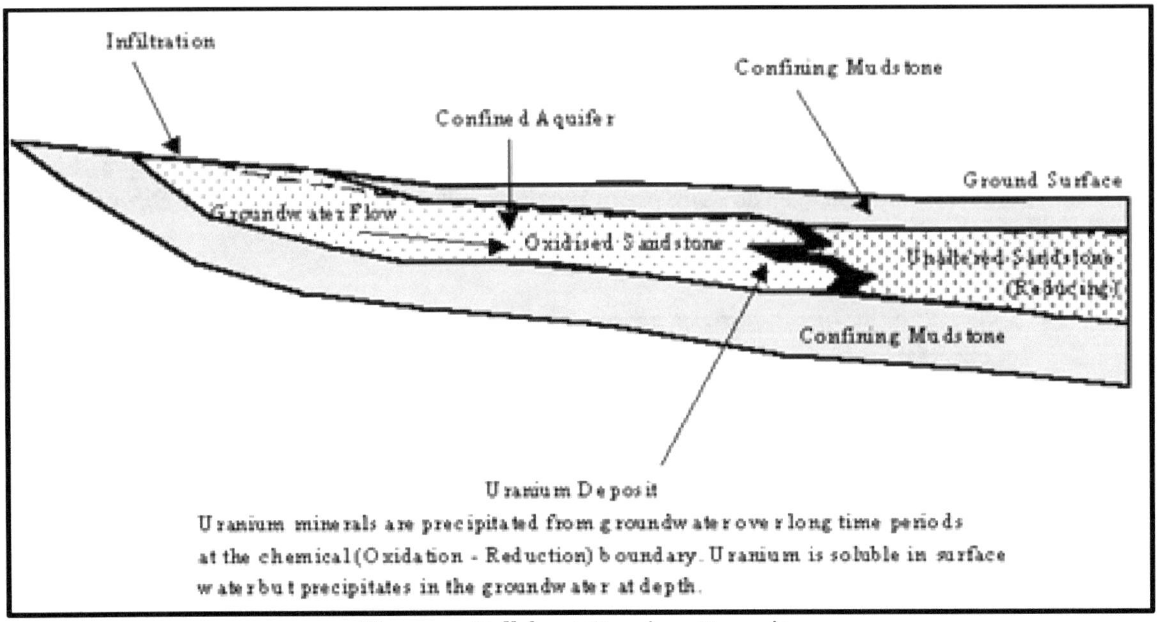

Fig. 3.54: Roll front Uranium Deposits

### 3.7.4.3 Unconformity-Related Uranium Deposits

Unconformity-related uranium deposits occur in close proximity to major unconformities, usually between quartz-rich basal sandstones/conglomerates, and underlying crystalline basement rocks. These are unique deposits that differ from the "sandstone-type" uranium deposits by the fact that they only occur along unconformities created mostly during the Proterozoic. They are mostly stratabound but sometimes exhibit structural control as they can be localized below, across and immediately above the unconformity (Fig. 3.55). Below the unconformity, the metasedimentary rocks which host the mineralisation are usually faulted and brecciated while the overlying younger Proterozoic sandstones are usually undeformed. Ore minerals are uraninite or pitchblende associated with strong quartz dissolution. They are rich and relatively high grade (up to 20% U) and constitute about one-third of the world's uranium resources, which include some of the world's largest deposits. The main deposits occur in Canada (the Athabasca Basin, Saskatchewan and Thelon Basin, Northwest Territories); and Australia (the Alligator Rivers region in the Pine Creek Geosyncline, Darwin area of Northern Territories, and Rudall River area, Western Australia).

The origin of the unconformity-related uranium deposits is attributed to remobilized basinal brines during diagenesis and probably with some circulating groundwater that leached uranium from the sediments, regolith and basement rocks, and depositing them within the sandstones and other structural openings (faulted basement and sediments) due to reduction linked to organic matter. The ore-forming solutions are slightly acidic and hot (160 to 220°C), at about 1kbar pressure, oxidized and Na-Ca-Mg rich (up to 6 molar Cl) basinal diagenetic brines generated within continental, oxidized, organic free, clastic formations. The source of the uranium is believed to be U-rich monazite and zircon in detritus and basement rocks. The deposits seem to show combined effects of syngenesis, diagenesis and epigenesis.

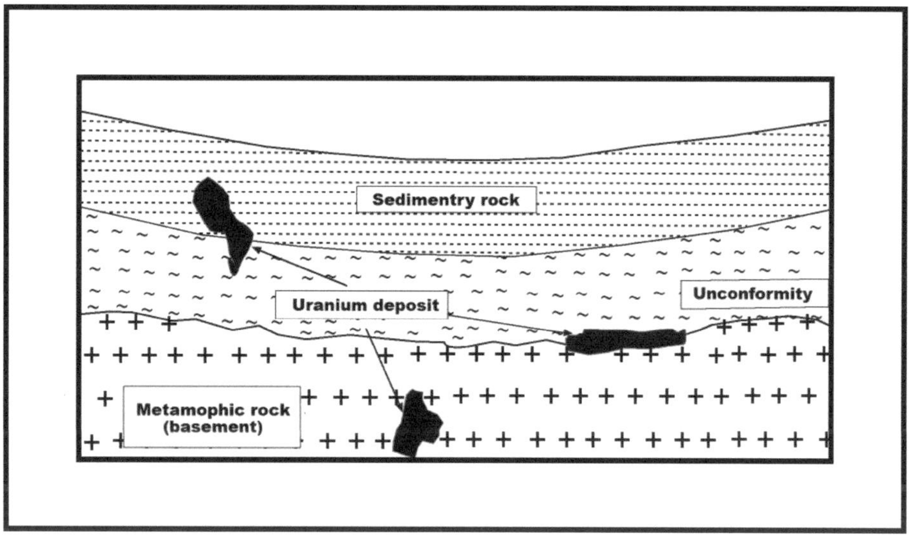


Fig. 3.55: Unconformity-related uranium deposits

### 3.7.5 Biogenic Mineral Deposits

Biogenic sedimentary mineral deposits are those deposits formed by the accumulation of the skeletal remains of organisms including plants, animals and microorganisms. These are mainly carbonates, siliceous deposits, phosphatic deposits, and plant organic remains as coals, tar sands and bituminous shales. Biogenic mineral deposits include non-metallic minerals and rocks, and solid mineral fuels.

- Calcium carbonates - Limestone and Dolomite
- Siliceous deposits - Chert, Diatomite
- Phosphate deposits
- Plant remains (Coal, Lignite)
- Marine organics (Tar Sand, Oil Shale)
- Biogenic Sulfur

#### 3.7.5.1 Phosphorite (Phosphate Rock)

Phosphorite, phosphate rock or rock phosphate are marine sedimentary rocks that contain high phosphate concentrations more than 4% $P_2O_5$, but usually in the range of 15–20% for commercial purposes. The phosphate is mostly contained as cryptocrystalline fluorapatite ($Ca_5(PO_4)_3F$) or as hydroxyapatite ($Ca_5(PO_4)_3OH$). Phosphorites are often associated with carbonate rocks, shales, and sandstones. They often occur in extensive layers, which may cover tens of thousands of square kilometers. Phosphorites usually show textures and structures similar to fine-grained limestones and sometimes may represent diagenetic replacements of carbonate minerals by phosphates. They also can be composed of peloids, ooids, fossils, and clasts that are made up of apatite. About 2% of world phosphate is derived from guano - the excreta of birds.

The origin of phosphorites is related to the precipitation of phosphorus in a variety of depositional environments that are very shallow, near-shore marine or low energy environments. This includes environments such as supratidal zones, littoral or intertidal zones, and most importantly estuarine. Currently, areas of oceanic upwelling cause the formation of phosphates because of the constant stream of phosphate brought from the large, deep ocean reservoir. This cycle allows continuous growth of organisms that utilize phosphate for shells... Significant deposits of phosphorites in the United States include the

Phosphoria Formation in Idaho and the Monterey Formation in California. Major deposits also occur in the Sechura Desert in Peru, as well as North Africa in Morocco and Mauritania.

#### 3.7.5.1 Coal Deposits

Coal is a black or brown combustible sedimentary rock formed from ancient vegetation which has been consolidated between other rock strata and transformed by the combined effects of microbial action, pressure and heat over a considerable time period. This process is commonly called 'coalification'. Coal is classified into four types based on the degree of coalification (carbon content) or the rank of coal: (i) Anthracite (86-97% C), (ii) Bituminous Coal (45-85% C), Sub-Bituminous Coal (35-45% C) and Lignite or Brown Coal (25-35% C). Coal occurs as layers or seams, ranging in thickness from millimeters to many tens of meters. Coal is composed mostly of carbon (50–98 per cent), hydrogen (3–13 per cent) and oxygen, and smaller amounts of nitrogen, sulfur and other elements. It also contains water and particles of other inorganic matter. When burnt, coal releases energy as heat which has a variety of uses. Coal is broadly separated into brown and black which have different thermal properties and uses.

Brown coal (lignite) has a low energy and high ash content. Brown coal is used to generate electricity in power stations located at or near the mine. Black coal is harder than brown coal and has a higher energy content. Thermal (steaming) coal is black coal that is used mainly for generating electricity in power stations where it is pulverized and burnt to heat steam generating boilers. Metallurgical (coking) coal is black coal with low ash that is suitable for making coke, which is used in the metallurgical production of pig iron. These coals must also have low sulfur and phosphorus contents, and are relatively scarce and attract a higher price than thermal coals. Coal is formed when dead plant matter decays into peat and is converted into coal by the heat and pressure of deep burial over millions of years.

## 3.8 Surficial Ore-Forming Processes

Surficial processes occur at or near the Earth's surface and involve weathering, leaching, infiltration and precipitation of minerals. In the concentration of minerals two surficial processes are important: chemical weathering and supergene enrichment. *Chemical weathering* involves the dissolution and removal of certain minerals or elements and produces the residual concentration of useful minerals without additional metal content. *Supergene enrichment* involves the oxidation and leaching of useful metals and precipitation and of secondary minerals with higher metal content in the subsurface.

### 3.8.1 Weathering

There are two main types of weathering; *physical* and *chemical* weathering. Both types may result in the concentration of useful economic minerals.

#### 3.8.1.1 Physical Weathering

Physical weathering and erosion produce the disaggregation of rocks through the effects of changing temperature sometimes assisted by water percolation. This process can facilitate the fragmentation of rocks along zones of weakness and possible concentration of heavy minerals such as zircon, rutile, chromite, diamond and gold. In temperate climates, the physical breakdown of rocks and minerals do cause disaggregation and the release of useful minerals that are subsequently concentrated by erosion and gravitational settling as eluvial placer deposits. In tropical environments, it is often difficult to separate the effects of physical weathering from that of chemical weathering because of the greater intensity of the latter.

### 3.8.1.2 Chemical Weathering

Chemical weathering involves the chemical breakdown of rocks and minerals mainly through the processes of oxidation and hydrolysis which produce selective dissolution and residual concentration of some useful minerals or elements. Water is the most important agent of chemical weathering; but other substances that may partake in the reactions include free oxygen, and inorganic and organic acids. Weathering as an agent of residual concentration of economic minerals is most effective in tropical and subtropical environments where high precipitation and deep-seated weathering can lead to the formation and accumulation of useful minerals or metals such *as bauxite (aluminum ore), kaolin and lateritic iron ore.*

### 3.8.1.3 Lateritic Weathering

Lateritic weathering is a very intense form of chemical weathering in tropical and semi-tropical environments where intense leaching and ferruginization produce thick weathering profiles or regolith often with Fe-rich caps. The most common minerals are clays and iron oxides. During lateritic weathering, the original body of rock is greatly reduced in volume by the process of hydrolysis and leaching, which removes ions from the original rock. Elements that are not leached from the rock may occur in higher concentration in the residual rock, and thereby accumulate as mineral deposits.

## 3.8.2 Supergene Enrichment

Supergene enrichment is a process that increases the concentration of metals in a primary mineralized body during chemical weathering. The chemical reactions usually involve hydrolysis and oxidation of sulfide and/or oxide ores at or near the surface accompanied by leaching and concentration in the subsurface by downward-percolating meteoric waters. Deposition of the leached elements usually occurs near the water table interface. Although supergene enrichment can apply to many transition metals such as Fe, Mn, Ni, it is copper sulfides (in the presence of pyrite) that are the outstanding example of ores enriched by this process. Several protore or low-grade primary deposits have been made more valuable by supergene enrichment. Supergene enrichment happens only when the structure and topography combine with atmospheric conditions to produce hydrodynamic groundwater flow systems that can drive large electrochemical cells for long periods, for millions to tens of millions of years for the process to achieve maturity.

A typical vertical profile of a supergene-enriched mineral deposit, consists of three zones (Fig. 3.56a): (a) primary or protore zone; (b) supergene enriched zone; and (c) leached (oxidized) zone which may be capped by an iron-enriched zone commonly referred to as *gossan* which is made up mainly of silica and iron oxides (hematite, goethite, limonite) commonly with a "*boxwork*' texture. Gossans may also contain relict constituents of the primary mineral assemblages such as malachite and azurite in copper deposits and cerussite and smithsonite in lead-zinc deposits.

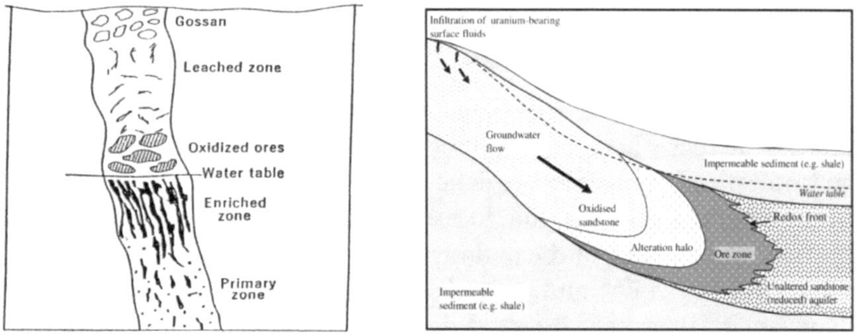


Fig. 3.56:( a) Zones of Supergene Enrichment (b) Meteoric Water Infiltration

#### 3.8.2.1 Meteoric Water Infiltration

Meteoric water infiltration is a form of supergene enrichment in which highly oxygenated groundwaters are capable of moving through sedimentary strata and leach elements that are later precipitated due to changes in the Eh-pH conditions, usually along a defined reduction front (Fig. 3.56b). The elements subject to this type of leaching are uranium, vanadium and molybdenum. This can lead to the formation of "infiltration mineral deposits" where meteoric water takes up a substance that is dissolved by weathering and gets it concentrated after considerable transport by infiltration in a different geological setting commonly in permissible clastic rocks.

## 3.9 Surficial Mineral Deposits

Surficial mineral deposits are those concentrations of useful minerals near or at the Earth's surface as a result of chemical weathering, and supergene enrichment. The difference between products of *lateritic weathering* and *supergene enrichment* is noteworthy. Lateritic deposits are concentrated by the leaching away of unwanted material and residual concentration of useful materials without any external contribution. In supergene enrichment, useful materials are removed and concentrated elsewhere, usually at depth or across a lateral front. Thus, a weathered profile may contain both lateritic and supergene deposits of a specific metal or material.
Three groups of deposits are formed by surficial processes:

1. Residual and Lateritic deposits,
    - Clay deposits
    - Bauxite deposits
    - Lateritic Iron deposits
    - Lateritic Manganese
    - Lateritic Nickel
    - Lateritic Gold
2. Supergene Enrichment deposits
    - Supergene Copper /Sulfides
    - Supergene Manganese
    - Supergene Niobium
    - Supergene Gold
3. Infiltration deposits
    - Roll front Uranium
    - Roll front Molybdenum
    - Karstic Bauxite

### 3.9.1 Residual and Lateritic Deposits

Residual deposits are accumulations of valuable minerals by chemical weathering after the undesirable constituents of rocks have been removed. The term 'lateritic deposit" is used to describe residual deposits in tropical and subtropical regions where the residual minerals contain iron oxides and hydroxides. The major requirements for the formation of residual deposits are: (a) parent rock of appropriate composition, (b) leaching of the undesirable constituents by intense chemical weathering and adequate drainage, and (c) low relief and crustal stability so that the residual accumulation is not eroded away as soon as it is formed. Residual deposits accumulate more in tropical and sub-tropical environments because of the effects of high precipitation and temperatures that accelerate intense weathering.

Aluminum and iron are the two most common elements concentrated by lateritic chemical weathering. After soluble elements are removed by hydrolysis, Fe and Al usually form hydroxides that are more stable under surficial conditions. Other transition metals such as Ni, Mn and occasionally, native metals like gold, may also be concentrated. Accumulation of these elements near the surface often produces lateritic ores. Common examples of residual deposits produced by weathering are: (1) bauxites, (2) lateritic iron ore, (3) lateritic nickel, (4) lateritic manganese, and 5) lateritic gold.

### 3.9.1.1 Bauxite and Lateritic Bauxite Deposits

Bauxite is the main ore of aluminum, and is found as a residual deposit formed from the decomposition of Al-rich parent materials. Bauxite ore consists of the minerals gibbsite (Al (OH) 3, boehmite (γ-Al (OH) 3) and diaspore (α-AlO (OH), and are sometimes mixed with kaolinite and the iron oxides, goethite and hematite.

Bauxite deposits are of four types: (a) *blanket bauxites* that commonly occur in the upper parts of thick surface (lateritic) blankets that cap plateaus in tropical areas for example, in Guinea, Ghana, Guyana and India: (b) *sedimentary bauxites* that occur as interlayered beds or lenses within clay or sedimentary sequence, are distinguished by pisolitic textures (Fig. 3.56b) e.g. deposits in Australia, South America and Malaysia: (c) *karst bauxites* that occur as pockets within limestones or karstified sediments e.g. deposits in the Caribbean islands (e.g. Jamaica), and, (d) *reworked bauxites* that are sedimentary reworked or detrital bauxites formed by erosion and redeposition. Most of the world's aluminium is mined from bauxite deposits. The world's largest producer of bauxite is Australia's Weipa Mine in northern Queensland, while other countries with large reserves include Guinea, Ghana, Brazil, Vietnam, and Jamaica.

*Lateritic bauxites* are iron-rich bauxites that are found within thick regoliths in tropical regions. They are formed by intense *in situ* weathering of aluminous silicate rocks including granites, syenite, trachyte, granite gneiss, sandstone, clays and shales. Most lateritic bauxites are relatively young because they form on erosion surfaces on top of plateaus and hills that are easily removed by erosion acting over long periods of time. Lateritic bauxites are highly variable in their nature and mode of occurrence. Textures of lateritic bauxites vary considerably, including massive, nodular, earthy and pisolitic. Typically, deposits are large, tabular or blanket types that are located close to the surface which facilitates simple strip-mining without the need for open pit development.

Profiles of lateritic bauxites vary depending on the climatic zone and type of parent material. A typical profile shown in Fig. 3.57(a) consists of a relatively thin soil layer overlying a horizon consisting of cemented pisolitic bauxite and underlying the cemented bauxite is a layer of loose pisolitic bauxite about 1-2m thick. The underlying substrate consists of a ferruginous layer overlying weathered sedimentary host rock. A profile from Northern Queensland, Australia shows the following horizons from the surface downwards: bauxite rubble, bauxite, bauxitic clay, clay and parent rock (Fig. 3.57).

Lateritic bauxites occur in many parts of the tropical and subtropical world including, India, Australia, West Indies, Brazil and West African countries (Guinea, Ghana, Nigeria and Cameroon). In West Africa, these lateritic bauxites are found as capping on old erosion surface or peneplain overlying a variety of aluminous rocks including shales, argillite, syenite, trachyte, pelitic schists and gneiss. In Guinea, one of the world's leading producers of bauxite with proven reserves of over 300 million tons and inferred reserves of over 40 billion tons, the blanket-type lateritic bauxites are found mainly in the western region (Boke, Daboia, Kindia, Fria) where they occur at low elevations (200 to 400m above sea level) in the flat topped plateaus with little or no overburden The source rocks include a series of argillites and formed from syenite and pelitic schists. A small sedimentary bauxite deposit is mined at Sangaredi.nine.

The ore minerals in the West African bauxites are predominantly gibbsite (Al $(OH)_3$ with minor boehmite (Al $(OH)_3$). (Gibbsite and boehmite are polymorphs). Alumina content ranges from 40 to 54% with low silica (<5%). At the largest deposit in Boke, a typical profile of the lateritic bauxite is 12 to 15 m thick regolith that consists of an upper iron-rich crust underlain by a softer bauxitic laterite that grades downward into a ferruginous bauxite layer, which is underlain by a layer of variegated clay and decomposed rock 10 m thick or more intervenes between the laterite and fresh rock. The bauxite consists chiefly of gibbsite and hematite with minor amounts of goethite, boehmite, quartz, and kaolinite.

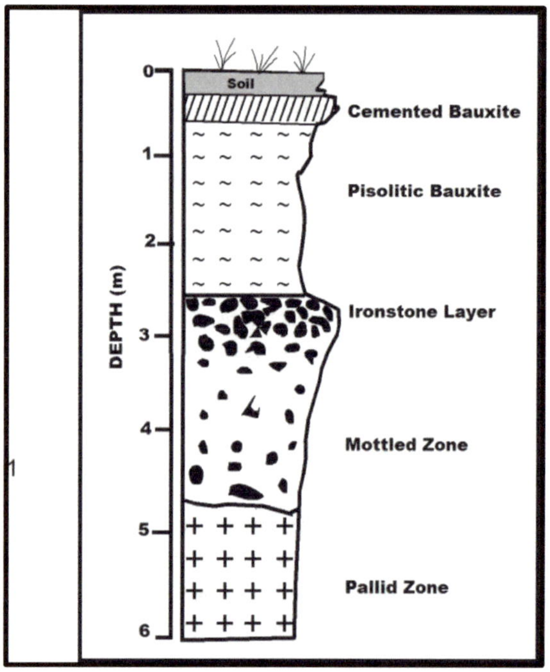

Fig 3.57: (a) Typical profile of lateritic (b) Pisolitic Bauxite Ore

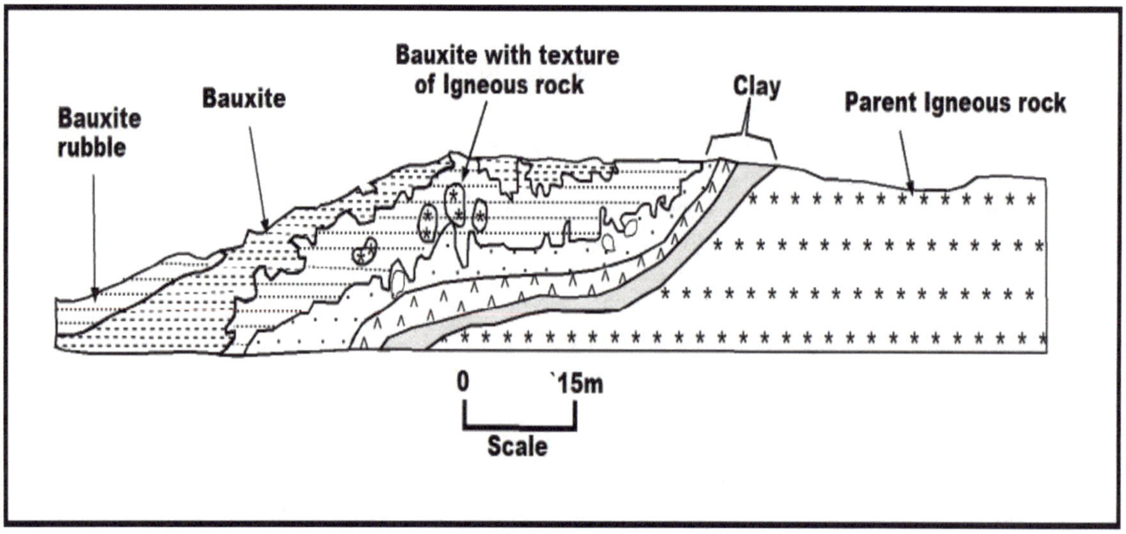


Fig. 3.58: Generalized profile of bauxite deposit in Queensland, Australia

The origin of lateritic bauxites is closely related to intense tropical weathering of aluminous parent rocks under conditions of high temperature and high rainfall which produce the leaching of silicates, iron oxides and kaolinite, and the precipitation of Al-hydroxides, usually gibbsite. The chemical changes associated with formation of lateritic bauxites include the removal of silica, alkalis, calcium, magnesium, iron and titanium, and concentration of alumina. The formation of bauxites has been attributed to a two-stage process: (a) chemical weathering and leaching of the parent rock to produce aluminum and iron-rich residue; and (b) further leaching of iron oxides and kaolinite to produce pure Al-hydroxides (gibbsite). Sometimes the bauxite is derived directly from underlying kaolinite (Fig. 3.58). The formation of lateritic bauxites is determined by the following conditions: (a) favorable parent rock with minerals that are soluble; (b) effective rock porosity that facilitates movement of water; (c) high rainfall with intermittent dry periods; (d) tropical warm climate; (e) low to moderate topographic relief; (f) good drainage; and (g) prolonged stability without much disturbance.

### 3.9.1.2 Lateritic Iron Ores

Lateritic iron ores are formed by intense chemical weathering involving leaching and concentration of residual elements in ferruginous host rocks to form economic mineral deposits. Suitable parent rocks are ferruginous sandstones, ultramafic rocks and low-grade banded iron formations (itabirites). The enrichment process involves the simple removal (dissolution) of silica and alumina in the primary ore accompanied by textural destruction, oxidation and the precipitation of residual iron oxides usually as soft or friable ore with enriched iron content. The ore grade of these deposits is in the range of 40 to 60% Fe depending on the original concentration in the parent material.

Lateritic weathering of various parent materials have been widely studied. In the case of low-grade banded iron formations known in Brazil as "itabirites", lateritic weathering under wet tropical conditions has resulted in well-developed weathering mantle, with a specific succession of horizons of different chemical, mineralogical and textural compositions capped by a ferruginous duricrust or canga. These deposits can be up to 30 meters thick with goethite and hematite as main ore minerals. A typical profile consists from top to bottom: (a) ferruginous cap (canga); (b) soft hematite of mostly hematite and goethite and no quartz, (c) weathered itabirite (friable and disaggregated), (d) slightly altered itabirite (minor changes) and (e) fresh itabirite (hard itabirite with hematite/magnetite) (Fig. 3.59).

Within the alteration profile there is desilicification and reduction in silica with consequent relative accumulation of iron content that changes from the original 40% up to 63 %. Lack of alumina in the itabirite and the intense leaching of silica inhibited the formation of clay minerals. The dominant iron minerals are goethite and hematite. This progressive lateritic modification tends to produce a varied surface textures, including massive, pisolitic, and botryoidal, or nodular aluminous/siliceous goethite and hematite, forming ferricrete or duricrust (Fig. 3.60). Examples of these deposits exist in Brazil, Guyana, Cuba, West Africa, India and Malaysia.

Apart from itabirites, lateritic iron ores are also derived from the weathering of ultramafic rocks which may contain enough iron to become potential parent rocks. A good example is the Conakry deposit in Guinea where a dunite is the source rock. The lateritic mantle is a duricrust about 6 m thick containing up to 50% Fe and 12% Al. In Nigeria, pisolitic or concretionary lateritic iron deposits have developed as thick caps on ferruginous sandstones with iron content of 40 - 50%Fe and 10 -12% Al. The minerals present include goethite, hematite and halloysite, with minor gibbsite.

The origin of lateritic iron deposits is attributed to the intense dissolution and removal of silica (and alumina) from the parent rocks and relative concentration of iron oxides by volume which results in iron enrichment.

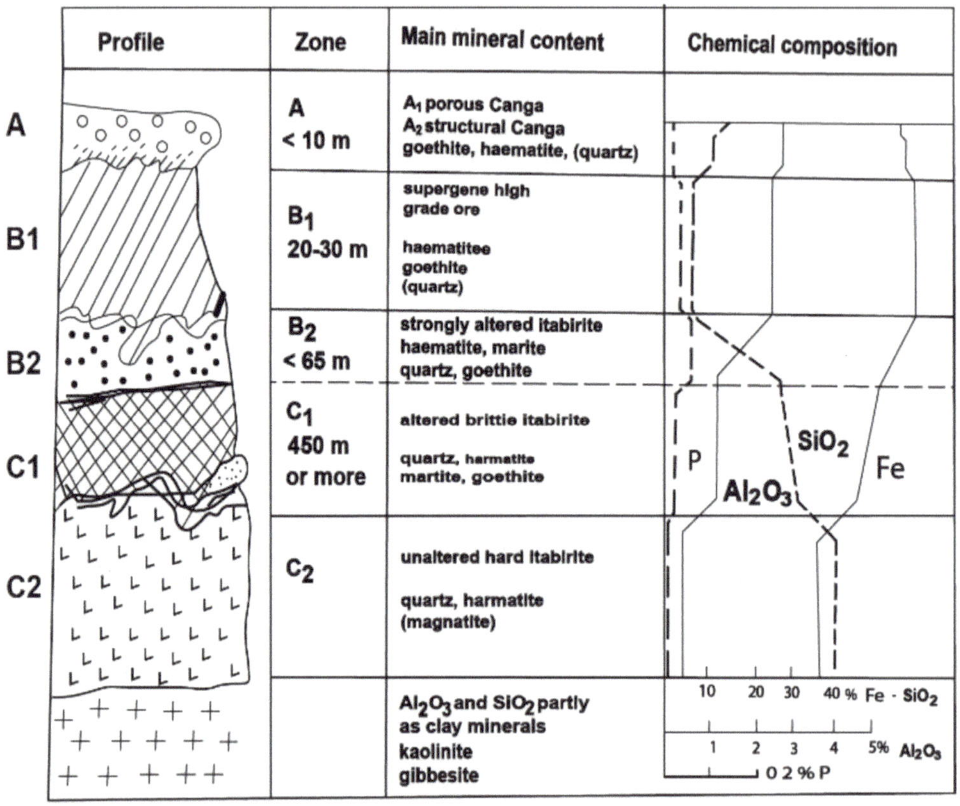


Fig. 3.59: Typical profile of lateritic iron ore, Brazil

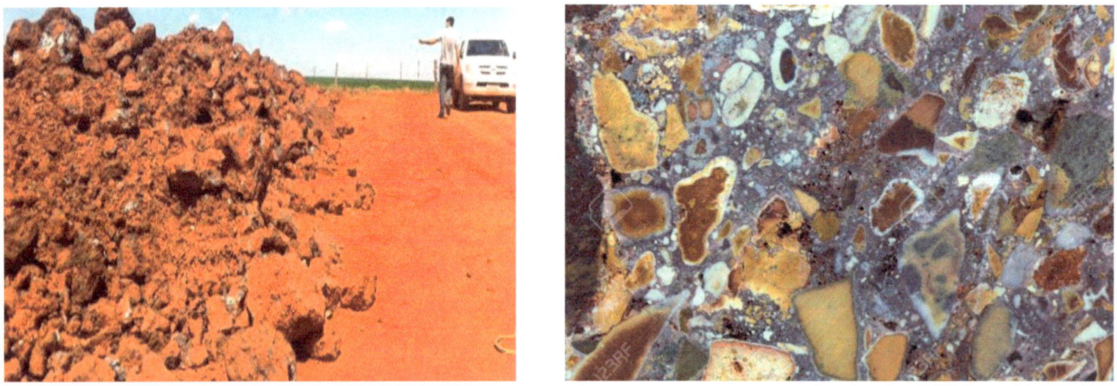

Fig. 3.60: (a): Lateritic Iron Ores, (b) Texture of Lateritic Iron, Central Africa

India **3.9.1.3 Lateritic Nickel Deposits**

Lateritic nickel ores account for over 60% of global nickel supply. They are formed from the intensive chemical weathering of olivine-rich ultramafic rocks such as dunite, peridotite and serpentinite (Fig. 3.60). The lateritization process removes the most soluble elements magnesium (Mg), calcium (Ca), and silicon (Si), and concentrates the least soluble elements (iron (Fe), nickel (Ni, manganese (Mn), and chromium (Cr). Mechanical weathering of the host material along fractures and faults increases the surface area exposed to the water driving the chemical weathering. This initial nickel content less than 0.1- 0.2% in the parent material may be enriched ten times in the course of lateritic weathering.

Two kinds of lateritic nickel ores have been distinguished: (a) *limonite types* and (b) *silicate types*. A third type which contains clay minerals (clay type) has been identified in some localities. Limonite-type lateritic ores (or oxide type) are highly enriched in iron due to very strong leaching of magnesium and silica. They consist largely of goethite and contain 1-2% nickel incorporated in goethite. The silicate type (or saprolite type) nickel ore formed beneath the limonite zone. It contains generally 1.5-2.5% Ni and consists largely of Mg-depleted serpentine in which nickel is incorporated. In pockets and fissures of the serpentinite rock, green garnierite a magnesium-based silicate can be present in minor quantities, but with high nickel contents of mostly 20-40% (Fig. 3.61b).

Typical lateritic nickel ore deposits are very large tonnage, low-grade deposits located close to the surface. They are typically in the range of 20 million tons and upwards at a grade of 1% Ni with some examples approaching a billion tons of ore. The ore deposits are restricted to the weathering mantle above the parent rock and tend to be tabular, flat and y large, covering many square kilometers of the Earth's surface. Examples of lateritic nickel deposits are found in Western Australia, Cuba, New Caledonia, Oregon, USA, Greece and USSR.

A typical lateritic nickel profile consists of a top ferruginous cap, underlain by the limonite zone followed by a saprolite (silicate) zone and underlain by weathered bedrock (Fig. 3.61a) and 3.62).

Fig. 3.61: (a) Laterite Nickel Profile (b) Laterite Nickel Ore with Garnierite

The origin of lateritic nickel deposits is similar to other lateritic iron and bauxite deposits. Chemical weathering is the most fundamental process in the formation of lateritic nickel deposits. The chemical weathering of the ultramafic rocks removes the most soluble elements magnesium (Mg), calcium (Ca), and silicon (Si)) and concentrates the least soluble elements iron, nickel and manganese. Two stages of weathering can occur: (a) the parent ultramafic rock may be altered to serpentinite, and (b) the serpentinite is subsequently weathered into a lateritic profile concentrating Ni.

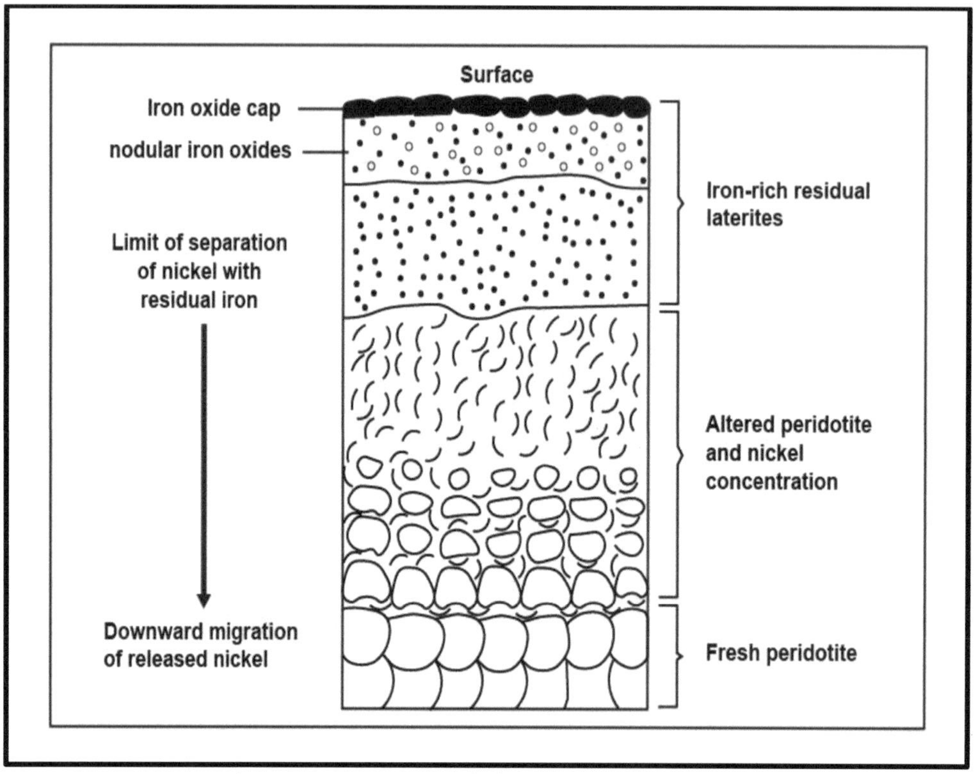


Fig. 3.62: Lateritic nickel ore profile, Australia

### 3.9.1.4 Lateritic Gold Deposits

Lateritic gold deposits are a relatively recent discovery, but they have acquired some importance, especially in Western Australia where large nuggets of gold have been found in laterites (Fig. 3.62). These deposits are formed by weathering of bedrock containing some fine gold, and chemical enrichment of gold occurs at the base of the lateritic weathering profiles. The source rocks are disseminated gold deposits or weathered ultramafic rocks or greenstones that contain gold specks. The gold is often deposited within iron oxides and/or aluminous concretions or hard crusts (Fig.3.63).

Examples of lateritic gold deposits can be found in Western Australia, Guyana, India and Brazil. In the Carajas region of Brazil, native gold is found within iron oxide-hydroxide nodules in the ferruginous zone where gold orebodies have been delineated. The Boddington bauxite mine in Western Australia was until its recent closure, the largest gold producer in Australia with annual production of 2400 kg. Sources of the gold at Boddington mine are quartz veins and hydrothermally altered rocks.

### 3.9.1.5 Lateritic Manganese Deposits

Lateritic weathering of manganese bearing metamorphic rocks (gondite) have produced some deposits in Brazil and many parts of Africa (Ghana, Ivory Coast, and South Africa). These concentrations within the lateritic profile consist of several Mn oxyhydroxides including manganite in association with Fe hydroxides. Lateritic manganese ores are often high-grade and of superior quality. Sources rocks for lateritic Mn include manganese quartzites, carbonates and volcano-sedimentary rocks that contain exhalative-sedimentary pre-enriched horizons.

Fig. 3.63: Lateritic Gold in Surficial Deposits, Australia

## 3.9.2 Supergene Enrichment Deposits

Supergene mineral deposits are those deposits formed by secondary enrichment near the earth's surface. Supergene processes which form this type of deposit are a remarkable form of weathering that leads to secondary enrichment of sulfide and oxide ores through the action of descending surface waters. This enrichment usually takes place by the alteration of primary sulfide or oxide minerals from the oxidized zone above the water table and the accumulation of the dissolved metals in the enriched zone below the water table. Supergene enrichment can occur in many types of ore deposits, but it is notably important in three categories of deposits: 1) sulfide deposits; 2) sedimentary Fe and Mn formations; and; 3) carbonatite Nb, Th and REE deposits.

### 3.9.2.1 Supergene Copper Deposits

Porphyry copper deposits are low grade disseminated mineralization of copper sulfides and pyrite in porphyritic granitic rocks. Secondary enrichment of porphyry copper deposits in the southwestern United States, Mexico, Peru, and Chile is an important factor in making those deposits economically viable ores. Enrichment blankets at many porphyry copper deposits form large laterally extensive zones (up to 30-40 square miles) that typically range from 200 to 1,000 feet in thickness, but thin laterally to 10 to 100 feet. The enrichment blanket at the Chuquicamata deposit in Chile is reported to attain thicknesses of up to 2,500 feet. Enriched sulfide ores are mainly composed of sooty to fine-grained chalcocite that coats and/or replaces pyrite and copper-bearing sulfides. (Fig. 3.64).

In volcanogenic massive sulfide deposits, copper and other metals are mobilized from primary massive sulfide ore and re-precipitated at depth. This precipitation of Cu sulfides with high Cu/S ratios can produce an economically significant copper enrichment at the reactive redox boundary between massive sulfide protore and downward penetrating fluids. In magmatic Ni-Cu bearing massive sulfides, supergene enrichment is recognized in many weathered profiles particularly in Western Australia, but such enrichment does not have a significant role in upgrading the metal tenor of the ores. The sediment-hosted copper deposits in Zambia also display supergene enriched zones characterized by widespread development of secondary copper minerals.

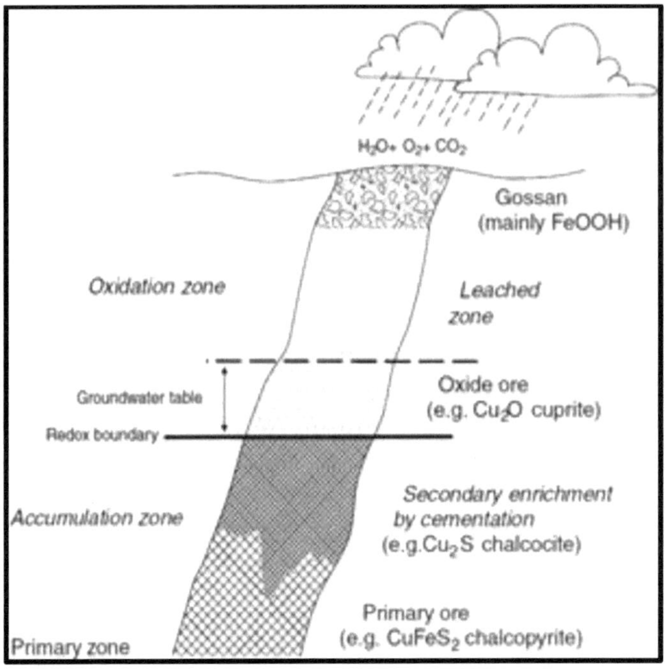


Fig. 3.64: Supergene enrichment of Copper ore from copper vein or porphyry copper pluton.

### 3.9.2.2 Supergene Iron Deposits

Many of the world's richest iron ores have undergone supergene enrichment of banded iron formations through the action of descending meteoric waters where the grade of iron ores have been increased two to three times. Prominent examples can be found in Western Australia and Brazil. The leaching process involves the removal of silica in the BIFs much more than the remobilization of Fe. The removal of silica produces more compact iron ores and the conversion of hematite to magnetite which enriches the Fe. In the supergene iron ores of Carajas, Brazil where primary itabirites with about 35% Fe have been enriched to up to 68% Fe. *Canga*, a cap rock consisting of detrital iron-rich fragments, cemented by limonite, shows an average grade of about 62% Fe. Zones enriched by lateritic weathering may merge with supergene enriched zones due to fluctuations in the water table, thus making the two related processes indistinguishable.

Enriched itabirite is typically disaggregated and porous, due to partial solution and removal of quartz; iron oxide is hydrated to varying degree, and enrichment is primarily residual. In the intermediate-grade ore with more than 57% Fe, leaching of silica is more complete, secondary enrichment is important, and nearly half of the iron is present as goethite. High-grade hematite ores with more than 66 percent Fe occur at the base of the supergene pile. These high grade hematite ores believed to be of supergene origin are the best quality iron ores. Examples are found in Carajás and Quadrilátero Ferrífero (Iron Quadrangle) provinces in Brazil and the Hamersley province in Western Australia.

### 3.9.2.3 Supergene Manganese Ores

Supergene manganese deposits are formed where chemical reactions taking place within tens of meters from the surface redistribute and concentrate manganese at lower depths within the weathering profile overlying manganiferous sediments and carbonates (Fig. 3.64). Sedimentary manganese deposits, especially those in carbonate rocks or with volcanic affiliation have been secondarily enriched to become ores by this process involving the leaching of primary Mn carbonates or silicates and conversion to Mn oxides such as pyrolusite and manganite, usually by oxidation and/or dissolution followed by migration and re-precipitation. A good example is the sedimentary manganese deposit at Groote Eylandt in Northern Australia where supergene enrichment has increased the ore grade from 15% to about 50%. Other examples include deposits in Brazil (Minas Gerais, Goias) where supergene ores occur at depth within the lateritic profile (Fig. 3.65). The most common minerals found on the deposit are a variety of Mn oxides, pyrolusite, cryptomelane and psilomelane with minor quartz, kaolinite, goethite and hematite. These combination of both primary and/or supergene minerals.

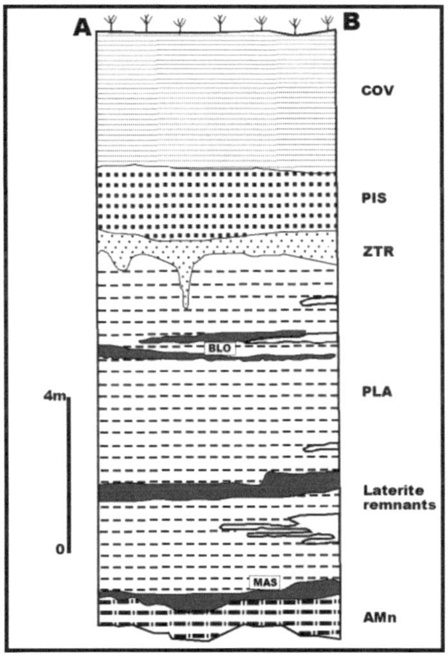


Fig. 3.65: Supergene Manganese Enrichment Zones (black)

### 3.9.2.4 Supergene Gold Deposits

Supergene gold deposits are rare but do occur within thick regoliths overlying disseminated or vein-type gold protore in which gold is remobilized and enriched by downward percolating acidic groundwater. Oxidative decomposition of sulfide minerals releases solid solution Au and provides physical and chemical opportunities for mobilization of micro-particulate gold. Examples are deposits in Otago, New Zealand.

### 3.9.2.5 Supergene Niobium Ores

Supergene niobium deposits occur in laterites formed from intense chemical weathering of carbonatites and secondary enrichment of disseminated niobium minerals, The process involves the decomposition of pyrochlore, apatite and magnetite, followed by removal of soluble carbonates and formation of secondary minerals, commonly Ba, Sr, K or Pb-bearing pyrochlore, and enrichment of Nb by the formation of Nb-bearing rutile, brookite, and anatase. The St. Honore carbonatite in Brazil hosts the largest niobium deposit in the world in which supergene enrichment occurs within a lateritic profile of up to 300 m deep. The Nb contents of the supergene deposits can be enriched up to 10 times or more above those of the primary carbonatite. Most primary deposits range in concentration from 0.1 to 0.3%Nb, compared to enriched values averaging 3% Nb. Examples of supergene Nb deposits are found mainly in Brazil (St. Honore, Araxa, Catalao, and Bingo). The Brazilian supergene deposits account for about 92% of the total worldwide production of Nb.

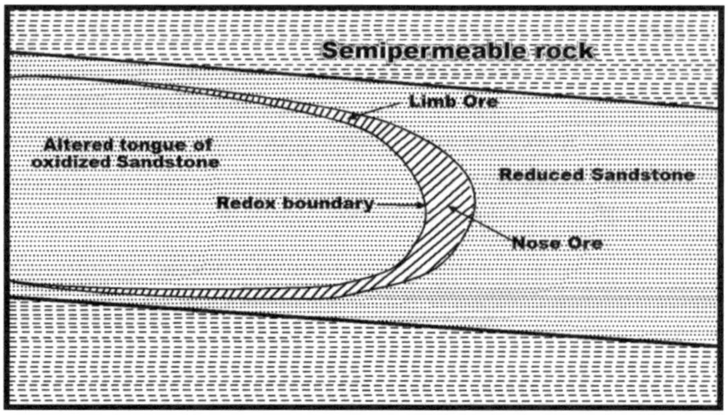

Fig. 3.66: Infiltration Roll-Type Uranium Deposits

### 3.9.3 Infiltration Uranium Deposits

Infiltration mineral deposits are formed where oxygenated meteoric water leaches and transports elements by infiltration through a permeable strata such as sandstones precipitating ores from solution due to reducing conditions. The so-called "roll front" uranium deposits are the most typical example of this process. Roll fronts mark the redox boundary in space and time where the infiltrating meteoric water carrying uranium lost its capacity to oxidize the percolated sandstone, and was forced to precipitate uranium from solution (Fig. 3.66). Uranium occurs in two valence states, the reduced form U4+ and the oxidized form U6+. The latter is highly soluble in oxidized fluids but becomes unstable under reducing conditions to precipitate as U4+. Common ore minerals include coffinite, uraninite, vanadium-rich clays and minor sulphides. Some deposits of metal such as molybdenum, copper, iron, vanadium, silver, and Pb-Z-Ba-F could have a similar genesis

## 3.10 Metamorphic Processes

Metamorphism is the process by which rocks and minerals are formed or transformed by the process of regional, dynamic and/or contact metamorphism. Temperature, pressure and fluids are the dominant agents of metamorphism, and their effects on pre-existing rocks and components can directly lead to the formation or accumulation of economic minerals and rocks. Indirectly, metamorphism can energize or generate ore-forming fluids by heating, devolatilization and remobilization of connate and meteoric fluids to leach and concentrate mineral deposits. Such processes and deposits are considered as hydrothermal, and were described in previous sections of this book.

The formation of mineral deposits in metamorphic environments is complex and not fully understood because it is difficult to absolutely differentiate what mineralization effects can be attributed to metamorphism compared to the pre-metamorphic conditions in the metamorphosed rocks. However, mineral deposits related to metamorphism can be formed in the following ways:

a. Crystallization or formation of new minerals or rocks e.g. talc, asbestos, graphite (metamorphic)
b. Recrystallization and/or deformation of existing mineral deposits or rocks e.g. marble, magnetite (metamorphosed or metamorphogenic)
c. Formation of new minerals due to reaction with magmatic fluids (contact metamorphic)
d. Dehydration and devolatilization of minerals during prograde metamorphism may produce fluids capable of leaching metals and concentration along permeable zones (Meta morphogenic).
e. Remobilization of heated formation or interstitial waters as metamorphic fluids (metamorphic-hydrothermal)

In the strict sense, only the first category of processes can be considered as truly metamorphic products because new economic minerals were formed by metamorphism. They are considered as real "metamorphic" deposits. The second category affects previously existing minerals that undergo metamorphic transformation which may produce only textural changes (increase in grain size) or changes in minerals with little or no change in chemical composition, e.g. change in iron oxides hematite to magnetite. They are considered as "metamorphosed" deposits. The third category most commonly involves the impact of heat and fluids from external (igneous) sources that interact with country rocks to form contact metasomatic "skarn" deposits. The fourth category generates fluids at low to medium temperature to form deposits that may be difficult to identify in the field, or differentiate from the fifth category of metamorphic-related hydrothermal deposits derived from heated connote brines.

Metamorphism affects existing rocks and minerals, and new minerals may be created or existing minerals become transformed by the effects of increased temperature, pressure and addition of fluids. Metamorphism can be regional or of local effect (contact); and both can produce mineral deposits.

## 3.10.1 Regional Metamorphism

Regional metamorphism is characterized by increased temperature and pressure over large areas of accumulated sediments that may lead to the dehydration and devolatilization of hydrous minerals. Prograde metamorphism due to increased temperature as metamorphic grade increases results in low-density minerals being replaced by minerals of high volumetric mass, while hydrous minerals are dehydrated and replaced by anhydrous ones and amorphous materials are crystallized. New minerals of economic value may be formed during this process through chemical reactions that take place. For example, minerals such as garnets, talc and graphite are formed, while existing minerals can be recrystallized by heat to enhance their economic value, such as increasing their grain size. Fine-grained calcite in limestone is recrystallized into coarse-grained calcite in marble. Hematite may crystallize as a new mineral from goethite, while hematite may recrystallize into a coarser-grained form of hematite called *specularite* or to magnetite during progressive metamorphism and reduced oxygen frugality. There are several aspects of regional metamorphism related to ore formation that will be explained below.

*Prograde metamorphism* may result in the dehydration and devolatilization of minerals which can generate mineralizing fluids capable of leaching metals from surrounding rocks to form mineral deposits which are called "metamorphogenic". Another concept of "retrograde leaching" has been proposed for this group of deposits which occurs after peak metamorphism, when a still-hot metamorphic complex is flooded and cooled by waters from outside; this will normally happen during uplift and extensional tectonics, and the resulting convection cells are open to the surface.

During *dynamic metamorphism*, a process known as *lateral secretion* can occur due to metamorphic reactions during structural deformation along fault zones (shearing) which can liberate mineral constituents such as quartz, sulfides, gold, carbonates, and oxides from deformed rocks, and channel these constituents as mineralization along zones of reduced pressure or dilation without fluid flow (Fig. 3.67a).

## 3.10.2 Contact Metamorphism and Metasomatism

Contact metamorphism occurs locally as a result of the intrusion of hot magma as plutonic rocks into sedimentary sequences (particularly limestones). Such interaction involving heat and magmatic fluids may produce chemical reactions that lead to the formation of mineral deposits, particularly if it involves the addition of metal-rich fluids from the intrusive body (metasomatism). Extensive replacement of surrounding rocks may take place, and mineral deposits may be formed, particularly for copper, tungsten and iron deposits. The ores and associated gangue mineral assemblages are called *skarn* deposits (Fig. 3.66c).

## 3.11 Metamorphic and Related Mineral Deposits

Mineral deposits related to metamorphic processes can be grouped into two types: *metamorphic* deposits which are new economic minerals formed directly by the effects of regional or contact metamorphism, and *metamorphosed* deposits which are existing economic minerals that are further concentrated or modified by metamorphism. A third group described as *metamorphogenic* by German geologists are difficult to identify in the field or easily differentiated from hydrothermal deposits in metamorphic origin, but they will be discussed briefly.

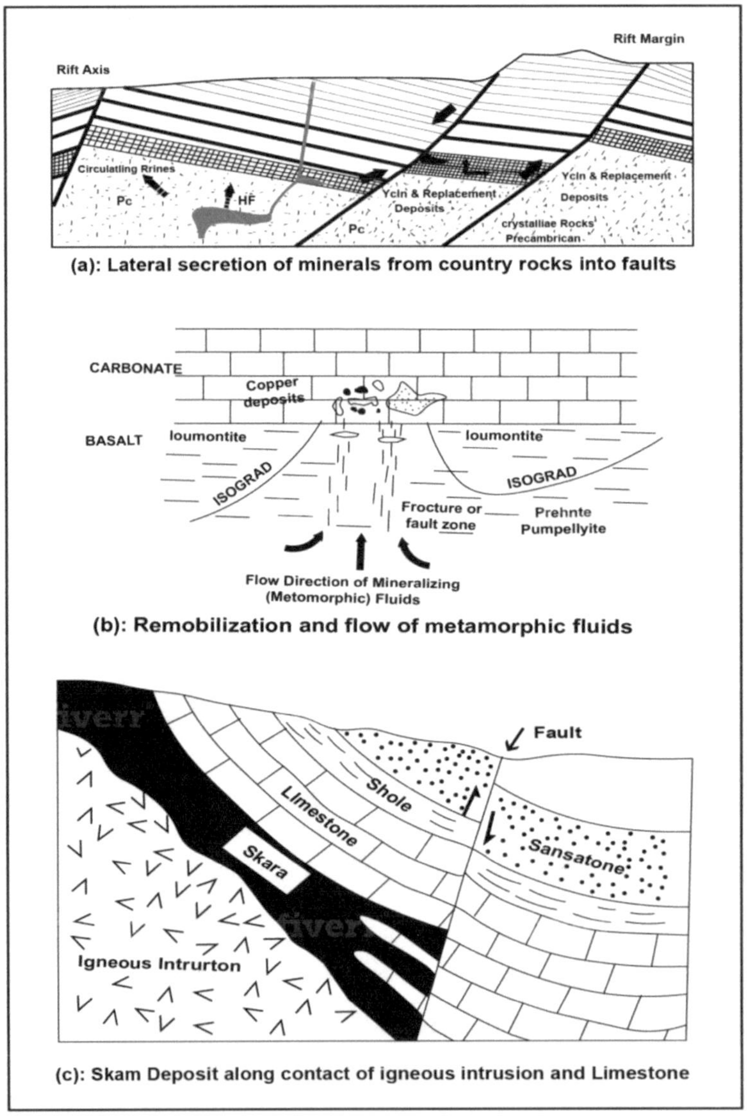


Fig. 3.67: Schematic diagram of metamorphic processes

Metamorphic related deposits are classified as follows:

1. Metamorphic Deposits
   - Industrial Minerals (Graphite, Asbestos)
   - Gemstones (Garnets, Zircon)
   - Contact metasomatic and Skarn Deposits
2. Metamorphosed Deposits
   - Iron Ores (Ferruginous Quartzites)
   - Manganese Ores (Gondite)
3. Metamorphogenic Deposits
   - Orogenic Gold?

## 3.11.1 Metamorphic Mineral Deposits

### 3.11.1.1 Industrial Minerals and Gemstones

Several industrial minerals and gemstones are formed during regional metamorphism of various rocks including mafic/ultramafic, pelitic and aluminous sediments. Such minerals include graphite, garnet, asbestos, talc, corundum, kyanite, sillimanite, and gemstones (e.g. ruby, sapphire, and emerald). Industrial rocks such as marble, slate and crushed aggregates of granitic gneisses and quartzites are of considerable economic value in building construction. Graphite is formed from metamorphism of coal, while talc and asbestos deposits are formed by regional metamorphism of ultramafic rocks. Several examples are found within the Precambrian greenstone and schist belts. Kyanite and sillimanite are formed from the metamorphism of aluminous sediments and shales. Marbles used as decorative stones are produced by regional metamorphism of limestones. Gemstones formed by regional metamorphism include varieties of garnets (spessartine, rhodonite, almandine, and grossular), corundum and spinels.

***Graphite Deposits***

Graphite, a crystalline form of carbon is an excellent example of economic minerals produced by regional and contact metamorphism of coal beds and highly carbonaceous sedimentary rocks at various grades of metamorphism. There are three major types of graphite deposits; (a) flake (crystalline) graphite; (b) amorphous (microcrystalline) graphite, and (c) vein (lump) graphite. Crystalline flake graphite deposits consist of flakes of graphite that occur in a variety of rocks including marble, paragneiss, iron formation, quartzite, and pegmatite and, in extremely rare cases, serpentinized ultramafic rocks

Amorphous (or microcrystalline) graphite deposits occur as intercalated lenses and beds within metamorphosed siliceous sediments such as quartz-mica schists, micaceous quartzites, or micaceous feldspathic quartzites, and rarely within marbles. They consist mainly of small graphite particles intergrown with other mineral impurities. Examples are found in Alabama and Texas, USA, where they are more than 100 feet thick; they range in length from a few feet to as much as half a mile. Amorphous graphite deposits.

Vein-type graphite deposits are the most economically important group that occurs as fissure-filling veins, pods and lenses, irregular bodies, stockworks, and saddle reef within metasedimentary rocks in the upper amphibolite and granulite facies grade of metamorphism. They are also found in zones associated with contact metamorphism by igneous intrusions. Vein graphite is composed of coarsely crystalline graphite-rich fragments that are typically 0.5 to 0.8 cm in diameter and are commercially described as "lumps". The origin of the vein graphite deposits is controversial, but they are believed to be products of high grade granulite facies metamorphism in the presence of a $CO_2$ rich fluid under low $fO_2$ conditions. Examples include extensive deposits in Sri Lanka, where graphite veins are currently mined.

### 3.11.1.2 Skarn Deposits

Contact metasomatic or skarn deposits are formed where a felsic magma (granite/granodiorite) intrudes sedimentary rocks, particularly carbonate rocks, resulting in the deposition of new metallic minerals with a unique assemblage of calc-silicate gangue minerals. This process is commonly accompanied by the discharge of acidic fluids from the intrusive body which facilitates chemical reaction with the sedimentary rocks and the ore deposition (Fig. 3.66c).

*Skarn* deposits are replacement deposits because they replace the carbonate host rocks. Common ore minerals in skarn deposits include both sulfides and oxides, such as molybdenite, chalcopyrite, sphalerite, galena, and oxides: cassiterite, wolframite, magnetite, scheelite, gold and hematite. Gangue minerals in skarns are unique, and include quartz, various garnets, amphiboles, pyroxenes, calc-silicates, olivines, talc, epidote, garnets and anhydrite. Skarn deposits are important as sources of iron, molybdenum, gold, tungsten, copper, lead, zinc, and tin. They constitute the world's premier sources of tungsten (more than 70% of world's tungsten production).

The mode of occurrence of ore minerals in skarns ranges from massive ores in some deposits to disseminated grains and veinlets of sulfides, molybdates, and tungstates in others. Reaction zones along the intrusive contacts are often narrow rims, often rich in Mn-silicates, calc-silicates and carbonates, formed between the intrusion and carbonate-rich host rocks. Replacement skarns are characterized by large areas of silicate replacement of carbonate rocks resulting from the passage of mineralizing solutions. Zoning of the silicate gangue minerals, and sometimes the metallic minerals is common in skarn deposits with a generalized pattern marked by an increase in the pyroxene to garnet ratio away from the igneous contact. Skarn deposits can be classified on the basis of the metal content e.g. skarn iron, skarn gold, skarn copper, skarn tungsten etc. Several examples of skarn deposits can be found in the western United States (Nevada, Arizona, New Mexico and California) including Battle Mountain (Au), Lower Fortitude (Au), Copper Canyon (Cu), in Nevada: and Twin Buttes (Cu), Christmas (Cu), in Arizona:

The origin of skarn deposits is connected with contact metamorphism and metasomatism that take place at 700-500°C takes place with some reaction with, and recrystallization of, carbonates to form calc-silicate minerals found as gangue. Acidic fluids from magmatic sources invade and react with limestones which results in the dissolution of the limestone and replacement by silicate minerals and sulfides of iron, copper, zinc, lead and silver, and oxides of iron, tin, and tungsten**.** Fluid inclusion and isotope studies indicate temperatures of formation from 225°C to more than 600°C, and a considerable degree of interaction of hypersaline magmatic fluids with convecting ground water.

## 3.11.2 Metamorphosed Deposits

Metamorphosed deposits are those mineral deposits that were originally sedimentary rocks but subjected to metamorphism that may have produced enhanced metal content and increased grain size. Such changes often increase the economic value of such deposits.

### 3.11.2.1 Ferruginous Quartzites

Ferruginous quartzites are good examples of iron ore deposits formed by the metamorphism of iron-rich sedimentary rocks. Other metamorphosed iron ores include banded iron formations (BIFs), granular iron formations (GIFs), and itabirites. Regional metamorphism often enhances the texture and grade of protore iron ores. Metamorphosed manganese ores are also common in metamorphic environments (Fig. 3.67). Most Archean iron-formations of the Algoma type are part of greenstone belts that have been deformed and metamorphosed. When BIFs are subjected to low-grade metamorphism, they retain most of the "banding" and a lot of their primary sedimentary features, but undergo some textural and mineralogical changes with increased crystallinity and more magnetite ore. They are called *"itabirite"* in Brazil to denote a metamorphosed iron formation composed of iron oxides (hematite, magnetite, martite) and abundant quartz, may be banded, schistose or compact. Itabirites have been described from Guyana and West Africa, and also occur in the Hamersley province of Australia

When ferruginous sandstones or shales are metamorphosed, they are locally described as either *ferruginous quartzites* (hematite-magnetite quartzites) or *ferruginous schists* (hematite schists) respectively, instead of 'banded iron formation' because they may not show continuous banding and are somehow different from the well-known BIFs of sedimentary origin in their mode of occurrence, lateral extent and nature of deposits. The ferruginous quartzites are not always banded, and if they are, the bands are not consistent in either width or lateral continuity because they are products of metamorphic differentiation and not sedimentary precipitation. The development of banding increases with metamorphic grade. Thus, these metamorphosed sediments may be described as *banded or granular hematite-magnetite quartzites.*

The ferruginous quartzites usually occur in medium to high grade metamorphic terrain in which iron-rich bands containing magnetite and/or hematite alternate with crystalline bands of quartz, with minor sillimanite. Examples of metamorphosed ferruginous quartzites occur widely in Precambrian shield areas of the world, and include several occurrences in India and central Nigeria near Okene, Kogi State, as well as the well-known iron ores of the Krivoi Rog basin and the Kursk magnetic anomaly in eastern Ukraine. Grades of iron ores range from 25 to 45% Fe.

The *ferruginous schists* are fine-grained quartz-hematite and hematite schists in which iron ore bearing units are associated with low-grade metamorphosed Precambrian metasediments and metavolcanics. They occur as bands and lenses about 1-30m thick, intercalated within low grade metamorphic rocks, mostly pellitic and semi-pelitic phyllites and schists. They are usually low grade and not mined. Examples are found in the Minas Gerais region of Brazil, Singhbhum district of India and those within the Schist Belts of northwestern Nigeria notably at Maru, Birnin Gwari and Muro.

#### 3.11.2.2 Metamorphosed Manganese Deposits

Metamorphosed manganese deposits are formed by the metamorphism of sedimentary manganese ores and manganiferous sediments, sometimes described as gondite. The grade of metamorphism can range from low-grade greenschist facies to high-grade granulite facies. The mineralogical assemblages include manganese oxides and manganese silicates. The texture of the manganese ores is enhanced by coarser and granoblastic texture as grade of metamorphism increases. Examples of these metamorphosed manganese ores are those found in India (Andhra Pradesh), and Namibia - within the Precambrian Damara metamorphosed manganese formations in Otjozondu. The Mn occurs in three stratigraphic zones associated with ferruginous quartzites and schists. Interbanded oxide and manganese silicate layers are represented by braunite,-jacobsite hausmannite, mineral assemblage, and by spessartite-rhodonite, diopside-acmite-quartz assemblage, respectively. The impact of metamorphism is to increase the manganese oxide content.

In Japan, manganese deposits of volcanogenic-sedimentary origin were later subjected to regional metamorphism. At Shiromaru mine near Tokyo, such metamorphosed deposits consist of braunite-rhodonite assemblage. At Muramatsu mine, Hagashaki Prefecture, similar manganese formations metamorphosed to amphibolite facies, exhibit a common mineral assemblage comprising braunite-rhodonite-piemontite-spessartite (Fig. 3.68).

Fig. 3.68: Metamorphosed Manganese Ore. Japan

### 3.11.3 Metamorphogenic Deposits

This class of metamorphic deposits was proposed by the German School of thought on economic geology led by WL Pohl. These deposits are purported to have been formed from fluids generated during prograde metamorphism of pelitic rocks by collecting water from the dehydration and devolatilization of sedimentary minerals. Such fluids definitely can be mobilized and produce lateral secretion of metals that are concentrated as mineral deposits. Such fluids are arguably different from the connate or sedimentary pore fluid that could also be heated and mobilized during metamorphism.

According to the theory, two end member models have been proposed to define a range of possibilities occurring in nature to generate the mineralizing fluid: (a) the prograde devolatilization model, and (b) the retrograde leaching model. In the first case, the mineralizing fluids are thought to have been generated and to have acquired their solute content within source rocks undergoing prograde metamorphism. Then, they flow towards lower pressure zones, possibly accompanied by further reactions with country rocks. Finally, changes in the chemical or physical environment along the flow conduits potentially induce mineral precipitation. Retrograde leaching, on the other hand, occurs after peak metamorphism, when a still-hot metamorphic complex is flooded and cooled by waters from outside; this will normally happen during uplift and extensional tectonics, and the resulting convection cells are open to the surface.

Metamorphogenic deposits cannot be easily recognized as such in the field, where they appear as epigenetic hydrothermal mineralization that may be difficult to differentiate from hydrothermal deposits formed by metamorphic fluids (described earlier in this chapter). Only the application of a broad spectrum of laboratory methods, such as isotope geochemistry and fluid inclusion study can allow a correct genetic diagnosis. A good example commonly cited are the orogenic *gold deposits* that are found in Precambrian greenstone belts in the form of structurally controlled quartz and quartz carbonate veins. This ore type has been described under the group of hydrothermal deposits.

## 3.12 Metallogeny and Plate Tectonics

Metallogeny is the study of the genesis and global distribution of mineral deposits in relation to major geotectonic features in the Earth's crust. It is widely known that the spatial distribution of mineral deposits in the Earth's crust varies considerably, both in time and space due to controls of certain deep-seated petrologic and tectonic features. This has led to the development of the concepts of "metallogenic provinces" and "metallogenic epochs". *Metallogenic provinces* or belts are geographic areas or regions characterized by a notable concentration of mineral deposits of a certain type or metal(s), while *metallogenic epochs* refer to periods in geological time when conditions were favorable for the formation of a group of mineral deposits. For example in Nigeria, there are metallogenic provinces for gold, in the Schist Belts, and for lead-zinc in the Benue Trough. Economic geologists are interested in metallogeny and metallogenic provinces in an attempt to achieve a better understanding of the nature and geo-tectonic settings of mineral deposits, and to use such knowledge in the exploration for mew deposits in a particular region or district.

The evolution of the Earth's crust is often marked by episodes of deformation and metamorphism of thick piles of accumulated sediments and volcanic rocks that are subjected to heat and pressure, igneous activities and topographic rise due to mountain building. Global tectonics produce the large-scale movements and processes that lead to the deformation of the earth's crust on a global basis. The prominent areas on the Earth's surface where global tectonics have been operative are known as "*orogenic belts*". They are regions of intense folding and faulting accompanied by granitic plutonism and metamorphism. These regions are often distinguished from the more stable basement areas known as "cratons". The formation and global distribution of mineral deposits have been linked to the evolution of orogenic belts. Prior to the 1960s, the prevalent tectonic theory in explaining the formation and evolution of orogenic belts was the "geosynclinal" theory which was based on "stabilitist" concept that has been discarded in favor of the more "dynamic" plate tectonics theory that has become widely accepted as providing an overarching framework to explain crustal evolution and metallogeny.

### 3.12.1 Plate Tectonic Theory

According to the plate tectonic theory, the Earth has a rigid outer layer, known as the lithosphere, which is broken up into a number of rigid plates; each plate moves relative to the others and interact along their boundaries, where they either converge, diverge, or slip past one another. Such interactions are thought to be responsible for most of Earth's major tectonic features such as orogenic belts, major faults, and metamorphic and igneous processes. Plate tectonics has provided a framework in which to describe the processes controlling creation and destruction of landforms, and the evolution of the Earth's crust, Moreover, plate tectonics explains the e distribution of global tectonic features that influence the distribution of mineral deposits. The driving force of plate tectonics is believed to be the convective circulation of heat from deep in the mantle to the lithosphere which is in part recycled back into the mantle. Extensional and compressional interactions at plate boundaries are the cause of profusely fertile metallogenetic systems where the metalliferous 'juices' emanate from deep within the mantle.

In plate tectonics parlance, crustal evolution is often considered in terms of a series of related processes known as the "Wilson Cycle" (Fig. 3.69), in which lithospheric crust is broken up by a rising plume of magma forming a domal uplift leading to the thinning of the overlying crust. As the crust continues to thin due to extensional tectonic forces, tensional faults and rifting develops as a rift basin widens to an ocean basin with sediments accumulating along its margins. Subsequently subduction is initiated on one of the ocean basin's margins and the ocean basin closes up by continental collision.

## 3.12.2 Mineral Deposits in Relation to Plate Tectonics

Plate tectonics provides a basis for understanding the global distribution and origin of mineral and energy deposits. The relationship of plate tectonics and mineral deposits can be viewed in three ways:

- Geological processes that occur due to energy released at plate boundaries control the process of mineral deposition.
- Mineral deposits form in particular tectonic settings which are governed by plate movements such as subduction zones and volcanic arcs.
- Reconstruction of fragmented continents can provide a useful basis for exploration for new mineral deposits.

The major tectonic settings governed by plate tectonics are:

- Intracontinental Rifts
- Oceanic Ridges and Basins (Divergent Plate Boundaries)
- Subduction and Collision Zones (Convergent Plate Margins)
- Cratonic Basins

Several requirements must be fulfilled in any tectonic setting for the production and accumulation of mineral deposits. A number of tectonic settings meet these requirements while others fail and may not be associated with mineral deposits. Figure 3.69 is a schematic diagram of mineral deposits in association with different plate tectonic settings.

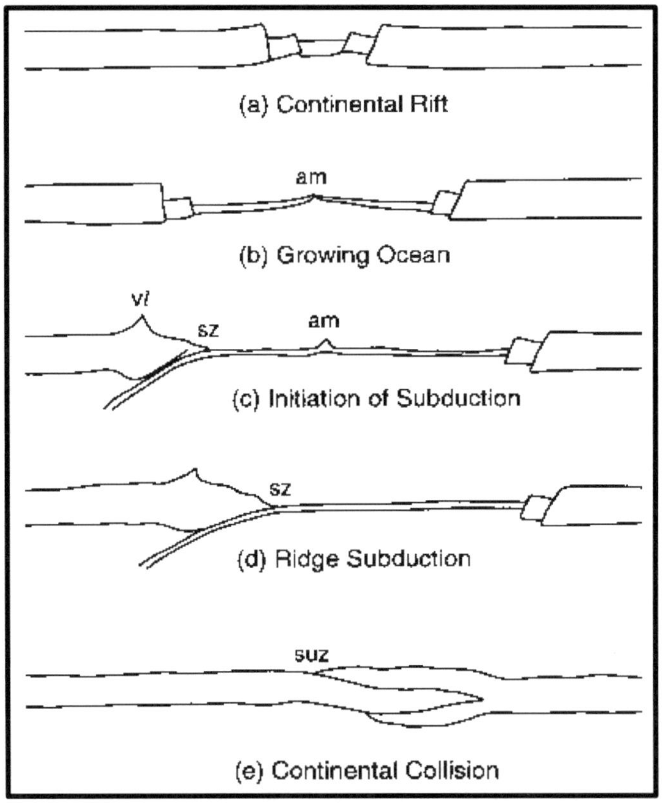


Fig. 3.69: Wilson's Cycle of Crustal Evolution

### 3.12.2.1 Mineral Deposits in Domal Uplifts and Rift Basins

Continental break-up is often initiated by the upwelling of mantle plume beneath the continental crust causing the overlying lithosphere to undergo regional uplift, thin, stretch and develop fractures that are invaded by magmas prying the continents apart, and as pathways for hydrothermal mineralization (Fig. 3.71). The magmas formed by this hot-spot activity produce high level alkaline granites, carbonatites and related volcanic rocks .Extensional failure of the lithospheric crust that occurs with triggers the development of a triple junction in which one arm of the rift fails, remaining as a fault-bounded sediment-filled basin known as an *aulacogen*, Thick sediments, evaporites and alkali basalts accumulate in aulacogens. The tectonic, igneous and sedimentary processes that accompany subcontinental mantle plume or hotspot activity provide favorable environments within which ore-generating systems can operate, and accumulate mineral deposits.

Mineral deposits associated with rift systems are found within alkaline intrusive rocks, fault zones, volcanic rocks and sedimentary rocks.

1. Granites intruded at the crustal doming stage are associated with tin, and tungsten and fluorite deposits e.g. Younger Granites of Nigeria.
2. Carbonatites derived from the mantle plume along rift margins may contain Nb, REE and Cu deposits (e.g. East African rift valley).
3. Evaporite deposits (halite and gypsum) accumulate in the rifts during the more advanced stages of continental separation.
4. Aulacogens that developed as failed rifts are characterized by the formation of vein-type Pb-Zn-Ag and barite deposits in limestones and shales during early and middle stages of rifting. (E.g. Benue Trough, Nigeria).
5. Advanced rift basins with thick sediments contain copper deposits e.g. Zambia
6. Strattabound Pb-Zn deposits in rift sediments associated with a juvenile ocean.
7. Geothermal fields that occur along the rift systems due to heat flow from the mantle may contain geo-epithermal mineralization, e.g. Ag and Hg deposits

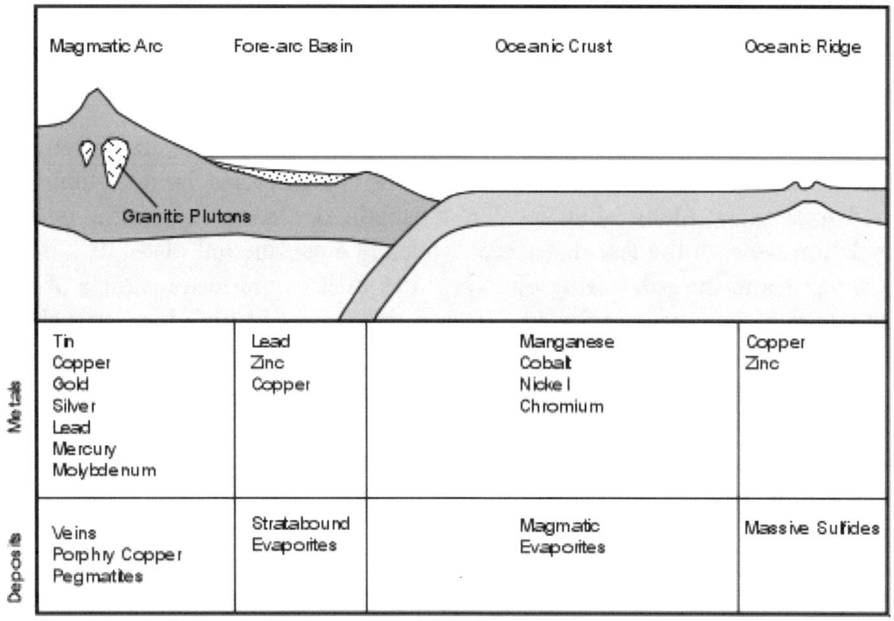


Fig. 3.70: Schematic diagram of mineral deposits in plate tectonic settings

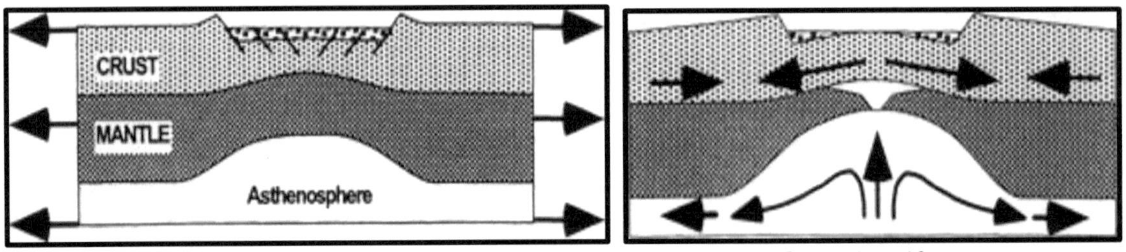


Fig. 3.71: (a) Passive intracontinental rift system (b) Active rift system

### 3.12.2.2 Mineral Deposits at Divergent Plate Margins

The opening of new oceans passes from a high heat-flow rift stage into a marine transgression and thermal contraction phase. Mid-oceanic ridges are developed as spreading centers that continue to enlarge into the ocean basin. Relatively shallow, epicontinental seas may form. As the young ocean widens, passive continental margins develop and creation of oceanic crust is accompanied by much volcanic outpouring of mafic basalts and ultramafic rocks. Mineral deposits are formed within the volcanic rocks and marine sediments by magmatic and hydrothermal fluids of connate and igneous origin (Fig. 3.71).

- Hydrothermal activity at the ridges gives rise to (a) Sulfide deposits and (b) Metalliferous sediments on the flanks of ridges. Important metallic deposits formed are Fe, Zn, Cu, Pb, Au and Ag. In the Red Sea metalliferous sediments containing Fe, Zn and Cu are being deposited.
- Manganese oxide deposits are important at some ridges e.g the TAG Hydrothermal field on the Atlantic Ridge.
- Ultramafic rocks in ophiolites containing asbestos, chromite and nickel ores. These are generally accessible in Phanerozoic orogenic belts to which sites they have been transported due to plate movement.
- Podiform chromite deposits associated with serpentinized ultramafic rocks.
- Cyprus-type volcanogenic massive sulfide deposits (Cu-Fe rich) are also associated with ophiolites and represent hydrothermal deposits formed at ocean ridges.
- Sedimentary exhalative (SEDEX) deposits of Pb-Zn along the edges of rift-related marine sediments

### 3.12.2.3 Mineral Deposits at Convergent Margins

As the ocean widens, it reaches the edge of another crustal plate, which leads to convergence along a zone where the two plates meet. Convergent plate margins are characterized by a tectonic plate subducting beneath a lower density plate. Along what is called a "subduction zone" where the denser plate, usually oceanic plate, is driven beneath the less dense plate, usually a continental plate. At a depth of about 100 miles (160 km), materials in the subducting plate begin to melt to produce magmas of basalt to andesite composition that rise to higher levels in the crus to form intrusive and high-level subvolcanic rock bodies. The trace of subduction on the seafloor is marked by deep oceanic trenches. Volcanic or Island arcs are developed on the overriding plate. Complete subduction of the oceanic crust may ultimately lead to continent to continent collision such as the Himalayas.

There are three main types of subduction zone configurations: (1) Ocean-Continent Plate boundary in which oceanic slab is subducted beneath continental slab (e.g. the highly significant Andean type collisional margins; (2) Ocean-Ocean Plate boundary in which two slabs of oceanic crust (e.g. Japan type plate margins); and (3) Continent-Continent Plate boundary involving two slabs of continental crust collide (e.g. Himalaya Mountain-type collisional margins) which results in thickened crust below collisional belts.

Various metallic mineral deposits are commonly found at convergent plate margins. At *ocean-continent* plate boundaries, the intrusive and volcanic rocks created by melting of subducted oceanic crust are the sites of the great porphyry Cu-Mo provinces of the world, while inboard of the volcanic arcs, significant Sn–W granitoid-hosted mineralization also occurs. The volcanic regions above the porphyry systems are also the sites of subvolcanic epithermal precious metal deposits of gold and silver (Au-Ag) (Fig. 3.72).

At *ocean-ocean* boundaries, porphyry Cu–Au deposits occasionally occur associated with the early stages of magmatism in these settings whereas the later, more evolved calc–alkaline magmatism gives rise to Kuroko-type massive sulfide (VMS) deposits such as those in felsic lavas in Japan. The back-arc basins are sites for the Besshi-type massive sulfides (Fig. 3.73). Along the Circum-Pacific Belt major metallic deposits occur in western North and South America, Japan, Philippines, New Zealand and Indonesia. More than half of the world's supply of copper comes from the porphyry copper deposits of this region.

In the zones of continent-continent collision, the thickened crust may melt and granites formed may contain tin-tungsten (Sn-W) mineralization (Fig. 3.74).

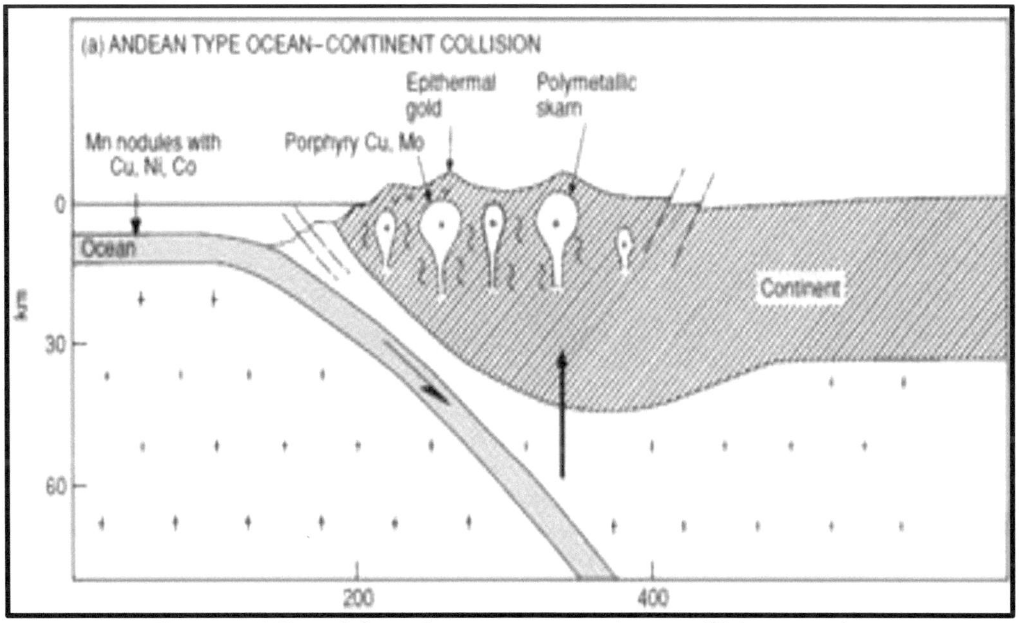

Fig. 3.72: Mineral deposits at ocean-continent collision margins

Important deposits associated with present and former convergent margins are:

- Base metals (Cu, Pb, Zn, Mo, and Sn)
- Precious metals (Au, Ag)
- Other metals (Sn, W, Sb, and Hg.

Plate tectonics plays an important role in the understanding the formation and global distribution of various metallic mineral deposits, particularly in relation to divergent and convergent plate margins. These tectonic processes which are known to have occurred in the Precambrian continue to repeat themselves throughout the Phanerozoic.

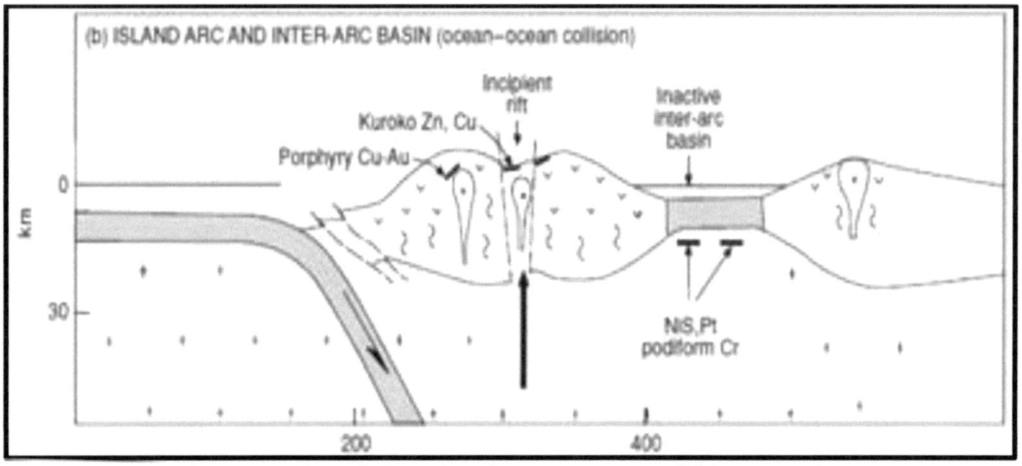


Fig.3.73: Mineral deposits of ocean-ocean collision margins

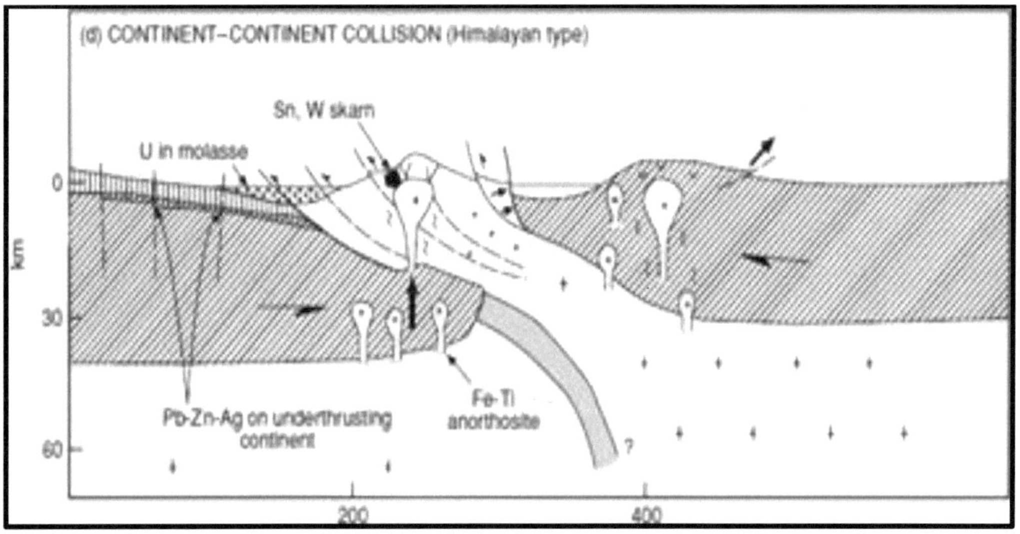


Fig. 3.74: Mineral deposits of continent-continent collision margins

## 3.13 Mineral Wealth of Africa

The mineral wealth of Africa refers to the totality of valuable mineral resources, tapped or untapped, that exist within the geographic boundaries of the continent. Mineral wealth provides the mineral raw materials that can be extracted and used for the economic benefit of the peoples of the various countries as perceived by their leaders. To be of benefit, minerals need to, first of all, extracted from the bowels of the Earth, and processed and used domestically or exported as commodities of trade to other countries to generate revenues. Such revenues are then utilized for the welfare and economic development of the nations involved. This cycle of conversion of mineral wealth to items of economic development and prosperity for a nation is often faced with challenges that may lead to rupture, and sometimes to total failure. Africa is widely believed to be the continent with the largest mineral wealth on Earth, while many of the African countries are among the poorest and least developed. Consequently, it can be reasonably assumed that the management of mineral wealth is still a major challenge in Africa and needs to be addressed.

## 3.13.1 Background

Africa is the second largest of the world's seven continents (after Asia), and covers a geographical area of approximately 30.million km2. The continent is well endowed with mineral resources, and believed to have the richest concentrations of metallic minerals and gemstones in the world. Africa's mineral wealth is partly attributed to its complex geology and multiple tectonic events that affected the prevalent Precambrian rocks that are known to contain the bulk of many of the world's valuable mineral deposits. Africa ranks first or second in the quantity of world reserves of bauxite, cobalt, industrial diamonds, gold, chromite, platinum, zirconium and phosphate rock, and a world leader in the production of several metals including gold, platinum, chromite, uranium, manganese, vanadium, tantalum and cobalt. There are vast resources of other minerals such as iron ores, bauxite, gemstones and rare earths that are being developed or yet to be developed. Africa is also rich in oil and gas, but these resources will not be examined here because they are not minerals. The value of mineral wealth that has been exploited in Africa in the past 100 years is estimated at over 10 trillion dollars.

Economic mineralization in Africa is unique in many ways, either in its exotic lithologies that are host rocks, or the very uncommon modes of their localization. For example, the Central African Copper Belt which stretches through Zambia into Congo (Kinshasa) is the largest and most prolific mineralized sediment-hosted copper-cobalt province known on Earth. The Bushveld Igneous Complex in South Africa is known as the largest and best developed layered mafic-ultramafic intrusive body in the world that is host to some of the most spectacular ore accumulations on the planet, including platinum, chromite, vanadium, copper, zinc, iron, asbestos, talc, mercury, gemstones, titanium and tin ores. The Great Dyke of Zimbabwe Is a very unique linear geological feature in south-central Africa that extends for a distance of over 500 km, and counts as a strategic economic resource with significant quantities of chrome, nickel and platinum.

The Witwatersrand Basin in South Africa, is the single largest repository of gold in the world containing about half of the world's gold reserve base, and has produced over 40,000 metric tons of gold, rich in uranium as a byproduct. The Southern Africa region also has one of the highest concentrations of kimberlite pipes in the world - a rock which is the sole source of primary and alluvial diamonds in Botswana, South Africa and Angola where 50% of global output of diamonds is located. The best quality diamond gems in the world have been sourced from African mines. How about carbonatites - the rare rock produced from carbonate magma? Africa contains over half of the world's known carbonatite intrusions - the preferred host rock for the rare earth elements which recently became critical components in high technology devices in high demand worldwide, for smartphones, digital cameras, computer hard disks, rechargeable batteries, flat screen televisions, computer monitors, and electronic displays.

Several African countries have generated substantial revenues from mining activities and mineral exports which constitute a major component of their gross domestic product (GDP) including Namibia, Ghana, South Africa, Botswana and Zimbabwe. While Africa's mineral wealth has been exploited over the past centuries by colonial governments and indigenous peoples, the continent still has a lot of resources untapped. In the last two decades, Africa has gotten a lot of attention and became a destination of choice for those international companies exploring for several minerals, particularly gold, diamonds and the valuable rare earth elements. Despite the growth in global mineral exploration and investment, large tracts of the continent are still relatively inaccessible or too remote for exploration companies. Other barriers to mineral resource development include but are not limited to a lack of financial resources, inadequate infrastructure, insecurity, illegal mining and poor governance.

While mineral resources are abundant in Africa, they are not evenly distributed across the continent. North Africa is known for its liquid fuels, phosphate and iron ore although the reserves are limited. In sub-Saharan Africa, metals and gemstones are concentrated in a few major areas, and over 75% of the continent's mineral production comes from three regions: Southern Africa for its gold, platinum, manganese, chromium, uranium, coal and diamonds, Central Africa for its copper, cobalt and gems, and Western Africa for its iron ore, bauxite, uranium and gold (and petroleum in Nigeria). Although Africa has tremendous mineral wealth, it still remains the continent with the most development challenges. According to the United Nations, over 80% of the 25 poorest countries in the world are located in Africa. It is apparent that Africa's mineral wealth is not making Africans wealthy.

## 3.13.2 Brief History of Mining in Africa

Historical records indicate that mining has a very deep root in Africa. The oldest mine ever discovered operated in Swaziland between 43,000 to 41,000 years ago (based on radiometric dating) where it was mined for hematite (ochre) - used for cosmetics and burial rights. In ancient Egypt, evidence indicated that iron ore was also mined about 6000 years ago for red hematite and black hematite (for cosmetics). Metal mining of copper and gold began in Egypt about 3500 years ago when they were used for making masks and other adornments. In subsequent centuries, the Egyptians worked gold mines in the Nubia region (south Egypt and Sudan), and gold became very valuable and was used as a major item of trade between Egypt and the former western African kingdoms of Ghana, Mali, and Songhai.

During the pre-colonial era, the fabled gold mines of south-central Africa attracted the Portuguese explorers in the late 1500s who established trading posts in today's Angola and Namibia, but failed to create any effective mining operations, and by the 1700s most European colonizers became focused on the more profitable slave trade. However, later in the 1880s after abolition of slavery, European explorers became engaged in a frantic race to conquer all of Africa and to exploit its natural wealth. The earliest known mining operations began in South Africa for copper mining in Northern Cape Province in 1854, followed by diamonds in 1867, and gold in 1870. For the next few decades, Europeans established mines throughout Africa. In south-central Africa, a chain of major copper mines stretched from what is now Zambia to present-day Congo (Kinshasa). In western Africa, colonial mining companies produced diamonds and gold in Ghana and diamonds in Sierra Leone, and tin, columbite and coal were extracted in Nigeria. In the 1950s, extraction of copper in Zambia and the DRC Congo reached very high levels followed by increasing production of iron in Liberia, bauxite in Guinea and uranium in Namibia and Niger.

Many of the colonial African countries gained independence in the 1960s, and the impact of colonial control on the mining industry continued in which mines are operated by European and some indigenous companies that pay a share of their profits and taxes to African governments. Some countries such as South Africa, Botswana Gabon and Ghana continued to make steady progress in mineral resource development while others like the DRC Congo, Liberia, Rwanda, and Sierra Leone were engulfed in civil strife, unrest and insecurity that has affected mining. However, in recent years, some countries such as Liberia, Guinea, Burkina Faso, Mali, Namibia Madagascar and Nigeria have settled down, and are attracting new investment from international companies from China, South Korea and Australia. Indigenization of mining companies is growing with occasional eruptions of labor unrest.

## 3.13.3 Geological Setting of Africa

The African continent is characterized by a geologically stable landmass of Precambrian basement rocks occupying over 70% of the land surface, and overlain by younger Phanerozoic sedimentary and volcanic cover rocks. The rocks can be divided into three main geo-tectonic domains: (a) *Cratons* (Archean to Early Proterozoic); (b) *Mobile Belts* (Mid - Late Proterozoic) and (c) *Cover Rocks* (Phanerozoic). There are four major cratons which are large masses of ancient rocks of Archean to Paleoproterozoic age that have remained stable since about 1.6 billion years (by) ago. They are the *Kalahari Craton, Congo-Tanzania Craton, West African Craton* and the *Trans-Saharan Metacraton* (Fig. 3.75). Each craton is made up of fragments of Archean (3.8-2.5 by) nuclei, often described as "shields", "blocks" or even as "cratons" surrounded or amalgamated by younger Early Proterozoic (2.5 -1.6 by) mobile (orogenic) belts. Therefore the cratons were created by a gradual process of continental growth or accretion in which new crustal materials are added to the older blocks. The cratons have remained mostly stable since 1.6 by ago.

The Archean rocks in the cratons consist predominantly of grey and banded gneisses, granite gneisses, various granitoids, and rather well-preserved and weakly metamorphosed volcanic rocks (greenstones) that are often intercalated with metasediments. The Early Proterozoic belts are characterized mostly by metavolcanic and metasedimentary rocks and granitoids in deformed orogenic belts (Eburnean progeny). Some parts of the cratons were covered by undeformed sedimentary sequences such as the Witwatersrand Basin in South Africa and the Franceville Basin of Gabon.

The cratons are separated from each other by a number of younger narrow mobile belts and sedimentary basins of Middle Proterozoic to Late Paleozoic ages, some of which were deformed by orogenic events during the Kibaran (1.2 by) and the Pan-African (900-500 my) orogenies. These fold belts, include the Mozambique belt along the east coast of Africa, the Damara and Katanga belts extending from Namibia into the Democratic Republic of the Congo and Zambia, the West Congo belt between Angola and Gabon, the Dahomey-Ahaggar belt between Ghana and Algeria, and the Mauritanide belt from Senegal to Morocco. The rocks consist mainly of metasediments and lesser amounts of metavolcanics intruded by orogenic granitoids, but also include extensive areas of rejuvenation of older rocks such as in the Trans- Sahara region. These mobile belts contain substantial deposits of copper-cobalt iron ores, manganese, nickel-copper, zinc-copper, gold, tin-tantalum and gemstones.

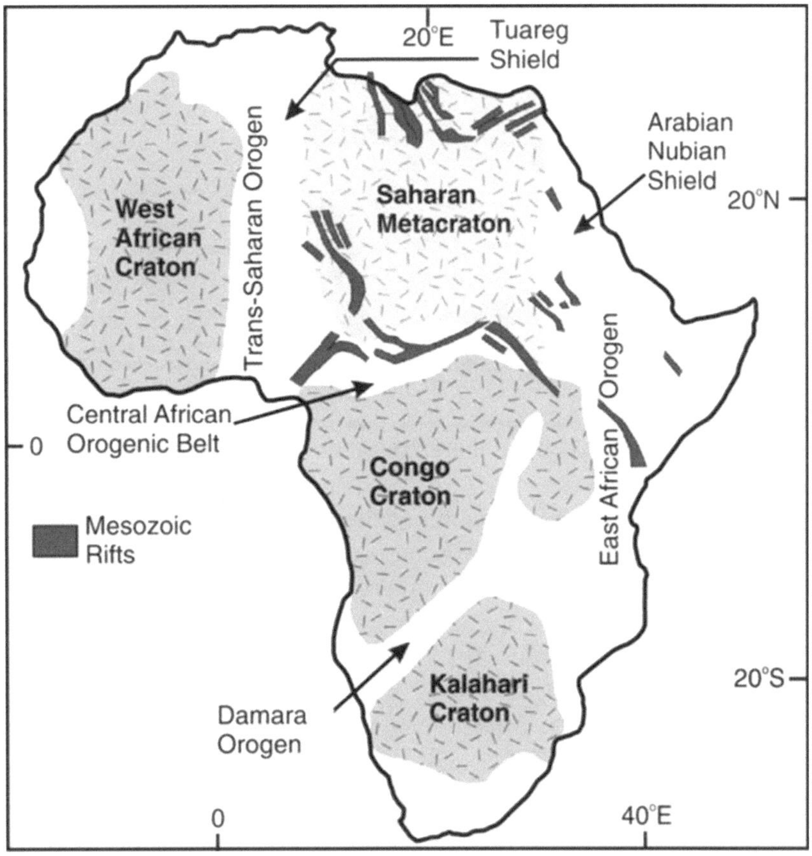


Fig. 3.75: Cratons and Mobile Belts in Africa

The Kaapvaal Craton which underlies most of southern Africa consists of the Archean Kaapvaal Shield and Zimbabwe Shield welded together by the Limpopo mobile belt, a zone of mostly reworked Archean rocks. During the Early Proterozoic, the southern part of the Kalahari craton was covered by thick sequences of supra-cratonic clastic sediments including the Witwatersrand and Ventersdorp-Transvaal basins. (Fig. 3.76) These are important host rocks for the extensive paleoplacer Witwatersrand gold and uranium deposits in South Africa, The Damara mobile of Pan African age in Namibia and South Africa separates the Kaapvaal Craton from the Congo-Tanzania Craton.

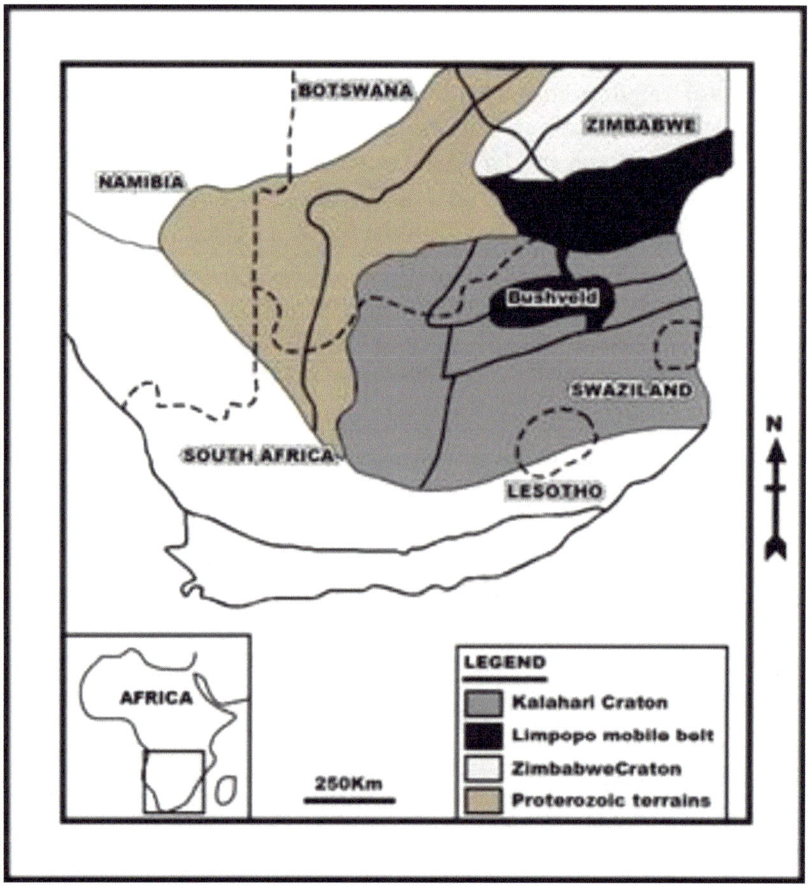


Fig. 3.76: The Kalahari Craton and its components

The Congo-Tanzania Craton (a combination of the Congo and smaller Tanzania cratons) underlies most of Central Africa is more complex because it comprises five Archean blocks welded together namely: Angola-Kasai Block, NE-Congo-Uganda Block, Cameroon-Gabon-Congo Block, Bangweulu Block and Tanzania Craton, flanked or joined by Paleoproterozoic orogenic belts (Rusizian, Ubendian and Usagaran Belts). The craton is truncated or flanked by the Mesoproterozoic (Kibaran and Irumide Belts) (Fig. 3.77). During the Late Proterozoic, several rifting events occurred along the southern margin of the Congo Craton, in the Lufilian and Zambezi Belts, with localized volcanism and deposition of clastic sequences of the Katanga Supergroup across Zambia and DRC Congo which are the host rocks to extensive copper and cobalt deposits. The Pan African orogenic belts surrounding the craton include the West Congo Belt to the west between Angola and Gabon, and the Central African Belt to the north between Congo and Cameroon-Chad and the Mozambique Belt to the east (Fig. 3.77).

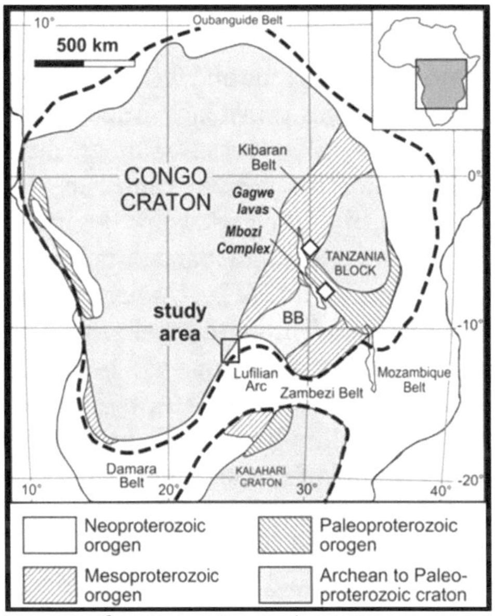


Fig. 3.77: The Congo-Tanzania Craton

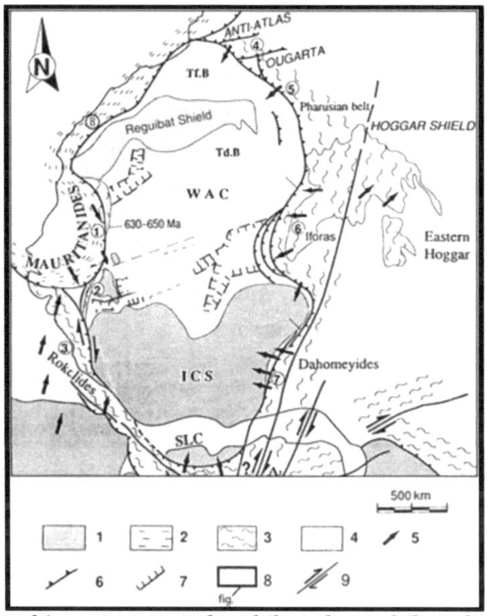


Fig. 3.78:The West African Craton and Mobile Belts Mobile Belts

The West African Craton which underlies most of west and northwest Africa comprises the Archean nuclei of the Reguibat Shield and the Man-Leo Shield that are surrounded and joined together by the Paleoproterozoic Birimian orogenic which consists of thick sequences of metavolcanic and metasedimentary rocks Fig. 3.78). Mid-Proterozoic sediments of the Tarkwaian conglomerates and quartzites lie to the east of the craton. The Dahomeyan-Ahaggar Belt to the west of Ghana and extending into Nigeria represents zones of deformed metasediments and reactivation of older rocks during the Pan-Africam Orogeny. The Birrimian rocks are the source of orogenic gold and manganese deposits in Ghana, Mali, and Burkina Faso, Tarkwaian conglomerate reefs of gold deposits in Ghana, bauxite and laterite iron deposits in Guinea and rutile and iron ore in Sierra Leone. Tin and tantalite deposits in Nigeria are associated with Pan-African granites and pegmatites.

The Trans-Sahara Metacraton is a large disjointed mass of mostly reworked basement rocks that underlies most of the Sahara and northeast Africa. The dominant rocks are Precambrian gneisses, granulites and granites covered by Phanerozoic sedimentary rocks in Niger, Algeria and Sudan.

After the Precambrian era, Africa's stable geological platform was punctuated during the Early Paleozoic by the formation of the Cape fold belt in South Africa, the Anti-Atlas belt in Morocco and the Mauritanides in Mauritania along the northeast borders of the older craton. Several broad areas of regional uplift and epeirogenic basins were developed, and the Karoo System of sediments and volcanic rocks were deposited in the Late Paleozoic as cover rocks over large areas in South Africa, Namibia, and East Africa. The breakup of Godwanaland in the Early Mesozoic led to the formation of the Atlantic Ocean accompanied by rifting and formation of several inland sedimentary basins and the deposition of thick marine and continental sedimentary sequences in North and West Africa. The Cenozoic also involved continuing sedimentary deposition in some of the inland basins, and structural deformation occurred during the Tertiary in the Atlas Mountains in the northwestern corner of the continent.

The most important geotectonic feature in Africa is the East African Rift system that stretches for 6500 km from the coast of Mozambique in the south to Turkey in the Middle East. The rift valleys were formed during the Cenozoic and are-associated with alkaline volcanics including kimberlites (source of diamond) and carbonatites (rare-earth metals), the rift system contains industrial minerals and salt lakes that contain common salt, soda-ash and potash.

## 3.13.4 Strategic Metals and Minerals of Africa

Africa is blessed with abundant mineral resources including precious metals, ferrous and ferro-alloy metals, base metals, solid energy minerals, industrial minerals and rocks, and precious gemstones. Africa seems to be favored by its unique geotectonic characteristics and the presence of extensive Precambrian igneous and metamorphic rocks. Approximately 80% of economic minerals worldwide are derived from such rocks. Africa produces more than 60 significant mineral and metal products, including what are described as "strategic minerals and metals" such as platinum, diamond, chromite, vermiculite, vanadium, iron ore, manganese, zircon, phosphate, titanium, gold, uranium, tin-tantalum, fluorspar, bauxite, copper and cobalt. It is the world's leader in the production and reserves of platinum, diamond, chromite cobalt, manganese, phosphate and gold. Africa's mineral wealth is exhaustive, and only those that are unique will be described in the following sections. Figure 3.79 is a distribution map of mineral resources in Africa. The centers of mineral concentration are in southern, eastern and western Africa, while large areas in the Trans-Sahara regions of Algeria, Chad, Libya, Sudan and Egypt are relatively devoid of metallic minerals, although some of those countries are rich in petroleum resources. The statistics of production used for the various metals and minerals are based on annual reports published by the United States Geological Survey and from Statista.com. Table 3.28 provides a summary statistics of global and African production and reserves as well as source countries in Africa.

### 3.13.4.1 Precious Metals and Minerals

The precious metals produced in large quantities in Africa are gold and platinum. Silver is only produced in small quantities as byproduct. Diamond is a very vital precious stone produced in Africa.

AFRICA'S MINERAL RESOURCES

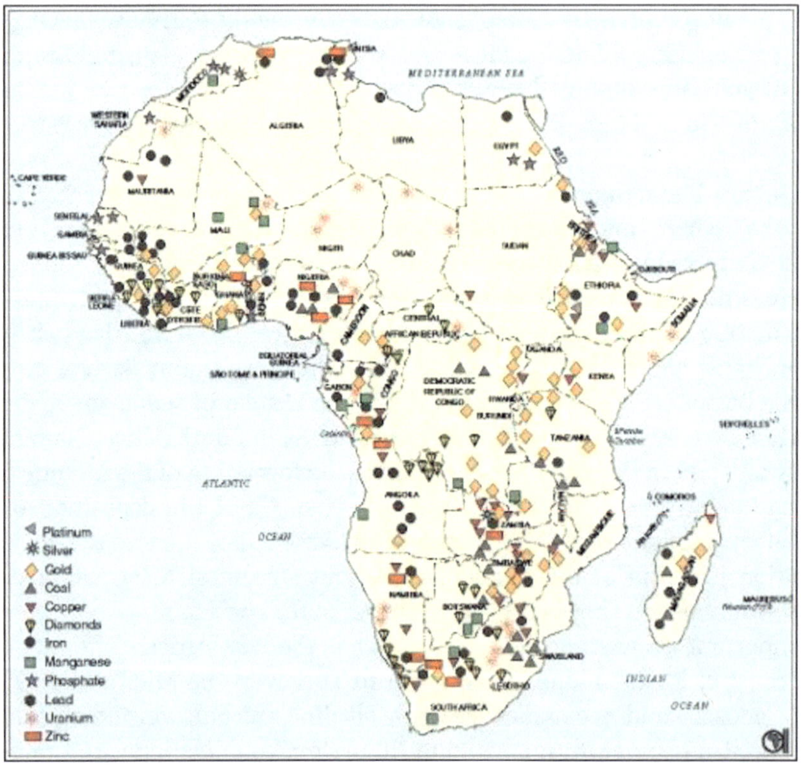


Fig. 3.79: Distribution of Mineral Resources in Africa

## Gold

Since the times of the Egyptian Pharaohs 3500 years ago, gold has been a commodity of value worldwide for use in jewelry and adornments. Gold is in very high demand worldwide, and global production in 2020 was estimated at 3250 metric tons with China (12%), Australia (10%), Russia (9%), USA (6%) and Canada (5%) as leading producers. Africa is the leading continent in gold output, contributing over 16% of world production in 2020, with North America 2$^{nd}$ (14%). Prior to 2007, South Africa was the global leader in gold production, but output has fallen steadily over the years from about 1000 tons in 1970, to 400 tons in 2001 and 120 tons in 2020. Grades remained the same at about 5 g/t. Gold production in Africa was about 512 tons in 2020 with over 60% from the West African region; from Ghana (28%), South Africa (23%), Mali (12%), Tanzania (10%), Burkina Faso (9%), Ivory. Coast (6%), Zimbabwe (5%), Guinea (3%), Senegal (2%), Nigeria (2%). About 45% of African gold production is attributed to artisanal and small-scale miners. African reserves of gold are not certain, but estimates indicate about 8000 metric tons accounting for 20% of global reserves. African countries export over 95% of their gold output mostly to Europe, Middle East and Asia. Gold price in 2020 was $56 million per metric ton compared to $8.5 million in the year 2000.

Gold mineralization in Africa occurs predominantly in Precambrian rocks and alluvial deposits derived from them. The four types of deposits are (a) orogenic gold; (b) sedimentary gold in pebble conglomerates; (c) BIF gold and (d) alluvial deposits. Orogenic gold is the most prevalent, and deposits have been exploited in the cratons in southern, eastern, and western Africa. Gold in banded iron formations is common in the Kalahari Craton in South Africa,

The Witwatersrand basin in South Africa is the world's largest repository of gold deposits that occur as paleoplacer deposits in Proterozoic pebble conglomerates. First discovered in 1867, over 40,000 metric tons of gold have been extracted from the Witwatersrand basin that still contains substantial gold reserves located deep (.3000 m) within the basin. In Ghana, gold reefs are found within the Mid-Proterozoic Tarkwaian pebble conglomerates east of the West African Craton.

## Platinum

Platinum is the world's second most expensive metal belonging to a group called the platinum group metals (PGM) consisting of iridium, osmium, palladium, platinum, rhodium, and ruthenium. The main uses for platinum are in jewelry, and electrical and dental industries. PGMs are indispensable in many industrial applications including in computer hard disks, mobile phones and glass, among others. Platinum alloys are used in capsules for radioactive power sources.

Africa dominates the world's production of platinum and palladium with major contributions from South Africa and Zimbabwe. They have a combined 85% of world production of platinum and 90% of world's reserves estimated as 70,000 tons. Other platinum producing countries are Botswana, Tanzania, Uganda, Gabon and Mozambique.

South Africa currently produces about 130 metric tons of platinum annually, representing about 72% of the platinum produced globally. It also produces 38% of the world's palladium, second only to Russia. South Africa is the only place in the world where platinum ores are mined directly from rocks. Elsewhere, most platinum is mined as a by-product of the refining of other metals, mostly nickel. All of South Africa's platinum ores are derived from the Bushveld Igneous Complex (BIC) that hosts the world's largest reserves of platinum and chromite. Within the BIC, platinum is located in two principal platinum layers or "reefs": the *Merensky reef* and the Upper Group 2 *UG2 chromitite reef*, both of which are about 1 m thick and occur consistently around the eastern and western sides of the BIC. Ore grades of the reefs are typically between 4-7 g/t. Zimbabwe produces about 15 metric tons annually with reserves of 1200 tons located within the Great Dyke. The price of platinum is $38:000 per kg ($38 million per ton).

### Diamond

Diamond is the most highly valued and popular gem that has been used for jewelry and adornments since historic times. Global production of natural diamonds from mines in 2019 was estimated as 137 million carats (Mc) with worldwide reserves of 1.2 billion carats. Africa is the world's largest producer of diamonds, producing as much as 50% of global production. Six of the seven leading producers of gem quality diamonds in the world are located in Africa, namely: Botswana, Angola, South Africa, Democratic Republic of the Congo, Namibia and Lesotho. They collectively produce over fifty million carats of diamond annually, and own some of the largest diamond mines in the world. In general, a 1 carat diamond costs between $1,800 and $12,000 depending on cut quality. In 2019, the amount of natural diamond produced in Africa was estimated as 67.5 million carats. To date, Africa has produced over 75% in value of the world's diamonds with more than 2.5 billion carats mined. In 2017, diamond exports from African countries were valued at $9.65 billion dollars in the global market. The USA is the world's top buyer of diamond jeweler, purchasing more than 40% of all stones.

Diamond mining in Africa is centered mostly on South Central Africa with minor contributions from West Africa. The major producing countries are Botswana (36%), DRC Congo (28%), Angola (15%), S. Africa (10%), Namibia (3%), and Zimbabwe (3%) Lesotho (2%), Sierra Leone (1%) Tanzania (1%) and others (1%). Most of the diamonds are produced directly from kimberlites and from alluvial deposits by dredging in Angola, Namibia and South Africa. Botswana stands out as the world's second-leading producer of diamonds on the basis of carat weight, and the leading producer on the basis of value, assessed on the larger size and higher quality of its diamonds compared to world leader, Russia. Botswana's Jwaneng mine is often referred to as "the richest diamond mine in the world.", and the mine in recent years has been producing about 15 million carats of high-quality diamonds per annum. South Africa which was once the world's leading producer of diamonds but now faced with dwindling resources, is still a major player in the gem trade with production of a few million carats annually.

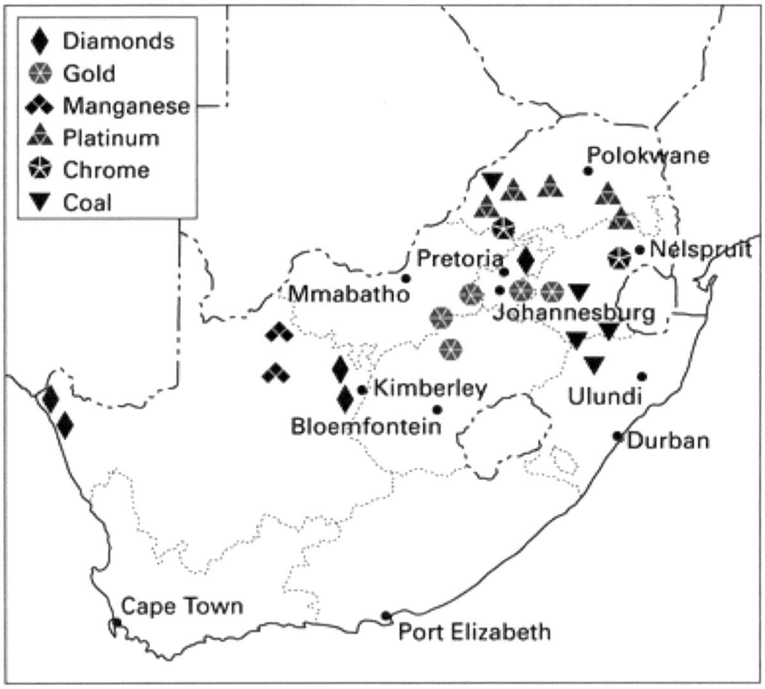


Fig. 3.80: Distribution of Precious and Ferro-Alloys, South Africa

Table 3.28: Summary Statistics on Mineral Reserves and Production in Africa

| Metal | Africa Reserves | World Reserves | Africa Production | World Production | Production by Country |
|---|---|---|---|---|---|
| Gold | 8000 tons | 36,000 tons | 516 tons | 3,250 tons | Ghana (28%), S. Africa (23%), Mali (12%), Tanzania (10%), B. Faso (9%), I. Coast (6%), Zimbabwe (5%), Guinea (3%), Senegal (2%), |
| Platinum | 64000 tons | 70,000 tons | 145 Mt | 170 | South Africa (90%), Zimbabwe (10%) |
| Iron Ore | 9 Bt | 180 Bt | 100 Mt | 2.5 Bt | South Africa (77%), Mauritania (13%), Sierra Leone (5%), Liberia (2%), Algeria (2%) |
| Chromium | 380 Mt | 800 Mt | 16.5Mt | 40 Mt | South Africa 9(67%), Zimbabwe (3%) |
| Cobalt | 3.8 Mt | 7 Mt | 109,000 | 140,000 | DR Congo (91%), Madagascar (3%), Zambia (3%), South Africa (2%) |
| Manganese | 240 Mt | 600 Mt | 9.8 Mt | 18Mt | South Africa (53%), Gabon (28%), Ghana (14%), Ivory Coast (4%) |
| Titanium | 150 Mt | 770 Mt | 2.6 Mt | 9.2 Mt | South Africa (35%), Mozambique (23%), Kenya (12%), Madagascar (11%), Senegal (11%) Sierra Leone (5%) |
| Diamond | 800 Mc | 1,200 Mc | 67 Mc | 137 Mc | Botswana (37%), Dr Congo (28%), Angola (15%), S. Africa (10%), Namibia (3%), Zimbabwe (3%) Lesotho (2%), Others (2%) |
| Bauxite | 75.5 Mt | 370 Mt | 9 Bt | 30 Bt | Guinea (94%), Sierra Leone (3%), Ghana (2%) |
| Copper | 40 Mt | 870 Mt | 2.3 Mt | 20 Mt | DR Congo (56%), Zambia (38%), South Africa (4%) |
| Uranium | 1.2 Mt | 6.14 Mt | 8,850 | 54,000 mt | Namibia (62%), Niger (34%), South Africa (4%) |
| Tantalum | 50,400 | 315,000 tons | 1420 | 1800 tons | DRC Congo (50%), Ewanda (35%), Nigeria (10.5%), Ethiopia (5%) |
| Fluorspar | 42 Mt | 310 Mt | 600,000 | 7 Mt | S. Africa (38%), Namibia (21%), Kenya (20%), Morocco (18%) |
| Phosphate | 36 Mt | 240 Mt | 56 Bt | 69 Bt | Morocco (60%), Egypt (8%), Tunisia (5%) South Africa (3%), Senegal (2.6%), Algeria (2%) Togo (1.3%) |
| Vermiculite | 20 Mt | 50 Mt | 240,000 | 500.000 | S. Africa (35%), Zimbabwe 12%), Uganda (12%), Egypt (3%) and Kenya (5%). |
| Zircon | 15 Mt | 74 Mt | 410,000 | 1.4 Mt | S. Africa (85%), Mozambique ( ), Kenya |
| Coal | 1.2 Bt | 1055 Bt | 290 Mt | 6.5 Bt | S.Africa (95%), Mozambique, (2.5%), Zimbabwe (1.2%), Botswana (1%) |

### 3.13.4.2 Ferrous and Ferro-Alloy Metals

Iron and the ferro-alloy metal (Cr, Mn, Co, Ti, and V) are the main raw materials for the iron and steel industry which has formed the backbone for the development of several **industries** in the global economy: including the defense industry, transportation and heavy engineering, energy and construction (including aeronautical and shipping construction industries. Africa has plentiful iron and steel metals necessary for industrial and economic development.

### Iron Ore

Primary iron (Fe) is the most significant metal on Earth and accounts for about 95% of the world's metal production... The term "iron ore" is used to connote an aggregate of iron minerals that are of commercial value containing an average 62% Fe. Iron Ore (Fe) is mined in about 50 countries worldwide and used to manufacture a variety of steel products and building infrastructure. Global economic growth is the primary factor that drives its supply and demand of iron ore. Iron ore production worldwide in 2020 was estimated as 2.5 billion tons (Bt), in which Africa produced 100 million tons. Global reserves of iron ore is estimated as 180 billion tons with Africa accounting for 5% of world reserves in South Africa, Liberia and Guinea.

Africa is blessed with substantial resources of iron ore, although several countries have vast deposits that are yet to be developed, such as Guinea and Liberia. Iron ore production in Africa is dominated by South Africa (77 Mt) and Mauritania (13 Mt). Other countries with significant production are Sierra Leone (4.7 Mt), Liberia (1.9 Mt) and Algeria (600,000 tons). Iron ore deposits in Africa are mainly of four types: (a) Banded iron formation; (b) Sedimentary Ironstones; (c) Lateritic iron ore, and (d) Magmatic deposits. Africa's most significant iron reserves are to be found in Western and Southern Africa. The dominant type of iron ore are the BIFs which are found within the Precambrian terrains. Currently, the bulk of iron ore mined in South Africa comes mainly from banded iron-formation within Proterozoic Transvaal sedimentary rocks in the Northern Cape Province which produces 88%, and the Limpopo Province (12%). The Bushveld Igneous Complex contains magnetite layers that are rich in titanium oxide, but the magnetite is more expensive to smelt. There are also vast low-grade deposits of Precambrian banded and oolitic iron formations in Pretoria.

The Precambrian terrains of western Africa contains extensive lateritic metamorphosed iron formations (itabirite).The most significant deposits are in Liberia in the Bomi Hills, Bong and Nimba ranges, and its extension into Guinea of the Nimba–Simandou Ranges, where high-grade hematite ores have been located. In North Africa, limited reserves of iron ore are concentrated in the Atlas Mountains and in the Western Sahara. Egypt, however, has medium-grade reserves, of which the most important are at Al-Baḥriyyah Oasis. The ore deposits in Morocco and Tunisia, which were once of considerable importance, have been severely depleted. Iron ore is a major raw material for steel production. Current price of iron ore is about $180 per ton.

### Chromium (Chromite)

Chromite is the main ore of chromium (Cr) and is directly obtained as mineral concentrate of chrome ore in mining operations. Chromium content of chromite ranges from 34% to 42%. Chromium is a very important metal, and 70% of its production is used as an alloying agent in the manufacture of corrosion-resistant stainless and hardened steel. World production of chromite in 2020 was 40 million tons. The five leading producers were South Africa with 16 Mt (40%), Turkey (16%), Kazakhstan (15%), India (10%) and Finland (6%). South Africa and Zimbabwe are the major producers of chromite ore in Africa. Chromite reserves in South Africa is 295 Mt which represents 42% of global reserves of 800 Mt. Zimbabwe produced 500,000 tons of chromite, with estimated reserves of over 80 million tons. Madagascar and Sierra Leone are the other countries in Africa with a significant production.

Chromite deposits in Africa occur within the two best known mafic layered intrusions: the Bushveld Igneous Complex of South Africa and the Great Dyke of Zimbabwe. Chromite occurs in the BIC as layers containing over 90% chromite with varying thickness ranging from a few centimeters to more than 2 m. twenty mines are exploiting chromite in the BIC. About 25% of the chrome ores are exported while the rest are used for domestic production of ferrochrome, an alloy of iron and chrome in local smelters. In the Great Dyke, chromite is found as seams or stratiform deposits within the mafic intrusion, Alluvial chrome deposits originating from the intrusion have accumulated in local stream valleys as thin beds that have yielded > 50% Cr. Minor production is associated with alluvial deposits derived from mafic intrusions in Sierra Leone and Madagascar. The price of refined chromium ranges from $7000 - $8000 per ton while chrome ore is about $150 per ton.

### Cobalt

Cobalt is considered as a strategic metal used in the production of corrosion-resistant and heat-resistant super-alloys, such as those used in military equipment and jet engines. Cobalt is also a component of rechargeable batteries. World production of cobalt in 2020 was 140,000 tons. Africa yields most of the global cobalt production, The Democratic Republic of Congo (DRC) is by far the world's largest producer, with annual production of 100,000 tons accounting for 70% of global production. Cobalt reserves of DRC are estimated as 3.6 million tons which is more than half of the global reserves of 7 million tons. Madagascar, Zambia and South Africa are the other African countries with some cobalt production in 2020 (3,500 3000 and 2600 tons respectively).

Cobalt is associated with copper as stratabound disseminated deposits in the Proterozoic sediments (mostly shales) of the Katanga Supergroup in the Central African Copperbelt that stretches for 800 km from Zambia into the Democratic Republic of Congo and Angola. Cobalt which occurs mainly as a sulfide mineral - carnotite ($CuCo_2S_4$) - is extracted as a byproduct of copper mining. The Copperbelt is one of the world's most prolific metallogenic provinces containing over 50% of the world's cobalt reserves with an estimated resource base of 5 million tons of 0.4% Co. Cobalt production in Madagascar and South Africa is in small quantities derived from the smelting of nickel ores associated with mafic intrusions.

Global demand for cobalt in jet aircraft and rechargeable (lithium-cobalt) batteries has continued to surge and is expected to double between 2020 and 2025 to 117,000 tons. China is the leading consumer of cobalt, driven by the demand from its rechargeable battery industry. Current price of cobalt is $80,000 per metric ton.

### Manganese

Manganese is a critical metal that is considered indispensable in the metallurgical production of steel because of its ability to remove unwanted oxygen and sulfur from iron ore melt. Manganese also serves as an alloying agent in steel and non-ferrous alloys of copper, aluminum, and nickel where it helps improve the strength, toughness, stiffness, hardness, and wear resistance in steels. Other uses of manganese include the production of chemicals and a component of alkaline manganese batteries.

Africa is the world's largest producer of manganese from several localities within the Precambrian terrains of southern and western Africa. Global production of manganese in 2020 was estimated as 18 million tons. South Africa is by far the world's largest producer of manganese with a production of 5.2 million tons in 2020 and reserves of 200 million tons. Other major producers of manganese in Africa are Gabon 3rd in the world with 2.8 Mt, Ghana 4th in the world with 1.4 Mt, and Ivory Coast (400,000 tons). Some manganese occurrences have been mined in Nigeria, Rwanda, Burkina Faso and Namibia. Africa is host to over 35% of the world's known land-based ore reserves of manganese, and produced about 54% of the 18 million tons (Mt) output in 2020.

Manganese deposits in Africa are associated mostly with lateritic weathering of Precambrian manganiferous sediments. The well-known South African manganese deposits occur mainly within the BIF-hosted Kalahari Manganese Field (KMF) where Proterozoic sedimentary banded iron-formation are interbedded with carbonate-rich manganese-bearing sediments enriched by lateritic weathering.

Manganese deposits in Gabon are mainly lateritic concretionary manganese ore developed on relatively unmetamorphosed and undeformed manganese-bearing black carbonaceous shales. The Mouanda deposit in southeastern Gabon is thought to be among the largest in the world. In Ghana, manganese is found as lateritic caps of high-grade oxide ores associated with chemical weathering of metamorphosed manganiferous sediments (phyllites) in the Proterozoic Birimian Supergroup. Elsewhere in western Africa, manganese deposits are found in Burkina Faso and Ivory Coast.

China is the main consumer of manganese for its steel production which is half of the world's output. The global demand for manganese is therefore overwhelmingly driven by steel output, accounting for well over 90% of global manganese consumption. While the production of steel may fluctuate, the demand for manganese for producing batteries is expected to grow considerably in the next decade. Current price of manganese ore is about $4.5 per metric tons.

### Titanium

Titanium (Ti) is a hard metal that is lightweight, high-strength and high corrosion resistant. It is used as a structural metal and alloy, and also converted to titanium dioxide powder which is the form in high demand by manufacturers. Some 94% of world demand for titanium is in the form of titanium dioxide powder, and about 6% as metal and metal alloys. The principal uses for titanium dioxide are as pigments for whitening, high quality paints, catalysts, ceramics, plastics, coated fabrics and textiles. Titanium metal and alloys are used mostly in aerospace applications. Ilmenite (FeTiO3) supplies about 90% of the world's demand for titanium, and the remaining 10% is derived from rutile (TiO2). World production of titanium ore in 2019 was approximately 8 million tons and the leading producers were China (30%), followed by South Africa (11%) and Australia (10%). World reserves are estimated as 820 million tons of combined ilmenite and rutile.

Africa is a major producer of titanium ore derived from beach placers along the shorelines of the following countries: South Africa, Mozambique, Sierra Leone, Senegal, Kenya and Madagascar. South Africa is the largest producer with over 820,000 tons of ilmenite and 110,000 tons of rutile. These economic minerals are derived from the extensive beach placer deposits located along the eastern, southern and northeastern coasts with smaller deposits along the west coast, north of Cape Town. South Africa holds about 18% of world titanium reserves. Mozambique, is host to the world's largest titanium placer deposit located 160 km from the city of Nampula. Annual production in 2019 was estimated at 600,000 tons of ilmenite and 10 tons of rutile. These deposits are unconsolidated which makes extraction relatively easy and profitable. Production from other African countries (Kenya, Madagascar, and Senegal) are a combined total of over 800,000 tons of ilmenite, while Sierra Leone produced 120 tons of rutile.

### Vanadium

Vanadium (V) is a soft metal used mostly as a ferroalloy for high speed tool steel and high-strength steel, and wear-resistant cast iron. Vanadium improves properties of ferrous alloys when it is present in small amounts. In 2019, approximately 90% of vanadium was recovered from magnetite and titano-magnetite ores, either from co-production or primary production. Global output of vanadium in 2019 was 73,000 tons, with contributions from only five countries namely; China (40,000 tons), Russia (15,000 tons), South Africa (8,000 tons), Brazil (7,000 tons) and the United States (500 tons).

South Africa is the only producer of vanadium in Africa and the third largest in the world. Vanadium is extracted as a byproduct of iron and titanium mining in the Bushveld Igneous Complex. Of the 38 million tons of vanadium reserves in the world, 12 million (32%) are located in the Bushveld Complex. The four vanadium mining operations in the Bushveld region contribute 25% of global vanadium production.

### 3.13.4.3 Base (Industrial) Metals

#### Aluminium (Bauxite)

Bauxite is the major ore of aluminum. Aluminium is remarkable for its low density and its ability to resist corrosion. Its alloys are ideal for use in aircraft, automobiles, beverage containers, and important in transportation and building industries. Of all the bauxite mined worldwide, about 85% is converted to alumina ($Al_2O_3$) for the production of aluminum metal, 10% for non-metal uses as various forms of specialty alumina, and the remaining 5% is used for non-metallurgical applications. Bauxite occurs predominantly within tropical and subtropical regions where intense lateritic weathering has accumulated vast deposits in or near the Earth's surface. The principal aluminum hydroxide minerals found in bauxites are gibbsite and the polymorphs boehmite and diaspore.

World bauxite production in 2020 was 370 Mt with the following five leading producers: Australia (30%), Guinea (20%), China (16%), Brazil (9%) and India (7%). Known global reserves are 30 billion tons located primarily in Guinea (25%), Australia (20%), Vietnam (12%), Jamaica (9%), Brazil (7%), and China (4%), Africa produced some 75 million metric tons of bauxite in 2020, and .contains some of the world's largest resources of bauxite, much of which are yet to be developed.. The four African countries that are actively mining bauxite with the following annual production are Guinea (72 Mt), Sierra Leone (2 Mt), Ghana (1.5 Mt) and Mozambique (10,000 tons).

Bauxites in Africa are of lateritic origin accumulating on elevated plateaus along a Tertiary erosion surface. In Guinea, bauxite forms blanket deposits overlying a series of Paleozoic argillite, siltstones and minor sandstone intruded by dykes. Alumina content varies from 40% to as high as 52% with an average of 46% $Al_2O_3$, and very low $SiO_2$ contents. The lateritic deposits in Ghana rest on phyllites, schists and metamorphosed lavas, greywackes and tuffs of the Proterozoic Birimian Supergroup and the Voltaian sediments. The Sierra Leone deposits located in the southern part of the Freetown Peninsular are products of weathering of Precambrian metasediments and dolerite dykes of the Kasila Group. Africa has extensive resources of bauxite particularly in Guinea and Ghana that have a promising future for development. Current price of crude-dry bauxite is $32 per ton, and $497 per ton for metallurgical grade alumina. Pure aluminium sells for about $2300 per ton.

#### Copper

Copper (Cu) and its alloys are the most widely used metals for industrial applications mainly by the wire and cable industry, taking advantage of its properties such as the electrical conductivity, corrosion resistance, and thermal conductivity. Copper occurs mostly in sulfide form, often in combination with other metals. Global production of copper in 2020 was about 20 million tons of which the five largest producers were Chile (5.6 million tons) (Mt), Peru (2.4 Mt), China (1.6 Mt), Democratic Republic of Congo (1.3 Mt) and the United States (1.3 Mt). Africa is a major world producer of copper contributing about 12% of world production derived mainly from three countries; Democratic Republic of Congo 56% (1.3 Mt), Zambia 38% (880,000 tons) and South Africa 4% (82,000 tons). Minor production accrues from countries in northwest Africa (Mauritania and Morocco), and as a byproduct of metal mining in Botswana, Namibia, and Zimbabwe.

Copper deposits in DRC Congo and Zambia are stratabound in Proterozoic sediments of the Katanga Supergroup in the Central African Copperbelt which contains one of the greatest concentrations of high-grade (3.5% Cu) copper resources in the world, with a resource base of over 150 million tons. Copper production in South Africa comes from the mining of a unique and rare deposit formed within a carbonatite pipe at the Palabora Mine. It is the leading producer of refined copper in South Africa, and accounts for 75% of the country's total copper production. Its open pit mine is one of the largest in the world, and production has transitioned into underground mining. Other sources of copper in South Africa are the polymetallic veins of Pb-Zn-Cu in the Black Mountain mine of Northern Cape Province and the Bushveld Igneous Complex where copper is produced as a byproduct of platinum mining in the Merensky Reef. Current price of refined copper is about $9,000 per ton.

**Tantalum**

Tantalum is a very hard, heavy and ductile metal that is highly resistant to corrosion. It is very conductive to heat and electricity and renowned for being nearly immune to chemical attack at room temperatures. Tantalum is considered a critical metal needed for applications in the electronics industry, in capacitors for medical appliances, mobile phones and in computer hard drives of which 35% is used by mobile phone manufacturers. Consequently, the global demand for tantalum has increased considerably since the mid-1990s. Global production of tantalum in 2020 was estimated at 1800 metric tons. The major sources are Central Africa (66%), Nigeria (18%), Brazil (6%) and Australia (5%), Africa is the leading producer with production from the following countries: DRC Congo, 700 tons (50%), Rwanda, 500 tons (35%), Nigeria, 150 tons (10.5%) and Ethiopia, 70 tons (5%). Other minor producers are Burundi, Uganda and Mozambique. World reserves are estimated as 315,000 tons with Africa being responsible for 16% - second only to South America. Tantalum in Africa is derived mostly from Precambrian granitic pegmatites and associated alluvial and eluvial deposits. Current price of tantalum oxide $Ta_2O_5$ is about $180,000 per metric ton.

### 3.13.4.4 Solid Energy Minerals

**Coal**

Coal is a combustible black sedimentary rock with high content of carbon and hydrocarbons. Coal is a solid fossil fuel and major energy source that is used worldwide. Coal is also an "industrial rock" that is used in the production of chemical fertilizers and several chemical products. Of all the fossil fuel sources, coal has become the fuel of choice because it is reliable, easily affordable and widely available. Globally, coal is a valuable commodity that is predominantly used to generate electricity, and also used by steel industries to extract iron from iron ore and in the production of cement. There are four types or rank of coal based on their degree of coalification (carbon content): anthracite (86-97% C), bituminous (45-86% C), semi-bituminous (35-45% C) and lignite (brown coal) (25-35% C).

Internationally, coal is the most widely used primary fuel, accounting for about 36% of the global total primary energy (electricity) consumption. However, the burning of coal in electric power plants is a major source of carbon dioxide ($CO_2$) emissions with impact on global warming. This has recently led to the decline in coal consumption relative to renewable energy sources. Nevertheless many countries still use a lot of coal, and even the United States still uses coal to generate more than 50% of its electricity. The growing implementation of clean coal technologies will also accelerate the use of coal as the primary source of electric power, leading to economic development, social progress and a higher quality of life: the economic miracles of the 21st century

Worldwide coal production in 2019 was about 7.4 billion tons (BT). China is the largest producer with 47% of global production and approximately 13% of the world's total reserves. Based on current estimated 1.1 trillion tons of proven coal reserves worldwide. Most of the reserves are anthracite and bituminous coal (70%).

Coal production in Africa accounts for about 5% of world production compared n to Russia (30%), North America (25%) and Asia/Australia (29%). South Africa is the leading producer with annual output of 278 Mt in 2019 representing over 95% of coal produced in the continent. Other countries with substantial production are Mozambique, 6.8 Mt (2.5%), Zimbabwe 3 Mt (1.2%), Botswana 2 Mt (.07%), Tanzania (304,000 tons), Niger (272,000 tons), Zambia (139,000 tons), and Nigeria (51,000 tons). In 2019, reserves of coal in South Africa amounted to approximately 9.9 billion metric tons, and almost 80% of the energy needs are met by use of coal because South Africa has no significant petroleum resources. Also, in addition to the extensive domestic use of coal, South Africa exports about 28% percent of its production as coal gas making it the fourth-largest coal exporting country in the world.

The largest coal deposits in South Africa occur within the sediments of the Ecca Group in the Paleozoic Karoo Basin. In recent years, Africa's energy needs are growing and this has led more countries, such as Botswana, Mozambique, Tanzania, Niger and Nigeria to turn to coal-powered plants. Zimbabwe has become the only other country actively coal mining in Africa on a large scale, with around 0.1% of the global reserve. Botswana has over 200 billion tons of coal reserves and the development of the coal industry has become a major priority. Mozambique is also expected to experience a sharp increase in coal production. Despite its oil wealth, Nigeria is using coal to generate power for cement and other industrial production.

### Uranium

Uranium is a radioactive element that has several applications in the military for its use in nuclear weapons, and for civilian use in power generation by nuclear reactors. Global production of uranium in 2019 was approximately 54,800 tons, the leading world producers were Kazakhstan (42%), Canada (13%), Australia (12%), Namibia (10%) and Niger (6%). Global uranium reserves recoverable at US$130/kg were estimated as 6.14 million tons. Australia has the largest reserves accounting for 28% and Africa for 17% of global reserves.

Uranium production in Africa in 2019 was a total of 8,850 tons (16% of world output), from three countries; Namibia (62%), Niger (34%) and South Africa (4%). Other African countries with uranium occurrences are Gabon and Nigeria. Namibia's annual production of 5500 tons is derived mostly from the Rossing uranium deposit which is localized within Precambrian pegmatites and alaskites. It is one of the largest known uranium ore deposits in the world, and is unique in its supposed metamorphic rather than igneous origin. Niger Republic produced 3000 tons in 2019 from several Paleozoic-Mesozoic sandstone-hosted (tabular and channel type) deposits around Arlit in central Niger. Uranium in South Africa is a byproduct of gold production in the Precambrian pebble conglomerates of the Witwatersrand Basin, and from sandstones of the Karoo Basin. Uranium which was discovered in the Karoo Basin in the 1960s is being mined from the Beaufort West district of the Karoo Province. South Africa's production has declined from 1000 tons in 2001 to about 350 tons in 2019. Current price of uranium is about $80,000 per metric ton.

## 3.13.4.5 Industrial Minerals and Rocks

Several industrial minerals and rocks are produced in Africa, and the following few selected for description are those in which Africa is a major or distinguished producer.

### Asbestos

Asbestos is a generic name for six naturally occurring asbestiform or fibrous minerals that belong to either the serpentine or amphibole group, namely: chrysotile (serpentine) and the other five amphiboles: crocidolite, amosite, tremolite, actinolite and anthophyllite. Chrysotile, crocidolite and amosite are the most common forms for commercial production. Chrysotile (white asbestos) is by far the most abundant and widely distributed of all the asbestiform minerals. Asbestos with its fibrous texture, has certain industrial properties that include excellent electrical insulation and high heat resistance.

For several decades, asbestos was widely used in the manufacture of building materials such as roofing and ceiling sheets, heat insulation, vehicle gaskets and brake linings, and in common household products. When asbestos became known as a potential health and safety hazard in the 1980s, its use and production were either banned or discouraged in several countries in Europe, USA and Southern Africa, However, it continues to be used in large quantities around the world and is still an important commodity in global trade. An estimated 1.5 million tons of asbestos continue to be used around the world each year, although this is significantly less than peak annual consumption of nearly 5 million tons three decades ago. India, China, Russia, and some developing countries are the major consumers of asbestos.

Production has declined considerably from about 4.3 million tons in 1979 to 1.3 million tons in 2019. Global reserves are estimated at over 250 million tons with largest reserves in Russia that holds 45% of global reserves. Russia is also the largest producer with 53% of the world's total, followed by Kazakhstan (16%), China (15%), and Brazil (11%). Asia consumes some 70% of the asbestos produced in the world with China, India and Indonesia as the largest consumers. In the past few years, despite the international campaign against asbestos, Russia has continued to increase its production of chrysotile asbestos, and even imported over 750 tons to the USA where it was not legally banned but with the permission of the Trump administration.

Fig. 3.81: Major Deposits of Some Industrial Minerals in Africa

In Africa, the leading producers of asbestos were South Africa and Zimbabwe. Until 2009 when asbestos mining was banned in South Africa, it was the world's leading producer of amphibole asbestos (amosite, crocidolite) and some chrysotile from mines in northeastern Transvaal and Northern Cape Province. At its peak in 1997, asbestos production in South Africa was about 380,000 tons, Zimbabwe 265,000 tons and Swaziland 25,000 tons. Currently, Zimbabwe is the only country that mines chrysotile asbestos, with annual production of 10,000 tons. The reserves in South Africa are estimated at 7.8 Mt, being the third-biggest in the western world after Canada and the USSR. Zimbabwe has reserves of about 5 million tons. Asbestos production value in Africa has been downgraded because of health reasons particularly in South Africa where the variety of asbestos produced (crocidolite and amosite) pose the highest health risk because of their short fibers. Chrysotile asbestos is still a resource that can be carefully extracted with minimal risk to health.

### Fluorspar (Fluorite)

Fluorspar or fluorite (CaF2) is an industrial mineral used for its chemical content mostly as a fluxing agent for smelting in the steel industry, and as additive in the ceramic production of decorative glassware and enamels. The purest grades of fluorite are a source of important industrial chemicals; hydrogen fluoride (HF) and hydrofluoric acid (70% HF), which are the intermediate source of most fluorine-containing fine chemicals, such as CFCs (chlorofluorocarbons), HFCs and HCFCs. These are used as industrial solvents, refrigerants, aerosol propellants and plastic foams. Global production of fluorspar in 2019 was 6 million tons, and China was by far the largest producer at 4 Mt (66%), followed by Mexico (1.2 Mt).

Africa is a major producer of fluorspar from several countries in south-central Africa and northern Africa accounting for over 10% of global production and 20% of the world reserves. South Africa is the continent's leading producer with 240,000 tons, followed by Namibia (132,000 tons), Kenya (130,000 tons) and Morocco (110,000 tons). Minor occurrences are known in Nigeria and Algeria. South Africa has the highest fluorspar reserves in the world accounting for 17.8%, followed by Mexico (13.9%). Fluorspar deposits in Africa are found in a variety of geologic settings including disseminations, stockworks and veins in silica-rich granites and alkaline intrusive rocks and carbonatites.

In South Africa, major deposits are located within the granites of the Bushveld Igneous Complex and some alkaline intrusives near Pretoria and Pilanesberg. Stockworks and veins of fluorspar in granite are mined extensively in South Africa, Fluorite veins in sediments are worked in South Africa, Morocco and Nigeria. Carbonate -replacement (manto} fluorspar deposits in limestones are very productive in South Africa, Kenya, and Morocco. Pipe-like bodies of fluorspar have been worked in South Africa and Namibia. Fluorite-bearing carbonatites are reported from Angola, Kenya, Morocco, Mozambique, Namibia, South Africa, Tanzania, and Uganda. Current price of fluorspar is $325 per metric ton.

### Phosphate

Phosphate is obtained from phosphorite or phosphate rock which is a chemical sedimentary rock composed mainly of phosphate minerals that occur as compact or nodular masses The dominant phosphate minerals are varieties of the mineral apatite: fluorapatite, ($Ca_5(PO_4)_3F$,) chlorapatite ($Ca_5(PO_4)_3Cl$), and hydroxylapatite, ($Ca_5(PO_4)_3OH$), usually admixed with other detrital materials. The phosphates originate from sources including shells of marine invertebrates containing calcium phosphate, and the bones and excrement of vertebrates such as birds. Phosphorites are deposited mostly in shallow marine or low-energy environments. Thick deposits occur as beds intercalated within limestone, shale and carbonaceous mudstone. Sedimentary phosphates constitute 75% of global phosphate, igneous phosphates from carbonatite 24% and guano - bird droppings 1%. The phosphate content of phosphorite varies greatly, from 4% to 35% phosphorus pentoxide (P2O5), but typical commercial phosphorites contain about 30% P2O5. About 90% of rock phosphate production is used for fertilizer and animal feed supplements while the rest 10% for industrial chemicals and other applications.

Phosphate rock has been known in Africa since at least 1873, when it was discovered in Algeria, followed by additional discoveries in northern Africa, notably in Tunisia, Egypt and, in the early part of the 20th century, in Morocco, Mauritania and South Africa. Deposits of Lower Tertiary age in northern Africa account for most of the commercial production of phosphate rock in Africa. Global production of phosphate in 2019 totaled 240 Mt, with China as the largest producer contributing 46%.

Africa produced about 15% of global phosphate in 2019. Morocco (and Western Sahara) is the largest producer with an output of 36 Mt (60%), followed by Egypt, 5 Mt (8%), Tunisia, 3 Mt (5%) South Africa, 2 Mt (3%), Senegal 1.6 Mt (2.6%), Algeria, 1.2 Mt (2%), and Togo (1.3%). Global reserves of phosphates are estimated as 69 billion tons, and reserves for Africa are 56 billion tons. Morocco with 59 billion tons alone accounts for 72% of global reserves. The vast phosphate deposits in Morocco are found within the Late Cretaceous, Paleocene and Eocene marine sediments. In Senegal and Togo, the phosphate beds are Eocene in age and occur as thin layers within the sediments. In central and southern Africa, phosphates of igneous origin occur in carbonatites in Uganda, Tanzania and South Africa (Palabora Complex).

The Palabora carbonatite is mined for copper, with apatite as a byproduct. Mabounie carbonatite in Gabon has considerable phosphate potential with an available annual production estimated at 2Mt/yr of high-grade (39% $P_2O_5$) concentration. Angola, Uganda, Tanzania, and Mozambique have great potential to develop several phosphate deposits that occur in carbonatite complexes. Minor amount of guano occurs in some coastal cavities of Cheringoma and Jofane limestones in Mozambique where it is used for local agricultural needs.

## Vermiculite

Vermiculite is a clay mineral formed by the weathering or hydrothermal alteration of ferro-magnesian mica (biotite and phlogopite). Vermiculite has a layered structure that contains water in between the layers, which undergoes significant expansion when heated. Expanded vermiculite has unique and versatile characteristics; it is lightweight and suitable for insulation, it is non-combustible, highly absorbent, pH neutral, inert, non-reactive to all but very strong acids, and compressible. These desirable characteristics allow expanded vermiculite to be widely used in industrial and commercial applications including: packaging, fire protection, refractory and high temperature insulation, lightweight concrete, plasters, and horticultural potting mixes.

World mine production of vermiculite was an estimated 500,000 metric tons in 2019. South Africa, the United States, Brazil and China were the leading producers. Africa accounted for 48% of global vermiculite in 2019, out of which South Africa's production of 180,000 tons represented 36% of global production and 72% of production in Africa. Other African countries with some production are Zimbabwe (12%), Uganda (12%), Kenya (5%) and Egypt (3%). Vermiculite reserves in the western world are estimated as 50 million tons, 50% in the United States, 40% in South Africa and 7% in Brazil. In South Africa, vermiculite occurs within the Palabora carbonatite where it is produced by the hydration of phlogopite (and biotite) by the loss of alkali and the addition of water through infiltration of water during weathering processes. In contrast, the vermiculites in the United States are products of hydrothermal alteration of mafic rocks.

## Zircon

Zircon ($ZrSiO_4$), a silicate mineral is the main source of zirconium. Zircon is also an industrial mineral as well as a gemstone. Zircon has some physical properties such as high bonding capacity, low acidity, low thermal expansion coefficient, high chemical resistance, and low wettability by molten metal, that makes it extremely suitable for use in metallurgical foundry and refractories (furnace linings and foundry bricks), and for casting applications and fabrications where its resistance to shock, abrasion and high temperature makes it valuable. Zircon dioxide (zirconia) powder is added to glazes for ceramic floor tiles and sanitary ware,

Most zircon is normally produced as a byproduct of the mining and processing of heavy-mineral sands where ilmenite is the principal ore mineral. World production of zircon mineral concentrates in 2019 was estimated as 1.4 million tons, of which the leading producers were Australia (550,000 tons), South Africa (350,000 tons), United States (100,000 tons), China (80,000 tons) and Mozambique (50,000 tons). World reserves of zircon are estimated at 74 million tons. The world's largest reserves are in Australia (64%), South Africa (19%) and India (5%).

Africa's production of zircon is very substantial representing about 30% of global output. South Africa is the second leading producer of zircon in the world, contributing 25% of global production and over 85% of African output, followed by Mozambique and Kenya. Other African producers of zircon in small quantities are Namibia, Sierra Leone, Senegal, Gambia and Nigeria. In South Africa, zircon is produced from the extensive beach placer deposits located along the eastern, southern and north-eastern coasts of the country, with two mining centers at Richard's Bay and Namakwa Sands. Smaller deposits are located along the west coast, north of Cape Town. Zircon (with ilmenite) are also mined from beach placers along the eastern coast of Mozambique and Kenya. Nigeria's zircons are of fluvial origin in the middle Benue valley.

### 3.13.4.6 African Gemstones

Gemstones are unique minerals (and rocks) that are valued for their beauty, durability, and rarity. They are usually cut and polished to make jewelry or other human adornments. Gemstones embedded in gold jewelry have been worn over the ages mostly by women who flaunt and relish their beauty. Africa is a "treasure trove" of precious and valuable gemstones. Long known for producing amazing crystals of diamond, Africa has also emerged as a major source of colored gemstones that are found in more than half a dozen different countries in southern and eastern Africa, stretching from Namibia in the southwest through Zimbabwe, Zambia, Malawi and Tanzania, to Kenya and Mozambique in Eastern Africa, and in the nearby island of Madagascar. All these producers are lined up along the Mozambique Belt - a gemstone metallogenic province. Nigeria has emerged in Western Africa as a country with a wealth of gemstones (Fig. 3.82).

Some of the colored gemstones that have become major articles of trade between Africa and Asia/Middle East are sapphire, ruby, emerald, tourmaline, garnets and tanzanite. The most important producers recently have been Tanzania, Mozambique and Madagascar. The varied gemstones from the mines of Africa are unique on their own and are appreciated worldwide by jewelers and consumers owing to their rarity, sheen, and extraordinary beauty. In fact, these gems are particularly intriguing and desirable as Africa's rich and valuable geological environment is known to produce roughly 19,000 variations of gemstones, which accounts for nearly 50% of the global gemstone turnover. Most of the flow of gemstones is towards Asia, Middle East and Europe.

Madagascar is famous for recent discoveries of ruby and sapphire. Mozambique has also become one of the prominent producers of Paraiba Tourmaline in a wide range of colors. Most of the global supply of Paraiba is now originating from Mozambique. Mozambique has also recently been recognized for the mining of high quality natural ruby stones, and some fine spessartite and red garnet gemstones in impressive sizes.

Tanzania is especially famous for its Tanzanite gemstone, but also produces very fine Tanzanian spinel in large sizes, and some uniquely colored unheated zircon. Other gemstones from Tanzania include the marvelous Rhodolite Garnet, the stunning Tsavorite Garnet, and several exclusive stones in various colors.

Nigeria, the populous West African nation is the one important gemstone producer that is not in the Mozambique Belt. Nigeria has produced respectable blue sapphire as well as large quantities of fine Paraiba Tourmaline and other varieties of tourmaline. Nigeria is also known for the fine Mandarin Spessartite, pyrope and almandine garnet, aquamarine and topaz.

### African Ruby and Sapphire

Ruby, a "bloody-red" variety of corundum is the second most valuable precious stone after diamond. Rubies can be valued at up to $1,500 a carat depending on its clarity, origin, and color. Myanmar (Burma) has for a long time dominated the world trade in rubies, until recent discoveries were made in Africa. The so-called "African Ruby" originating mainly from Mozambique is of the highest quality and has compensated for the demand for clear ruby specimens. More recently, there have been more discoveries of top-grade ruby found in eastern Mozambique, including some material of excellent color that does not require heat treatment. Other sources of African ruby are the neighbouring Kenya, Tanzania, and Madagascar which is known to also contain high quality ruby. These ruby deposits are lined up in the so-called Pan-African Mozambique Belt of Eastern Africa, which is now recognized as a gemstone metallogenic belt, a unique area rich in corundum gems.

Sapphire is the royal-blue variety of corundum. It is a beautiful gem that is in high demand worldwide. One carat of sapphire ranges in price from $700 to $1200 depending on quality. In 1998 large deposits of high quality sapphire were discovered mostly as xenocrysts in alkali basalts in the Limagne Rift in northern, central, and eastern Madagascar. A combination of size and quality of the gems have made the Madagascar sapphire popular worldwide. Good quality sapphire is also found in the alkali basalts of the Mambilla Plateau of Nigeria and Cameroon, and new discoveries were made recently at Aksum in Ethiopia.

### African Paraiba Tourmaline

Tourmaline is a boron aluminium silicate that occurs in a variety of colors (pink, red, green, blue or multicolor). The term "Paraiba tourmaline" is generally used as a trade name for this copper-bearing elbaite tourmaline that is found in only three countries in the world (Brazil, Mozambique and Nigeria). They are found in association with granitic pegmatites. Some Paraiba tourmaline appear in electric blue-green colour that is both unusual and stunningly beautiful and was first reported from Brazil in 1988, and in Nigeria around 2000 and Mozambique in 2005. It is valued for its glow that is not seen in other gemstones due to the presence of copper. Mozambique has recently become famous for its high quality Paraiba tourmaline, and most of the world supply now originates from Mozambique rather than the original source in Brazil. Nigeria produces large quantities of Paraiba tourmaline which is the most popular gemstone in Nigeria and substantial quantities are produced in Oyo, Kaduna and Nasarawa states.

### Tanzanite

Tanzanite is a trade name for gem-quality specimens of the mineral zoisite (epidote group) with a unique blue color (blue zoisite). It is named after the country of its original discovery, Tanzania where the only known deposit of commercial importance is mined near the base of Mount Kilimanjaro at the city of Arusha in northern Tanzania. Tanzanite is a relatively new gemstone discovered in the 1960s that has become the second most popular and probably the most valued blue gem in the market due to its glitter and rarity. Apart from its strong presence, tanzanite also stands apart among other gemstones because it exhibits trichroism - showing three different colors of blue, burgundy and violet when viewed from three different directions. Tanzanite is located within the Mozambique Belt in Tanzania, and is produced only in a few small mines. Reserves are however expected to last for at least 20 years. These mines have also produced pink, yellow and green zoisite as well as a host of other minerals including chrome diopside, tsavorite garnet, blue apatite, and the mint green form of grossularite garnet known as "merilani mint garnet" that has become a favorite of gem collectors. One carat of tanzanite is worth about $500.

### African Garnets

Africa has become a major source of excellent and beautiful garnets with great color and clarity. Garnets found in Africa include: *red* almandine, *orange* spessartine, *pink-red* rhodonite, *green* tsavorite, *yellow* 'mali', and *blue* garnet. They are found mostly within the metamorphosed Precambrian gneisses and pegmatites. The first blue garnets known in the world were found at Bekily, in Madagascar in 1998. It is a garnet that resembles sapphire and has also been found in Kenya. *Rhodonite* is probably the world's finest and most valued of the garnets because of its reddish-purple color, rarity and luster. *Tsavorite* with its bright green color is sometimes used in place of emerald. Tanzania and Kenya are the leading producers of high quality tsavorite garnet and rhodolite gems. *Spessartine*, the orange garnet is beautiful but scarce. Nigeria is known to contain some of the finest specimens of spessartine from Oyo State, and a little bit of rhodolite, pyrope and almandine garnet. These gem deposits come from colluvial and alluvial deposits derived from mostly granitic pegmatites. Some rhodolite and spessartine also occur in Mozambique. Madagascar is a producer of red almandine garnet. Yellow garnet named "Mali" was first discovered in 1994 in the West African country of Mali which is still its only known source. This rare gemstone has great brilliance and a high dispersion and is in high demand but with a limited supply of just a few crystals at a time.

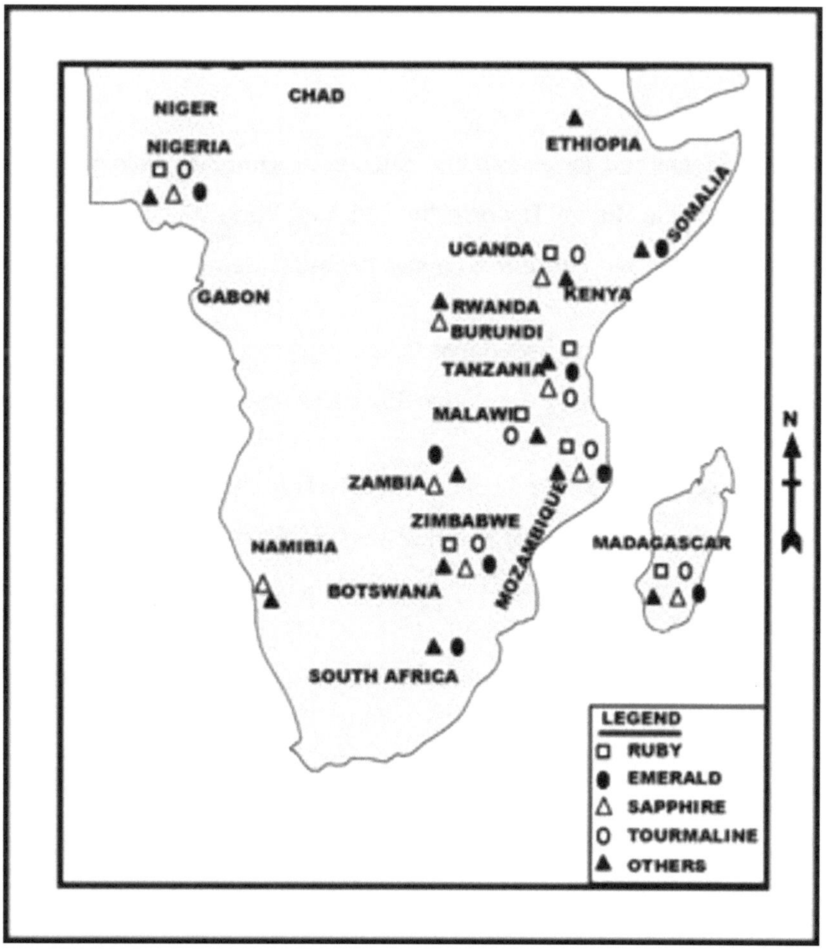


Fig. 3.82: Distribution of Major Deposits of African Gemstones

## 3.14 Review Questions

1. List the main types of magmatic deposits and explain their processes of formation
2. What are hydrothermal deposits? Describe the four main sources of ore forming solutions.
3. Attempt a classification of hydrothermal deposits according to (a) morphology and (b) temperature of deposition.
4. Describe the sources, composition and transportation of hydrothermal solutions.
5. Differentiate between the following pairs of terms: (a) stratiform and stratabound; (b) epigenetic and syngenetic; (c) potassic and argillic alteration; (d) mesothermal and epithermal deposits
6. What are sedimentary mineral deposits? Discuss the geological processes that lead to the formation of sedimentary ore deposits. Cite examples.
7. Describe the main features and differences between volcanogenic massive sulfide (VMS) and sedimentary exhalative (SEDEX) deposits, Give example.
8. What are the differences between lateritic weathering and supergene enrichment? List two examples of deposits formed by each process.
9. Define metallogeny and metallogenic provinces. Explain the relationship between plate tectonics and mineral deposits
10. What are "metamorphic" and "metamorphosed" mineral deposits? Describe their mineralogical characteristics and give examples.

## References

Arndt N, Ganino C (2012) Metals and society: an introduction to economic geology. Springer, Berlin,

Bateman, A. M. (1950). Economic Mineral Deposits, 2nd ed. New York: Wiley, 916p.

Brown, A.C., (1992, Sediment-hosted stratiform copper deposits: Geoscience Canada, v. 19, no. 3, p. 125-141.

Carlson, CA. (1991). Spatial distribution of ore deposits: Geology, v. 19, p. 111-114.

Carne RC, Cathro RJ. (1982) Sedimentary exhalative SEDEX) zinc-lead-silver deposits, northern Canadian cordillera. CIM Bull 75:66–78

Cox, DP. and Singer, DA., *eds.* (1986) Mineral deposit models: U.S. Geological Survey, Bulletin 1693, 379 p.

Dahlkamp, F.J. (1993), Uranium ore deposits: Springer-Verlag, 460 p.

Edwards, R., and Atkinson, K. (198) Ore deposit geology and its influence on mineral exploration: Chapman and Hall, London, 466 p.

Evans AM (1993) Ore geology and industrial minerals. An introduction. Blackwell Publishing Ltd., Oxford, 389 p

Gross, G.A., (1980,) A classification of iron formations based on depositional environments: Canadian Mineralogist, v. 18, p. 215-222.

Groves, D.I., Goldfarb, R.J., Robert, F., and Hart, C.J., (2003) Gold deposits in metamorphic belts: overview of current understanding, outstanding problems, future research, and exploration significance: Economic Geology, v. 98, p. 1-30.

Groves DI, Bierlein FP (2007) Geodynamic settings of mineral deposit systems. J Geol Soc Lond. 164:19–30

Guilbert, J.M., and Park, C.F., Jr., (1986) The geology of ore deposits: W.H. Freeman and Co., 985 p.

Haggerty, S.E (198), Diamond genesis in a multiply-constrained model: Nature, v. 320, p. 34-38.

Hayes TS, Cox DP, Piatak NM, Seal RR, II (2015) Sediment hosted stratabound copper deposit model: U.S. Geological Survey Scientific Investigations Report 2010–5070–M, 147 p-

Hedenquist, JW and Lowenstern, JB., (1994) The role of magmas in the formation of hydrothermal ore deposits: Nature, v. 370, p. 519-526.

Hutchinson, CS. (1982) Economic deposits and their tectonic setting: John Wiley, 365 p.

James, H and Trendall, AF. 1982) Banded iron formations: distribution in time and paleoenvironmental significance, *in* Holland, H.D. and Schidlowski, M., *eds.*, Mineral deposits and the evolution of the biosphere: report of the Dalhem workshop: Springer Verlag, p. 199-218.

Jenkin GRT, Lusty PAJ, McDonald I, Smith MP, Boyce AJ, Wilkinson JJ (2015) Ore deposits in an evolving Earth: an introduction. In: Jenkin GRT, Lusty PAJ, McDonald I, Smith MP, Boyce AJ, Wilkinson JJ (eds) Ore deposits in an evolving Earth, v. 393. Geological Society, London, Special Publications, pp 1–8

Jensen MLY, Bateman AM (1979) Economic mineral deposits. Wiley, New York, 593 p

Kesler, SE. (1991), Mineral resources, economics, and the environment: Macmillan College Publishing Company, 391 p.

Kirk, J., Ruiz, J, Chesley, J., Walshe, J., and England, G. 2002), A major Archean gold- and crust-forming event in the Kaapvaal Craton, South Africa: Science, v. 297, p. 1856-1858.

Kirkham, RV., Sinclair, WD., Thorpe, R., and Duke, JM., *eds.* (1995) Mineral deposit modeling: Geological Association of Canada, Special Volume 40, 798 p.

Lowell, JD and Guilbert, JM (197) Lateral and vertical alteration-mineralization zoning in porphyry ore deposits: Economic Geology, v. 65, p. 373-408

Misra, KC (2000) Understanding mineral deposits: Kluwer Academic Publishers, 845 p.

Mitchell, AHG and Garson, MS (1981), Mineral deposits and global tectonic set- tings: Academic Press, Geology Series, London, 405 p.

Naldrett, AJ. (2004) Magmatic sulfide deposits. Geology, Geochemistry and Exploration: Springer, Berlin, Heidelberg, New York, 727 p.

Ohmoto, H (1996), Formation of volcanogenic massive sulfide deposits: the Kuroko perspective: Ore Geology Reviews, v. 10, p. 135-177.

Pirajno, F., (199) Hydrothermal mineral deposits. Principles and fundamental concepts for the exploration geologist: Springer Verlag, Berlin, 709 p.

Pirajno, F. (2000) Ore deposits and mantle plumes: Kluwer Academic Publishers, Dordrecht/Boston, 556 p.

Pirajno, F., (2004), Hotspots and mantle plumes: global intraplate tectonics, magmatism and ore deposits: Mineralogy and Petrology, v. 82, p. 183-216.

Pohl WL (2011) Economic geology: principles and practice In: Metals, minerals, coal and hydrocarbons – Introduction to formation and sustainable exploitation of min-eral deposits. Blackwell Publishing Ltd., Oxford, 663 p

Revuelta, MB (2016) Mineral Resources: From Exploration to Sustainability Assessment. Springer Publishers, 641 p.

Robb, (2004), Introduction to ore-forming processes: Blackwell Publishers, 373 p.

Sangster, DF. (1990) Mississippi Valley-type and Sedex lead-zinc deposits: a comparative examination: Canadian Institute of Mining and Metallurgy, Transactions, v. 99, p. B21-B42.

Sawkins FJ (1984) Metal deposits in relation to plate tectonics. Springer, Berlin, 325 p

Stanton RL (1972) Ore petrology. McGraw-Hill, New York, 713 p.

## Pictures of African Gemstones

Botswana Diamond

Mozambique Ruby

Madagascar Sapphire

Rhodolite Garnet

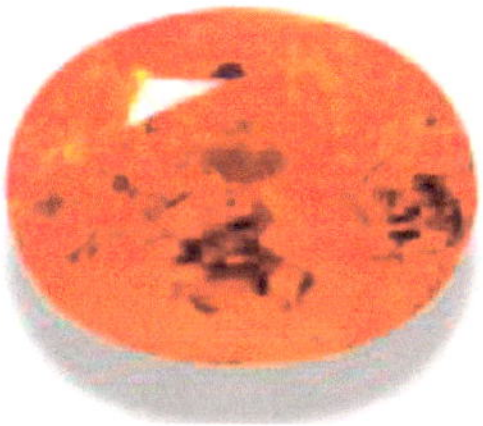

Nigeria Spessartite

Tsovite

Mali   Garnet

Blue Garnet

Mozambique Paraiba

Tanzanite

Nigeria Topaz

South Africa Zircon

# Chapter 3 Photographs of Metallic Mineral Deposits

## Magmatic Deposits

Plate 3.1: Disseminated Diamond in Kimberlite

Plate3.2: Cassiterite in Greisenized Granite

Plate 3.3: Disseminated Nickel Sulfide in Komatite

Plate 3.4: Platinum in Sulfide Rich Reef, Stillwater

Plate 3.5: Stratiform Chromite in Peridotite

Plate 3.6: Stratiform Magnetite in Anorthosite

Plate 3.7: Pegmatite with Gemstone

Plate 3.8: Mineralized Pegmatite Body

# Hydrothermal Deposits

Plate 3.9: Greisen-Quartz Vein

Plate 3.10: Gold-Quartz Veinlet

Plate 3.11: VMS Deposit in Tuffs

Plate 3.13: Porphyry Copper Ore

Plate 3.14: Porphyry Molybdenum

Plate 3.15: Porphyry Copper Mine

Plate 3.16: MVT Pb-Zn Ore in Carbonate

Plate 3.17: MVT Banded Ore

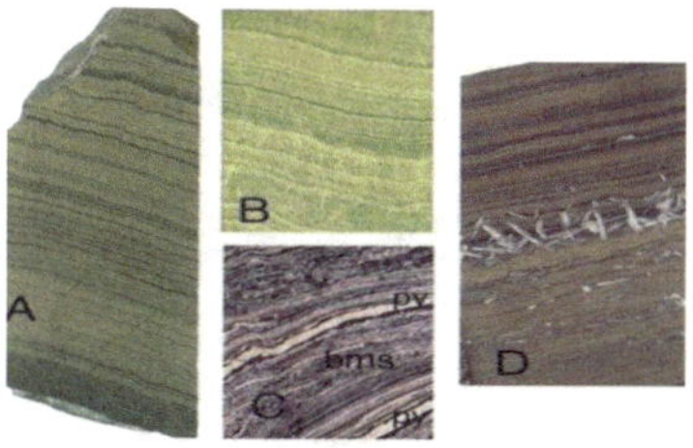

Plate 3.18: SEDEX Drill Cores

Plate 3.19: SEDEX Fine Banded Ore

Plate 3.20: Carlin Gold Disseminated

Plate 3.21: Mississippi Valley Type

Plate 3.22: Epithermal Gold-Silver

Plate 3.23: Geo-Epithermal Silver Ore

Plate 3.24:-Epithermal Mercury

## Sedimentary Deposits

Plate 3.25: Lake Superior Type BIF

Plate 3.26: Algoma Type BIF

Plate 3.27: Rapitan BIF

Plate 3.28: Sedimentary Ironstone

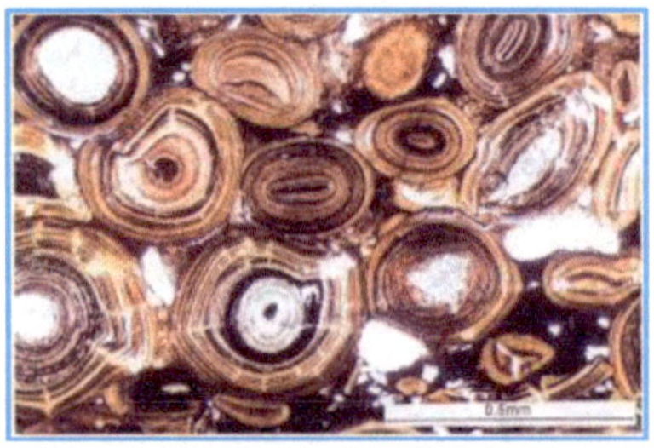
Plate 3.29: Oolitic-Pisolitic Texture

Plate 3.30: Bedded Oolitic Ironstone

Plate 3.31: Sedimentary Copper, DRC

Plate 3.32: Sedimentary Cobalt, ZAM

Plate 3.33: Sedimentary Cu-Ag Ore

Plate 3.34: Sandstone Uranium

Plate 3.35: Sedimentary Uranium

Plate 3.36: Radioactive Pebble Conglomerate

Plate 3.37: Gold in Conglomerate

Plate 3.38: Phosphorite Beds

Plate 3.39: Gypsum Evaporite Drposiy

## Metamorphic Mineral Deposits

Plate 3.40: Ferruginous Schist

Plate 3.41: Ferruginous Quartzite

Plate 3.42: Talc Deposit

Plate 3.43: Skarn Copper Veins

Plate 3.44: Skarn Garnet (Grossular)

Plate 3.45: Skarn Gold

## Residual and Supergene Deposits

Plate 3.46: Lateritic Bauxite, Guinea

Plate 3.47: Pisolitic Bauxite, AUST

Plate 3.48: Bauxite and Kaolin

Plate 3.49: Lateritic Iron Ore

Plate 3.50: Supergene iron Gossan

Plate 3.51: Lateritic Manganese Ore

Plate 3.52: Lateritic Gold

Plate 3.53: Lateritic Nickel

Plate 3.54: Lateritic Niobium

# Chapter 4
# Economic Geology of Metals

## 4.1 Introduction

Descriptive economic geology provides a useful perspective on the geological characteristics of mineral deposits. Economic minerals are of three main types: *metallic, non-metallic* and *mineral fuels*. Metallic minerals are made up of one or more metals that are extracted from ores. Metals are chemical elements or a class of solid substances characterized by high electrical and thermal conductivity as well as by malleability, ductility, and high reflectivity of light. Metals have been of historic value to humankind, and the growth of human civilization has relied heavily on the discovery of various metals. Prehistoric man used metals to build tools and weapons, but as our knowledge of metallurgy developed, metals became the backbone of industrial revolution. This was pivoted on their uses in industrial and technological applications, as well as their essential role in the advancement of agriculture, transport and mining that are so relevant in today's modern society. Six metals, namely; gold, silver, lead, copper, tin and iron were used since prehistoric times. Other important metals were subsequently discovered in the 18$^{th}$ and 19$^{th}$ centuries, including aluminium, beryllium, cadmium, chromium, magnesium, molybdenum, nickel, palladium, platinum, titanium, vanadium and zinc. Compared to other materials, metals are almost irreplaceable because of their unique properties which include as follows (1) they have much higher fracture toughness than other materials of which steels are the toughest known; (2) they are more conductive than ceramics and polymers; and (3) they have the best known overall mechanical properties at high temperatures up to a few hundred degrees.

Metals can be classified in many different ways, including by geochemical abundance, geochemical affinity, mineralogical association and the way they are used. In terms of geochemical abundance, there are five geochemically abundant elements; aluminum, iron, magnesium, manganese, and titanium that constitute more than 0.1 percent by weight of the Earth's crust. The geochemically scarce metals, which embrace all other metals including copper, lead, zinc, gold, and silver, constitute less than 0.1 percent and include most of the precious metals of highest value. In describing mineral production statistics, the abbreviation that are used for *metric tons* is (mt), *million tons* (Mt) and *billion tons* (Bt).

For the purpose of describing the economic geology of metals, they will be classified into seven groups:

1. Precious metals (Au, Ag, PGM)
2. Ferrous and Ferro-Alloy metals (Fe, Mn, Ni, Cr, Co, Ti, V, W, Mo, Nb-Ta)
3. Base (Industrial) metals (Cu, Pb, Zn, Sn)
4. Light metals (Al, Mg)
5. Minor Metals And Elements (Li, Be, Sb, Bi, REE, U, Th)

# 4.2 Precious Metals

Precious metals are the most valuable, naturally occurring metals of economic value that are particularly corrosion-resistant, ductile and of high luster. They are both rare and expensive. The three precious metals are gold (Au), silver (Ag), and the platinum-group elements (PGE).

## 4.2.1 Gold

Gold is a precious metal of considerable importance and high economic value. Throughout history, apart from copper, gold was the earliest metals to be conscientiously sought after by humans, and was always valued for wealth, adornment, and used as a strong and reliable currency. Gold is a metal as well as a mineral that appears most commonly in native form, and as tellurides. Gold is used worldwide in several applications including for jewelry (83%), for gold coins and medals (6%), electrical equipment (6%), dental implants (3%) and a variety of industrial applications including architectural glass and plating (2%).

In 2020, world gold production was 3250 metric tons (mt). The five leading producers were: China (380 mt), Australia (320 mt), Russia (300 mt), USA (200 mt) and Canada (180 mt). Africa contributed about 8% of world production in 2020. Major producers are Ghana (140 mt) and South Africa (120 mt). However, because of widespread illegal mining and smuggling of gold, Africa's production may be underestimated. The world's consumption of gold in 2020 was approximately 4300 tons which includes contribution from already existing stock. Current prices of gold ranges between $1300 and $1900 per ounce. Global world reserves of gold are estimated as 54,000 metric tons (mt).

Table 4.1: Common Gold-bearing Ore Minerals

| **Ore Mineral** | **Formula** | **Gold Content (Wt. %)** |
|---|---|---|
| Native Gold | Au (Ag, Cu, Hg, Pd) | 70 - 100 |
| Sylvanite | (Au,Ag)Te4 | 42 |
| Calaverite | AuTe2 | 25 |
| Petzite | Ag3AuTe2 | 25 |
| Krennerite | (Au,Ag)Te2 | 44 |
| Nagyagite | AuPb(Sb,Bi)Te2-3S6 | 19 |
| Aurostibite | AuSb2 | 42 |
| Fischesserite | Ag,AuSe2 | 30 |
| Uytenbogaardt | Ag,AuS2 | 34 |
| Maldonite | Au2Bi | 65 |

### 4.2.1.1 Mineralogy and Geochemistry

Gold is one of the two metals (and platinum) that occur predominantly in the native form. Native gold is the primary ore mineral in gold deposits in association with tellurides, especially *calaverite* (AuTe2), Other tellurides containing silver (Ag) are *sylvanite*, (Au,Ag)Te4, *krennerite*, (Au,Ag)Te2 and petzite (Ag3AuTe2). The antimonide, *aurostibite*, AuSb2, occurs in some gold deposits. Table 4.1 provides a list of common gold minerals and their Au contents.

The natural abundance of gold in most igneous and sedimentary rocks is very low, usually in the range of 1 to 5 parts per billion (ppb) or 0.001 to 0.005 part per million (ppm or g/ton). However, during magmatic crystallization, Au generally fractionates into the felsic and residual melt with increasing concentrations that may form mineralization in pegmatites. In hydrothermal solutions, Au is transported as

Au-Cl and Au-S complexes, and tends to be concentrated in the vapor phase of fluids at high temperatures and pressures. Gold-arsenic (Au–As) and Au–Sb associations are common in gold deposits. Gold is precipitated from solution due to changes in physicochemical conditions such as temperature, pressure, oxygen fugacity, and sulfur fugacity. The close association of Au with felsic magmas and their associated magmatic-hydrothermal solutions is consistent with the association of gold deposits with felsic intrusive and volcanic rocks, and quartz veins. In orogenic belts, gold veins tend to be associated with metamorphosed mafic volcanic rocks derived from the mantle.

In the surficial environment, Au is almost inert and very resistant to chemical decomposition. After the enclosing minerals have been broken down, Au survives as discrete particles that may accumulate in soils, and because of its specific gravity becomes concentrated as a heavy mineral in stream and beach placer deposits. In lateritic weathering, gold may concentrate as detrital grains within the regolith, and may be remobilized as chloride and organo-metallic complexes and concentrated during supergene processes, usually under acidic conditions.

### 4.2.1.2 Deposit Types

Gold occurs in igneous, sedimentary and metamorphic environments, and several deposit types have been recognized, including intrusion-related (magmatic), hydrothermal, sedimentary, metamorphic and lateritic deposits. Global distribution of gold production according to deposit types indicate that placer deposits (including Witwatersrand) contribute about 32%; mesothermal gold deposits (including orogenic gold) 40%; epithermal deposits (including Carling type) 15%, porphyry and skarn deposits (10%); and magmatic deposits (3%). In economic terms, it is difficult to specify what concentration of gold makes a "gold deposit", but under current mining practices, a gold deposit must have grades exceeding 1 gram per ton (g/t) (1 ppm), or at least contain a significant proportion of ore exceeding that grade. As a byproduct of the mining of other metals, gold concentration of 0.2 to 0.5 g/t can be extracted.

***Magmatic (Intrusion-Related) Deposits***

- Gold in coarse-grained granitic rocks, pegmatites and albitites (e.g. Nastamine, Zimbabwe) and albitites (e.g. Bhukia, India). Gold has a close affinity with felsic and alkali magmas and may be enriched in residual fluids during fractional crystallization. Gold occurs in association with base and precious metals as fine disseminations and sulfide inclusions in coarse-grained granitic pegmatites
- Gold in high-level porphyry dykes, sills, and stocks (e.g. Golden Sunlight, Montana). High level intrusions of calc-alkaline and alkali porphyries and dikes are host to n disseminated/stockworks that contain Au and Cu-Au in close association with vein deposits.
- Gold in carbonatites and carbonatite-related deposits. Carbonatites and related rocks (e.g. Palabora, South Africa contain gold (>0.5 g/t) in association with Cu mineralization

***Contact Metasomatic Deposits***

- Auriferous contact metasomatic or gold skarn type deposits (Dolcoath mine, Montana). Intrusive-related contact metasomatic deposits include an example of the contact orebody which consists of chalcopyrite and an auriferous sulfide and telluride of bismuth intergrown with silicates and calcite.
- Gold skarns in which gold is the major economic metal are found in regions with porphyry copper deposits (e.g. Canty, British Columbia and Fortitude, McCoy, Nevada).

### *Hydrothermal (Vein-type) Deposits*

A substantial portion of the world's gold deposits are of hydrothermal origin where they occur in a variety of vein/lode deposits that are formed at various temperatures ranging from <150°C to >500°C, and at varying depths of emplacement.

- *Hypothermal* porphyry Au, Cu-Au deposits (300° -500°C) (e.g. Porcupine, Ontario, Canada). Hypothermal Au deposits often have a close connection of the hydrothermal deposits with igneous bodies. Au and Au-Cu deposits formed at high temperature and substantial depth characterized by finely disseminated gold in veinlets and stockworks. Breccia pipes associated with porphyries often contain native Au and Au tellurides finely disseminated within gangue minerals filling open spaces in the breccia pipes (e.g. Golden Sunlight, Montana).
- *Mesothermal* quartz-auriferous vein deposits (200° -300°C). Auriferous quartz veins/lodes formed at intermediate temperature and depth (2-5km) in fissures or fractures are widespread in sedimentary and metamorphic terranes of orogenic belts. They are known for their large size and continuity at depth, and constitute a major source of the world's gold production. Typical mineralization includes the sulfides chalcopyrite, sphalerite and galena with quartz-carbonate gangue. Examples include the gold deposits of the Sierra Nevada, California, Nova Scotia, Canada and the Birimian Au deposits in Ghana, Burkina Faso and Mali, West Africa.
- *Epithermal* disseminated Au-Ag deposits (50°-200°C) formed at low temperature and shallow depth that occur as very fine dissemination in carbonaceous sediments (e.g. Carlin type, Nevada).
- *Geothermal-Epithermal* deposits (50° -250°C). Low sulfidation Au-Ag deposits in volcanic rocks and pyroclastics associated with hotspring activity (e.g. El Oro, Mexico and Walhi, New Zealand).
- *Volcanogenic Massive Sulfide* (VMS) deposits contain varying amounts of gold, although the concentrations are usually <2 g/t, but there are some gold mines with 100 g/t. Only about 5% of VMS deposits worldwide contain mineable gold (e.g. LaRonde Penna, Quebec, Canada).

### *Sedimentary Deposits*

- Paleo-placer (old pebble conglomerate) deposits (e.g. Witwatersrand, South Africa; Jacobina, Brazil). Gold is very resistant to weathering, and tends to accumulate as detrimental fragments in recent and old placer deposits which both account for over 60% of the world's current gold production. Paleoplacer gold occurs as grains and fracture-fillings in the matrix of the conglomerates. Average Au content is about 5 g/t.
- Eluvial-Alluvial placer deposits (e.g, Niger State, Nigeria). Eluvial gold in regoliths and alluvial (stream) placers in recent drainage systems are major sources of gold worldwide.

### *Surficial Deposits*

Gold in weathering profiles have been concentrated as lateritic (detrital) deposits and supergene enriched saprolite supergene deposits.

- *Lateritic* deposits (e.g. Western Australia, Guyana, India and Brazil). Disseminated gold and auriferous veins subjected to lateritic weathering are concentrated in the gossans and regolith. In the Carajas region of Brazil native gold is found within iron oxide-hydroxide nodules in the ferruginous zone where gold orebodies have been delineated. The Boddington bauxite mine in Western Australia was until its recent closure, the largest gold producer in Australia.
- *Supergene* (Saprolite) deposits occur at depth in regoliths in which Au is mobilized and concentrated. Examples include supergene ores developed over orogenic gold deposits in Otago, New Zealand and several deposits in Western Australia.

### *Metamorphic Deposits*

Metamorphic Au deposits are formed through the passage of metamorphic fluids during prograde metamorphism of volcanic and sedimentary rocks in orogenic belts. These deposits are also considered as hydrothermal.

- Orogenic gold in greenstone belts (e.g. Tarkwa, Ghana; Kalgoorlie, Australia). Orogenic gold in greenstone belts and other metamorphic terranes are mostly of Precambrian age and may represent the most widespread type of gold deposits. Most of the "mesothermal" gold vein deposits described earlier belong to this category.
- Gold in Banded iron formation (BIF) (e.g. Blue Dot, South Africa; and Homestake, South Dakota). Banded iron formations (BIFs) of oxide and carbonate facies, are host rocks to some gold deposits in cross-cutting auriferous quartz/quartz-carbonate veins/veinlets or as fine disseminations associated with pyrite and arsenopyrite
- Gold in metamorphosed black shale and greywacke (e.g. Bendigo, Victoria, Australia). Minor gold has been found in metamorphosed and usually deformed black shale and greywacke. The gold veins occupy fractures and saddle reefs, with high grade ores that are associated with pyritic and graphitic zones.

### 4.2.1.3 Exploration for Gold

Exploration and discovery of gold deposits are quite challenging because gold as a mineral is very resistant to chemical decomposition in the surface environment, and shows no distinguishing physical properties that can be directly detected by geophysical techniques except when it is associated with other sulfide minerals or iron oxides. Because of the occurrence of gold as discrete grains in surficial systems, its geochemical dispersion is very limited except for the use of associated elements as "pathfinder". Indirect methods can be very useful in identifying lithologic units and structural features that are associated with gold mineralization but may not lead directly to gold discovery. In normal circumstances, particularly in "greenfield" exploration, the search for gold deposits begin with target selection, followed by a combination of reconnaissance exploration involving remote sensing, airborne geophysical surveys, field reconnaissance, geologic mapping, and prospecting.

Remote sensing can assist in identifying lineaments and other structural features such as faults and fractures where spectral analysis can identify zones of hydrothermal alteration. Airborne magnetic and electromagnetic surveys are also helpful in identifying lithologic features and the presence of magnetic minerals. Several ground geophysical methods commonly used to locate metallic deposits can be used for gold, including electrical resistivity, self-potential and induced polarization as well as very low frequency (VLF) electromagnetic and magnetic methods particularly if sulfides and iron oxides are associated with gold mineralization. Brownfield exploration for gold has become a popular approach to exploration in many developing countries of Africa where metallogenic provinces have been identified.

Geochemical exploration for gold at reconnaissance and detailed stages can utilize stream sediments and soils respectively as the appropriate medium in search of anomalous concentrations of certain elements that are pathfinders to gold, including arsenic, antimony, selenium, copper, mercury, galena, sulfur and tellurium. Panning of stream sediments can reveal particles of gold that may indicate the presence of gold and show a pattern of increase towards a source area. Advance exploration involving drilling in known areas of mineralization has been successful in making new discoveries or extensions of known mineralization. For example in Nigeria where Thor's Exploration, a Vancouver-based exploration company discovered new gold reserves at the old Iperindo Mine near Ilesha, SW Nigeria.

### 4.2.1.4 Gold Production in Africa

Gold is a precious mineral resource that is valued by most African countries. It is so important that its acquisition has led to civil war in some countries in Africa. Africa is a major producer of gold, contributing more than 16%% of global production in 2020, ahead of North America (19%). Prior to 2007, South Africa was the global leader in gold production, but production has fallen steadily over the years from about 1000 tons in 1970, to 400 tons in 2001 and 120 tons in 2020. Grades remained the same at about 5 g/t. The production decline has been attributed to low ore reserves, increased production costs in mines that are getting old and deep, and labor unrest. Currently a large proportion of gold production in Africa comes from the West African region; from Ghana, Mali, Burkina Faso, Guinea, Ivory Coast, Nigeria and Senegal. Other countries with substantial gold production are Tanzania and Zimbabwe in East Central Africa although production has declined over the years. Ghana has now overtaken South Africa as Africa's leading gold producer. The West African region has become the leading gold producing region and a major destination for investors involved in gold exploration in several countries including Ghana, Mali, Burkina Faso, Guinea, Senegal, Ivory Coast and Nigeria (Fig. 4.1).

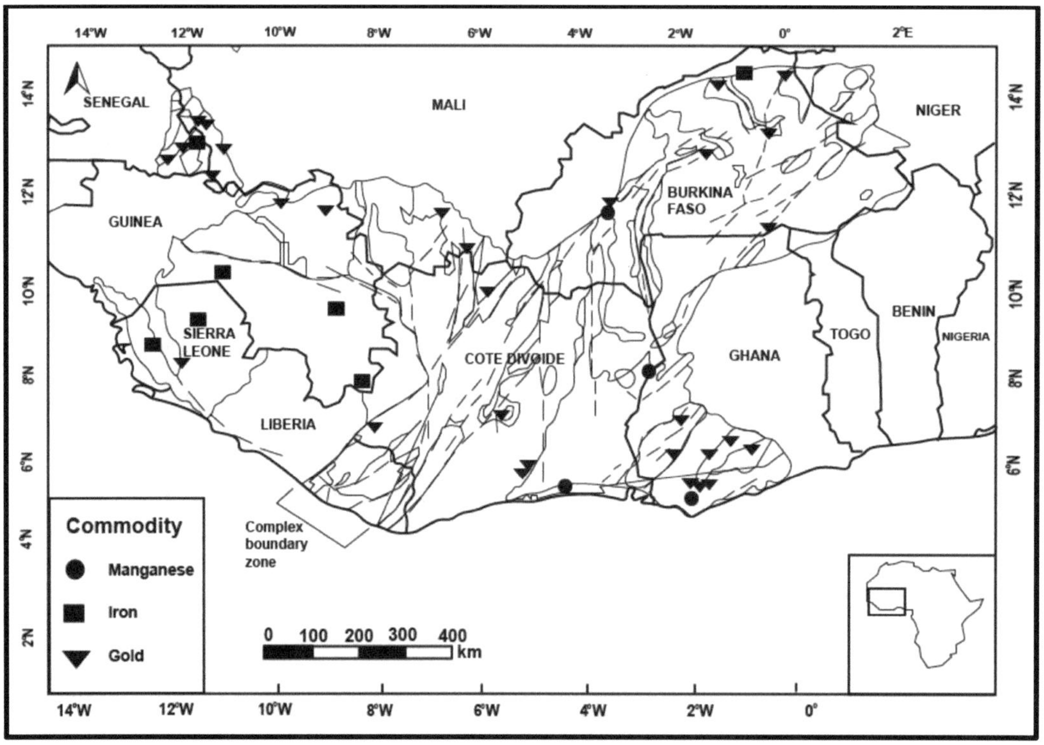


Fig. 4.1: Major Gold, Iron ore and Manganese in West Africa

### *Ghana*

Ghana began gold production in the mid-19th century when the country was known as the "Gold Coast". Production has doubled in the last decade from about 70 tons in 2010 to about 140 tons in 2020, surpassing South Africa. Gold deposits are of three types: (a) orogenic gold characterized by mesothermal quartz-vein deposits hosted in metavolcanic rocks and metasediments in the Proterozoic Birimian greenstone-granite belts, (b) paleo-placer gold deposits hosted in Tarkwaian conglomerates and (c) alluvial placers found in recent drainage systems near coastal areas.

Most of the gold deposits in Ghana are confined to the southern region within two NE trending belts: Sefwi-Bibiani belt and Ashanti belt which hosts the very important deposits at Obuasi and Prestea. Gold mineralization occurs as vein systems occupying fractures and shear zones in greenstones, mafic volcanics and granitoids. Gold in the Tarkwaian paleo-placers is found within the matrix of the Tarkwaian conglomerates with minor hydrothermal remobilization. The Tarkwa gold mine in the Ashanti region is the major deposit developed in the conglomerate, with grades as high as 15-45 g/t. Alluvial gold placers are being mined and dredged including the extensive Dunkwa Goldfields, lying in the Offin River. Lateritic weathering has enriched some of the primary deposits (e.g. Konongo goldfields).

### *South Africa*

South Africa was for a long time, the world's leading producer of gold but production has plummeted from about 1000 tons in 1970 to a little over 110 tons in 2020. The decline has been partly attributed to the increasing cost of mining diminishing reserves, and the on-going issue of social upheaval and political instability. The bulk of South Africa's gold occurs as paleo-placer in quartz-pebble conglomerates and quartzites of the Archean Witwatersrand Basin where gold has been mined for more than 100 years, and still possess the greatest unmined reserves in the world (Fig. 4.2). Since the inception of mining in 1887, the Witwatersrand Basin has yielded over 50,000 tons of gold. Among the major gold mines are the East Rand and TauTona mines which are the deepest gold mines in the world (3.6 km deep). Apart from the paleo-placer deposits, the Archean Amelia greenstone belt contains substantial gold in quartz-carbonate veins associated with oxide facies banded iron formation at the Blue Dot mine where three stratabound orebodies; Goudplaats, Abelskop, and Bothmasrust have been mined.

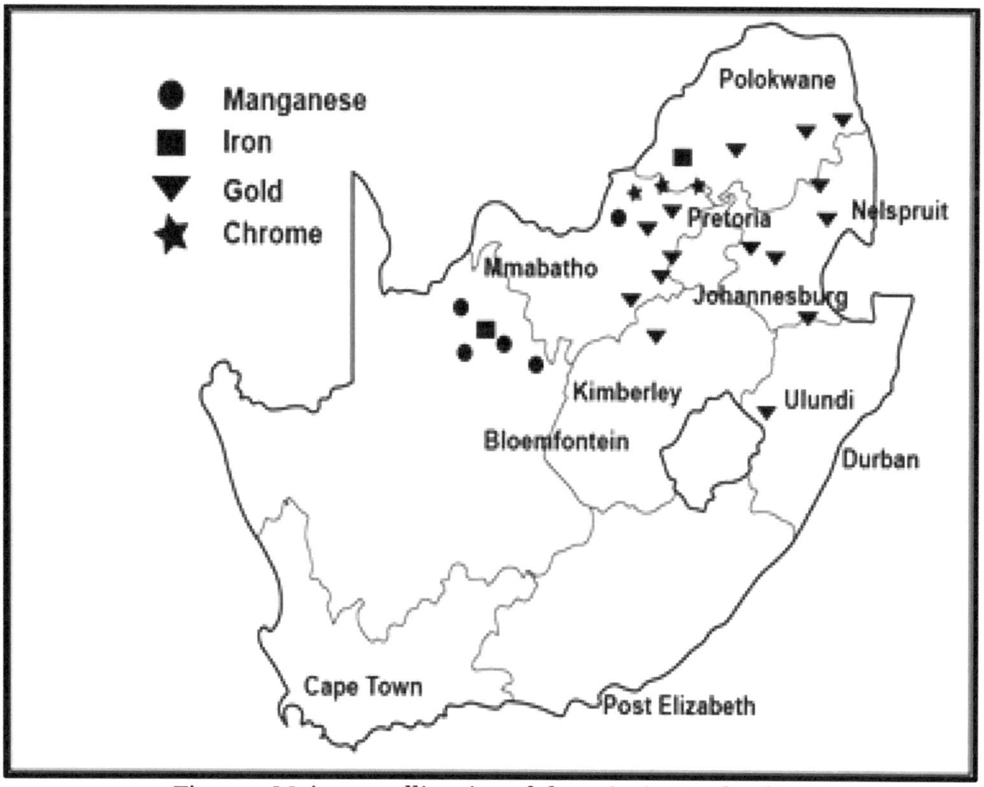


Fig. 4.2: Major metallic mineral deposits in South Africa

### *Mali*

Mali is a landlocked country in sub-Saharan West Africa that has emerged as a leading producer of gold since the 1980s (Fig. 4.1). Current annual gold production is about 60 tons, and Mali is the third largest gold producer in Africa. The geology is characterized by the presence of Proterozoic Birimian sequence in the south and southwest areas containing gold deposits, and the reworked gneisses of the Tuareg Shield to the north and northeast. Three types of gold deposits have been identified: (1) auriferous pyrite-quartz veins and lodes (orogenic gold) with native Au and sulfides in Birimian metasediments (greywackes and marbles); (2) stratabound mineralization in tourmalinized quartzites and marbles (gold mainly in sulfides) in Birrimian metasediments; (3) alluvial and lateritic/eluvial placer. Birimian rocks occur mainly as elongated N-S trending inliers of Archean-Proterozoic rocks in the western and southern parts of the country in two separate belts: (a) the Kenieba Valley on the border between Mali and Senegal, and (b) the Tabakoto-Baroya region in south Mali.

In the Kenieba valley in west Mali, there are a number of major gold mines on the border with Senegal which are localized within a Birrimian weakly metamorphosed sediments including sandstone, calcareous greywackes and marbles intruded by synorogenic diorites. These rocks are different in lithology from the predominant Birimian metavolcanics (greenstones). The major mines are Yatela, Sadiola, Loulo (Randgold) and Segala Mines. Sadiola is the most important lode deposit and largest gold mine of the country, with an annual production of about 12 tons at 3 g/t Au and proven reserves of 240 tons of gold. The Loulo mine is a stratabound gold mineralization that occurs in slightly metamorphosed tourmalinized sandstone bed, 10 to 22 m thick, with disseminated Au-bearing pyrite and stockworks of quartz-carbonate veinlets mineralized over a length of about 800 m. Estimated reserve is 8.2 t of gold with an average grade of 4 g/t Au.

In the Tabakoto-Baroya region of south Mali, gold vein deposits are less prominent but include the Morila and Syama mines, and the recently-discovered, world-class Kobada deposit with an estimated resource base of 62 tons of gold. The annual gold production at the Syama Mine is about 6 tons with estimated reserves of 143 tons Au in sulfide and oxide (lateritic) ores. The oxide ore, in a deeply weathered regolith extending to 35-40 m below the surface is of an average grade of 3.2 g/t reaching 30g/t in places due to supergene enrichment.

### *Tanzania*

Tanzania is the $4^{th}$ largest producer of gold in Africa with annual production of 51 tons in 2019. There are several gold mines; including Buzwagi, Geita, Liuka, Golden Pride, North Mara, Tulawaka and Bulyanhulu. The most productive gold deposits are located within the Archean Nyanzian greenstone belts which occur as narrow belts in the northern and northeastern regions of the country where most of the gold is found. The deposits occur in the form of massive or disseminated Fe-sulfide-rich bodies in banded iron formation (BIF) or tuffs, and as primary gold in quartz veins along shear zones in the metavolcanics and metasediments. They are relatively shallow deposits that are mined in open pits except for the Bulyanhulu deposit. Mineralization is mostly interpreted as epigenetic, although for some deposits in the BIFs, a syngenetic origin is possible. The top four producers are Geita, Golden Pride, Liuka and Buzwagi. Alluvial deposits up to 5 m thick are mined for gold, with yields that range from 5-30 g/t.

### *Burkina Faso*

Burkina Faso is a land-locked country that lies north of Ghana. It is Africa's $5^{th}$ largest producer of gold with annual production of 46 tons in 2019. Most of the gold deposits are localized within the Early Proterozoic Birimian greenstone-granite belts, with minor occurrences in the Tarkwaian clastics. There are several operating gold mines and prospects mainly in the south and northeast regions, including Youga, Poura,

Kiaka, Taparko, Essakane and Inata. The Youga Mine orebodies in southern Burkina Faso, close to the border with Ghana are epigenetic, mesothermal orogenic-type gold deposits localized within Tarkwaian arkosic sandstones and Birrimian greenstones. Mineralization consists of native Au with sulfides. The auriferous quartz veins vary from 1 to 6 m in thickness, and contain visible gold and crystals of pyrite, galena, and chalcopyrite. Extensive lateritic and supergene enrichment have enhanced ore grade. Ore grade averages 2.7 g/t with estimated reserves of 6.6 million tons (Mt).

There are several producing mines in Burkina Faso. The Inata Mine in the south consists of en echelon quartz veins along major shear zones between metasedimentary and metavolcanic units. The ore reserve is approximately 120 Mt of gold with an average grade of 1.2 g/t. The Poura mine, in the south-central region consists of auriferous quartz lodes in Birrimian greenstones with estimated reserves of 22 Mt averaging 17 g/t. The Taparko Mine, located in the northeast region, consists of a series of mineralized quartz veins in greenstones. Ore grade is about 3 g/t with reserves of 12.6 million tons of ore. The Essakane Mine located in the far northeastern part of the country, is a large, low-grade deposit with estimated reserves of 155 million tons grading 1 g/t. The world-class Kiaka gold prospect located in the south of the country is hosted by Birrimian greenstone and metasediments, and has proven reserves of124 Mt at 1.09 g/t Au. Several lateritic/eluvial gold deposits, overlying mineralized quartz veins contain gold in the range of 1 to 10 g/t.

### *Ivory Coast*

Ivory Coast is an aspiring gold producer that mined a record 32 tons of gold in 2019, up 35% from 24 tons in 2018. The rise in production was partly attributed to the start-up of production in the northern mine of Sissingue. Gold deposits are found within the Birrimian greenstone-granite belts that underlie the entire country. Mineralization occurs mainly as primary quartz veins and extensive laterite/saprolite deposits. The Ity Mine in western Ivory Coast is the oldest mine in the country and has produced over 30 tons of gold since its inception in 1991. Primary gold mineralization occurs as veinlets and, mostly, skarn along the contact between granitic intrusion and metasediments. However, most of the economic gold mineralization is associated with the thick saprolite zone.

The Tongon mine in the northern edge of the country is the biggest gold mine with annual production of 7.5 tons of gold mainly from laterites has estimated reserves of 98 tons. The gold mineralization at Bonikro Mine, north of Abidjan is hosted partly within a small granodiorite intrusion that extends into surrounding basalts to the south, and is localized along a shear zone. Gold occurs associated with quartz and pyrite, with the highest gold grades occurring around the intersection of the shear and the granodiorite. Overall, the deposit has an average grade of about 2g/t Au.

### *Zimbabwe*

Zimbabwe is a significant gold producer in Africa although its mining industry has under-performed in recent years. Gold production has dropped from 33 tons in 2018 to 27 tons in 2019. There are some major gold mines and several hundred small mines operated by artisanal and small-scale miners. Some of the important gold mines are Freda Rebecca, Rio Zim, Falcon, Pickstone Peerless, Eureka and Sabi. More than 90% of gold deposits in Zimbabwe are associated with Archean greenstone belts which are some of the richest in the world. Other gold deposits occur in the Limpopo Mobile Belt in the south of the country and in the Proterozoic Piriwiri rocks in the northwestern part of the country.

Gold deposits are of two types: orogenic vein-type and disseminated stratabound deposits. The orogenic gold deposits occur as quartz veins filing fractures or as impregnations in shear zones. All the Archean rock types within this greenstone belt are known to host gold mineralization. More than 50% of the known gold occurrences are in the ultramafic and mafic metavolcanics and about 25% are hosted in felsic volcanics and intrusives. Vein-type gold deposits consist of quartz, carbonates and sulfides, while the stratabound deposits are hosted as sulfide bands in banded iron formation and disseminations in associated volcaniclastic and clastic rocks.

Table 4.2: Gold Production in Africa and Deposit Types

| | **Country** | **Production 2019 ( tons)** | **Major Mines** | **Deposit Type** |
|---|---|---|---|---|
| | Ghana | 142 | Tarkwa, Damang, Chirano, Ahafo Prestea, Asanko, Nzema, Bibiani, Iduapriem, Obuasi, , Konongo | Orogenic gold<br>Paleo-placer<br>Alluvial & Lateritic |
| | South Africa | 120 | East Rand, TauTona, South Deep, Kromdraal, Mponeng, Kloof , | Paleo-placer<br>BIF gold |
| | Mali | 61 | Morila, Sadiola, Yatela, Syama, Loulo, Segala, Kobada | Orogenic gold<br>Lateritic/Eluvial |
| | Tanzania | 51 | Buzwagi, Geita, Luaka, Golden Pride, North Mara, Tulawaka, Bulyanhulu, | Orogenic gold<br>BIF gold<br>Stratiform sulfides |
| | Burkina Faso | 46 | Kiaka, Youga, Taparko, Poura, Essakane, Inata, | Orogenic gold<br>Alluvial placer |
| | Ivory Coast | 32 | Ity, Tongon, Angovia, Sissingue, Yoaoure, Kouakounianou, Bonkro, | Orogenic gold<br>Lateritic/Saprolite<br>Skarn gold |
| | Zimbabwe | 26 | Freda Rebecca, Rio Zim, Falcon, Pickstone Peerless, Eureka, Sabi, | Orogenic gold<br>BIF gold |
| | Guinea | 16 | Lero-Karya, Kart, Siguiri, Kiniero | Orogenic gold<br>Lateritic/saprolite<br>Alluvial gold |
| | Senegal | 12 | Sabodala, Mako, (Boto) | Orogenic gold |
| | Nigeria | 10 | Segilola | Schist Belt Gold<br>Alluvial Gold |

***Guinea***

Guinea produces about 20 tons of gold annually from orogenic gold deposits and eluvial and alluvial placers, with estimated reserves of 700 tons. The gold deposits are localized in shear zones within the Birimian greywacke and metasediments in the greenstone belts of the Siguiri Basin in the northeastern part of the country (Upper Guinea). There are three types of deposits: (a) auriferous quartz veins and stockworks, (b) lateritic and/or supergene deposits, and (c) alluvial placers. Major gold mines are the Lero-Karta, Kart. Siquiri and Kimero mines. The Lero Mine, the largest producer, generates about 12 metric tons of gold of 2.6g/t grade. The Kart open-pit mine located close to Lero is estimated to contain about 185,000 tons of gold with 4.6g/t grade. The Siquiri Mine located on the bank of River Niger in northeast Guinea, produces about 7 tons of gold per year. Mineralization occurs as a series of quartz veins hosted by metasediments that have been deeply weathered to a depth of 100 m in places to form laterite and saprolite or supergene-enriched gold mineralization. The Kimero Mine in the central part of Guinea consists of primary quartz veins that produce about 2 Mt of gold at an average grade of 3.12g/t.

### *Senegal*

Senegal is an emerging gold producer within the Birrimian greenstone gold province. The largest gold deposit in the country was discovered in Diabougou village in eastern Senegal near the Mali border in 2006 which started a gold rush by artisanal miners and corporate mining companies. Annual production is currently estimated as 12 tons, with an estimated resource base of 200 tons. Most of the gold mineralization occurs along the eastern Senegal-Mali structural corridor ("Kenieba Window") where a major mineralized fault zone transects Birrimian metavolcanics, metasediments, schists and metagreywacke (turbidites). The mineralized zones contain tourmalinized quartz-carbonate veins associated with pyrite, pyrrhotite and arsenopyrite. They are very similar to those of the adjoining Kenieba Valley in Mali. Major gold mines are the Sabodala, Mako and Boto mines.

The Sabodala Mine is located within the Mako Mafic Volcanic Belt in the West Birimian province. The host rocks are low-grade metamorphosed mafic volcanic rocks inter-layered with ultramafic units and local sedimentary horizons. Gold mineralization is associated with quartz, quartz-albite-carbonate-pyrite and ankerite/siderite alteration veining. Annual production is about 4 tons at an average grade of 2.4 g/t. The Mako Mine is situated within the same Kenieba Window geological setting as the Sabodala deposit. Gold mineralization is hosted within a felsic volcanic unit sandwiched between two basalt sheets and associated with disseminated and veinlet pyrite, quartz-carbonate veining and zones of hydrothermal brecciation within the felsic unit. Annual production is 4.3 tons averaging 2.2 g/t. The recently discovered Malikounda deposit is the largest gold deposit so far, with estimated reserves of 20 tons averaging 1.80 g/t, with projected annual production of 4 tons.

## 4.2.2 Silver

Silver (Ag) is one of the precious metals that is highly valued for its rarity, economic importance and unique physical properties. Silver is white, soft, ductile, malleable, resistant to oxidation, and has the highest electrical and thermal conductivity of all metals. It is used in coins, jewelry, and tableware, and has many industrial applications including electrical and electronic products, mirrors, and photography. Silver occurs in nature in one of four forms: 1) as a native element; 2) as a primary constituent in silver minerals; 3) as a natural alloy with other metals; and, 4) as a trace to minor constituents in the sulfide ores of other metals. Silver is obtained mainly as a by-product from copper, lead, zinc, and gold ores. In fact, the economic viability of many base metal and gold deposits relies on by-product silver. Most of the silver produced worldwide is a product of inclusion in other minerals, notably galena, sphalerite, chalcopyrite, bornite and tetrahedrite. Global production of silver in 2019 was 27,000 metric tons (mt}. The top five producers are Mexico (6300 mt), Peru (3800 mt), China (3600 mt), Russia (2100 mt) and Poland (1700 mt). Silver is one metal that is not produced in Africa in substantial quantities. The world reserves base for silver is estimated as 500,000 tons.

### 4.2.2.1 Mineralogy and Geochemistry

Silver is found rarely as a native metal but occurs as sulfides, tellurides and arsenides in combination with Au, Pb, Cu, and Hg. More than 40 such minerals are known in nature, but the major ones are Ag metal, electrum (Au-Ag alloy), *argentite* (Ag2S), *cerargyrite* (AgCl); various sulfides of arsenic, antimony, and silver, such as *pyrargyrite* and *tetrahedrite*, and in minor amounts in galena (PbS) and pyrite (FeS2e) (Table 5.3). *Argentite* is the most common Ag mineral found in mineral deposits. Telluride minerals of Ag such as *hessite* (Ag2Te) and *empressite* (AgTe) are not an important source of Ag although they occur rarely in some deposits. *Freibergite* is found rarely as an Ag mineral in Cu-Ag-Pb deposits where it may occur together with galena intergrowth or grain coating. *Friebergite* forms a solid solution series with tetrahedrite

$(Cu,Fe)_{12}Sb_4S_{13})$. Silver replaces copper in the freibergite-tetrahedrite series, so that when Ag increases, Cu content decreases

Silver concentrations in crustal rocks are very low, usually in the range of 0.02 to 0.1 ppm. Highest values are found in organic-rich sedimentary rocks such as black shales. In magmatic-hydrothermal environments, Ag is higher in mafic than felsic rocks due to its affinity for sulfides. During magmatic differentiation, Ag concentrations may increase in hydrothermal solutions in the presence of chloride and sulfide complexes. In hydrothermal systems where mixing of magmatic-hydrothermal and meteoric waters and boiling may occur at shallow depths, Ag may be transported in substantial concentration in a vapor AgCl phase to form minerals such as chlorargyrite (AgCl). In the surface environment, Ag is relatively soft and does not accumulate as placer mineral, although during lateritic weathering it can accumulate as secondary minerals within ferruginized caps.

Table 4.3: Selected Silver Minerals and Silver Content

| | Mineral | Chemical Composition | Silver Content |
|---|---|---|---|
| | Native silver | Ag (with Au, Cu, Hg, As, Sb) | Up to 100% Ag |
| | Argentite/Acanthite | $Ag_2S$ | 87% |
| | Proustite | $Ag_3AsS_3$ | 65% |
| | Pyrargyrite | $Ag_3SbS_3$ | 60% |
| | Stephanite | $Ag_5SbS_4$ | 68% |
| | Pyrargyrite | AgCl | 75% |
| | Hessite | $Ag_2Te$ | 52% |
| | Empressite | AgTe | 46% |
| | Freibergite | $[(Ag,Cu)_{12} (Sb,As)_4S_{13}]$ | 20-50% |

#### 4.2.2.2 Types of Deposits

While silver is extracted as a major commodity in some deposits, over 70% off world production is extracted as a byproduct from various polymetallic deposits of Cu, Pb and Zn. The main types of silver deposits include contact metasomatic, epithermal, mesothermal, porphyry-type, sediment-hosted Cu deposits with minor concentrations from volcanogenic massive sulfides (VMS) and sedimentary exhalative deposits (SEDEX).

***Contact Metasomatic and Skarn Deposits***

- Silver commonly in association with Zn and Pb is found in contact metasomatic and skarn deposits originating from granitic intrusion into limestone host rocks (e.g. Western Cordilleras in British Columbia, Canada; Montana and Alaska, USA).
- Mineralization occurs as replacement bodies, containing silver, lead, zinc and pyrite with minor amounts of Mn, Cu, As, Sb and Au. Gangue minerals are quartz and carbonates. Silver is often concentrated in tetrahedrite associated with galena (e.g. Tintic, Utah).

***Hydrothermal Deposits***

- *Porphyry-Type (Sn-Ag)* deposits characterized by felsic subvolcanic porphyry intrusions that host tin deposits (*porphyry tin*) and produce substantial silver, sometimes in association with lead and zinc (Sn-Pb-Zn-Ag). Tin-silver veins occur as late-stage veins in these deposits. Silver is also recovered as a by-product of porphyry copper and gold deposits.

- *Epithermal (Au-Ag) or Ag* deposits in volcanic rocks that are related to fossilized geothermal systems are the main source of silver worldwide (e.g. Veladero, Argentina and Bolivia). They account for about 20% of global production. Mexico, the world's leading producer is endowed with this type of deposits with the largest and richest silver mine (Fresnillo district). They occur as vein, stockwork, disseminated, and replacement deposits. The grade of silver ranges from 500 to 2000 g/t.
- *Polymetallic Silver Vein Deposits*: Silver-bearing veins were the traditional sources of silver in the early days of silver mining. Most of the veins are mesothermal, based on their mineral assemblage of quartz-carbonate veins. There are three vein types: (a) Ag veins (e.g. Cornwall, UK); (b) Ag-Pb-Zn veins (e.g. Silver Valley, Idaho; Benue Trough, Nigeria), and (c) polymetallic Ag-Ni-Co-As-Bi-(U) veins e.g. Echo Bay, Canada).

***Sedimentary Deposits***

- *Sediment-hosted Cu-Ag Deposits:* Silver is found in copper-bearing stratabound deposits containing disseminations of copper sulfides in a variety of cl sedimentary rocks including black shale, sandstone and limestone (e.g. White Pine, Michigan; Kuperschiefer, Germany). Silver occurs in its own minerals mainly native Ag, but the most important silver carriers are base-metal sulfides - bornite and chalcopyrite.

#### 4.2.2.3 Silver Deposits in Africa

Africa is not a major producer of silver. The only known major silver deposits are in Morocco in the Anti-Atlas region of North Africa. The Imiter Mine is a hydrothermal vein deposit in Late Proterozoic felsic volcanic rocks. Current estimated resources at Imiter are 8,000 tons of silver metal. The Zgounder deposit is also a silver-rich hydrothermal vein deposit in the same area of Morocco that has been exploited since the 12thcentury and operated on and off for a long time but was closed since 1990 due to low silver prices and recently reopened under new management in 2014. From 1982 to 1990, operators of the underground mine extracted 500,000 tons of ore grading 330 g/t using a cutoff grade of 125 g/t Ag. Zgounder has proven and inferred reserves of 6 million tons of silver at 343 g/t.

Numerous mineral deposits of various types produce silver as a by-product of both base metal and gold mining. Notable examples include:

1. the lead-zinc deposits of Nigeria's Benue Trough where Ag is extracted from galena;
2. the cobalt-nickel-arsenide (copper-gold-silver) hydrothermal vein deposits Bou Azzer, Bleida,Tiouit, and Bou Madine in the Anti-Atlas region of Morocco
3. the Mississippi Valley-type massive lead-zinc sulfide districts of Beddiane and Toussit-Bou Bekkar in Morocco and the Peeing district in South Africa;
4. sediment-hosted and variably metamorphosed lead zinc deposits such as those in the Aggeneys and Gamsberg districts of the Namaqua Mobile Belt of South Africa and the Rosh Pinah deposit in Namibia;
5. gold deposits of the Witwatersrand in South Africa;
6. the Phalaborwa carbonatite deposit of South Africa;
7. polymetallic veins in granite-greenstone belts in South Africa, Namibia, Zimbabwe, Ethiopia, Egypt, and many others.

## 4.2.3 Platinum Group Metals

Platinum is a precious metal that is part of a family of six platinum group metals (PGM) namely: platinum (Pt), palladium (Pd), rhodium (Rh), iridium (Ir), osmium (Os), and ruthenium (Ru). Platinum and palladium are actually the most commercially important of the PGM with largest utilization in the automotive industry where they are applied to decrease harmful emissions from vehicle systems Platinum is usually found alongside palladium and other metals, notably base metals such as copper, nickel, and chrome. The PGMs have similar physical and chemical properties and are found together in the same mineral deposits. They also show similar properties including high density, high electrical conductivity, high melting points, and low reactivity. The uses of platinum and palladium are diverse, including as catalytic converters for pollution abatement and as components of laboratory equipment and for production of industrial chemicals. Platinum alloys are used in capsules for radioactive power sources which provide space probes with electrical energy. Platinum has earned the 'honor' of being the most expensive metal on Earth; it is more expensive than gold, and is becoming increasingly used in jewelry, decoration and adornments.

Global production of platinum and PGMs (platinum group) in 2020 were 170 and 470 metric tons respectively. The world's five largest producers of platinum based on 2020 statistics are South Africa (130 mt), Russia (22 mt), Zimbabwe (15 mt), Canada (10 mt) and USA (4 mt). The world resources base for platinum in mineral concentrations that can be mined economically are estimated as 100,000 tons, the vast majority being located in South Africa. Platinum occurs mainly as native platinum (usually tiny particles but sometimes nuggets), as an alloy with other precious metals, and as traces in sulfides or chromite. Typical mining grades are between 3 and 10 g/t, and even lower if nickel or chromium can be mined at the same time. Platinum, palladium and rhenium are the most common of the PGMs.

### 4.2.3.1 Mineralogy and Geochemistry

The minerals of PGM are *native pla*tinum and alloys with other PGM and/or Fe and sulfides *cooperat*e (PtS) and b*raggite* ((Pt,Pd,Ni)S). Of less importance are the telluride and arsenide minerals (Table 4.4). Platinum group metals have a close affinity with iron as indicated by the common occurrence of alloys of PGM with iron (up to 20% Fe) and their close association in mafic and ultramafic rocks. The abundance of PCM in crustal rocks is very low (Pt 0.005 ppm, Pd 0.01 ppm), and ultramafic and mafic rocks tend to have higher concentrations of platinum, while felsic rocks contain no PGMs. Consequently, most of the platinum deposits are found in ultramafic/mafic rocks such as the Bushveld Igneous Complex.

In hydrothermal systems, PGMs in similarity to Au and Ag, are transported as chloride and sulfide complexes, but rarely are there enough concentrations by fractional crystallization to form hydrothermal deposits. Black shales tend to accumulate platinum (as well as gold), such as in the Kupferschiefer Copper Shale. Metamorphic fluids are known to have formed Pt-Au quartz-sulfide veins in metasediments (schists) in parts of the Southern Alps, mostly derived from leaching of sediments.

In the surficial environment, platinum just like gold is very resistant to chemical dissolution and therefore concentrates as heavy minerals in stream placers where precious nuggets have been found. Lateritic weathering produces residual PGM in the ferruginous caps (gossans) and in lower lateritic zones within the regolith. However, the concentrations are rarely high enough to be extracted at a profit. Just like gold, platinum metals may be weakly remobilized, and as the solubility of the different platinum metals s is not the same, weathering of primary PGM ore often leads to relative enrichment of Pt over Pd thereby creating some supergene concentration of platinum deeper in the regolith, usually beneath the water table.

Table 4.4: Selected Platinum Minerals

| | Mineral | Chemical Composition | PGM Content |
|---|---|---|---|
| | Native Platinum | Alloyed with PGM & Fe | Up to 100% Pt |
| | Cooperite | PtS | 82% Pt |
| | Braggite | (Pt,Pd,Ni)S | 59% Pt; 20% Pd |
| | Sperrylite | PtAs2 | Up to 54% Pt |
| | Moncheite | (Pt,Pd)(Te2,Bi) | 37% PGE |
| | Laurite | RuS2 | 61% Ru |

#### 4.2.3.2 Deposit Types

Platinum group metals are found almost exclusively within ultramafic and mafic rocks of layered intrusive complexes commonly associated with chromium and nickel ores. Such an example is the Bushveld Igneous Complex of South Africa that contains over 75% of the world's platinum, 54% of palladium, and 82% of rhenium resources. Apart from the magmatic PGM deposits, there are other minor contributions from hydrothermal veins and alluvial placers. Some podiform chromite deposits are associated with elevated platinum concentrations while small ultramafic intrusions (e.g. Alaska–Urals) usually contain platinum only in the subeconomic range.

***Magmatic Deposits***

Magmatic deposits of PGM occur as products of early and late crystallization in ultramafic and mafic magmas by processes of dissemination and gravitational settling, or late crystallization involving liquid immiscibility and injections.

- *Disseminations* of platinum- enriched chromite in small intrusions of dunite (e.g. Urals-Alaska) that are subeconomic but contribute to the formation of unique placer deposits such as in the in the Urals, Koryakia and Aldan Shield (USSR).
- *Stratiform* layers of platinum-rich sulfide- and chromite-bearing mafic and ultramafic rocks (platinum reefs) produced by gravitational crystal settling (e.g. Merensky reefs, Bushveld) or immiscible liquid segregation or injection (e.g. Vlakfontein, South Africa and Sudbury, Canada).

***Contact Metasomatic Deposits***

- Contact metasomatic deposits of platinum along the contact between pyroxenite intrusives and surrounding host rocks of dolomite and banded iron formation (e.g. Potgieter Mast, South Africa). Enrichment of platinum has been observed along the contact of the reefs with surrounding rocks.

***Hydrothermal Veins***

- Hydrothermal quartz veins containing platinum have been reported from the Bushveld and Sudbury Igneous Complexes. In the Waterberg deposit, platinum-bearing quartz veins occur near the southern boundary of the northern limb of the Bushveld Complex. The veins occupy a fault zone between Rooiberg felsic rocks and Triassic sandstones.

***Sedimentary Deposits***

- Alluvial placer deposits containing platinum alloys derived from ultramafic rocks, dunite, serpentinite and pyroxenite occur in a few localities in the Urals, Alaska and Colombia. In Colombia, most of the platinum placers contain gold. Approximately 20% of PGM produced in Russia is derived from alluvial placers in the Urals region where they are shed from small ultramafic bodies.

### 4.2.3.3 Exploration for Platinum

Platinum deposits are almost exclusively associated with layered ultramafic and mafic intrusions. Based on the concept of ore-rock association, target selection should focus on identifying such potential host rocks followed by geophysical and geochemical exploration. Prime targets in host rocks are sulfide-rich (Ni-Cu) sections and areas with chromitite and magnetite layers in layered intrusive complexes. Applicable geophysical methods include induced polarization to detect sulfides and magnetic and electromagnetic methods for chromite and magnetite. Field reconnaissance methods using heavy mineral sampling along stream beds by panning can be helpful in identifying potential source areas upstream. Regional and detailed geochemical surveys of stream sediments and analysis for Cu, Ni and Cr, and pathfinder elements such as arsenic and tellurium can help delimit potential target areas in sulfide-associated platinum deposits. Detailed soil surveys analyzing for several ore and associated elements are applicable to define potential mineralized zones.

### 4.2.3.4 Platinum Deposits in Africa

Africa dominates the world's production of platinum and palladium with major contributions from South Africa and Zimbabwe with a combined 85% of world production of platinum and 90% of world's reserves. Other platinum producing countries are Botswana, Tanzania, Uganda, Gabon and Mozambique.

***South Africa***

South Africa currently produces about 130 tons of platinum annually, representing about 72 percent of the platinum and 38 percent of all the palladium in the world, all from the Bushveld Igneous Complex. In addition, the Bushveld contains at least 70,000 tons of PGM reserves which accounts for 88% of world reserves. There are two principal platinum layers or "reefs" in the Bushveld Complex, the *Merensky reef* and the Upper Group 2 *UG2 chromitite reef*, both of which are about 1 m thick and occur consistently around the eastern and western sides of the Bushveld. A third PGM-rich layer, the Platreef, varying between 5 and 90 meters is only found in the northeastern edge of the complex (the Potgietersrus "Limb"). Ore grades of the reefs are typically between 4-7 g/t. There are several platinum mines in South Africa but the leading producers are the Rustenburg mine, Mogalakwena mine, Marikana mine, Bathopele mine, Khomanani mine, and others. In recent years, platinum mining has declined due to the fact that the richest deposits have been mined and the cost of production is going up steeply.

***Zimbabwe***

Zimbabwe is 2nd in Africa and 3rd in the world in platinum production that comes mainly from the Great Dyke intrusive complex. The country operates eight mines that produce about 15 tons of platinum annually. The Great Dyke is a unique geological feature that runs through the center of Zimbabwe for about 550 km in a north-south direction. The PGM occurs in a layer known as the Main Sulphide Zone, which is typically about 3 m thick, although the economically mineable zone may be only 1m thick depending on ore grade and metal prices. Ore grades are generally below 4 g/t, of which about 55 per cent is platinum and the rest is palladium and rhenium. Only four mines are currently in active production in Zimbabwe. The Ngezi Mine located about 150 km south of Harare (capital city) is Zimbabwe's principal PGM producer while Mintesa Mine, in the southern part of the Great Dyke, is the oldest mine. Others are Mimosa, s and Unki mines.

***Other Countries***

Other countries including Tanzania, Botswana, Gabon, Namibia and DR Congo with intrusion-related Ni-Cu sulfides and stratabound Cu-Co-Ag deposits that may contain PGM as byproducts of mining and smelting. The Kalahari Copperbelt of Namibia and Botswana is a broad region with stratabound copper-silver deposits hosted in Late Proterozoic metasedimentary rocks. Platinum group metals in copper-silver mineralization are associated mainly with molybdenite, bismuth-bearing minerals and copper sulfides.

## 4.3 Ferrous and Ferroalloy (Iron and Steel) Metals

Ferrous metal is a metal that contains iron in its composition which implies that iron is a large percentage of the elemental composition. There are various types of iron, namely: pure iron, cast iron, wrought iron, pig iron, carbon steel, and stainless steel. Steel or carbon steel is obtained by blowing oxygen through molten iron, thereby reducing its carbon content to 2%. The properties of steel can be adapted by alloying it with other metals, - the so-called steel metals or ferroalloys.

Ferroalloys are metals that are added to iron in high proportion to enhance or impart certain distinctive properties to materials or manufactured products. The main ferroalloy metals are manganese (Mn), chromium (Cr), nickel (Ni), Tungsten (W), molybdenum (Mo), cobalt (Co), and vanadium (V). The ferroalloys are very important in the iron and steel industry where they are used to improve properties such as tensile strength, ductility, fatigue, strength, and corrosion resistance.

### 4.3.1 Iron Ore

Primary iron (Fe) is the most significant metal on Earth and accounts for about 95% of the world's metal production. It is a very important and widely used commodity as raw material for the manufacture of iron and steel products. The term "iron ore" is used to connote an aggregate of iron minerals that are of commercial value. Iron can be found in trace or minor amounts almost everywhere on the Earth's surface, but economic concentrations of iron (iron ores) are usually produced by geological processes. Iron occurs very rarely in the native form except in meteorites and some rare volcanic rocks, but it is a major constituent of rock-forming minerals where it appears as silicates, carbonates, and oxides. Metallic iron is relatively soft and easily susceptible to oxidation and corrosion. Other useful physical properties include malleability, ductility and conductivity.

The predominant use of iron is in the form of cast iron and steel of various types, such as stainless steel. Carbon Steel which is also known as structural steel is a major staple in the construction industry where it is used in the construction of buildings, skyscrapers and bridges. Ferrous metals are also used in shipping containers, industrial piping, automobiles, railroad tracks, and many commercial and domestic tools. Most ferrous metals are magnetic which makes them very useful for motor and electrical applications.

Global resource base of crude iron ore is estimated to exceed 800 billion tons containing more than 230 billion tons of iron. World production of iron ore in 2020 was 2.5 billion tons with proven reserves of 170 billion tons. The five leading producers are Australia (930 million tons) (Mt), Brazil (480 Mt), China (350 Mt), India (210 Mt), and Russia (99 Mt). Other countries in the top ten include South Africa which is $6^{th}$ in the world and $1^{st}$ in Africa with an annual production of 77 Mt, and Ukraine, Canada, the United States, and Kazakhstan. China and India are the world's leading importers of iron ore. The world's resources are dominated by low-grade ore, though current world production of iron ore is largely from deposits of high-grade ore that are composed of either rock of massive hematite, magnetite, or pisolitic goethite.

#### 4.3.1.1 Mineralogy and Geochemistry

The four main ore minerals of iron are hematite, magnetite, goethite/limonite and siderite. The important minerals are shown in Table 4.5. *Hematite* ($Fe_2O_3$) with 69.9% Fe is the most common mineral in iron ores worldwide and occurs as a primary as well as secondary mineral often enriched by supergene processes. Magnetite ($Fe_3O_4$) is found in lesser amounts particularly in igneous and metamorphic environments where it can be coarsely crystalline due to the effects of elevated temperatures of crystallization. It occurs mostly in magmatic, contact metasomatic, and metamorphosed BIFs or banded magnetite quartzites.

Goethite and limonite which are reddish or brownish–yellowish respectively are earthy products of chemical weathering of iron minerals and made up of aggregates of different iron hydroxides. They are found mostly as secondary minerals in weathered ferruginous caps and regolith, and as minerals in oolitic ironstones. Typical ore grades in iron ores range from 25% to 60% Fe which may be further enriched to over 65% by lateritic weathering and supergene processes. The cut-off grade for mining of iron ores is 32% Fe. A large portion of the world's iron ore resources are low-grade ores (<30% Fe). Deleterious impurities in iron ores such as high phosphorus, sulfur and silica and alumina that reflect the presence of quartz and clay impurities may affect their quality and ease of beneficiation.

Table 4.5: Major Minerals of Iron and Metal Content

| | Mineral | Chemical Composition | Iron Content |
|---|---|---|---|
| 1 | Magnetite | $Fe_3O_4$ | 72.4% |
| 2 | Hematite | $Fe_2O_3$ | 69.9% |
| 3 | Goethite | $Fe_2O_3H_2O$ | 62% |
| 4 | Limonite | $(Fe,O,OH,H_2O)$ | 59.8% |
| 5 | Siderite | $FeCO_3$ | 48.2% |
| 6 | Chamosite | $(Fe^{2+},Mg)_5Al(AlSi_3O_{10})(OH)_8$ | 28.6 |
| 7 | Pyrite | $FeS_2$ | 46.67 |
| 8 | Ilmenite | FeTiO4 | 30.8 |

Iron is the third most abundant element in the Earth's crust, after silica and alumina. Concentrations of Fe are higher in ultramafic and mafic rocks than in intermediate and felsic rock due to the relative abundance of Fe-bearing silicate minerals. Iron occurs in two oxidation states, the divalent or ferrous form and the trivalent or ferric form. The occurrence of Fe in varying oxidation states affects its stability in the surficial environment. Reduced iron is more mobile than oxidized iron which is considered immobile, and this difference has resulted in the concentration of oxidized secondary minerals as duricrust and ferruginous caps in lateritic weathering of Fe-bearing rocks. The solubility of iron in natural water is affected by the prevailing oxidation state eH and hydrogen ion concentration (pH). Also important is its tendency of iron in solution to form complex ions with inorganic as well as with organic material. Iron will remain in solution in a reduced state (ferrous iron but will precipitate if the ferrous iron is oxidized to ferric iron,

In the surficial environment, weathering of iron-bearing rocks often leads to residual enrichment and formation of Fe hydroxide minerals, goethite and limonite.as in lateritic iron ore deposits. Some oxidized iron can be mobilized in the form of colloidal or solid oxy-hydroxide particles, adsorbed on suspended clay flakes, and concentrated by organic matter. The stability and formation of iron minerals are strongly controlled by pH and Eh conditions. Iron is mobile in slightly acidic meteoric water, and can be mobilized by downward percolating groundwater and precipitated as enriched iron ore at depth in the weathered regolith as supergene deposits.

#### 4.3.1.2 Deposit Types

Iron deposits are widely distributed worldwide and the types of deposits are diverse. However, there are six main types of iron-ore deposits that currently provide most of the world's iron production (1) Banded Iron Formations (BIF) (with high-grade hematite and magnetite ores)' (2) Ferruginous Quartzites (Hematite-

Magnetite Quartzites); (3) Magmatic magnetite and titano-magnetite ores (4) Contact metasomatic/skarn magnetite deposits, and (5) Oolitic-pisolitic hematite ironstones; and (6) Lateritic and supergene ores. BIFs constitute over 90% of iron ore produced worldwide and represent over 60% of global reserves of iron ore.

***Magmatic Deposits***

- Stratified magnetite and Ti-magnetite deposits in layered igneous complexes (e.g. Bushveld, South Africa; Duluth, Michigan). Formed in mafic intrusive complexes by magmatic crystallization and separation of immiscible iron magma that crystallized as massive magnetite or Ti-magnetite ores.
- Volcanogenic magnetite and magnetite-apatite deposits (e.g. El Laco, Chile; Kiruna, Sweden). Magnetite ores are associated with mafic volcanic rocks.

***Contact Metasomatic Deposits***

- Skarn Iron deposits (e.g. Urals, USSR; Hebei, China). These deposits are linked to metasomatism at the contact of intermediate to felsic intrusions with carbonate rocks. The deposits are usually small to medium-sized (<100 Mt), and are mined for high-grade iron ore of magnetite and hematite requiring little or no beneficiation
- Metasomatic siderite deposits (e.g. Erzberg, Austria). Massive siderite deposits replacing carbonate rocks. Ore grade is low, about 40% Fe.

***Hydrothermal Deposits***

- Epigenetic vein deposits of apatite-iron oxide (e.g. El Romeral, Chile). Formed as magnetite-apatite ore in veins associated with dioritic intrusives.

***Sedimentary Deposits***

- Banded Iron Formations (unmetamorphosed) (e.g. Lake Superior, Michigan) Sedimentary deposits consisting of rhythmic alternating thin layers of iron oxide and chert. Consist mainly of magnetite and hematite ores of Lake Superior or Algoma type BIF.
- Oolitic-pisolitic ironstones (e.g. Lorraine, France; Alabama, USA). Iron-rich clastic sedimentary iron ores with oolitic and pisolitic texture. The ores contain hematite, magnetite and goethite with average iron concentrations of 42% Fe.

***Metamorphic Deposits***

- Hematite-Magnetite Quartzites (e.g. Krivoy Rog, Ukraine; Andhra Pradesh, India). These are high-grade metamorphosed iron-bearing clastic sediments that are converted to homogeneous or banded hematite and magnetite quartzites.
- Metamorphosed BIFs (e.g. Abitibi, Canada; Tamil Nadu, India). Metamorphosed BIFs with metamorphic features and recrystallization.

***Surficial Deposits***

- Lateritic iron ores (e.g. Minas Gerais, Brazil). Deposits are formed by chemical weathering of iron-bearing rocks in tropical and subtropical environments. Ore minerals include hematite, goethite and limonite.
- Supergene iron ores (Hamersley, Australia; Carajas, Brazil). Deposits formed by supergene processes which increase the grade, texture and quality of iron ores derived from weathered BIFs, oolitic ironstones and iron-rich ultramafic rocks. The iron minerals in these deposits are mostly hematite and goethite.

### 4.3.1.3 Iron Ore Deposits in Africa

Africa is blessed with substantial resources of iron ore, although several countries have vast deposits that are yet to be developed, such as Guinea and Liberia. Iron ore production in Africa is dominated by South Africa (77 Mt) and Mauritania (13 Mt). Other countries with significant production are Sierra Leone (4.7 Mt), Liberia (1.9 Mt) and Algeria (600,000 tons). (Fig. 4.3). Vast deposits have been identified for development in Guinea and Liberia.

***South Africa***

South Africa is the world's sixth-largest iron ore producer and the third largest iron ore exporter after Brazil and Australia. Its annual production in 2019 was 77 Mt, with estimated reserves of 1 billion tons. Production of iron ore has declined in recent years due to rising operational costs. Currently, the bulk of iron mined in South Africa comes mainly from banded iron-formation within Proterozoic Transvaal sedimentary rocks in the Northern Cape Province which produces 88%, and the Limpopo Province (12%). Most iron ore production comes from five mines namely the Sishen, Khumani, Beeshoek and Kolomela Mines in the Northern Cape, and the Thabazimbi Mine in the Limpopo Province. The total mineable iron ore reserves for the five largest BIF-hosted iron ore deposits were estimated to be approximately 1.51 Bt Nearly 70% of the South African iron ore operations are conducted by Kumba Iron Ore Limited which is one of the world's top suppliers of exported iron ore. Kumba owns three of the five major mines, Kolomela, Sishen and Thabazimbi.

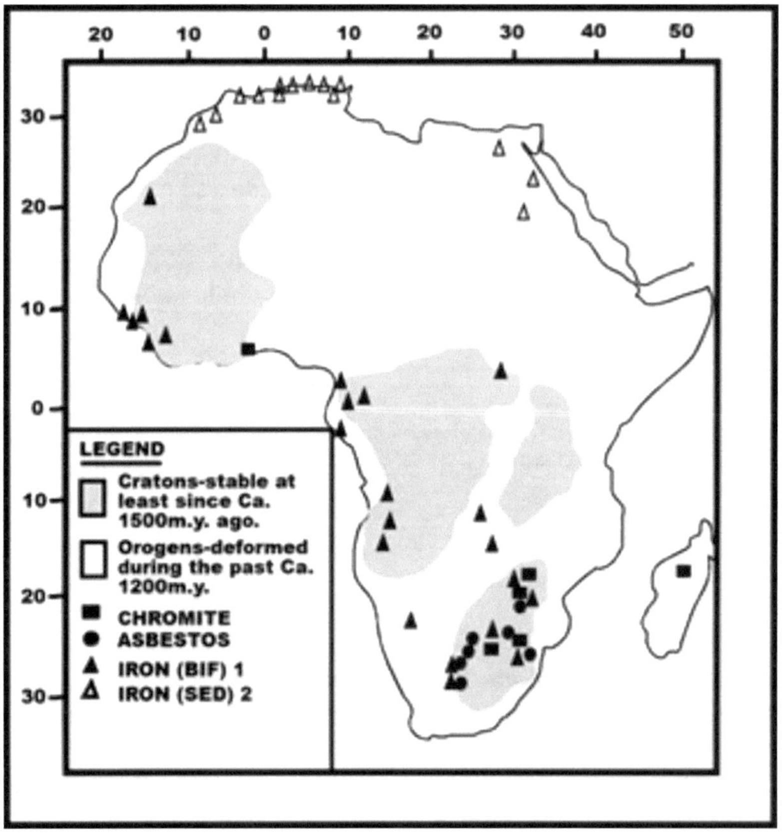

Fig. 4.3: Major Iron and Chrome Deposits in Africa

There are five types of iron ore deposits in South Africa: (a) BIF of Algoma type which occurs in the Archean greenstone belts; (b) BIF of Lake Superior type which includes the high-grade deposits at Thabazimbi and Sishen; (c) oolitic hematite iron silicate ores of the Clinton type, which includes large deposits in the Proterozoic Pretoria Group in the Transvaal basin; (d) magmatic magnetite ore deposits in the Bushveld Complex and elsewhere; and (e) small isolated siderite mudstone (blackband') deposits in the Karoo Sequence, Most of South Africa's iron ores are mined from a variety of high-grade hematite and magnetite BIF deposits, mostly of Lake Superior type that all occur within the Griqualand West and Transvaal areas of the Late Archean to Early Proterozoic Transvaal Supergroup, and in units of the Postmasburg Group in Griqualand West. These ores have been affected by extensive supergene enrichment which has produced the largest reserves of high-grade (>60% Fe) iron ores in South Africa, such as the Sishen, Khumani, Beeshoek and Sishen-South (Welgevonden) deposits. In addition, there is evidence of enrichment of iron attributed to hydrothermal activity and/or contact metamorphism of the BIFs. Magnetite BFF and other low-grade deposits exist in large quantities but are not currently extracted because of the higher costs of beneficiation of magnetite-bearing deposits.

Within the Bushveld Igneous Complex and similar intrusions, extensive magnetite seams contain 11 to 17% titanium oxide (TiO2) which makes them quite unsuitable for smelting to iron by conventional (blast furnace) methods because of the high melting point of titanium. However, some of the lower seams contain significant quantities of vanadium, up to about 1, 8 % V2O5, which are mined as a source of this metal and for the recovery of the iron as a profitable byproduct for steel alloy production by direct furnace reduction.

The *Sishen Mine* is by far the largest iron mine in South Africa and is among the world's largest open-pit mines. It is located in the Northern Cape Province about 280 miles north of Kimberley. The mine produces around 27 million tons (Mt) per annum of 65% Fe hard hematite ore with a high lump proportion, and low impurity levels of generally less than 0.05% P, 2.5% SiO2 and 1.2% Al2O3. Sishen represents one of the largest iron ore reserves in South Africa and in the world with an estimated resource base of 2.43 billion tons of ore grading 58.6% iron metal. The currently mineable reserves are 130 Mt of high-grade ore (60%Fe) and 1.3 billion tons of low grade (30-50% Fe) ore. The Sishen deposits are sedimentary BIFs hosted by the Palaeoproterozoic Transvaal Supergroup within the Griqualand West sub-basin. Four ore types occur, namely: laminated; massive; brecciated and conglomeratic ore.

### *Mauritania*

Mauritania in northwest Africa lies north of Senegal and borders Morocco and Algeria to the north. It is the second largest iron ore producer in Africa and the fifteenth largest in the world with an annual production of 15 Mt and crude ore reserves of about 1.1 billion tons in 2019. Iron production comes mainly from the Zouerate district in northern Mauritania where high-grade hematite ores are produced from banded hematite quartzites (also called banded iron formations). The Société Nationale Industrielle et Minière (SNIM), a public/private joint venture is the sole iron ore producer in the country. The company owns and operates the three iron ore mines—the Guelb el Rhein Mine, the Kedia Mine, and the M'Haoudat Mine in Zouerate. Three types of iron ore deposits have been recognized in Mauritania: (a) Banded hematite quartzites in Proterozoic metasediments or metamorphosed Lake Superior type BIF; (b) Algoma type magnetite BIF associated with Archean greenstone belts of the Reguibat Shield; and (c) oolitic-pisolitic (sedimentary) ironstones of Paleozoic age.

The Zouerate region where the producing mines are located contains eight major iron deposits in four steeply-dipping lenticular orebodies of banded hematite quartzites consisting of parallel ferruginous layers associated with siliceous clastic sediments including conglomerate, sandstone, quartzite and mica schist. The richer hematite ores have ore grades >65% Fe suitable for direct shipping ore while the siliceous ores contain 55% Fe. Proven and probable reserves of direct-shipping ore total 300Mt, and those requiring beneficiation 3,000Mt. Total resource base is estimated as 2400 Mt. Other hematite deposits with good prospects are in the development pipeline including the Legleitat iron ore deposit with an estimated resource base of 12.2 Mt of iron ore containing 59% Fe that has the potential to support the production of 1 Mt/yr of direct-shipping grade iron ore for 10 years

In the central area of the Reguibat Shield, magnetite ores of Algoma type BIF occur within the deformed Archean basement rocks with two massive orebodies at Guelb Rhein and Oum Arwagen. However, the magnetite ores are low grade with approximately 37% Fe. The apparent depletion of the high-grade ores in recent years has resulted in a number of new projects focused on the lower grade magnetite ores. The Guelb el Rhein deposit contains several hundred million metric tons of magnetite ore grading 37% Fe. Large deposits of untapped oolitic-type iron ores are also present in Phanerozoic sedimentary rocks in eastern Mauritania.

### *Algeria*

Algeria located in North Africa is the leading iron ore producer in the Arab world with an annual production of about 800,000 metric tons in 2019. Production has fluctuated over the years reaching between 1 million and 2 million tons in the 1990s to 2010s. Currently, iron production is derived mainly from two iron ore facilities at Ouenza and Boukhadra mines in the province of Tébessa in eastern Algeria, which contain estimated reserves of approximately 100 million tons of ore grading 45 to 60% Fe. Lower grades occur at depth. There are two other small mines at Anini near Sétif and Rouina in Aïn Defla province with estimated reserves of about 3 Mt and 3.5 Mt of iron ore respectively. The Quenza and Bou Khadra iron ores occur as metasomatic deposits within Early Cretaceous (Aptian) reef limestones. The Quenza Mine accounts for 75 percent of total iron production, while the Bou Khadra Mine provides about 20%.

### *Sierra Leone*

Sierra Leone in West Africa was once a leading producer of iron ore in Africa with an annual production of 4.7 Mt in 2017. However, mining operations have been disrupted since 2018 by legal and ownership tussle between the private mining companies and the Sierra Leonean government. Iron ore was produced from two locations, the Marampa mine, and Tonkolili mine.

The *Marampa* mine is hematite-specularite iron ore mine located approximately 150km north-east of Freetown. The deposit lies within the Marampa Schist formation, a Late Proterozoic sequence of steeply-dipping and folded metasediments comprising mica schist, ferruginous schist, quartzites and metavolcanics. The iron-bearing quartz-hematite schist is up to 65m thick containing low-to medium-grade ore with iron content averaging 37% Reproduction at the mine reached a target of 1.5 Mt per annum before the complete shutdown in 2019 due to dispute with the government.

The *Tonkolil*i iron deposit occurs in the Archean Sula Mountain-Kangari Hills greenstone belt in the northern part of the country. The iron occurs as residual or secondary ore caps resting on Precambrian banded iron formation associated with metavolcanic rocks. The caps are as a result of tropical lateritic weathering and leaching of underlying quartz-magnetite rocks and the iron-rich amphibolites. Two well-developed weathering zones exist within the deposits and are associated with the development of lateritic regolith and supergene enrichment zones overlying the primary BIF. These zones feature a surface hard cap about 60 m and a saprolite zone overlying the BIF material. The Tonkolili mine contains 11.5 billion tons of magnetite ore with 500 Mt average grade of 56.3% Fe in the lateritic enriched zones and 31% Fe in the quartz-magnetite ore. . It is considered to be one of the world's largest magnetite deposit

The Bagla Hill iron ore deposit occurs in the southeastern part of the country and is a metamorphosed banded magnetite quartzite made up of alternating layers of magnetite and quartzite. Reserve estimate was 384 million tons of primary ore (magnetite) at an average grade of 18.1% Fe, and about 90 million tons of hematite ore at 34.3% iron.

### *Liberia*

Liberia and Guinea are two West African countries with vast resources of BIF iron ore in Early Proterozoic greenstone belts. However due to political, social and financial reasons, and lack of foreign investment and transport infrastructure, these huge reserves have not been fully developed. In the 1970s, Liberia was the leading producer of iron ore in Africa, but production gradually declined from a peak of over 4 Mt to about 2 Mt per annum. Most of the iron in Liberia and Guinea are BIFs associated with the Birrimian supracrustal greenstone belts that overlie the Archean basement. Currently, there are three major operating mines in Liberia; Yekepa, Bong and Bomi mines.

The Nimba Range BIFs located near the north-central border with Guinea were first to be mined from 1963 until the 1990s when they were exhausted. Production peaked at 12 million tons in 1974 but subsequently declined due to market conditions and exhausted reserves. The Yekepa mine is located about five km s west of the now exhausted Nimba Range deposit. The production which started in 1985 is from high-grade oxidized ore, with an average iron content of 60–62%. Total ore reserves at Yekepa are 501million tons at 48.3% Fe. The mine produced 4.9 Mt at its peak in 2014.

The Bong Mine is located about 110 km NE of Monrovia within the Bong Range. The mine was operated by the Bong Mining Company between 1965 and 1990 when 158 million tons of medium to high-grade iron ores were extracted. It is currently being exploited by a company China Union which exported its first ore of 50,000 tons in 2014. Other mines like the Bomi Hills mine and Mano River mine situated north of Monrovia in western Liberia extract ore from smaller orebodies that have been worked intermittently for high-grade direct shipping magnetite ore in addition to magnetite concentrate beneficiated from itabirite.

### *Guinea*

Guinea has the world's largest untapped iron deposits located mainly in the southern and central parts of Guinea: the Mount Simandou and Mount Nimba region (in the southeast), the Zogota region in the southeast; as well as the Kalia region in Central Guinea. Of great importance is the world-class Simandou BIF deposit in south-east Guinea which has reserves of over 1.8 billion tons grading 65.5% Fe and planned production of 100 million tons per annum for a period of more than 40 years—this is the sixth largest iron ore deposit in the world. Production has commenced in the Simandou mine after much turmoil and many mining rights disputes over the years. The mine is being operated by a consortium of Rio Tinto and Chinese partners. The mine is expected to produce up to 95 Mt per annum of ore grading 65% Fe.

## 4.3.2 Chromium

Chromium (Cr) is a steely-gray lustrous that is known to have high corrosion resistance. When polished, it gains a very shiny surface, which is used to plate other metals so as to form a protective and attractive covering. Chromium is mined from chromite ores, and 70% of production is used as an alloy in the manufacture of corrosion-resistant stainless steel and other steel alloys with exceptional hardness, durability and tempering properties, for example for aerospace components.

World production of chromium in 2020 was 40 million tons. South Africa was the largest producer of chromium in the world in 2020 with production amounting to 16 million metric tons (Mt), equivalent to 40% of world production, followed by Turkey with 10 Mt. Other countries are Kazakhstan, India, and Brazil. Zimbabwe is a notable producer of chromium from the Great Dyke with estimated production of 350,000 tons in 2020. China is the world's leading consumer of chromium mostly for steel production.

#### 4.3.2.1 Mineralogy and Geochemistry

The only ore mineral of chromium is *chromite* ((Fe,Mg)(Cr,Al,Fe)2O4). The mineral chromite contains only about 25% Cr2O3; with varying amounts of the following: 6–18% FeO, 0–22% MgO, 0–62% Al2O3 and 0–30% Fe2O3. Some chromites contain minor amounts of Mn, Ti, V or Zn. In some mineralized environments such as the Bushveld Igneous Complex in South Africa, chromite may contain enough Pt and Ni to be extractable as a byproduct. The cut-off grade for mining chromium ore is 20% chromite.

Chromium is one of the metals that are closely associated with Fe. The crustal abundance of Cr is about 100 ppm. It is strongly enriched in ultramafic and mafic rocks in which peridotite contains an average of 2700 ppm Cr, gabbro and basalt 200–500 ppm, and granite only 20 ppm. During magmatic crystallization, Cr enters the early crystallizing minerals mostly as chromite or in substitution for Fe in mafic minerals. Chromium occurs in three oxidation states; Cr+6, +3, +2. In natural environments, the most stable oxidation states of chromium are Cr+3 and Cr+6. Chromium distribution is also controlled by redox processes. Oxidation of Cr+3 to Cr+6 occurs in the oxidizing environment during the weathering of Cr-bearing rocks. In soils, and in surface water and groundwater. Cr+6 is the most mobile of Cr as the chromate anion (CrO4)2. Organic matter and other reductants favor the conversion of Cr+6 to Cr+3 and immobilization of hexavalent chromium which leads to enrichment in organic-rich sediments.

#### 4.3.2.2 Deposit Types

Most chromite deposits of economic value are derived from magmatic crystallization and segregation in ultramafic and mafic igneous rocks. There are minor occurrences of placer and lateritic deposits, but most of the world's chromite resources are located in stratiform chromite deposits, such as the Bushveld Complex in South Africa.

***Magmatic Deposits***

- *Disseminated* chromite in ultramafic bodies and serpentinites (e.g. Philippines; Great Dyke, Zimbabwe). Chromite disseminations occur within dunites, peridotites, and serpentinites in greenstone belts. They are usually subeconomic, and not a significant source of chromium.
- *Podiform* chromite in ophiolites and Alpine-type mafic peridotite intrusions (e.g. Turkey, Kazakhstan). Podiform chromite deposits that occur as small masses and lenses ("pods") are an important source of chromite, particularly as the primary source for both high-chromium, low-aluminum ore, used in metallurgical applications. They are found in alpine-type peridotites formed in oceanic crust and tectonically emplaced as ophiolites along continental margins. Podiform chromite mines have produced 57%of the world's total chromite production.
- *Stratiform* chromite in layered igneous complexes (e.g. Bushveld Igneous Complex, S. Africa; Stillwater Complex, USA)). Stratiform deposits are formed as large sheet-like orebodies, in layered mafic to ultramafic igneous complexes. They contain about 98% of the worldwide chromite reserves.

#### 4.3.2.3 Exploration for Chromium

Chromium is always found in ultramafic/mafic rocks and serpentinites. Target areas should focus on such rocks so that exploration can involve combined geological, geophysical and geochemical methods. Chromite and associated minerals, magnetite, ilmenite and Cr-spinel are heavy minerals that are dense and relatively magnetic, and can be detected in the host rocks by magnetic, electromagnetic and gravity methods through aerial and ground surveys. Moreover regional and detailed stream sediment can be applied using a suite of "pathfinder elements" including Cu, Ni, Co and As (arsenic). Both the chromite ores and ultramafic host rocks have distinct geochemical signatures of value in mineral exploration. Podiform-type chromite may create some challenges for exploration because of their smaller size and intense structural deformation of ophiolite host rocks.

### 4.3.2.4 Chromite Deposits in Africa

South Africa combined with Zimbabwe account for 90% of global chromite reserves and produce 50% of the world's chromite ore which are derived from two of the world's best known mafic igneous complexes: the Bushveld and the Great Dyke. Other African countries that produce chrome ore are Sierra Leone and Zambia (Fig. 4.3).

#### *South Africa*

South Africa is the world's largest producer of chromium and ferrochrome. Ferrochrome (FeCr) is a type of ferroalloy (an alloy of Fe and Cr) containing about 50-70% Cr by weight and used in stainless steel production The country holds about 70% of the world's total chrome reserves, most of it derived from the Bushveld Igneous Complex (BIC), and produces 50% of the world's chromite ore and 75% of the world's ferrochrome. The annual production of chrome in 2020 was 16 million tons (Mt).

The Bushveld Igneous Complex is the world's largest layered mafic-ultramafic complex, occupying an area of about 66,000 km2 in the northeastern part of South Africa, and hosts the largest chromite and PGE resources in the world. The Bushveld Complex has the form of an enormous, irregularly shaped saucer about 370 km across with its center buried underground but its rim exposed. Chromite occurs in the Bushveld in layers varying in thickness from a few centimeters to more than 2 m. These chromite layers are confined essentially to one portion of the complex called the "Critical Zone", which has been traced laterally around the complex; the major deposits are in the western and eastern Bushveld, where the resources are vast but are of lower grade (45% Cr2O3). Chromite is mined primarily from the UG2, and LG and MG chromitite seams of which the UG2 also contains significant amounts of PGM's. Thus several platinum mines produce chromite as a byproduct. There are several primary chrome mines, specifically maintained to provide chromite feed to the developing ferrochrome industry. Most of South Africa's chrome mines are developed along the Eastern BIC, in the Steelpoort Valley.

#### *Zimbabwe*

Zimbabwe is the second largest producer of chrome in Africa. The country has a significant 19% of global chrome ore reserves of about 10 billion tons. The country's chrome ore is of a higher grade than in other parts of the world, with an average chrome to iron ratio of 2:1. The chromium reserves are second only to South Africa, but remain largely untapped due to poor regulations and lack of foreign investment in the context of political instability. Annual chrome production has fluctuated over the years, from a low of 250,000 tons to a high of 750,000 tons. Chrome is mainly mined along the Great dyke of Zimbabwe, a unique geological structure that stretches for approximately 500 km N-S in the heart of the country.

Chromite occurs as seams or stratiform deposits within the ultramafic/mafic intrusion, especially in the Darwendale, Chegutu, Kadoma, Kwekwe and Lalapanzi areas. Most of the deposits are small and are worked by small mining operators and the tributers who are in a lease agreement with the three mining companies who hold the mineral titles: Zimalloys, Zimasco and Maranatha Ferrochrome. The Great Dyke has in places generated some alluvial chrome deposits that have accumulated in local stream valleys as thin beds that have yielded > 50% Cr.

Other small chrome deposits are found in the Archean greenstone belts outside the Great dyke in south-central Zimbabwe where Cr occurs as podiform lenses in serpentinites, e.g. in Zimasco mine at

Shurugwi. The Mashava, chrome is found in greenstone remnants within the Proterozoic Limpopo mobile belt on its southern flank south of Mberengwa.

## 4.3.3 Manganese

Manganese (Mn) is a silvery metallic element that has a close affinity with iron, and has similar physical and chemical properties. Manganese is not found in the native state but occurs mainly as oxides, hydroxides, silicate and carbonates. About 90% of manganese produced is used in the steelmaking industry (metallurgical) by removing oxygen and sulfur from iron ore melt, and also as a constituent in steel and non-ferrous alloys (copper, aluminum, nickel). Manganese helps improve the strength, toughness, stiffness, hardness, and wear resistance in steels. Other uses of manganese include the production of chemicals and alkaline manganese batteries.

Global production of manganese in 2020 was estimated as 18 million tons, and the five leading producers are South Africa (6.2 Mt), Australia (2.9 Mt), Gabon (2.3 Mt), China (1.8 Mt), (Brazil, 1 Mt). Ghana is the 10th largest producer with 0.39 Mt. Global resource base of manganese is estimated at 600 million tons.

### 4.3.3.1 Mineralogy and Geochemistry

The main minerals of manganese are oxides, hydroxide, silicate and carbonates but rarely as a sulfide (Table 4.6). The most important manganese ore is pyrolusite ($MnO_2$). Other economically important manganese ores usually show a close spatial relation to the iron ores. Land-based resources of manganese are large but unevenly distributed. About 80% of the known world manganese resources are in South Africa. Manganese is the 12th most abundant of the crust's elements, with an average concentration of 0.1% (1000 ppm). Its geochemical behavior and distribution are affected by the close association with iron. Mafic rocks, such as basalt, typically contain about 0.2% Mn, whereas granitic rocks contain about 0.06 %. Manganese ores contain a minimum of 15 percent manganese, but most ores are of considerably higher grade, ranging up to about 50% Mn. The cut-off grade is about 35%.

Table 4.6: Manganese Minerals and Metal Content

| | Mineral | Chemical Composition | Manganese Content |
|---|---|---|---|
| | Pyrolusite | MnO2 | 63% |
| | Braunite | 3Mn2O3. MnSiO3 | 64% |
| | Hausmannite | Mn3O4 | 72% |
| | Rhodochrosite | MnCO3 | 49% |
| | Cryptomelane | K[(Mn4,Mn2)8O16 | 57% |
| | Manganite | MnO(OH) | 52% |
| | Psilomelane | $(Ba,H_2O)_2Mn_5O_{10}$ | 47% |

Manganese occurs in three oxidation states, Mn2+, Mn3+, and Mb4+. In the surface environment, Mn is quite mobile and the solubility and mobility in various solutions is strongly controlled by the acidity (pH) and oxidation potential (Eh) of the environment. In general, manganese solubility is highest in more acidic (lower pH) conditions and less oxidizing (lower Eh) conditions in which dissolved manganese exists primarily in the most soluble +2 oxidation state. The transport of large quantities of manganese to sites of ore formation, generally requires the existence of oxygen-poor (low Eh) ocean water or surface water. Manganese tends to accumulate in soils and sediments at high redox potential.

#### 4.3.3.2 Deposit Types

Manganese deposits are classified into four general types: hydrothermal, sedimentary, metamorphic, and supergene. The most important sources of manganese are the sedimentary deposits and their metamorphosed equivalents. The sedimentary manganese deposits are the most abundant and economically relevant group of deposits, and can be divided into volcanogenic and non-volcanogenic subtypes. The following are the deposit types of manganese ores.

***Hydrothermal Deposits***

- Epigenetic veins of Mn carbonates and oxides produced by hydrothermal solutions and hot spring related-hydrothermal activity (e.g. Butte, Montana; Franklin, New Jersey).

***Sedimentary deposits***

- *Volcanogenic* deposits are generally small in size and associated with volcano-sedimentary sequences of primarily Phanerozoic age (e.g. Hokkaido, Japan).
- *Non-volcanogenic* deposits occur in clastic, carbonate, or iron-formation-dominated sedimentary sequences of various ages. They can be subdivided into oolitic oxide and carbonate ores.

(i) Oolitic manganese deposits ((e.g. Nikopol, Ukraine; Groote Eylandt, Australia).
(ii) Siderite and carbonate-rich Mn deposits (e.g. Karoo, South Africa)

***Metamorphosed Deposits***

- Associated with Archean and Proterozoic greenstone belts and Algoma-type iron formations. They are products of low-grade regional and contact metamorphism of manganiferous sediments (e.g. Postmasburg, S. Africa; Minas Gerais, Brazil)

***Surficial Deposits***

- Lateritic deposits (e.g. Nsuta, Ghana; Mounda, Gabon). Formed by lateritic weathering and residual concentration of secondary manganese minerals
- Supergene deposits (e.g. Kalahari, South Africa; Groote Eylandt, N. Australia). Formed by mobilization of Mn in weathered regolith and enrichment by supergene process.

#### 4.3.3.3 Exploration for Manganese

Manganese is a metal that is still in high demand because of its importance in the iron and steel industry. Exploration for manganese will definitely involve a combination of surface geological investigations, geophysical surveys and subsurface drilling. In selecting exploration targets, the focus should be on areas with potential for iron and manganese such as greenstone and schist belts as well as shallow marine sedimentary basins. Geophysical methods particularly magnetic and electromagnetic (because of association with Fe) can help in defining drilling targets. Although Mn is a very common element in the surficial environment, detailed geochemical surveys involving soil and stream sediments can be helpful in defining anomalies.

#### 4.3.3.4 Manganese Deposits in Africa

Africa is the world's largest producer of manganese from several localities within the Precambrian terrains of southern and western Africa. Global production of manganese in 2020 was estimated as 18 million tons. South Africa is by far the world's largest producer of manganese with a production of 5.2 million tons in 2020 and reserves of 200 million tons. Other major producers of manganese in Africa are Gabon 3rd in the

world with 2.8 Mt, Ghana 4th in the world with 1.4 Mt, and Ivory Coast (400,000 tons). Some manganese occurrences have been mined in Nigeria, Rwanda, Burkina Faso and Namibia. Africa is host to over 35% of the world's known land-based ore reserves of manganese, and produced about 54% of the 18 million tons (Mt) output in 2020.
Manganese deposits are mainly of sedimentary and supergene origin comprising four major types, namely banded iron formation (BIF)-hosted, black shale-hosted, oolitic and supergene/karst-hosted deposits. South Africa is the world's largest producer of manganese, followed in third place by Gabon. Ghana is also a notable producer of manganese. Minor manganese occurrences have been reported in Nigeria, Rwanda, Burkina Faso and Namibia (Fig. 4.4).

### *South Africa*

South Africa produced 6.2 million tons of manganese in 2019, and is by far the world's leading producer (35%) and exporter of manganese. It has the largest world manganese reserves with over 370 million tons and is credited with more than 80% of all world manganese resources which are located predominantly in the giant Kalahari manganese field in the Northern Cape Province (Fig. 4.4). The manganese deposits in South Africa are mainly of two types: *sedimentary* (BIF hosted and oolitic) and *supergene*. Prograde regional metamorphism has affected some manganese within the BIFs to create "metamorphosed" deposits. There are four main subtypes based on the type of host rock: banded iron formation (BIF) hosted, black shale hosted, oolitic ironstone and supergene/karst hosted.

The BIF hosted Kalahari Manganese Field (KMF) contains world-class deposits, worked by numerous mines that are confined within the Proterozoic Hotazel Formation which is a stratigraphic sequence composed of sedimentary banded iron-formation interbedded with carbonate-rich manganese-bearing sediments of the Postmasburg Group in the Transvaal Supergroup of Northern Cape Province. The area is largely covered by Kalahari sand deposits. The KMF deposits are exposed in a 90-km-long N-S belt over an area of about 400 km2. It contains three distinct manganese units, which are interlayered with iron-rich units over a thickness of approximately 150 m. The manganese- and iron-rich horizons are flat-lying and particularly fine-grained and up to 10m thick. Production comes from secondarily enriched high-grade deposits with an average of 48% Mn. The supergene deposits are accumulations that occur within karst-fill features in the dolomites of the Campbell and Subgroup in the Postmasburg Fe-Mn field, Principal ore minerals are braunite, cryptomelane, hausmannite, manganite, and rhodochrosite.

In some areas of the Kalahari Basin, low-grade metamorphic ores, with various oxide and carbonate ore minerals have developed, and the ores upgraded during the Late Proterozoic by a major hydrothermal-metamorphic event in the northwest part of the basin, producing some of the exceptional variety of Mn minerals found in the Kalahari basin, such as at the Mamatwan mine with 20-40% Mn. The ores from Wessels mine (so-called Wessels-type ores) are coarser, and of higher Mn content (30-44% Mn).

### *Gabon*

Gabon located on the central-western coast of Africa ranks third in the world with reserves of 45 million tons and annual production of 2.3 million tons in 2019. The Moanda mine is a key manganese operation in the country and the world's second-largest producer of high-grade manganese ore (2$^{nd}$ to KMF), The manganese deposits of Gabon are mainly lateritic concretionary manganese ore developed on relatively unmetamorphosed and undeformed manganese-bearing black carbonaceous shale unit in the argillaceous Bangombe Formation of the Paleoproterozoic Francevillian Supergroup overlying Achaean basement in the northwestern part of the Congo Craton The resource base is estimated as 325 Mt at about 49 wt% Mn, with regional exploration potential for another 100 to 200 Mt.

The Mn deposits that are up to 100 m thick occur on the surface of flat plateaus that are old Cenozoic lateralized erosion surfaces, and the manganese ore deposits are represented by manganocretes developed in areas where this surface overlie a manganiferous black shale. The black shale contains manganese carbonates, and the supergene processes have upgraded the manganese content from 15 wt% in the black shales to about 45 wt% in the ores The deposits are large and shallow and amenable to open-pit mining. The origin of the high-grade ores is related to the lateritic weathering of rhodochrosite-bearing black shale.

The Moanda mine near Franceville is the major manganese operation in the country and the world's second largest producer of highest grade manganese ore. Reserves are estimated at 250 million tons with a metal content of 48%–52% Mn.

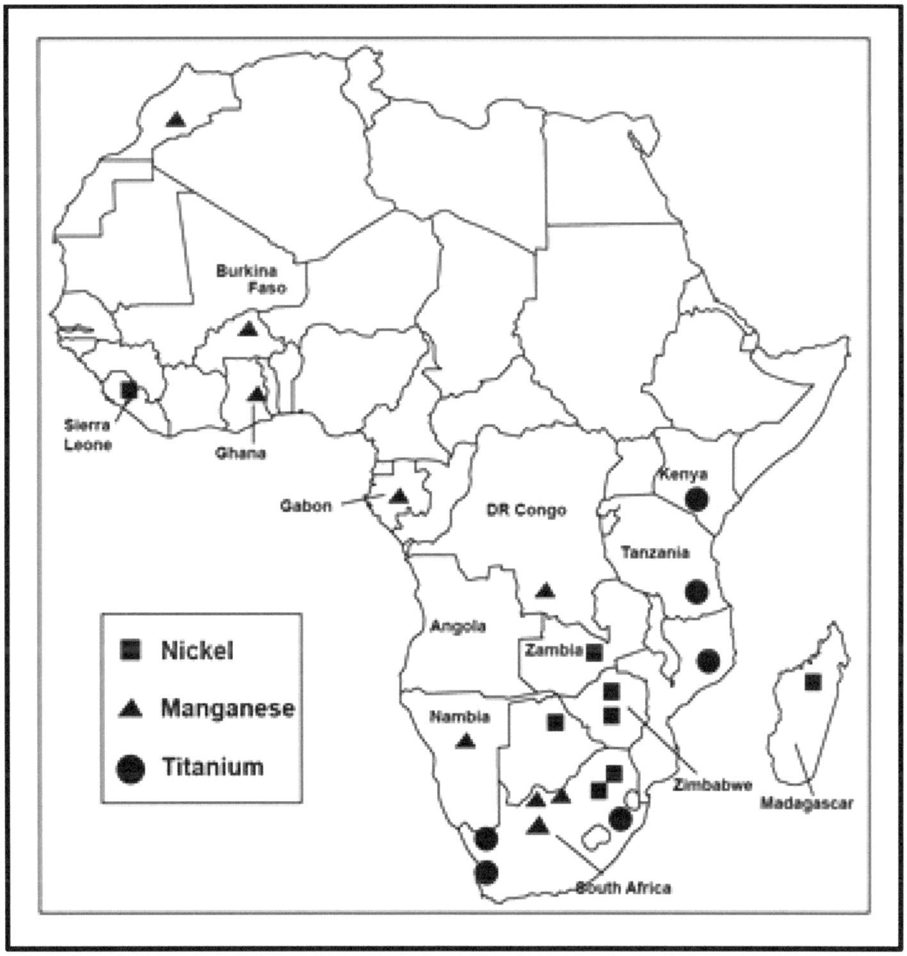


Fig. 4.4: Manganese, Nickel and Titanium Mines in Africa

### *Ghana*

Ghana is the tenth world's largest producer of manganese with about 400,000 tons in 2019. The manganese deposits are lateritic caps of high-grade oxide ores associated with chemical weathering of metamorphosed manganiferous sediments (phyllites) in the Proterozoic Birimian greenstone belts of the West African craton. Manganese ore is found as a series of elongated flat or irregular-shaped bodies, on the top of a series of hills. They are contained in the laterite profile overlying an extensive horizon of manganese-rich meta-sediments (metamorphosed shale) that seem to exist at a particular level in the stratigraphy of the Upper Birimian Group. The supergene Mn (IV) oxide ore caps, which are seldom more than 10-20 Mt in size, constitute premium high-grade metallurgical ores with high Mn/Fe ratios and 46-50% Mn. The ores are massive, patchy, or bedded accumulations of psilomelane, with numerous veinlets and cavity fillings of pyrolusite.

The *Nsuta* mine near Tarkwa in the country's Western Region is the only operational facility where mining has taken place since 1916. The high-grade lateritic oxide ores became exhausted in 1997, and since then low-grade carbonate (rhodochrosite) ore, with a grade of 28-30 wt% Mn, has been mined. The ore minerals are *nsutite* (Nsuta MnO2) which is the most abundant oxide component of the ores, which occurs mainly as replacements and cavity fillings in deeply weathered Precambrian phyllites. Other minerals include cryptomelane, pyrolusite, and lithiophorite. The 30-40 m thick manganese carbonate bed at Nsuta is interbedded with fine-grained black carbonaceous shale that grades downwards below and upwards above, the ore bed into a succession of greywacke and shale of turbidite origin. The manganese originated as a metamorphosed manganiferous sediment; as low-grade gondite which has been concentrated on an old erosion surface by lateritic weathering.

## 4.3.4. Nickel

Nickel (Ni) is a lustrous silvery-white metal that is hard, ductile and malleable. Primary nickel can be used as an alloy, such as ferro-nickel, or with chromium and other metals to produce stainless and heat-resistant steels for use as pots, pans, kitchen sinks, etc. Sixty five percent (65%) of nickel is used to manufacture stainless steel, and strengthen them to withstand corrosive environments and extreme temperatures. Twenty percent of produced Ni is used in other steel and non-ferrous (including "super") alloys, often for highly specialized industrial, aerospace, and military applications. The rest are used in electroplating and for coins and a variety of chemicals. In its various applications, it can often be recycled and the secondary nickel is then, commonly used as supplementation to newly mined material

World production of nickel was estimated as 2.5 million tons in 2010, with the following countries as the world's five largest producers: Indonesia (800,000 metric tons (mt), Philippines (420,000 mt), Russia (270,000 mt), New Caledonia (200,000 mt) and Australia (180,000 mt). Other notable producers are Canada, China, Brazil and the United States. Africa produced about 1% of world production mostly from southern Africa. Identified global resource base for nickel averaging 1% or greater is estimated at least 150 million tons (Mt), while identified reserves are 80 to 100 million tons. Australia has the largest reserves of about 20 Mt.

### 4.3.4.1 Mineralogy and Geochemistry

The most important ore mineral in magmatic deposits is pentlandite that is mostly intergrown with pyrrhotite (FeS) (Table 4.7). Garnierite is a mixture of nickel-rich silicates (Ni-serpentine, Ni-talc, Ni-smectite, Ni-chlorite, and others) found in laterites. Nickeline is widespread in hydrothermal deposits.

Crustal abundance of nickel averages 80 ppm. Nickel shows close affinity with Fe, and Mg in silicate melts. Concentrations are elevated in ultramafic/mafic rocks than in felsic rocks. Average contents in igneous rocks are; ultramafics 1000- 6000 ppm, basalt 160 ppm, and granite 6 ppm. During magmatic crystallization, sulfide deposits are generated by extraction of chalcophile metals such as Ni, Cu, Co, and PGM into a coexisting Fe-sulfide magma by liquid immiscibility which are subsequently crystallized or injected as massive sulfide deposits. In the surficial environment, Ni occurs in only one oxidation state Ni+2, and is released during lateritic weathering to accumulate in Fe-rich layers as a substitute for Fe in limonite and goethite. The most common Ni-bearing minerals in laterite deposits are garnierite and nickeliferous limonite/goethite. Where intense lateritic weathering involves acidic solutions or acid groundwater, Ni may be mobilized to accumulate in saprolite zones deeper in the weathering profile by supergene enrichment.

Table 4.7: Nickel Minerals and Metal Content

| | Mineral | Chemical Composition | Nickel Content |
|---|---|---|---|
| | Pentlandite | (Ni,Fe)9S8 | 35% |
| | Nickeline | NiAs | 44% |
| | Millerite | NiS | 65% |
| | Gersdorffite | NiAsS 3 | 35% |
| | Rammelsbergite | NiAs2 | 28% |
| | Ni- Skutterudite | (Co,Ni)As3 | 28% |
| | Ni Garnierite | Ni,FeSiO4 | 25–30% |
| | Nickeliferous Goethite/limonite | (Fe,Ni)O(OH) | Variable |
| | Siegenite | (Ni,Co)3S4 | 43% |

### 4.3.4.2 Deposit Types

Most economic nickel deposits occur in two geological environments: *magmatic* and *laterites*. Lateritic nickel accounts for about 40% of nickel production, and over 70% of global resources. Minor contributions of nickel as byproduct also come from hydrothermal polymetallic veins and sedimentary exhalative (Sedex) deposits.

***Magmatic Deposits***

- *Volcanic* Fe-Ni-Cu sulfide deposits in ultramafic (komatite) lavas (e.g. Kambalda, Australia). Formed by crystallization of immiscible sulfide melt in ancient ultramafic lava flows
- *Plutonic* Fe-Ni deposits in mafic intrusions (e.g. Sudbury, Canada; Stillwater Complex, USA). Formed by fractional crystallization within a magma chamber of immiscible sulfide melt as distinctive ore-rich layers within layered mafic intrusive complexes. Sulfide ore grades range from 0.15% to around 8% Ni, but 93% of known deposits are in the range 0.2–2%

***Hydrothermal Deposits***

- ***Veins*** of nickel sulfides in structural openings are found rarely in gabbroic intrusions (e.g. Zambia).

***Lateritic Deposits***

- *Latecitic deposits* which are Concentrations of nickel silicate and hydroxide formed by tropical and subtropical weathering of Ni-bearing ultramafic rocks with average nickel content of 1–1.6%. During the process of weathering, silica and other elements such as Mg and Ca are removed while the residual Ni and Fe are concentrated. The principal Ni minerals are garnierite (hydrous nickel silicate) and nickeliferous limonite. Lateritic nickel ores supply over 40% of world production. (e.g. Indonesia, Philippines and Australia)
- *Saprolite deposits* formed by supergene enrichment may occur in a saprolite zone deeper in the soil profile.

### 4.3.4.3 Exploration for Nickel

There are two main types of nickel deposits of interest in mineral exploration: magmatic and lateritic. They are both associated with ultramafic rocks. Exploration for nickel should be generally focused on finding ultramafic rocks with potential for laterite development. Modern geophysical methods used at various levels (regional, district and detailed) can be helpful during exploration for Ni-Co laterite deposits. At a regional scale, aeromagnetic, radiometric, gravity, and optical remote-sensing techniques provide target areas with major fault zones (locale for intrusive slivers), ultramafic/mafic bodies and related Ni-Co laterite deposits. This can be followed up by geological mapping, and the use of ground-based, magnetic, gravity, electromagnetic, electrical, (induced polarization), and ground-penetrating radar methods that can be useful in delineating the weathering profile of the Ni-Co laterite deposits. Geochemical stream sediment and soil surveys using geochemical indicators, e.g. Cu, Ni, Co and Cr can be very useful in defining orebodies.

### 4.3.4.4 Nickel Deposits in Africa

Africa is a relatively small producer of nickel accounting for less than 1% of world production but has substantial unproven resources in laterites which are widespread in equatorial Africa. Current production of nickel is derived mainly as a byproduct of mining platinum group metals from layered igneous complexes such as the Bushveld of South Africa and Great Dyke of Zimbabwe, as well as from "stand alone' nickel in ultramafic bodies intercalated within Archean greenstone belts. Lateritic nickel deposits in the tropical regions are known to have developed over exposed dunite and peridotite lithologies in the basal sequence of layered and other smaller intrusions.

Nickel production in Africa was concentrated mainly in Botswana, South Africa and Zimbabwe (Fug. 4.4), along what is described as the East African Nickel Belt (EANB), but in recent years Madagascar has emerged as the leading producer with annual production of 58,000 tons in 2019, followed by South Africa (52,000 tons), Botswana (28,000 tons), and Zimbabwe (19,000 tons). The known nickel deposits of the EANB all occur in mafic-ultramafic intrusive rocks of the Mid-Proterozoic Kibaran igneous event of mafic-ultramafic bodies. Laterite deposits are developed over exposed dunite and peridotite intrusives.

All of the primary Ni producers in Africa are of relatively low nickel concentrations usually less than 1.5% Ni, and thus their operations are adversely affected by fluctuating world Ni prices. Several large nickel deposits have been discovered elsewhere in Africa including Burundi, Tanzania and Ivory Coast, but until recently, their development has been hindered by political risk and limitations in energy and transport infrastructure.

***Madagascar***

Madagascar, a large island off the east coast of Africa has emerged as a leading producer of lateritic nickel. The Ambatovy mine situated in the eastern part of the island is one of the world's largest lateritic nickel mines with an annual production capacity of 60,000 tons of nickel. Current production is estimated as 58,000 tons in 2019. The deposit consists of two large weathered ultramafic lateritic orebodies located three miles apart overlying the Atampombato ultramafic-mafic intrusive complex that cuts across N-S trending Archean gneisses and migmatites. The igneous complex consists of multiple intrusions of gabbroic to syenitic rocks with two small outer ultramafic bodies. The lateritic deposits are found in weathered regolith ranging in thickness from 20 to 100 m with an average of 40m. The residual Ni along with Co and Fe are concentrated near the top ferricrete limonitic layer while an enriched saprolite layer is developed near the base of the weathered profile. Proven and inferred reserves are estimated as 180 million tons of ore ranging from 0.07% to 1.04% Ni.

***South Africa***

Most nickel production in South Africa comes from magmatic sulfide ore deposits, where it is produced as a byproduct of platinum mining at several mines within the Bushveld Igneous Complex. Concentration of Ni is in the well-known Merensky reef, where the ore mineral is pentlandite with an average nickel content of

0.24%. Because of the size of the Bushveld Complex, it is believed that the Ni resources are large when mined from the platinum reefs.

Outside the Bushveld, the largest individual resource of nickel as a primary product is that of the Nkomati mine in Mpumalanga, near the southeastern coast. The deposit is located within the Uitkomst layered ultramafic-mafic body of Bushveld age intruded into the basal sediments of the Transvaal Supergroup, Nickel occurs as pentlandite in lenticular masses within several layers of gabbro and peridotite. Because ore grades of 0.3 to 0.7% Ni are marginal for profitable production, at the current low Ni prices, the Nkomati mine has faced operational losses and temporary shutdown. Currently, two new nickel mining projects of similar mineralization as Nkomati are in the pipeline, namely the Zebediela nickel sulfide project in Limpopo, and the Burgersfort nickel project, in Mpumalanga, both may increase production by up to 20,000 tons per annum.

### *Botswana*

Nickel production in Botswana is derived from two operating facilities that exploit magmatic Ni-Cu deposits; the Tati nickel mine and the Selebi-Phikwe copper-nickel mine where Ni is considered a byproduct of copper mining. Production has fluctuated over the years due to low nickel prices and disputes between mining companies and the Botswana government. Both deposits are located in eastern Botswana near its border with Zimbabwe.

The Tati mine exploits the Phoenix deposit which is a relatively small deposit (up to 4.5 Mt of ore at 2.05% Ni and 0.85% Cu) localized within low- to medium-grade metamorphosed gabbro–norite intrusion situated within or at the periphery of the Tati Archean greenstone belt. Magmatic sulfide mineralization comprising masses of pyrrhotite, chalcopyrite, and pentlandite generally occurs in disseminated, massive, and vein like forms throughout the gabbro-norites. Proven reserves at Tati are estimated as 180,000 tons of 0.28%Ni and 0.21%Cu.

The Selebi-Phikwe mine is located in eastern Botswana, about 200 kilometers (km) south of Francistown. The deposit is hosted by Proterozoic high-grade metamorphosed gabbro-norites, pyroxenites and peridotite lenses intruding gneisses. Mineralization occurs as massive lenses and disseminations. Estimated reserves are 124, 000 tons of 0, 23%Ni and 0.27% Cu.

### *Zimbabwe*

In Zimbabwe, nickel occurs in two geological environments: Archean komatiite lavas and layered/unlayered ultramafic/mafic complexes. Within the Archean greenstone belts, Ni occurs as sulfide lenses in rocks of komatiite composition which occur as ultramafic bodies intercalated within metasedimentary and metavolcanic sequences in central and northeastern Zimbabwe. One of such operating facilities is the Trojan mine which is localized within piles of ultramafic lava metamorphosed in the lower amphibolite facies with lensoid bodies of serpentinite. The ore of pentlandite which appears as massive lenses is about 250 m long and 30 m wide. There are several other ore bodies containing massive and disseminated ore.

Nickel also appears in layered and non-layered mafic-ultramafic intrusive bodies such as. Empress and Madziwa Great dyke where Ni is produced as a byproduct of metal mining. It is also found in nickel laterite e.g. the northern part of the Great Dyke, and also as a hydrothermal vein along shear zones. . Most of the Zimbabwean primary nickel mines have been depleted or temporarily closed due to political and economic factors, but they still contain at least 0.4 Mt of Ni as defined mineral resources

### *Zambia*

In northwestern Zambia, a new type of hydrothermal Ni sulfide deposit in gabbro emplaced along major regional fault zones is being developed, which contains approximately 0.5 Mt of Ni with an average grade of about 1%. The latest total indicated and inferred resources for the deposits, at a cut-off grade of 0.6% Ni, are

10.3Mt at 1.2% nickel, 0.2% copper, 0.07% cobalt, 0.6g/t palladium and 0.3g/t platinum. The Munali nickel mine located 75 km south of Lusaka in southern Zambia is now operational and exporting high grade +10% Ni concentrate. Production ramp up is now on track to produce 3,300 tons of Ni in 2020 building up to 4,000 tons in 202.

## 4.3.5 Titanium

Titanium (Ti) is a hard silvery-grey metal that is lightweight, high-strength and high corrosion resistant. It is used as a structural metal and alloy for parts in high-speed aircraft, spacecraft and missiles. About 65 percent of titanium metal is used in aerospace applications; the remaining 35 percent is used in medical, sporting goods, and other applications. Titanium oxide is used as a whitening, brightening and opacifying agent. High-quality white paints usually contain significant amounts of titanium dioxide. Ilmenite (iron titanium trioxide) supplies about 90% of the world's demand for titanium. World ilmenite resources total about 1 billion tons of titanium dioxide). Identified world resources of other titanium minerals (primarily rutile, titanium dioxide) total about 230 million tons of contained titanium dioxide.

Global production of "sponge" titanium in 2016 was estimated as 170,000 tons of which the largest producers were China (60,000 tons), Japan (54,000 tons), Russia (38,000 tons), Kazakhstan (9,000 tons) and Ukraine (7,000 tons). Titanium sponge is a porous, brittle form of titanium, while titanium ore is titanium oxide usually in mineral concentrates. South Africa and Mozambique are the leading producers of titanium ore in Africa. China has the world's largest titanium reserves, with a total of 200 million tons identified to date, which make up 28.9% of the world total

### 4.3.5.1 Mineralogy and Geochemistry

The two most important ore minerals of titanium are ilmenite ($FeTiO_3$) and rutile ($TiO_2$). Other minerals are listed in Table 4.8. Titanium is a relatively abundant metal with a crustal abundance of 0.6%. During magmatic processes, Ti follows Fe in magmatic crystallization where $Ti^{4+}$ is predominantly partitioned into Fe-Ti or Fe oxides such as ilmenite and magnetite, or into one or more of the $TiO_2$ phases, rutile, anatase and brookite. Titanium may also substitute for $Mg^{2+}$ or $Fe^{2+}$ in silicate minerals (pyroxene, amphibole, mica and garnet), leading to enrichment of Ti in amphibole and mica. Titanium is enriched in mafic (>1% $TiO_2$) and ultramafic (>2% $TiO_2$) rocks relative to felsic and alkaline rocks (0.2% $TiO_2$).

Titanium is relatively immobile during prograde metamorphism except at high grade granulite facies. In normal hydrothermal processes, titanium is not involved in sulfide mineralization, but contact metasomatism may produce magnetite-rich rocks containing appreciable Ti. In sedimentary rocks, the concentration of $TiO_2$ is determined by the abundance of detrital oxides and silicates, such as chlorite and clay minerals, and diagenetic phases, such as anatase Titanium concentrations in shale, sandstone and carbonate rocks average 4600, 1500 and 400 ppm respectively. In soils and stream sediments a large proportion of the Ti is held in heavy minerals, such as rutile, ilmenite and sphene, all of which are relatively insoluble. Some Ti may be taken into solution in stream water, through the weathering of ferromagnesian minerals and authigenic phases, such as anatase, but secondary dispersion is generally restricted by adsorption to clays.

Table 4.8: Titanium Minerals and Metal Content

| | **Mineral** | **Chemical Composition** | **Titanium Content** |
|---|---|---|---|
| | Rutile | $TiO_2$ | 60% |
| | Anatase | $TiO_2$ | 60% |
| | Ilmenite | $FeTiO_3$ | 35% |
| | Brookite | $TiO_2$ | 60% |
| | Perovskite | $CaTiO_2$ | 35% |
| | Titanite/Sphene | $CaTiSiO_5$ | 18% |

### 4.3.5.2 Deposit Types

Economic concentrations of titanium are found in two main deposit types: *magmatic* and *placer* deposits. Other minor sources include metamorphic, metasomatic and hydrothermal deposits. Ilmenite is the main ore of magmatic and hydrothermal sources while rutile is the dominant ore in sedimentary, metasomatic and metamorphic sources.

***Magmatic Deposits***

- Disseminated ilmenite in mafic/ultramafic lavas and amphibolites.
- Iron-titanium oxide (Ti-magnetite and ilmenite) deposits in layered and massive mafic igneous complexes (e.g. Bushveld Igneous Complex, S. Africa; Lac la Blache, Canada)
- Iron-titanium and titanium oxide (ilmenite, hemo-ilmenite) deposits in anorthosite intrusive complexes (e.g. Allard Lake, Canada; Rogaland, Norway)
- Apatite-ilmenite-magnetite in carbonatites and alkaline igneous complexes.

***Contact Metasomatic Deposits***

- Formed by alteration of existing ilmenite-bearing rocks by fluids from mafic magmas converting ilmenite to rutile by removal of Fe (e.g. Kagero, Norway).

***Hydrothermal Deposits***

- Hydrothermal veins of rutile formed in shear zones along the contact of anorthosite gabbro intrusions with country rocks (e.g. Roseland, Virginia)

***Placer Deposits***

- Beach and alluvial deposits of heavy minerals containing ilmenite and rutile derived from the weathering of metamorphic and igneous rocks (e.g. southern coast, Australia; Ghangbama, Sierra Leone). Much of the high-grade ilmenite and rutile used today is extracted from unconsolidated placer deposits.

***Sedimentary Deposits***

- Iron-titanium oxides in older, lithified heavy-mineral beach sand deposits that may have undergone extensive post-depositional alteration and consolidation (e.g. Karoo. South Africa and Sierra Leone, West Africa).

***Metamorphic Deposits***

Both contact and regional metamorphism can transform Ti-minerals to rutile deposits.

- *Contact* metamorphism where country rocks intruded by and reacted with titanium-, vanadium-, and niobium-bearing fluids from adjacent alkali-rich rocks to form brookite ($TiO_2$) e with high amounts of vanadium and niobium (e.g. Magnet Cove, Arkansas).
- *Regional* high grade metamorphism (eclogite facies) of Ti-rich sediments results in the formation of rutile deposits (e.g. Piampaludo, Italy).

***Lateritic and Supergene Deposits***

- Formed by the lateritic weathering of carbonatites. Weathering of perovskite, magnetite, and titanite in titaniferous carbonatite, resulting in development of anatase (e.g. Tapira, Brazil).

### 4.3.5.3 Titanium Deposits in Africa

Titanium in Africa is produced mostly from beach placer deposits along the shorelines of South Africa, Mozambique, Kenya Sierra Leone and Madagascar (Fig. 4.4). Because these deposits are unconsolidated makes recovery of valuable heavy minerals such as rutile and ilmenite quite profitable.

***South Africa***

South Africa is the largest producer of titanium ore in Africa. In 2018, it produced over 500,000 tons of ilmenite and 100,000 tons of rutile. These economic minerals are produced from the extensive beach placer deposits located along the eastern, southern and northeastern coasts with smaller deposits along the west coast, north of Cape Town. Titanium minerals are recovered at three major mines namely, Richard's Bay, situated along the Indian Ocean coastline in northern KwaZulu-Natal, Hillendale, 20 km south-west of Richards Bay and Namaqua Sands mine on the west coast. All these deposits are heavy mineral beach sand deposits that are mined by dredging or hydraulic mining methods. Other sources of titanium ore in South Africa include the primary byproducts of mining Ti-Fe ores in the Bushveld Igneous Complex and the Palabora carbonatite.

The Bushveld contains the world's largest resource of titaniferous magnetite in the Upper Zone of the Rustenburg Layered Suite. Between 10 and 30 titaniferous magnetite-bearing layers have been identified although not currently mined for their titanium content due to the cost of mineral processing from hard rocks. These titaniferous magnetite layers may become economically viable when the ilmenite beach placers are exhausted in the future. Titanium is also produced as a byproduct of copper mining in the Palabora carbonatite complex.

***Mozambique***

Mozambique, along the eastern coast of Africa is host to the world's largest titanium placer mineral deposit, located 160 km from the city of Nampula. The Moma mine contains a number of beach sand deposits, which include Namalope, Congolone, Nataka, Pilivili, Mualadi, Martini, Marrua, and Quinga. Heavy surface mineralisation occurs in the sand units near Namalope flats and Tupuito high dune area. The sand deposits occur in a sequence of fine to medium grained sands and silty sands associated with clay units. The mineralisation contains about 85% ilmenite, 5% zircon and 2% rutile. Annual production in 2018 was estimated as 600,000 tons of ilmenite and 10 tons of rutile. Ore reserves are estimated as about 26 million tons of ilmenite, 1.8 Mt of zircon and 0.55 Mt of rutile Annual production in 2018 was 600,000 tons of ilmenite and 10 tons of rutile.

***Kenya***

Titanium ore in Kenya is produced mainly from the placer Kwale Sands on the southeast coast of the country. This is the first large mining development in Kenya in recent years. Annual production in 2018 was estimated as 330 000 tons of ilmenite and 100 tons of rutile. The Kwale mine features a high grade titanium ore body in beach sands with a high value mineral assemblage, rich in rutile. Mining initially commenced on the Central Dune orebody before transitioning to the South Dune orebody in June 2019. The Kwale operation currently accounts for approximately 65% of Kenya's mining industry by mineral output value.

***Sierra Leone***

Sierra Leone located along the coast of West Africa is one of the world's leading sources of rutile and has the largest natural rutile reserves in the world, estimated as about 500 million tons. The deposits are found in sands and gravels of the Lanti River south of the Gbangbama region in the southern province. Four groups of deposits are known to be distributed around the country: the Gbangbama, Sembehun, Rotifunk and the Kambia deposits. Annual production of rutile in 2018 was about 170,000 tons. Ore grade of rutile deposits ranges from 0.5 to 2%$TiO_2$. The Rotifunk Titanium Prospect in northern Sierra Leone is being developed for mining. The deposit is sedimentary in origin and occurs in clastic sedimentary rocks. The ore mined is composed of ilmenite, rutile and zircon. Annual production will be 56,000 tons of ilmenite and 83,000 tons of rutile for a mining life of 12 years.

## 4.3.6 Vanadium

Vanadium (V) is a silvery white soft metal that is used as a ferroalloy for high speed tool steel, high-strength steel and wear-resistant cast iron. Vanadium oxide is also utilized as a catalyst in sulfuric acid production. Global production of vanadium in 2019 was 73,000 tons, with contributions from only five countries namely; China (40,000 tons), Russia (15,000 tons), South Africa (8,000 tons), Brazil (7,000 tons) and the United States (500 tons). In 2019, approximately 90% of vanadium was recovered from magnetite and titano-magnetite ores, either from co-production or primary production. Co-production derived from iron slag processed for steel production remains the main source of vanadium, accounting for 71 % of 2019 global supply. Primary production from primary ores accounted for 18% of global supply in 2019. Secondary production of vanadium by recovery of material from fly ash, petroleum residues, alumina slag, and from the recycling of spent catalysts used in some crude oil refining accounted for 11 % of global supply. However, because vanadium is usually recovered as a byproduct or co-product from other mining operations, demonstrated world resources of the element are not fully indicative of available supplies.

### 4.3.6.1 Mineralogy and Geochemistry

The most important ore minerals from which vanadium is derived are *magnetite* and *titano-magnetite* that sometimes contains over 1% vanadium (vanadium-titanium magnetite). Other minerals containing vanadium are shown in Table 4.9. *Coulsonite* is a vanadium magnetite in which vanadium substitutes for iron in magnetite. *Patrónite* and carnotite are the only minerals ever mined as primary ores of vanadium, but are no longer sources of vanadium. Because of a mutual chemical affinity, some vanadium-bearing minerals contain uranium. These include *carnotite*, a hydrous potassium uranium vanadate, and *tyuyamunite*, a hydrous calcium uranium vanadate.

In nature, vanadium exists in three oxidation states; $V^{+3}$, $V^{+4}$ and $V^{+6}$. Because of similarities in ionic sizes, vanadium has a close geochemical affinity with iron and titanium in igneous rocks and minerals. At high temperatures, vanadium is mainly in the trivalent ($V^{+3}$) state and substitutes for $Fe^{+3}$ in magnetite, but with increased oxygen fugacity, vanadium changes to $V^{+4}$ and substitutes for $Ti^{+4}$ in titano-magnetite and ilmenite-hematite. Crustal abundance of vanadium is 150 ppm. In igneous rocks, vanadium is preferentially included in magnetite, titanium minerals and chromite, compared to mafic silicates. Consequently, average vanadium contents decrease from mafic rocks (250 ppm) to granites (20 ppm).

During fractional crystallization, vanadium does not fractionate much into residual fluids and rarely forms in hydrothermal and pegmatite deposits.

Chemical weathering of primary minerals liberates vanadium which is hardly soluble and is mainly dispersed by mechanical processes and not transported to depositional basins by runoff. However the high concentrations of vanadium in carbonaceous shales, oil shales, coals and petroleum suggest either an affinity of vanadium for organic matter or the favorability of conditions that preserve organic matter to partition dissolved vanadium into the sediment. Studies indicate the latter is the preferred explanation. Vanadium from several sources in oceanic basins (sea water, hydrothermal, volcanic etc) are precipitated from sea water by reduction of V+5 along with organic matter. During diagenesis some vanadium-organic complexes migrate with petroleum out of carbonaceous rocks, Vanadium is an important trace element in petroleum

### 4.3.6.2 Deposit Types

The most important deposits of vanadium are the magmatic magnetite and titano-magnetite deposits associated with layered mafic intrusions such as the Bushveld igneous Complex. In addition to primary deposits, current world production is also produced from recycled iron slags, petroleum ash and coal ash. The main types of primary ore deposits are:

Table 4.9: Vanadium Minerals and Metal Content

| | Mineral | Chemical Composition | Tungsten Content |
|---|---|---|---|
| | Coulsonite | FeV2O4 | Up to 4.8% |
| | Montroseite | VOOH | 42% |
| | Carnotite | K2(UO2)2(VO4)2·3H2O | 20% |
| | Tyuyamunite | Ca(UO2)2(VO4)2·5–8H2O MnWO4 | 20% |
| | Vanadinite | Pb5(VO4)3Cl 19% | 19% |
| | Patronite | VS4 | 28% |

***Magmatic Deposits***

- Titaniferous magnetite is the most important source for vanadium, accounting for 85 percent of its world production. It's typically mined in South Africa and China before being processed for vanadium extraction

***Sedimentary Deposits***

- Sandstone-type uranium deposits commonly contain vanadium that occur in the mineral carnotite (e.g. Colorado, USA). This deposit type accounts for about 2% of world production. Vanadium concentrations in sandstone-hosted uranium-vanadium deposits generally stand at 1% V2O5 or higher, although some deposits have grades of over 2.5% V2O5.
- Vanadiferous Black Shales that are rich in vanadium s are found mainly in late Proterozoic and Phanerozoic marine successions. Concentrations in these vanadiferous black shales regularly exceed 0.18 percent V2O5, and can be as high as 1.7 percent V2O5 (e.g. Hubei Province, China).
- Stone Coal (a type of hard compact coal) is an important vanadium-bearing resource in China and the gross reserves of vanadium in stone coal accounts for more than 87% of the domestic reserves of vanadium. .Average grade of vanadium extracted from these coals is 1.07% V2O5.

#### 4.3.6.3 Vanadium Deposits in Africa

***South Africa***

South Africa with annual production of 8,000 tons of vanadium in 2019 is the leading producer in Africa and the world's third largest. Vanadium is mined as a byproduct of iron and titanium mining in the Bushveld Igneous Complex. Vanadium occurs in quantities varying between 0.3 and 2% in layers of titanium-rich magnetite in the Upper Zone of the Bushveld Complex and in breccia pipes. Of the 38 million tons of vanadium reserves in the world, 12 million are in the Bushveld Complex. The four vanadium mining operations in the Bushveld region contribute 25% of global vanadium production.

### 4.3.7 Tungsten

Tungsten (W) is a greyish-white metallic element with high electrical resistance and the highest melting point among pure metals. It is used as a ferroalloy in the manufacture of hard and heat-resistant steels and in superalloys to form wear-resistant coatings. The high electric resistance makes tungsten usable as electrodes, heating elements and field emitters, and as filaments in light bulbs and cathode ray tubes. The high density is exploited in armor-piercing ammunition and in sports (e.g., arrowheads in archery). Today, the majority of tungsten is used in manufacturing hard metals, such as tungsten carbide used for grinding and cutting tools.

China is by far the world's leading producer and exporter of tungsten. Production among the world's five largest producers in 2019 was China (70,000 metric tons), Vietnam (4,800 tons), Russia (1,500 tons), Bolivia (1,200 tons) and Rwanda (1,100 tons). Global reserves of tungsten is estimated at around 3.5 million tons.

#### 4.3.7.1 Mineralogy and Geochemistry

The main ore minerals of tungsten are wolframite ($(Fe,Mn)WO4$) and scheelite ($CaWO4$). A list of the other tungsten minerals is shown in Table 4.10. Scheelite is the most abundant tungsten mineral and is present in approximately two thirds of known tungsten deposits. Stolzite occurs as a result of hydrothermal alteration of scheelite. Tungstenite ($WS_2$) is the most common example of the sulfide group of tungsten minerals. It has the same hexagonal layered structure as molybdenite ($MoS_2$).

The crustal abundance of tungsten is about 1.5 ppm. Tungsten shows a close association with tin (Sn) and molybdenum (Mo), as they occur together in many deposit types. Average concentration of tungsten in granitic rocks is 2 ppm and 0.5 ppm in mafic rocks. During magmatic crystallization, tungsten fractionates into the residual fluids along with silica and alkalis which may end up in pegmatites or hydrothermal solutions. Hydrothermal transport of tungsten is mostly by acidic solutions. At temperatures of 300-500$^{o}$C, the reaction of the acidic hydrothermal fluids with carbonate rocks during contact metamorphism results in the formation of calcium feldspar (plagioclase) and precipitation of metals as contact metasomatic deposits and skarn minerals. In the surficial environment, the geochemical behavior of tungsten is similar to that of molybdenum. It is soluble and mobile in soils derived from the weathering of ores and rocks. It may be toxic with long term residency as tungstate particulates.

#### 4.3.7.2 Deposit Types

Almost all primary tungsten deposits are associated with granitic intrusions or with medium- to high-grade metamorphic rocks. Most (i.e. more than 70% of intrusion-related deposits were formed during the Mesozoic period (251-65 million years), although other ages are also known. Tungsten deposits are usually located within or near the areas affected by subduction-related convergent plate tectonics, particularly in the Circum-Pacific belt. The world's largest accumulation of tungsten deposits is found in Eastern Asia, extending from Korea, Japan and China into Vietnam and Thailand. The various deposit types are skarns (40%), hydrothermal veins, breccia and stockworks (35%), porphyry tungsten (15%), disseminations/greisens (5%), stratabound deposits (3%), and pegmatites (2%).

Table 4.10: Minerals and Metal Content of Tungsten

| | Mineral | Chemical Composition | Tungsten Content |
|---|---|---|---|
| | Wolframite | (Fe,Mn)WO4 | 61% |
| | Scheelite | CaWO4 | 64% |
| | Ferberite | $FeWO_4$ | 61% |
| | Hubnerite | MnWO4 | 61% |
| | Stolzite | PbWO4 | 40% |
| | Tungstenite | WS2 | 74% |

***Magmatic Deposits***

- Disseminated deposits consist of tungsten minerals disseminated in altered (greisenized) granite. Tungsten generally occurs as wolframite; scheelite may be important in some deposits (e.g. Krasno, Czech Republic).
- Pegmatite deposits (e.g. Okbang deposit in South Korea).

***Contact Metasomatic Deposits***

- Skarn deposits of scheelite (tungsten skarn) formed by the replacement of carbonate rocks along the contact with granitic intrusions. Mineralization might be mono-metallic tungsten (almost exclusively as scheelite) or polymetallic often with Mo or base metals: Pb, Zn, and Cu (e.g. Cantung, Canada; Shizhuyuan, China).

***Hydrothermal Deposits***

- Hydrothermal quartz-wolframite veins, greisens and stockworks within granitic intrusions or surrounding rocks. Rarely scheelite veins occur (e.g. Gilanwe, Rwanda: Panasqueira, Portugal; San Fix, Spain).
- Porphyry tungsten deposits of wolframite and/or scheelite (with Mo) dissemination, quartz veinlets and stockworks associated with granite porphyries. Mineralized breccia zones, either irregular or pipe-shaped, may also be present (e.g. Xin Luo Keng, China: Logtung, Yukon, Canada).

***Sedimentary Deposits***

- Stratabound tungsten deposits in metasedimentary rocks (e.g. Mittersill, Austria, and possibly Damingshan, China,).
- Placer deposits of scheelite or wolframite in alluvial, eluvial, and, in some cases, marine sediments usually not too far from source areas (e.g. Heinze Basin, Maymar).

### 4.3.7.3 Exploration for Tungsten

In the exploration for tungsten, the selection of target areas should be based on geological criteria, with favorable intrusive rocks being the principal focus of investigations. The main exploration methods for tungsten include geological prospecting with the aid of an ultraviolet (UV) lamp), alluvial heavy mineral prospecting, and stream-sediment and soil geochemistry, and some geophysical techniques. Selection of an exploration target should be based on geological criteria, with favorable granitoids being the primary consideration. The conspicuous fluorescence of scheelite in UV-light has been used worldwide for many discoveries by locating outcropping orebodies and alluvial trails of scheelite Therefore, the most efficient method for defining tungsten targets is by stream sediment heavy mineral evaluation and chemical analysis of fine alluvial sediments and soils that may contain scheelite It is well known that scheelite and wolframite particles are resistant enough to make alluvial prospecting a very effective technique In soils, tungsten dispersion is limited due to resistance to secondary dispersion or mobilization, although such limitations may be useful in detailed delineation of ore bodies.

#### 4.3.7.4 Tungsten Deposits in Africa

Rwanda and Zimbabwe are notable producers of tungsten in Africa. Most of the deposits are hydrothermal in origin.

***Rwanda***

Rwanda, a small landlocked country in Central Africa is the world's fifth-largest producer of tungsten with an annual output of 1100 tons in 2019. Most of the tungsten is derived from polymetallic vein deposits of tin-tungsten-tantalum within Proterozoic metasediments in the Kibaran metallogenic belt that extends from Uganda into the Democratic Republic of Congo. They are mostly linked to Kibaran granites associated with the Mesoproterozoic Kibaran orogeny. There are several tungsten occurrences in Rwanda within the so-called "tungsten belt" including Rutsiro, Nyakabingo, Gifurwe and Bugarama deposits. The mineralized tungsten (with Sn-Ta) occurs as series of quartz and ferberite ($FeWO_3$) veins (up to 1 m wide) localized within a black schist-dominated formation of Proterozoic Rwandan Supergroup metasediments that include carbonaceous and pelitic schists and quartzites. Most of the mining of tungsten is carried out by artisanal and small-scale miners which makes production estimates unreliable.

***Zimbabwe***

The RHA open-pit mine in the Kamativi Tin Belt of northwestern Zimbabwe produced tungsten starting in 2015 from a small but high-grade deposit containing a series of parallel ore-bearing quartz-tourmaline veins in garnet-mica schists of the Tshontanda Formation in the Zambezi Mobile Belt. The ore minerals are wolframite, ferberite and minor scheelite.

### 4.3.8 Molybdenum

Molybdenum is a silvery-white metallic element that is ductile and highly resistant to corrosion, and has one of the highest melting points of all pure elements (only the elements tantalum and tungsten have higher melting points). It is used principally as an alloying agent in cast iron, steel, and super alloys to enhance hardness, strength, toughness, and wear and corrosion resistance. Electrical conductors made of molybdenum are used in thin-film transistors (TFTs) (e.g., for flat screens), thin-film solar cells, and halogen lamps. The largest world producers in 2019 were China (130,000 tons), Chile (54,000 tons), United States (44,000 tons), Peru (28,000 tons) and Mexico (19,000 tons). Identified molybdenum resources of the world are about 18 million tons, of which 5.4 million tons are located in the United States. Most molybdenum production is derived as a main product and byproduct of copper mining in porphyry type deposits and porphyry skarns.

#### 4.3.8.1 Mineralogy and Geochemistry

Molybdenite ($MoS_2$) with 66% Mo is the main primary mineral of molybdenum. Other minerals include wulfenite ($PbMoO_4$) with 26% Mo, and powellite ($CaMoO_4$) with 48% Mo. Crustal abundance of molybdenum is 1-2 ppm. Felsic and mafic rocks show little difference in molybdenum content with granitic rocks averaging 1 ppm, basalts, 1ppm; gabbros, 0·6 ppm and ultramafic rocks, 0.2 ppm. Shales poor in organic matter contain about as much molybdenum as igneous rocks (1 ppm) whereas carbonaceous and pyritic shales are much richer in molybdenum with contents as high as 100 ppm. Chemical weathering of molybdenite ores, releases Mo in the environment where it is transported via association with organic matter and iron and manganese hydroxides into aquatic systems where it has long residence times.

#### 4.3.8.2 Deposit Types

Molybdenum is closely associated with intermediate to felsic intrusive and subvolcanic porphyries where it occurs as disseminations, hydrothermal veins, stockworks, greisens and skarns. Most of the world's molybdenum. Types of deposits are listed as follows:

***Magmatic Deposits***

- Pegmatite bodies and dukes containing molybdenite in association with cassiterite (Abitibi, Quebec; Sabshendo, China).

***Contact Metasomatic Deposits***

- Skarn Mo- or Mo-Cu deposits along the contact of intrusive rocks and carbonate country rocks (e.g. Tibet, China)

***Hydrothermal Deposits***

- *Porphyry* molybdenum and Cu-Mo deposits in which molybdenum and copper sulfides occur as disseminations, quartz veins and stockworks in quartz monzonite and alkali granites and porphyries (e.g. Climax, Colorado; Butte, Montana). These deposits contribute more than 50% of world production and 75% of China's production.
- *Hydrothermal* veins of quartz-molybdenite and other sulfide minerals occupy fractures and faults in a variety of igneous and metamorphic rocks (e.g. Tyrol, Austria). Sometimes, molybdenum is associated with tungsten and tin in greisens (e.g. Xing Luo Keng, China).

#### 4.3.8.3 Molybdenum Deposits in Africa

Africa is not a major producer of molybdenum because the majority of global production comes from porphyry deposits that formed mostly along Mesozoic-Cenozoic convergent plate margins around the Pacific Ocean. The only known copper-molybdenum mine is in Namibia where it is attributed to the so-called "world's oldest porphyry copper deposit" of the Archean age.

***Namibia***

The Haib copper-molybdenum mine is a large porphyry deposit located at the southern end of Namibia and close to the South African border. The measured and indicated resources of the mine are estimated to be 456.9 million tons (Mt) grading 0.31% copper mined at the rate of 8.5 Mt per annum. Molybdenum is a byproduct of copper mining. The Haib deposit is regarded as the oldest known porphyry-type deposit hosted by a Proterozoic porphyritic, quartz-feldspar porphyry containing identifiable quartz veins and typical hydrothermal alteration. The host rock is part of the Richtersveld geological sub-province comprising Proterozoic volcano-sedimentary sequence in the Namaqua-Natal Province in the northern corner of South Africa.

## 4.3.9 Niobium-Tantalum

Niobium and tantalum are two metals that are almost always found together in nature because of their similarity in physical and chemical properties. Niobium is a lustrous, gray, ductile metal with a high melting point, relatively low density, and superconductor properties. Tantalum is a very hard, heavy, grayish-silver, ductile, and easily fabricated metal. It is highly conductive to heat and electricity and renowned for being nearly immune to chemical attack at room temperatures.

Approximately 80% of the world's produced niobium is used to manufacture high strength low-alloy steels. Niobium enhances the steels' mechanical strength, toughness, high-temperature strength, and corrosion resistance for use in pipelines, transportation, and structural applications. Appreciable amounts (>20 percent) of niobium are used in nickel-, cobalt-, and iron-based superalloys for high-temperature applications in jet engines, gas turbines, rocket subassemblies, turbocharger systems, and combustion equipment. Niobium alloys are also used to manufacture superconducting magnets for medical hardware, such as magnetic resonance imaging (MRI) and nuclear magnetic resonance (NMR) instruments.

Tantalum is mostly used in the electronics industry in capacitors for medical appliances such as hearing aids, cell phones, hard drives and pacemakers, airbag protection systems, ignition and motor control modules. In its pure form, tantalum is ductile and can be drawn into fine wire, enabling its use as filaments. Tantalum's low mechanical strength and high biological compatibility allow it to coat stronger substrates, like stainless steel, for medical applications.

World annual mine production of niobium since 2008 has been steady at about 60,000 tons while tantalum has declined from 1400 tons in 2005 to about 800 tons in 2017. Brazil, Canada, and Australia are the leading global producers of niobium and tantalum mineral concentrates. Brazil produces the greatest amount (90%) of niobium mineral concentrates while Australia leads in the production (50%) of tantalum mineral concentrates. The main source of Nb in Brazil and Canada is the pyrochlore group minerals associated with carbonatites as disseminations and lateritic/sedimentary deposits. Global production of tantalum in 2020 was estimated at 1800 metric tons. The major sources are Central Africa (66%), Nigeria (18%), Brazil (6%) and Australia (5%), Africa is the leading producer with production from the following countries: DRC Congo, 700 tons (50%), Rwanda, 500 tons (35%), Nigeria, 150 tons (10.5%) and Ethiopia, 70 tons (5%). Other minor producers are Burundi, Uganda and Mozambique. Mining is through artisanal and small scale mining which makes estimates of production not very reliable. World reserves are estimated as about 315,000 tons with Africa being responsible for 16% second only to South America.

### 4.3.9.1 Mineralogy and Geochemistry

Niobium and tantalum do not occur as free metals in nature, but form mainly oxide minerals and rare silicates (Table 4.11). Niobium and tantalum also substitute for major ions in a number of other minerals, in which they typically have low concentrations. The vast majority of the economically important species are oxides that form part of a solid solution series between niobium and tantalum end members of columbite (niobite) and tantalite. The columbite-tantalite mineral group is the most common group of tantalum- and niobium-bearing minerals.

Niobium is primarily derived from the complex oxide minerals of the pyrochlore group ((Na,Ca,Ce)2 (Nb,Ti,Ta)2 (O,OH,F)7 ), which are found mainly in carbonatites and peralkaline igneous rocks. Tantalum is derived mainly from the mineral tantalite ((Fe,Mn)(Ta,Nb)2 O6 ), which is found as an accessory mineral in rare-metal granites and pegmatites. Wodginite (Ta,Nb,Sn,Mn,Fe)O2 is also an important source of tantalum.

The average crustal abundance of niobium is 8 ppm and tantalum is 0.7 ppm. .Their large ionic size and high charge reduce their ability to substitute for more common elements in rock-forming minerals and make them practically immobile in many aqueous solutions. They show close affinity with the alkali metals and tend to fractionate into the residual fluids of granitic melts. Although niobium and tantalum generally show coherent geochemical behavior, some chemical processes are able to separate them, which results in preferential enrichment or depletion of one or the other. Niobium tends to be enriched in peralkaline, Na-rich melts while tantalite has preference for potassium -rich melts in pegmatites.

In the surficial environment, both niobium and tantalum are resistant to weathering and tend to accumulate in eluvial and alluvial placers. However the pyrochlore group of minerals breakdown into secondary minerals during lateritic weathering and are subject to mobility and concentration in supergene deposits.

Table 4.11: Niobium and Tantalum Minerals and Metal Content

| | Mineral | Chemical Composition | Niobium Content (%) | Tantalum Content (%) |
|---|---|---|---|---|
| | Columbite | $Fe,Mn)(Nb,Ta)_2O_6$ | 78.72 | n.a. |
| | Tantalite | $Fe,Mn)(Ta,Nb)_2O_6$ | n.a. | 86.17 |
| | Pyrochlore | $Na,Ca)_2Nb_2O_6(O,OH,F)$ | 75.12 | n.a. |
| | Microlite | $Na,Ca)_2Ta_2O_6(O,OH,F)$ | n.a. | 83.53 |
| | Tapiolite | $Fe,Mn) (Ta,Nb)_2O_6$ | 1.33 | 83.96 |
| | Ixiolite | $Ta,Nb,Sn,Mn,Fe)_4O_8$ | 8.30 | 68.96 |
| | Wodginite | $Ta,Nb,Sn,Mn,Fe)O_2$ | 8.37 | 69.58 |
| | Loparite | Ce,La,Na,Ca,Sr)(Ti,Nb)O3 | 16.15 | n.a. |
| | Lueshite | $NaNbO_3$ | 81.09 | n.a. |
| | Euxenite | $Y, Ca, Ce, U, Th)(Nb, Ti, Ta)_2O_6$ | 47.43 | 22.53 |
| | Struverite | $Ti,Ta,Fe)O_2$ | 11.32 | 37.65 |
| | Ilmenorutile | $Fe_x(Nb,Ta)_{2x}4Ti_{1-x}O_2$ | 27.9 | n.a. |

### 4.3.9.2 Deposit Types

Niobium and tantalum mineral deposits are most commonly associated with igneous rocks, including granites, pegmatites, syenites and carbonatites. Some secondary deposits, where niobium- and tantalum-bearing minerals have been concentrated by weathering and sedimentary processes, are also known. Primary niobium and tantalum deposits occur mostly as magmatic disseminations in alkaline granites, pegmatites and carbonatites, and can be divided into three main types, on the basis of the igneous rocks with which they are associated.

***Magmatic Deposits***

- Carbonatites and associated rocks. Disseminated pyrochlore in carbonatite igneous rocks in alkaline igneous complexes (e.g. St. Honore, Brazil & Canada). Reserves estimate of 23.5 tons of 0.59% Nb2O5.
- Alkaline to peralkaline granites and syenites. Magmatic disseminations and hydrothermal concentration of Nb and lesser Ta in silicic alkaline granites and syenites (e.g. Pitinga, Brazil; Ilimausaaq, Greenland). 100 Mt of 0.1-1% Nb2O5 and <0.1% Ta2O5.

- Granites and pegmatites of the LCT group that are enriched in lithium, cesium and tantalum (e.g. Yichun, China; Beauvoir, France; Tanco, Canada). Mostly tantalum deposits disseminated in granites and pegmatites.
- Contact metamorphic deposits in which Nb-Ta minerals occur in the contact aureole of igneous intrusives (e.g. Rwanda deposits).

***Sedimentary Deposits***

- Carbonatite-sourced sedimentary deposits overlying carbonatite intrusions in which niobium is enriched (e.g. Tomtor, Russia).
- Placer deposits (alluvial) of tantalite derived from the weathering of pegmatites and columbite from weathering of granites and transportation to drainage systems (e.g. Nigeria, Malaysia and Congo).

***Lateritic and Supergene Deposits***

- Niobium deposits produced by lateritic weathering and supergene enrichment of pyrochlore (e.g. Araxa, Brazil).

### 4.3.9.3 Exploration for Niobium Deposits

Niobium and tantalum are closely associated with pegmatites and alkaline granites. Initial exploration typically involves a search for rare-metal peraluminous granite intrusions that can serve as the source of pegmatites. Based on extensive geochemical studies, the rare metal pegmatites with mineralization potential may be identified by elevated concentrations of alkali elements such as Li, Rb, Cs and Nb/Ta ratios. Pegmatites with potential to host a tantalum deposit generally have low Nb:/Ta ratios in bulk chemical analyses, as well as abundant tantalum-rich oxide minerals. Radiometric surveys can aid in exploration for LCT-type pegmatites. Geochemical surveys of heavy minerals in stream sediments and soils are also useful tools in defining geochemical anomalies.

### 4.3.9.4 Niobium-Tantalum Deposits in Africa

Current world production of niobium is dominated by two countries: Brazil with 90% and Canada with 8%. The remaining 2% is shared among several countries including Burundi, China (Yichun Mine), Ethiopia (Kenticha Mine), Nigeria, Russia (Lovozero Mine), Rwanda, and Uganda (Fig. 4.5). Prior to the discovery of Nb in carbonatites in the 70s, Nigeria was the world's leading producer of columbium (now niobium) from columbite in alkaline granites of the Jos Plateau.

On the other hand, African countries are substantial producers of tantalum and coltan in the columbite group minerals primarily within rare earth metal pegmatites and granites enriched in lithium, cesium and tantalite (LCT). The pyrochlore group minerals are of no significance in the production of Ta in African deposits. Collectively, African operations are currently responsible for more than half of the world's primary Ta. The leading producers are Democratic Republic of Congo, Rwanda, Mozambique, Ethiopia), Nigeria, Namibia, Zimbabwe and Burundi. The estimated Ta resources of Africa are >50,000 tons of contained tantalum ore (Ta2O5), representing 16% of the world resources. Currently, the economically most important mining districts in Africa are located within the Kibaran Belt (mainly Rwanda and the DR Congo), around the Kenticha mine (Ethiopia) and in the Alto Ligonha Province (Mozambique). Contribution to Ta production from Nigeria is derived mostly from weathered pegmatites and alluvial placers along the Pegmatite belt of Central Nigeria. Overall production from pegmatites and related placer deposits is mostly from hundreds of artisanal and small-scale workings with only a handful of semi-industrial Ta mines.

### *Democratic Republic of Congo (DRC)*

The Democratic Republic of Congo is believed to be the leading producer of tantalum in Africa and one of the world's largest. However, because its society is fractured from decades of civil war and the use of the "coltan" (the local name for columbite-tantalite mining), as a source for funding of militias, it has been difficult to evaluate mineral production and sources. Tantalite is regarded as one of the "conflict minerals" from DRC and is subjected to sanctions by the international trade community.

More than a million local artisanal and small scale miners extract tantalum metal from coltan ore in eastern Congo, from where the mineral is smuggled through Rwanda to East Asia. Most of the tantalum ore is derived from the pegmatites of the Proterozoic Kibaran granites within the Kibaran orogenic belt. The tantalite produced is the Fe-rich variety as opposed to the Mn-rich variety found elsewhere in Africa. Production of coltan ore is estimated at about 250 tons per annum. The nature of mining in the Congo makes it impossible to know what tantalum reserves the country possesses, but they are believed to be substantial and will last for decades to come.

### *Rwanda*

Rwanda is a small country in the eastern border of DRC. The country is a major producer of columbo-tantalite contributing about 18% of current global production. It is a more peaceful country than DRC with a better organized mining industry. Columbite-tantalite (coltan) occurs mainly in the granitic pegmatites in the Karagwe-Ankole Belt sequence of metasediments that is an extension of the Kibaran Belt to the west. The Nb-Ta-Sn bearing pegmatites are found at the middle and outer contact metamorphic aureole of the associated granite plutons, usually in lenticular and concordant bodies that are up to 10 m wide and 250 m long (e.g. Ntunga Nb-Ta-Sn pegmatite). The mined coltan ore has an average tantalum pentoxide (Ta2O5) content of 26% and niobium pentoxide (Nb2O5) of 18%. Most of the production is derived from artisanal and small scale workings, although some companies and cooperatives are getting involved in mining activities. Estimated production of columbite-tantalite is estimated as 220 tons. However, because a substantial amount of tantalite produced in DRC is channeled through Rwanda, it is difficult to have a realistic estimate of the true production from mines in Rwanda.

### *Mozambique*

Mozambique, located on the eastern coast of Southern Africa contains tantalum reserves estimated by the United States Geological Survey (USGS) as 3,200 metric tons with annual production of 90 tons. Most of the tantalum is found within the Pan-African Mozambique Belt in which synorogenic granites and pegmatites are host to Nb-Ta mineralization. The Alto Ligonha pegmatite belt hosts historic tantalum mines, including the Marropino, Morma and Muiane mines. The columbite-tantalite of the Alto Ligonha pegmatite belt is Mn-rich compared to the Fe-rich variety in the Kibaran belt. The Marropino mine is the most important mining operation located in the northern part of Mozambique in the Zambezia Province. Reserves are estimated as 21.7 million tons of ore grading 0.02% tantalum. The Morma mine also located in the Zambezia Province contains reserves of 7.5 million tons of ore grading 0.07% tantalum.

### *Nigeria*

Before the discovery of pyrochlore as the main source of world supply of niobium, Nigeria was the world's leading producer of columbite as a source of niobium. Most of the columbite was derived from decomposed biotite granites and alluvial deposits in the Jurassic Younger Granite Province of the Jos Plateau. Resources of columbite have dwindled over the years. Most of the current production of tantalum and coltan originate from decomposed pegmatites and alluvial deposits associated with the Pan-African Older Granite pegmatites mostly in Central Nigeria along a well-defined pegmatite belt in Nasarawa, Kaduna and Kogi

States. Most of the mining is undertaken by artisanal and small scale miners thereby making correct estimates of production of tantalum and coltan ores very difficult. The proportion of tantalite in the ores range from 10% to 60% $Ta_2O_5$. Estimated annual production is about 80 to 100 tons of tantalum. Nigeria harbors substantial resources of tantalum along the pegmatite belt but community conflicts and insecurity bedevils the country's mining industry.

***Ethiopia***

Ethiopia has in recent years emerged as a notable producer of tantalum in Africa with an output of about 70 metric tons derived mainly from the Kenticha mine located in the Oromia Region of southern Ethiopia. Tantalum-niobium mineralization occurs in zoned beryl-columbite-bearing pegmatite bodies containing black tourmaline, greenish and bluish beryl, Fe-columbite (magnetic) and Mn-tantalite (non-magnetic) associated with biotite granites that were deformed and metamorphosed to upper greenschist and amphibolite grade within an extension of the N-S trending Pan-African Mozambique orogenic belt. The Kenticha pegmatite rarely carries cassiterite but has spodumene (lithium pyroxene). The estimated reserves of primary ore are e 2400 metric tons at a grade of 0.02% $Ta_2O_5$ with a ratio of 1:1 up to 3:1 of Ta and Nb (60%Ta to 20% Nb). Spodumene, a source of lithium is mined as a byproduct.

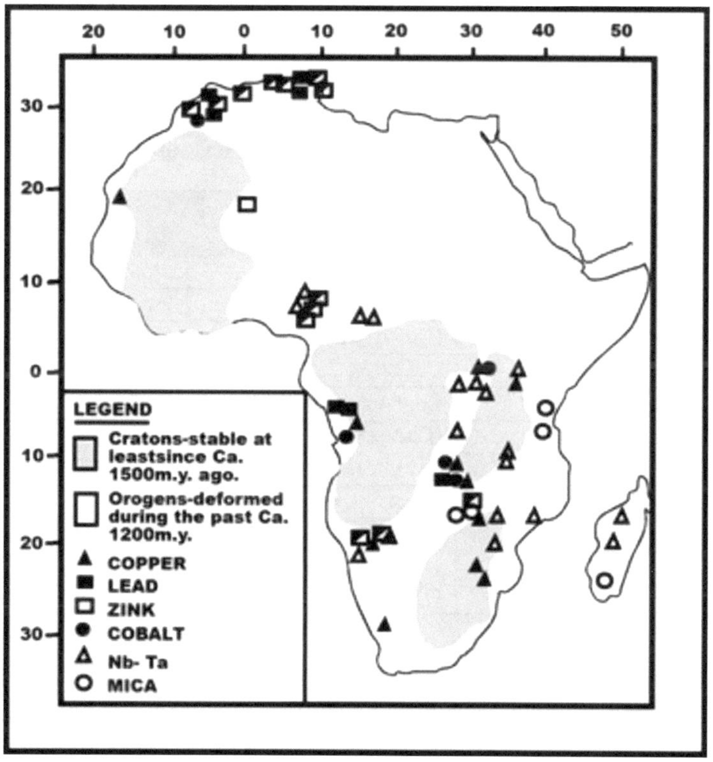


Fig. 4.5: Major Base Metal Deposits in Africa

## 4.3.10 Cobalt

Cobalt (Co) is a bluish-gray, lustrous, brittle and very hard metallic element that has close affinity with nickel and iron. It is used as a ferroalloy in the production of corrosion-resistant and heat-resistant super-alloys especially, for jet engines. Cobalt is an ingredient in the production of lithium-cobalt batteries as well as an important color pigment for glass, glazing, and lacquers. Cobalt is mostly mined as a byproduct of either copper or nickel. China is the leading consumer of cobalt, driven by the demand from its rechargeable battery industry.

World production of cobalt in 2020 was 140,000 tons. The Central African Copper Belt yields most of the global cobalt production. The Democratic Republic of Congo (DRC) is by far the world's largest producer, accounting for roughly 71 percent of global production with an output of about 100,000 tons in 2020. Other leading producers are Russia (5,900 tons), Cuba (4900 tons), Australia (4,700 tons) Philippines (4,600 tons) and Madagascar (3,300 tons). The total global cobalt reserves amount to 7 Mt tons, and the DRC has the largest cobalt reserves in the world, at some 3.6 million tons as of 2019, which is nearly half of the world's reserves of the metal. Australia, in second place, holds 1.2 million metric tons of the global cobalt reserves (17%). Cobalt is mostly mined as a byproduct of either copper or nickel. China is the leading consumer of cobalt, driven by the demand from its rechargeable battery industry.

### 4.3.10.1 Mineralogy and Geochemistry

There are several minerals that contain cobalt, although many are rare or unique to individual localities (Table 4.12). Cobaltite (CoAsS) is the main mineral of cobalt, and cobalt also occurs in copper and nickel minerals. In mineral deposits, cobalt forms minerals as sulfosalts, sulfides, arsenides and oxides including safflorite (CoAs), glaucodot ((Co,Fe)AsS) and skutterudite ($CoAs_3$) minerals. The mineral cattierite (CoS2) is similar to pyrite and occurs together with vaesite (NiS2) in the copper deposits. Geochemical behavior of cobalt is similar to nickel, and they are closely associated in igneous, metamorphic and surficial environments.

Table 4.12: Cobalt Minerals and Metal Content

| | Mineral | Chemical Composition | Cobalt Content |
|---|---|---|---|
| | Cobaltite | CoAsS | 35% |
| | Skutterudite | (Co,Ni)As3 | 23% |
| | Linnaeite | (Co,Ni)3S4 | 58% |
| | Carrollite | CuCo2S4 | 29% |
| | Erythrite | Co3(AsO4)·2−8H2O | 37% |
| | Safflorite | (CoAs2)5 | 21% |
| | Glaucodot | ((Co,Fe)AsS) | 27% |
| | Cattierite | CoS2 | 48% |
| | Spherocolbaltite | CoCO3 | 49% |
| | Asholite | Ni,Co,Ca,Mn·$(OH)_2(H_2O)$ | |

### 4.3.10.2 Deposit Types

Economic concentrations of cobalt can be found in four different geological settings. Cobalt is almost always a by- or co-product of mining for other base metals, chiefly nickel and copper. However, the vast majority of the world's cobalt resources are associated with magmatic nickel-copper sulfide deposits, sediment-hosted copper-cobalt deposits and lateritic deposits.

***Magmatic Deposits***

- Sulfide deposits formed from immiscible liquid sulfide phase that concentrated metallic elements including Ni and Co in ultramafic-mafic igneous rocks (e.g. Sudbury, Canada; Kambalda, Australia),

***Hydrothermal/Volcanogenic Deposits***

- Deposits formed by precipitation of minerals from hydrothermal fluids passing through volcanic host rock (e.g. Bou Azzer, Morocco; Keretti, Finland),

***Sedimentary Deposits***

- Sediment-hosted sulfide deposits precipitated in shallow marine sediments and converted to sulfides and sulfates during sedimentary diagenesis (e.g. DR Congo; Mt. Isa, Australia)

***Laterite Deposits***

- Formed by tropical chemical weathering of cobalt silicates and sulphides in ultramafic rocks causing cobalt to become enriched in residual weathered rocks (e.g. New Caledonia).

#### 4.3.10.3 Cobalt Deposits in Africa

Most of the cobalt in the world is mined in Africa as a by-product of platinum-nickel and copper mining operations. The Democratic Republic of Congo is the largest world and African producer of cobalt with annual production in 2019 of 56,000 tons followed by Zambia (6000 tons) Madagascar (4700 tons), South Africa (2900 tons), and Zimbabwe (400 tons) (Fig. 4.5).

***Democratic Republic of Congo and Zambia***

The Democratic Republic of Congo (DRC) and Zambia both share a common copper-cobalt metallogenic province. The Central African Copperbelt that extends from the northern part of Zambia into the southern part of Congo-(Kinshasa)) is the largest sediment-hosted copper district in the world notable for its cobalt content. The copper belt is approximately 500 km long and 30–50 km wide, and is estimated to contain a total resource of more than 4.9 billion tons grading 3.37 percent copper (165 Mt Cu metal and 4.4 Mt Co) . The ore, contained within fine-grained clastic sedimentary rocks of the Katanga Supergroup that formed about 1 billion years ago in an intracratonic or epicontinental setting, occurs as single layers and in multiple stacked layers and in multiple stacked layers. Individual ore zones may measure several kilometers long and tens of meters in thickness. Copper and cobalt have been produced from the Central African Copperbelt since the early decades of the 20th century from a multitude of large and small mines along the 500-km length of the belt.

Among the major cobalt producers in DRC are the Kufundwa mine (46,000 tons), Mutanda mine (23,000 tons), Kamoya South mine (11000 tons), Mukondo mine (8000 tons), Tenke Fungurume mine (8000 tons) and Kamoya Central (7000 tons), However, mining and exploration activities have been seriously compromised since the latter part of the 20th century because of economic and political instability and international claims of human rights violations in the region.

In contrast, since 2000 when the re-privatization of the copper mines in Zambia was implemented, production of copper and cobalt have again begun to increase, and exploration efforts are renewed. Many prospects are being developed in Zambia, but the DRC is subject to international boycott due to human rights and child labor violations.

***Madagascar, South Africa and Zimbabwe***

Cobalt production in these countries is a byproduct of nickel production derived from lateritic ores in Madagascar (Ambolevy mine), and from magmatic sulfides in layered igneous complexes and ultramafic-mafic intrusives within Archean greenstone belts of South Africa and Zimbabwe.

# 4.4 Base (Industrial) Metals

Base metals are non-ferrous metals of widespread industrial utilization including copper, lead-zinc, and tin. They were named "base metals" because alchemists of the past believed they were common and inexpensive metals, as opposed to the "noble" metals, gold and silver. They are non-ferrous, industrial metals that are distinguished from the precious or noble metals, by oxidizing or corroding relatively easily and reacting variably with diluted hydrochloric acid (HCl). Base metals such as copper and zinc have become vital materials in industrial production and are widely used in modern electronic communication and information technology.

## 4.4.1 Copper

Copper (Cu) and its alloys exhibit many desirable properties. It is ductile, malleable, hard, tough, strong, wear, and corrosion resistant. It also has high-tensile strength, fatigue strength, and thermal and electrical conductivity. The production of cop-per is mainly utilized by the wire and cable markets, taking advantage of properties such as the electrical conductivity, corrosion resistance, and thermal conductivity. Excellent malleability, ductility, and resistance against atmospheric attack distinguish copper metal and its alloys (e.g., tin or zinc); copper also shows strong antibacterial properties. Other applications include structural and aesthetic uses.

Copper is mainly used as an electrical conductor, in building materials and constituent of various metal alloys, such as brass (copper and zinc) and bronze (copper and tin). Copper sulfate is used as a pesticide and fungicide. Copper is found in nature as a native metal and as sulfides in combination with iron and sulfur. . Global production of copper in 2020 was about 20 million tons of which the five largest producers were Chile (5.6 million tons (Mt), Peru (2.4 Mt), China (1.6 Mt), Democratic Republic of Congo (1.3 Mt) and United States (1.3 Mt). Africa is a major world

producer of copper contributing about 12% of world production derived mainly from three countries; Democratic Republic of Congo 56% (1.3 Mt), Zambia 38% (880,000 tons) and South Africa 4% (82,000 tons). There are other notable producers including Zambia with 0.8 Mt. Global copper consumption is about 23.8 Mt and is steadily increasing at an annual rate of about half a million tons. China is the world's leading consumer of copper consuming about 51% of world production.

### 4.4.1.1 Mineralogy and Geochemistry

The most important copper ore minerals are sulfides chalcopyrite and bornite. More than 70% of producing copper mines exploit chalcopyrite ore. Copper also occurs as native Cu and oxides/hydroxides (Table 4.13). During weathering and oxidation of copper sulfides, the common minerals in the enriched zone include chalcocite, bornite and digenite. Minerals in the oxidized zones include carbonates malachite, azurite, chrysocolla, cuprite, tenorite, native copper and brochantite. Exploitable grades of copper depend on the size or tonnage of mining operations. In small deposits, mineable ore grade range ranges from 3-5% Cu, whereas in large tonnage operations, the cut-off grade is 0.5% Cu, but can be as low as 0.3% if molybdenum and gold are byproducts.

Crustal abundance of copper is an average of 68 ppm, while abundance in mafic rocks and felsic rocks are 100 ppm and 10 ppm respectively. The two principal oxidation states of copper are Cu+1 and +2 although some +3 complexes are known. Copper occurs rarely as native metal usually in volcanic rocks, but has a very strong affinity for sulfur and can precipitate from magmatic sulfide melts by liquid immiscibility and from magmatic-hydrothermal solutions. Among sediments, shales have the highest copper contents with mean value of 70 ppm due to affinity for organic matter content, and carbonates have the lowest (6 ppm). Acidic and sulfur-poor ore-forming hydrothermal solutions transport copper mainly in the form of chloride complexes, such as (CuCl) at more than 250°C and as $CuCl_2$ at lower temperatures.

Studies have shown that at high concentrations of reduced sulfur, copper is effectively transported, in the form of $CuS\text{-}(HS)_2$ which is only possible if solutions are alkaline. Typical copper concentrations in ore-forming hydrothermal solutions ranges between 100 and 500 ppm. Sulfur-rich vapors segregating from magma may contain up to 1% Cu. This may explain the high concentrations of copper in volcanogenic massive sulfides and high-sulfidic porphyry copper deposits. In the surficial environment, weathering and oxidation of copper sulfides produce secondary minerals in the ferruginized caps, and supergene enrichment near the water table due to reduction which produces enriched sulfides such as chalcocite and covellite.

Table 4.13: Copper Minerals and Metal Content

| | Mineral | Chemical Composition | Copper Content |
|---|---|---|---|
| | Native Copper | Cu | 100% |
| | Chalcopyrite | $CuFeS_2$ | 34% |
| | Bornite | $Cu_5FeS_4$ | 63% |
| | Covellite | Covellite | 66% |
| | Chalcocite | $Cu_2S$ | 80% |
| | Digenite | $Cu_9S_5$ | 79% |
| | Enargite | $Cu_3AsS_4$ | 48% |
| | Tetrahedrite-Tennantite | $Cu_{12}(As,Sb)_4S_{13}$ | 35 -50%Cuu (with Fe,) |
| | Azurite | $Cu_3(CO_3)_2(OH)_2$ | 55% |
| | Malachite | $Cu_2CO_3(OH)_2$ | 52% |
| | Cuprite/Tenorite | CuO | 88% |
| | Dioptase | $CuSiO_2(OH)_2$ | 45% |
| | Chrysocolla | $(Cu,Al)_2H_2Si_2O_5(OH)_4 \cdot n(H_2O)$ | 33% |
| | Brochantite | $Cu_4(SO_4)(OH)_6$ | 56% |
| | Cubanite | $CuFe_2S_3$ | 23% |

#### 4.4.1.2 Deposit Types

There are various types of deposits that generate copper ores including magmatic, contact metasomatic, hydrothermal, porphyry-type, volcanogenic massive sulfides (VMS), sedimentary exhalative (Sedex), sedimentary precipitates and oxidized residual and supergene-enriched deposits. About 75% of the copper produced worldwide comes from porphyry copper deposits, followed by sediment-hosted deposits

***Magmatic Deposits***

- Native copper in lavas (e.g. Keweenaw; Michigan). Disseminations and impregnations of native copper in lava flows.
- Copper is associated with nickel in layered mafic intrusions (e.g. Sudbury, Canada). Stratiform Ni-Cu deposits consist of massive to stringer and disseminated sulfides, usually located in the basal portions of the layered intrusions. Copper grades range from 7-10% Cu.
- Copper in carbonatites (e.g. Palabora, South Africa). Copper veins, stringers and disseminations in carbonatites. Ore grade is around 1% Cu. Ore minerals include chalcopyrite, bornite and cubanite replacing magnetite. Major source of copper in South Africa.

### *Contact Metasomatic Deposits*

- Skarn deposits (e.g. Bingham, Utah; Ertsberg-Irian Jaya, Indonesia) Skarn deposits form when carbonate-rich rocks are k invaded by magma) or hydrothermal fluids from igneous intrusions. Skarn deposits are usually small except when combined with porphyry copper deposits. Pyrite-chalcopyrite assemblages are most common in the skarn deposits such as at Ertsberg-Irian Jaya, Indonesia.
- Porphyry skarn deposits are associated with subvolcanic porphyries (e.g. Twin Buttes in Montana). Bornite-chalcocite mineral associations are the dominant Cu-Fe sulfides in porphyry skarns.

### *Hydrothermal Deposits*

- Porphyry-type deposits (e.g. Bingham, Utah; Bisbee, Arizona; Ely, Nevada; La Caridad, Mexico). Copper, Cu-Mo and Cu-Au occur as disseminations, stringers and stockworks associated with high-level porphyry intrusions. Over 75% of world copper production derives from this deposit type. Most deposits are found around the Circum-Pacific region. High tonnage but low-grade with 0.5-1.5% Cu.
- Hydrothermal vein deposits of various types occur in a variety of igneous and metamorphic rocks as well as sedimentary rocks, including polymetallic veins such as Pb-Zn-Cu (Black Mountain, S. Africa) copper–tin veins (e.g., in Cornwall, UK), chalcopyrite siderite veins, and cordillera-type veins in the Sierra Nevada, USA and hydrothermal iron oxide-copper-gold (IOCG) deposits with high magnetite content (e.g. Olympia Dam, Australia). Cu veins in Cretaceous sediments (e.g. Benue Trough, Nigeria)
- Volcanogenic Massive Sulfides (e.g. Noranda, Canada; Cyprus; Kuroko, Japan). These deposits consist of generally conformable, **massive** bodies of massive sulfides, usually of copper with zinc and lead (Pb) hosted in submarine mafic to felsic volcanic rocks. They are a major source of world copper production. Ore grade ranges between 2 and 5% Cu.
- Sedimentary exhalative deposits (e.g. Rammelsberg, Germany). These are polymetallic ores formed by precipitation of metals into seawater sediments usually of Pb-Zn-Ag but with little copper as chalcopyrite. Copper concentrations are usually less than 1&, but recovered as a byproduct of Zn-Pb-Ag mining.

### *Sedimentary Deposits*

- Sediment-hosted deposits (e.g. Central Africa copper belt, Zambia; Kuperschiefer, Germany). Stratabound copper deposits that occur as finely disseminated sulfides in fine-grained to coarse sediments (shale, sandstone and dolomite) of which chalcocite and bornite are the dominant copper minerals with lesser amount of native Cu. They contain large copper resources second only to r porphyry type deposits. Grade of copper ranges from 3-7%.

### *Surficial Deposits*

- Supergene enrichment deposits associated with the oxidation and leaching of copper and deposition beneath the water table as chalcocite and covellite. Oxidized zone may contain copper hydroxides, (e.g. Red Mountain, Arizona)

### 4.4.1.3 Exploration for Copper Deposits

Exploration for copper involves the application of a wide range of geological, geochemical and geophysical methods particularly because of the association of copper with several other metals such as Pb, Zn, Mo, Au, As etc. Porphyry copper deposits are the most common source of copper followed by sedimentary deposits.

Exploration efforts should start with targeting copper mineral districts using geological, remote sensing and magnetic methods that can detect potential large scale targets or host rocks followed by regional geochemical surveys involving rock, stream sediment, water and soil sampling using Cu and a variety of indicator elements. Ground-based geophysical methods such as induced polarization may be used to detect sulfide bodies. Advanced exploration using spot drilling to investigate subsurface anomalies provides useful information about wall rock alteration that point to proximity of sulfide mineralization.

#### 4.4.1.4 Copper Deposits in Africa

Africa is a major world producer of copper ore estimated annually as around 3 million tons derived mainly from three countries; Democratic Republic of Congo and Zambia in the Central African Copper Belt and South Africa. However, there are many other countries that produce copper in smaller quantities including Botswana, Namibia, Mauritania, Morocco, Zimbabwe and Tanzania (Fig. 4.5). The dominant deposit types are sediment-hosted and magmatic deposits.

***Democratic Republic of Congo (DRC)***

DRC is the largest copper producer in Africa with 1.5 million ton output in 2019. Most of the copper deposits occur within the Katanga portion (aka Katanga Copper Belt) of the Central African Copper Belt that extends from Zambia into the southeastern region (Katanga Province) of DRC. The Katanga copper belt is about 250 km long and 70 km wide. The copper-cobalt ores occur in Proterozoic sediments of the Katanga Supergroup of the Lufilian Arc. Over 70 orebodies have been identified within four large mining centers. The sulfide deposits are located in two stratigraphic levels forming the "lower" and "upper" orebodies, with a total cumulative thickness varying between 15 and 55 m. Total resources of the Copperbelt in the DRC and Zambia are estimated at more than 150 million tons of copper and 8 million tons of cobalt. The Katanga copper belt includes some of the highest grade copper deposits in the world with grades generally above 5% Cu.

There are several mines operating in the region and several deposits are in the process of development. Gécamines (La Générale des Carrières et des Mines), a state-owned mining company is a majority shareholder of several mines including Kambove, Kipushi, Kolwezi and KOV which is an opencast mine and the largest high-grade copper operation in the world. The average grade of copper at KOV is 4.2%.The largest copper mines and their annual production are Kinsenda mine (650,000 tons), Kamfundwa mine (400,000 tons), Tenke Fungurume (110,000 tons) Kipol mine (100,000 tons), Frontier mine (84,000 tons) and Kamoya Central mine (80,000 tons).

***Zambia***

Zambia is the second largest copper producer with production in 2019 of 950,000 tons. All the deposits are located within the Zambian portion of the Central Africa Copperbelt. The Zambian Copperbelt forms the southeastern extension of the Katanga Copper belt and part of the 900-km-long Proterozoic Lufilian Arc that contains one of the world's largest accumulations of sediment-hosted stratiform copper mineralization. Just like DRC, there are several stratabound or stratiform deposits in Zambia which occur in two major economic concentrations of copper and cobalt, hosted within a "lower orebody" (copper only) and "upper orebody" within the arkosic and shale units of the Lower Roan Group of Katangan Supergroup. The ore mineral assemblage includes pyrite, bornite, chalcopyrite, and chalcocite, although in the lower orebody, sulfide phases are partially or completely replaced by malachite and copper oxides. Economic ore grades range between 3.5-10% Cu.

Five big mines dominate Zambia's copper production: Lumwana (122,000 tons), Kansanshi (270,000 tons), Monani (120,000 tons), Nkana and Konkola. The balance of Zambia's copper production comes from a handful of smaller mines including Nchanga, Mufulira, Kitwe, Lubambe, Chibuluma and Luanshya owned by Chinese companies. The Nkana mine located 1 km southwest of Kitwe is one of the largest in Africa. The mine is underground as well as open pit and has been in operation since 1932. Copper and cobalt mineralization occurs within the mineralized Ore Shale Formation about 2-20 m thick, with chalcopyrite and bornite as the main ore minerals. The Mufulira mine is the largest underground copper mine in Africa.

***South Africa***

South Africa is a notable producer of copper but production has fluctuated over the years ranging between 130, 000 and 75,000 tons but was 82,000 tons in 2019. South Africa's copper resources originate from three main sources; (a) magmatic/hydrothermal copper from the Palabora carbonatite in Limpopo, (b) copper in the Black Mountain polymetallic veins of Zn-Pb-Ag-Cu, in Northern Cape, and (c) copper as byproduct of platinum mining in the layered Bushveld igneous complex.

The Palabora Cu-Fe-P deposit is located in Limpopo Province, South Africa, where it is associated with a 2 km2 carbonatite core within a pipe-like igneous complex (Palabora Igneous Complex) composed predominantly of dunite, pyroxenite and apatite-rich pegmatitic pyroxenite. There are several phases of carbonatite dyke emplacement within the complex. Mineralization consists of magnetite and copper emplaced in two stages. The first involved the main magnetite (weak copper) phase, with a high Ti-magnetite, interpreted to be orthomagmatic within the alkaline complex and early carbonatite. The second hydrothermal phase, associated with the late transgressive carbonatite dykes, introduced the bulk of the copper mineralization with lesser magnetite.

Mineralization involved early magmatic dissemination and late-stage hydrothermal veins. Bornite is the principal sulfide with some chalcocite in the disseminated ore while chalcopyrite dominates in the late-stage stringers and veinlets. However, the main ore and highest Cu grades of up to 1% Cu are in the transgressive carbonatite where chalcopyrite predominates with lesser bornite and minor cubanite. Both chalcopyrite and bornite replace magnetite. Ore reserves are estimated as 245 Mt @ 0.7% Cu. Annual production of Palabora mine is approximately 62,000 tons.

Other areas with copper mineralization are the polymetallic veins of Pb-Zn-Cu in the Black Mountain mine of Northern Cape Province are a source of copper in chalcopyrite which by itself is not in viable quantities. In the Bushveld Igneous Complex, copper is produced as a byproduct of platinum mining in the Merensky Reef. Platinum occurs along with nickel, and copper sulphides, most commonly pyrrhotite, pentlandite, and chalcopyrite, which grade about 0.1 to 0.2% Ni and 0.1 to 0.15 % Cu.

## 4.4.2 Lead and Zinc

Lead (Pb) and zinc (Zn) are two metals that are closely related and often occur together in mineral deposits, so it can be justified to consider them together. There are few ore deposits that contain only lead or zinc, and most mines produced both metals. Lead is a silvery-blue, soft, dense, malleable and very corrosion-resistant metal. Its most important commercial use is in the manufacture of lead-acid storage batteries. Lead is commonly utilized in alloyed form, which increases its low-tensile strength. When added to metal alloy, lead improves their machinability. Other uses of lead include the manufacture of ammunition, electrical, electronic and communication applications.

Zinc is a blue-grey metal that is brittle at ambient temperatures but is malleable at 100 to 150°C. Because of its anticorrosion properties, more than one-half of all zinc metal is used for galvanizing followed by the production of zinc-based alloys, brass, and bronze. Steel coated with zinc (galvanized steel) exhibits high levels of corrosion resistance. This application is responsible for around 50% of total demand. Zinc-based alloys are also used in die casting, ranging from automotive components to toys and models

Global lead production originates from both primary (mining of ores) and secondary (recycled) sources. Currently, secondary lead production accounts for more than half of all lead produced throughout the world which is estimated as 12 million tons. The five countries with the largest mine-produced lead in 2019 were China (2.1 million tons), Australia (430,000 tons), Peru (290,000 tons), United States (280.000 tons), and Mexico (240,000 tons). For zinc, global production is estimated as 11 million tons with the following countries as the largest producers; China (4.3 million tons), Peru (1.4 million tons), Australia (1.3 million tons), India (800,000 tons) and the United States (300,000 tons).

### 4.3.2.1 Mineralogy and Geochemistry

The major ore mineral of lead is galena, and zinc is sphalerite. Wurtzite is a polymorph that has the same chemical composition as sphalerite but different structure. Both occur in colloform sphalerite. Galena is a common source of silver, with contents ranging from 0.01 to >1% Ag. Silver substitutes for lead in the crystal lattice or forms minute inclusions of silver minerals such as acanthite and tetrahedrite. Sphalerite often contains up to 5% of cadmium. There are secondary minerals of lead-zinc mostly carbonate and sulfate that form due to oxidation and chemical weathering (Table 4.14).

Crustal abundance of Pb averages 15 ppm and Zn is 80 ppm. The concentration of Pb in rocks generally increases during magmatic crystallization, and is higher in felsic than mafic igneous rocks with mean values of 20 ppm in granites, 5 ppm in gabbros and 2 ppm in basalts. Zinc on the other hand, substitutes strongly for Fe and Mn in igneous rocks resulting in higher Zn values in mafic rocks (100 ppm) compared to felsic rocks (40 ppm). In hydrothermal solutions, both Pb and Zn are transported as chloride or bisulfide complexes. In fluids that are acidic with high salinity and oxidation state, metal-chloride complexes are a preferred transportation mechanism. Lead and zinc are precipitated as sulfides due to sudden changes in pH, cooling and dilution or mixing with other fluids especially those carrying reduced sulfur.

In the surficial environment, the behavior of Pb during weathering contrasts sharply with that of Zn. Galena is more stable than sphalerite, because a thin film of nearly insoluble lead minerals (cerussite and anglesite) quickly develop on all exposed surfaces. During weathering, galena is slowly oxidized by atmospheric oxygen to either the sulfate (anglesite) or the carbonate (cerussite). These secondary minerals become quite insoluble in natural waters and tend to concentrate in surface residual deposits. The Pb content of weathered lead ore commonly remains as high as that of the original unweathered ore. In contrast zinc sulfides are oxidized to soluble secondary products that leach away from the weathered ore. The mobilization and transport of Zn in acidic solution (both vertically and laterally), and re-precipitation as secondary, non-sulfide zinc produces enriched supergene zinc deposits. The low Pb content in supergene non-sulfide ores is a result of the relative immobility of the metal in drainage waters with low pH, While these two metals are closely related, they tend to be 'odd couple' in the surficial environment.

### 4.4.2.2 Deposit Types

Lead and zinc ores occur together in several deposit types. However, the most important sources are all of hydrothermal origin: volcanogenic massive sulfide (VMS); sedimentary exhalative (SEDEX), Mississippi Valley-type (MVT) and polymetallic vein deposits. Lead and zinc are also found in chimneys and mantos, skarns and as impregnations in sandstones but contribute little to world production. Minimum ore grades are between 3 and 10% lead with zinc respectively.

Table 4.14: Lead and Zinc Minerals and Metal Content

| | Mineral | Chemical Composition | Metal |
|---|---|---|---|
| | Galena | PbS | 86% Pb |
| | Cerussite | $PbCO_3$ | 77% Pb |
| | Anglesite | $PbSO_4$ | 68% Pb |
| | Sphalerite | ZnS (with Fe, Mn, Cd, etc.) | 38-66% Zn |
| | Wurtzite | ZnS (with Fe, Mn, Cd etc.) | 38-66% Zn |
| | Smithsonite | $ZnCO_3$ (often with Cd) | 52% |
| | Hemimorphite | $Zn_4Si_2O_7(OH)_2{\cdot}H_2O$ | 54% |
| | Willemite | $ZnSiO_4$ | 59% |

***Hydrothermal Deposits***

- *Volcanogenic massive sulfide* (VMS) deposits are formed on the ocean floor where cooling magma, either as volcanic rocks erupted on the ocean floor or emplaced at depth in shallow intrusions, supplies the heat needed to generate and mobilize hydrothermal solutions of mixed magmatic and marine sources. The ores are a combination of Zn-Pb-(Cu-Au) (e.g. Kuroko, Japan;
- *Sedimentary exhalative* (SEDEX) deposits are formed by hydrothermal solutions released into a water reservoir (usually the ocean) or water-rich sediments resulting in the precipitation of stratiform ore within deeply buried basins. They are found mostly within clastic sediments. The ores dominated by Pb-Zn-(Ag) with little or no Cu (e.g. Broken Hill, Australia; Red Dog, Alaska).
- *Mississippi Valley type* (MVT) deposits are epigenetic stratabound deposits hosted mostly by carbonate rocks (dolostones) by hydrothermal solutions of diagenetic connate brine origin mobilized during diagenesis (e.g. Pine Point, Canada)
- *Hydrothermal* veins of quartz-carbonate Pb-Zn-Ag-Ba deposits formed as fissure-filling in sedimentary and metamorphic rocks. Although Pb-Zn veins are widespread, they currently contribute little to world production (e.g. Benue Trough, Nigeria).

### 4.4.2.3 Exploration for Lead-Zinc Deposits

Lead-zinc deposits occur in a variety of forms and geological environments. Some are structurally controlled while others are controlled by sedimentary or stratigraphic features. In the search for lead-zinc deposits, regional areas of interest may be targeted by a combination of geology, geochemistry and geophysics. While some types of lead-zinc deposits are difficult geophysical targets, on a regional scale, a combination of gravity and aeromagnetic surveys, used in conjunction with geochemistry, effectively focus exploration into favourable structural and lithological settings. Detailed gravity and seismic surveys may be used to delineate areas with host carbonates in regions of shallow cover. Remote sensing can be useful in identifying structural features that control Pb-Zn mineralization. Geochemical surveys using stream sediments and soils can be used for district-scale or orebody exploration particularly in tropical environments where Pb is relatively immobile in soils but dispersive in stream sediments.

### 4.4.2.4 Lead-Zinc Deposits in Africa

Africa is not a major producer of lead and zinc, although several countries in North Africa, West Africa and Southern Africa produce small quantities of both metals including Morocco, Tunisia, Algeria, Burkina Faso, Nigeria, Eritrea, South Africa, Namibia and Botswana (Fig. 4.5).

### *North Africa*

The Atlas belt across North Africa (Morocco-Algeria-Tunisia) is a known world-class province for lead-zinc deposits. The zinc-lead deposits of the Atlas belt are broadly of Mississippi Valley Type (MVT), low-temperature carbonate-replacement deposits formed within the Mesozoic-aged broad carbonate shelf sedimentary sequence deposited on the southern margin of the Tethys Ocean. Jurassic carbonate rocks in the High Atlas Mountains in the Touissit-Bou Beker area of northeastern Morocco and at El Abed in adjacent western Algeria host the largest known MVT deposits in Africa. The ore district includes five major deposits, four in Morocco and one in Algeria, and more than 30 subeconomic occurrences. The mined ores average 7% Pb, 3% Zn, and 600 g/t Ag.

In Algeria, the most significant MVT lead-zinc deposits have been mined at El Abed near the Algerian-Moroccan border, at Kherzet-Youcef in the Sétif region of central Algeria and near Behala in northern Algeria. Algeria has the potential capacity to produce up to 220,000tons per year of mined zinc following the development of the VMS deposit in the Tala Hamza zinc mine near Behala, with estimated reserves of 65 million tons grading 6.3% zinc and 1.8% lead, In Tunisia there are several Pb-Zn occurrences of which the most important prospect is the Djebba MVT Zn-Pb deposit located in the Atlas Zn-Pb metallogenic belt. Estimated resource base is 2.7 million tons of 6.1% Zn and 3.3% Pb.

### *West Africa*

Lead-zinc massive sulfide (VMS) deposits occur within the Paleoproterozoic Birimian greenstone belts of Burkina Faso. The Perkoa zinc deposit is located in the Sanguie Province, approximately 120 km west of the capital city of Ouagadougou. The host rocks are felsic to intermediate series of volcanic and volcaniclastic rocks within the Birimian Supergroup. Base metal mineralization is scarce within the Birrimian sequence, and the Perkoa deposit represents the only significant zinc-silver massive sulfide mineralization discovered in the s region. Mining commenced in 2013 with the most recent annual production of 165,000 tons of zinc concentrate in 2018. Ore reserves are estimated as 2 million tons averaging 12.8% Zn

In Nigeria, several Pb-Zn-Ba veins have been mined within the Cretaceous rocks of the Benue Trough at Abakaliki, Isiagu, Zurak and Gombe. Production is small and rarely exceeds a combined value of 10,000 tons of galena concentrate with valuable silver.

### *South Africa and Namibia*

Zinc-lead-copper deposits variably of sedimentary exhalative (SEDEX) type are hosted by a series of pelitic and psammitic metasediments of the Proterozoic Bushmanland Group in the Northern Cape Province. The most prominent deposit is exploited at the Black Mountain mine in the Aggeneys which has produced significant quantities of Zn, Pb, Ag and Cu. Production has fluctuated over the years but currently produces about1.5 Mt of copper-lead-zinc-silver ore per year. An unusual association of carbonate-hosted MVT-type and vein-type lead-zinc mineralization with hydrothermal gold mineralization has been described in proximity to the Bushveld Complex in South Africa. Small VMS deposits occur in greenstone belts in South Africa, such as theMaranda Cu-Zn deposit in the Murchison greenstone belt and the BienVenue Zn-Ba-Pb-Ag-Audeposit in the Barberton greenstone belt have been exploited to varying extent.

Namibia produces Zn and Pb from Neoproterozoic zinc-lead-copper mineralization at the Rosh Pinah mine in the Damara belt. The complex lead-zinc-copper mineralization of the Tsumeb mine, also in the Damaran of Namibia is of VMS type. Other small VMS mineralization occurs in the Gorob, Matchless, and Otjihase deposits in Namibia

## 4.4.3 Tin

Tin is a silvery white metallic element that is soft, malleable, corrosion-resistant, non-toxic and chemically stable. Its main uses are in making tin plates for tin cans. Tin has many important uses as an alloy. It can be alloyed with lead and with copper to produce bronze, and for brass and solder. Tin-coating of other metals is also used especially, for tableware and food containers. World production of tin is 310,000 tons in 2019; the largest producers are China (85,000 tons), Indonesia (84,000 tons), Myanmar (54,000 tons), Peru (18,000 tons), Bolivia (17,000 tons) and Brazil (17,000 tons).

### 4.4.3.1 Mineralogy and Geochemistry

The only mineral of commercial importance as a source of tin is cassiterite ($SnO_2$).The other tin mineral that is found less commonly in tin ores is stannite ($Cu_2FeSnS_4$). Very rare minerals that contain some tin are cylindrite, franckeite, canfieldite, and teallite (Table 4.15).

The abundance of Sn in the Earth's crust is estimated as 2 ppm. Concentrations of Sn in ultramafic rocks averages 0.15 ppm, mafic rocks 1.2 ppm, and granitic rocks 4.5 ppm. In magmatic rocks, Sn is known to be closely associated with felsic or granitic rocks particularly peraluminous more than peralkaline type. Consequently, tin-bearing granites are considered as products of partial melting of continental crustal materials.

During magmatic crystallization, under reducing conditions, tin as $Sn^{2+}$ is a highly incompatible element in granitic systems, and fractionates into the residual melt. But as temperature decreases and oxygen fugacity increases, the solubility of Sn in granitic melt decreases, which in the latest stages of magmatic evolution, tin which is oxidized largely as $Sn^{+4}$ has effectively decreased the saturation concentration of tin and may lead to the crystallization of cassiterite as a late-stage accessory mineral of primary origin. This is the source of cassiterite dissemination in granitic rocks. The late-phase crystallization may be accompanied by late-stage addition of albite (albitization) with elevated Na/K ratios and enhanced values of Li, Rb, F and Nb in the residual fluids. This explains the presence of disseminated well-formed crystals of cassiterite in albitized peraluminous biotite granites such as those of the Younger Granites of Jos Plateau, Nigeria.

Hydrothermal fluids that form tin deposits in granites are usually acidic with high salinity. Precipitation of cassiterite ($Sn^{4+}$) is attributed to destabilization of the chloride complexes due to several reasons including sudden decreases in any of the following; temperature pressure, pH and oxygen fugacity or mixing with fluids of different composition. Greisenization is a common alteration in hydrothermal tin deposits.

Table 4.15: Tin Minerals and Metal Content

| | Mineral | Chemical Composition | Tin Content |
|---|---|---|---|
| | Cassiterite | $SnO_2$ | 78% |
| | Stannite | $Cu_2FeSnS_4$ | 27% |
| | Tealite | $PbSnS_2$ | 30% |
| | Cylindrite | $Pb_3Sn_4FeSb_2S$ | 28% |
| | Franckeite | $(Pb,Sn)_6Fe^{++}Sn_2Sb_2S_{14}$ | 13% |
| | Canfieldite | $Ag_8SnS_6$ | 10% |

In the surficial environment, tin is very resistant to chemical decomposition, and accumulates as heavy minerals in eluvial, alluvial and marine placers. About 50% of world tin production is derived from placer deposits.

### 4.4.3.2 Deposit Types

The most important tin deposits are currently from secondary placers which account for over half of the world's production. Primary tin deposits of various types are granite-related, and found in greisens, granitic pegmatites, and disseminations in highly evolved granite intrusions, skarns, porphyry tin deposits and polymetallic veins emanating from granites (Sn-W-Zn; e.g. Cornwall, England).

***Magmatic Deposits***

- Dissemination of cassiterite (and columbite) in intensely albitized cupolas of peraluminous granites (e.g Younger Granites, Nigeria). Source of alluvial tin ore.
- Stanniferous pegmatites of the quartz-microcline with albite variety (e.g. Wamba-Jemaa, Nigeria: Musha, Rwanda). Cassiterite is irregularly disseminated in the pegmatite body.

***Contact Metasomatic Deposits***

- Skarn and metasomatic deposits formed by interaction of tin-bearing fluids from granites with carbonate rocks (e.g. Dachang, China).

***Hydrothermal Deposits***

- Porphyry tin deposits (e.g. Bolivia). Sulfide-cassiterite deposits commonly known as tin porphyries mined more for their silver than for the tin they contain.
- Hydrothermal greisen-quartz-cassiterite (+wolframite) veins formed at high temperature in association with stanniferous granites. Major source of primary tin deposits (e.g. Cornwall, England; Ririwai, Nigeria; Herberton, Queensland).
- Hydrothermal tin-silver deposits associated with rhyolitic volcanics and subvolcanic stocks (e.g. Pirquitas, Argentina; Potosi, Bolivia).

***Sedimentary Deposits***

- Placers (eluvial and alluvial) deposits (e.g. Nigeria, Malaysia

### 4.4.3.3 Exploration for Tin

Economic cassiterite deposits are currently found mostly as secondary (placer) deposits, Exploration for such deposits have traditionally utilized panning of stream sediments to find traces of cassiterite. Recent exploration efforts have been focused increasingly towards finding primary deposits and placers that are covered or buried by more recent sediments or by volcanic rocks such as those buried by basalts in Nigeria's Jos Plateau. Remote sensing and geophysical methods can be helpful in locating buried stream beds.

The search for primary tin deposits can utilize geochemical methods, particularly rock geochemistry in identifying granites with tin-bearing potential within tin metallogenic provinces. Various remote sensing techniques are useful in the exploration for primary tin deposits, particularly when combined with airborne geophysical methods for the definition of broadly favourable areas with stanniferous granites). For direct target identification, geochemical surveys using stream sediments and soils can be useful tools. .

### 4.4.3.4 Tin Deposits in Africa

Tin resources in Africa are located primarily in central and western Africa, although the total output is not significant on a global scale. The leading African producers are Democratic Republic of Congo (10,000 tons), Nigeria (7,500 tons), Rwanda (3,800 tons) and South Africa (1,500 tons) (Fig. 4.6). Historically, Nigeria was once the sixth largest producer of tin in the world, but production capacity has fallen due to dwindling resource base and poor mining practices.

***Democratic Republic of Congo***

Historically, the Democratic Republic of Congo (DRC) has been well known for tin production in Africa since the early 20th century. However most of the output was through informal mining by artisanal and small scale miners in the densely forested eastern part of the country. For a long time, civil war and insecurity prevented any effort to organize and develop mining activities until the last decade. The main focus of mining is the Bisie tin mine in North Kivu, a remote province in the eastern region of the DRC. It is one of the world's largest t tin deposits. Mineralization occurs as polymetallic veins of Sn-Cu-Zn-(Pb-Ag), and blocks or disseminations of cassiterite within chlorite schists of the Kibaran orogenic belt comprising Proterozoic metasedimentary units of pelitic schist, carbonaceous schist, quartz-sericite schist, mica schist, and amphibolite rock types. Two high-grade tin deposits have been delineated: the Mpama North and Mpama South. The Mpama North deposit is rich in tin-copper mineralization, while the Mpama South hosts silver-zinc-lead mineralization. At Mpama North, mineralisation occurs in narrow veins, blocks or disseminations of cassiterite and is identified within chlorite schist.

As of 2017, the proven and probable ore reserves at the Bisie tin project were estimated to be 4.67 Mt, grading 3.58% Sn. The mine is estimated to contain 167,300 tons of tin. Current annual production is about 8000 tons, with plans for rapid expansion.

***Nigeria***

Prior to the discovery of crude oil in the early 60s, Nigeria was one of the world's largest producers of tin ore, but due to several other factors including political instability, privatization policies and the lure of cheap oil revenues, tin production fell drastically to less than 2000 tons. However, in recent years due to efforts to diversify revenue sources, production has risen to about 8000 tons annually. The major source of cassiterite and associated columbite-tantalite (coltan) are alluvial and eluvial placer deposits and decomposed Jurassic Younger Granites around the Jos Plateau, and pegmatite areas of Kaduna and Nasarawa State. The pegmatites are associated with the Pan=African syntectonic Older Granites of the Precambrian basement complex. There is little or no derivation of tin from hydrothermal vein deposits in the Jos Plateau. Mining is of tin is mostly by artisanal and small scale operators.

***Rwanda***

The mining of cassiterite (tin ore) and tantalite (coltan) are the two largest contributors to the mining sector and the economy of Rwanda because of demand and monetary value. The country produces around 4000 tons of tin ore for export in 2019. The tin and associated Ta and W occur as primary mineralisation in the form of pegmatite, quartz veins, and greisen, but also as secondary mineralization in alluvial or eluvial deposits. Primary quartz veins and pegmatites are localized within the Rwandan metasediments of the Kibaran mobile belt in central Rwanda, and are believed to have originated from the syntectonic Kibaran or Pan-African granites that occur in adjoining areas. Mining of cassiterite is undertaken mostly by artisanal and small scale producers. Several mining companies such as the Rutongo Mines Ltd produce an average of about 1000 tons of cassiterite per annum. Locally mined cassiterite had an average tin content of 63% and an average tantalum content of between 1% and 2%.

### *South Africa*

South Africa was once a significant producer of tin in up to the 1990s when the world price of tin crashed and most production was shut down. Since 2015, plans are ahead to revive tin mining.

All the deposits of commercial importance are associated either with the Cape granites near Cape Town and Stellenbosch or with the younger granites of the Bushveld Igneous Complex in the Transvaal. Cassiterite occurs as disseminations in granites and as veins and lodes in both the granites and surrounding country rocks. Of the six separate tin fields in South Africa, Rooiberg, Zaaiplaats, and Nylstroom deposits have provided well over 90 per cent of the total tin produced in the Republic of South Africa. The two major producers, Rooiberg Tin Limited (formerly Rooiberg Minerals) and Zaaiplaats. Bushveld Minerals has planned to extract cassiterite from the previous mine dump to be recycled sandwiched between 6 000 t of contained tin in coarse granite at Groenfontein and 12 500 t of contained tin at Zaaiplaats. Both primary deposits are envisaged as future open pit quarry-type mines that are earmarked for collective development in the next four to five years.

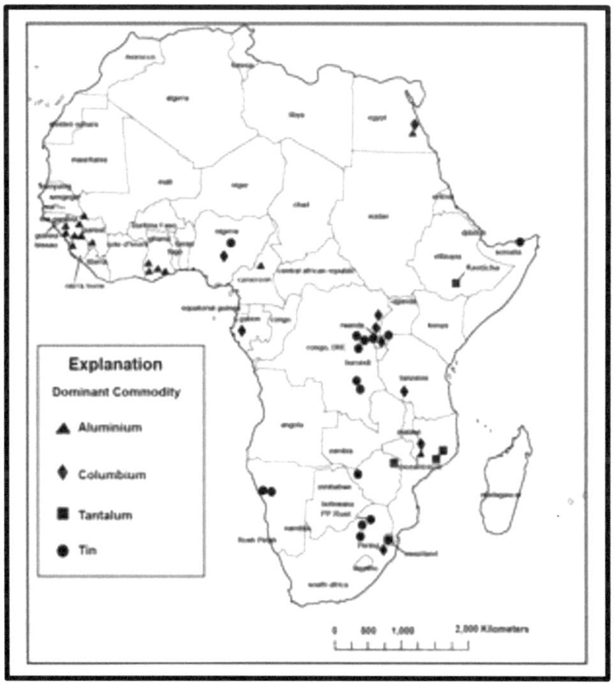


Fig. 4.6: Tin, Niobium-Tantalum and Aluminium Deposits in Africa

## 4.5 Light Metals

Light metals are characterized by low specific gravity. Aluminum and magnesium are the best known examples widely used in industry. Aluminum is the most important of the non-ferrous metals, and is commonly produced from bauxite, which is a loose soil or a hard rock with 30–65% $Al_2O_3$. About 95% of bauxite produced is processed into aluminum metal. The remaining 5% serves as an industrial raw material for numerous special products such as abrasives, Portland cement, technical ceramics, glass, chemicals, paints, and refractories. Favorable attributes of aluminum metal such as lightweight, strength, and excellent corrosion resistance (Lu 2010) allow its use in many applications from building air frames to food packaging.

Magnesium is the other light metal, but will not be described in detail. Magnesium, just like calcium, is easily dissolved and transported in surface and groundwater. Various raw materials, natural and industrial brines, and seawater are used for the production of magnesium and magnesium compounds. Applications of the extremely light magnesium metal employ either the pure metal or aluminum alloys. Magnesium-aluminum alloys are mainly used in making beverage containers. About 40% of magnesium is used for die casting in the car industry in order to reduce weight and fuel consumption. Other sectors include the space, air-craft, and chemical industry. Magnesium is mainly utilized as magnesium oxide in applications such as refractory material (e.g., furnace linings for the production of iron and steel), glass, and cement

## 4.5.1 Aluminium (Bauxite)

Aluminium (Al) is a silvery-white, soft, very light but strong and ductile metal Aluminium is remarkable for its low density and its ability to resist corrosion Aluminium and its alloys are ideal for use in aircraft, automobiles, beverage containers, and important in transportation and building industries. Aluminum makes up 8% of the Earth's crust; and is the most abundant metal element. Although aluminum is a common element in many rocks, it is only concentrated for mining only in bauxite ore made up of aluminum hydroxide. Of all the bauxite mined worldwide, about 85% is converted to alumina (Al2O3) for the production of aluminum metal, 10% for non-metal uses as various forms of specialty alumina, and the remaining 5% is used for non-metallurgical applications such as pigments for plastics and special calcined alumina for the production of ceramics, refractories, and abrasives,

World bauxite production in 2020 was 370 Mt with the following five leading producers: Australia (30%, Guinea (20%), China (! 6%), Brazil (9%) and India (7%). Known global reserves are 30 billion tons located primarily in Guinea (25%), Australia (20%), Vietnam (12%), Jamaica (9%), Brazil (7%), and China (4%), Africa produced some 75 million metric tons of bauxite in 2020, and .contains some of the world's largest resources of bauxite, much of which are yet to be developed.. The four African countries that are actively mining bauxite with the following annual production are Guinea (72 Mt), Sierra Leone (2 Mt), Ghana (1.5 Mt) and Mozambique (10,000 tons). There are vast resources of bauxite in Africa that are yet to be developed due to lack of financial resources, energy sources and transportation infrastructure.

### 4.5.1.1 Mineralogy and Geochemistry

Bauxite ore is the main source of aluminum and contains the aluminum minerals *gibbsite, boehmite* and *diaspore* that are polymorphs. Other related minerals are *alunite*, a sulfate and nepheline, a silicate containing over 35% Al (Table 4.16).

Aluminum is a lithophile major element found mostly in primary form as a component of feldspars, felspathoids and clay minerals. With an abundance of 8%, aluminum is the third most abundant element in the Earth's crust after oxygen (47%) and silicon (27%). In spite of its abundance and enrichment, economic concentrations that can be exploited are rare. Viable aluminum deposits are only found as bauxite formed by lateritic weathering in humid tropical and subtropical areas of the world particularly in the equatorial regions.

Bauxite formation is a process involving the chemical breakdown of aluminum-bearing minerals and rocks, and removal of unwanted elements mainly silica and iron, and concentration of alumina as a residual deposit. In sedimentary environments, alumina is transported to sedimentary basins to form extensive clays and shales, and rarely as economic deposits.

Table 4.16: Aluminum Minerals and Metal Content

| | Mineral | Chemical Composition | Al Content |
|---|---|---|---|
| | Gibbsite | Al(OH)3 | 65% |
| | Boehmite | g-AlO9OH) | 85% |
| | Diaspore | g-AlO9OH) | 85% |
| | Alunite | KAl3[(OH)6(SO4)2 | 37% |
| | Nepheline | (Na,K)[AlSiO4] | 36% |

### 4.5.1.2 Deposit Types

The main ore of aluminum is bauxite which is a surficial deposit that forms by weathering of aluminum-bearing rocks in an area that has a tropical climate with alternating wet and dry seasons, and a terrain with high relief and good drainage, all of which promote chemical leaching and the resulting concentration of aluminum. The aluminum is concentrated when silicate minerals (rock-forming minerals that contain silica) and clays are decomposed, and silica and other components are removed, leaving a soil relatively enriched in aluminum. Most tropical bauxites occur at high elevations or plateaus where they are associated with old erosion surfaces. Bauxites are of various types and are classified as follows:

***Types of Bauxite Deposits***

- *Lateritic Bauxite* formed by tropical weathering of aluminous igneous, metamorphic and sedimentary rocks (e.g. Boke, Guinea). Textural varieties include pisolitic, massive, earthy, and nodular. Gibbsite or mixture of gibbsite and boehmite are the main ore minerals. Lateritic bauxite accounts for about 85% of the world's bauxite.
- *Karst Bau**xite*** formed in limestone terrain by weathering of residual or transported materials such as clays or volcanic tuff respectively (e.g. Jamaica). Ore minerals are mainly gibbsite in younger (Cenozoic) deposits and boehmite in older (Paleozoic) deposits. Karst bauxite represents 14% of global bauxite.
- *Sedimentary Bauxite* formed from detrital Al-rich material that is transported, deposited and lithified as sediments (e.g. Ghana).

### 4.5.1.3 Bauxite Deposits in Africa

Africa contains some of the world's largest resources of bauxite. The four countries that are actively mining bauxite are Guinea, Sierra Leone, Mozambique and Ghana. Other countries with substantial resources yet to be fully developed are Tanzania, South Africa, Egypt and Cameroon (Fig. 4.6). Combined bauxite production in Africa in 2019 was estimated as 60 million tons (Mt,)

***Guinea***

Guinea is the largest producer of bauxite in Africa, and contains over 30% world's reserves. It produces about 60 million tons of bauxite ore annually which accounts for over 95% of African bauxite production. The reserves are mainly located in the following regions: Boké, Kindia Fria, Boffa in *Lower Guinea*; Tougué, Pita, Mali, Mamou, Dalaba in the Fouta Djallon Province in *Middle Guinea*; and Dinguiraye, Dabola and Siguiri in Upper Guinea. The most important deposits are found in the Boke region in western Guinea (Lower Guinea). Bauxite reserves are estimated at over 40 billion tons, 23 billion tons of which are located in the Boke region. Bauxite production is sourced from three open pit mines in the Boke region: Sangaredi, Kindia and Fria. The lateritic bauxite deposits of Guinea occur at low altitudes from about 300 to 400 m above sea level in the flat topped plateaus with practically negligible soil cover and overburden. The lateritic bauxite is underlain by a series of Paleozoic argillite, siltstones and minor sandstone intercalations that are widely intruded by sills and dykes of Mesozoic dolerites.

Alumina content varies from 40% to as high as 52% with an average 46% $Al_2O_3$. One of the unique features of Guinea bauxite, particularly in Boke bauxite belt is their relatively high chemical quality with respect to fairly low silica content that does not increase with lowering of alumina content. The total silica content ranges from about 1-2%, of which half is in the form of "reactive" (soluble) silica which makes them ideal for processing in the low temperature alumina refinery. Bauxite deposits, occurring near the coastal belt of Guinea, sometimes have relics of andalusite and pyrophyllite minerals, which reduces the available alumina content in the ore.

### *Ghana*

Mining of bauxite began in Ghana (then Gold Coast) in 1941, and by 1960, production level had reached 160,000 tons annually, but production had dwindled over the years due to several factors including lack of investors, fluctuating prices, and inadequate power supply and infrastructure challenges. However prospects for development in recent years have improved considerably with participation of Chinese companies. Bauxite production in 2019 was 1.4 million tons.

There are several bauxite deposits in Ghana and the main occurrences are in four groups: (1) Sefwi Bekwai (also known as Awaso) in the *Western Region*; (2) Nyinahin in the *Ashanti Region*; (3) Atewa Range (near Kibi) and (4) Mount Ejuanema both in the *Eastern Region*. Current production of bauxite comes from Awaso in the Western Region of Ghana. The lateritic bauxite deposits of Ghana are mostly *blanket* deposits that are associated with old erosion surfaces and occur as cappings on the flat tops of hills or mountains whose elevations generally exceed 600 m above sea level. The deposits in the Ashanti and Western Regions rest on phyllites, schists and metamorphosed lavas, greywackes and tuffs of the Proterozoic Birimian Supergroup while the Ejuanema deposits and others in the Eastern Region were formed on flat-bedded shales, mudstones with sandstones of Carboniferous Voltaian System. In the formation of the bauxites, the presence of kaolinitic clay beneath the bauxite indicates gibbsite did not form directly from alteration of primary minerals but through clay minerals. Out of the bauxite deposits in the Eastern Region, only the Kibi and Nyinahin deposits are of such sizes as to warrant long term industrial investment.

The *Sefwi-Bekwai* deposit in the Western region is the principal producer of bauxite in Ghana. It consists of a lateritic bauxite layer about 5- 20 m thick forming a blanket deposit that caps flat hills in altitudes of 300-600 m. The ore mineral is mostly gibbsite with minor boehmite. Alumina ranges from 50-55%with about 20% iron. The Nyinahin bauxite deposit in the Ashanti Region which is being developed was discovered in 1922, and has a 30 km strike length (NE-SW) and 5 km wide corridor. It occurs on ten hills of which three are being developed. The bauxites are 4.5m to 25m thick and covered on the average by thin layers of overburden and topsoil. Measured reserves on 3 hills as 100 million tons at 44.5 % $Al_2O_3$ and 2.5 % $SiO_2$. The ore minerals are gibbsite (32-84%) and boehmite (1-10%)

In 2019, Ghana signed a memorandum with China to further explore Ghana's deposits of bauxite at two locations; Aweso in the Western Region, and Atewa in the Eastern Region of Ghana. The potentials for further exploitation of Ghana's bauxite resources are huge.

### *Sierra Leone*

Most of the country's bauxite deposits are found in the south on the Freetown Peninsular and in the north at Kamakwie and Makumre. In 2018, Sierra Leone experienced a rise of more than 22% in its bauxite exports, with the total volume recorded at 1.99Mt – before falling by 13% in 2019 to 1.73 million tons. Production target in 2020 was 2 Mt. Bauxite deposits are of lateritic origin derived from the weathering of Precambrian gneisses and metasediments of the Kasila Group along a Tertiary erosion surface.

### *Mozambique*

Mozambique started producing bauxite in the 1990s. The Mina Alumina mine near Manica, in the Manica Province on the border with Zimbabwe produces aluminum ore from the deposit that comprises several small, about 1m thick bodies of bauxite with gibbsite nodules intermixed with aluminous clays. The rock containing the deposits is weathered to depths of more than 30 m. Mining of bauxite has been going on commercially in the area since the 1990s, and production in recent years has reached the level of about 10 000 tons annually. This bauxite ore is exported to Zimbabwe where it is used as a raw material for the production of aluminium sulphate. Bauxite reserves have been estimated at 10 Mt and those of white kaolinitic clays are assumed to be from 10 to 15 Mt. The overall grade of the bauxite ore is relatively low with alumina content of about 40-45% Al2O3, although the gibbsite nodules in the bauxites are as high as 60% alumina with low silica.

### *Tanzania*

Bauxite deposits in the Usambara Mountains of northeastern Tanzania occur as remnants of residual deposits on two plateaus in Lushoto and Usambara Mountains. The parent rocks for the deposits are mainly granulites and feldspathic gneisses of the Neoproterozoic Mozambique belt. The plateaus represent a well preserved Late Cretaceous-Lower Tertiary old erosion surface. The bauxite deposits contain mainly $Al_2O_3$ (40-69%), $Fe_2O_3$ (3-10 %) and low $SiO_2$ (0.16-7%). Gibbsite makes up to 98% of the bauxite ore and is accompanied by goethite in the ore. Boehmite occurs in small amounts and is usually accompanied by hematite. Kaolinite is the sole clay mineral encountered in the bauxite ore, suggesting a development of the bauxite deposits on a well-drained peneplanation profile. Ore reserve estimate of bauxite at Lushoto is about 37 million tons. Mining licences have been granted for the exploitation of the Lushoto bauxite deposit.

## 4.5.2 Magnesium

Magnesium (Mg) is one of the common elements in the Earth's crust. It constitutes 2% and is the fourth most abundant metallic element after silicon (27%), aluminium (8%) and iron (5%). Average MgO content in common rocks amounts to 6.7% in basalt, 3.3% in andesite and 0.4% in rhyolite; average dunite contains 38% MgO. Magnesium has a low density (only two-thirds that of aluminum), and because of this property, it has found extensive use in the aerospace industry. However, because the pure metal has low structural strength, magnesium is mainly used in the form of alloys—principally with 10 percent or less of aluminum, zinc, and manganese—to improve its hardness, tensile strength, and ability to be cast, welded, and machined. Casting, rolling, extruding, and forging techniques are all employed with the alloys, and further fabrication of the resulting sheet, plate, or extrusion is carried out by normal forming, joining, and machining operations. Magnesium is the easiest structural metal to machine and has often been used when a large number of machining operations are required. Magnesium alloys have a number of applications: they are used for parts of aircraft, spacecraft, machinery, automobiles, portable tools, and household appliances.

The minerals of magnesium are magnesite (MgCO3), dolomite (CaMg(CO3)2), brucite (Mg(OH)2) and forsterite (Mg2SiO4). Sources of magnesium metal are seawater, natural brines and residual brines from potassium salt processing; and mafic rocks dunite, serpentine, dolomite, brucite and magnesite rock. Seawater with 0.3% MgCl2 or 0.86 g/L Mg is an unlimited source of magnesium. Some playa lake brines (Dead Sea, Great Salt Lake) contain recoverable magnesium as one of several products. The world's annual production is estimated as 1 million tons, and the leading producers are China (80%), Russia (7%), USA (5%), Israel, Israel (3%) and Kazakhstan (2%).

## 4.6 Minor and Specialty Metals

The term 'minor and specialty metals' is to describe a group of metals that are of limited or special uses such as in nuclear energy or high technology applications . These include rare earth elements (REE), uranium, thorium, lithium and zirconium. This selection of metals in this category are based on the author's discretion and may not be consistent with the view of others.

### 4.6.1 Rare Earth Elements

The rare-earth elements (REEs), also called the rare-earth metals are 15 elements that range in atomic number from 57 (lanthanum) to 71 (lutetium). In chemistry parlance, they are commonly referred to as the "lanthanides." and two more elements, yttrium (atomic number 39) and scandium (atomic number 21) are included in the group as rare-earths because they share chemical and physical similarities and close affinities with the lanthanides. Although REEs are not rare in terms of average abundance in crustal rocks, the economic concentrations of REEs as ore deposits are relatively very few worldwide. The REEs have certain chemical and physical properties that make them indispensable and non-replaceable in many high-technology applications. For this reason, they are in high demand in the defense, energy, industrial, and military technology industries which have made the REEs critical or strategic minerals for the world's superpowers. REE consumption is also growing due to their increasing usage in products that affect our daily lives, such as hybrid cars, catalytic converters, wind power generators, household appliances, industrial motors, MRI machines, iPods and computer hard drives, and green energy technology.

Global production of REEs in 2020 was about 300,000 tons of which the leading producers were China 62% (140,000 tons), United States 12% (38,000 tons), Burma 10% (30,000 tons), Australia 9% (17,000 tons) and Madagascar 4% (8000 tons). There is a bit of a crunch in the supply chain because of its dominance and control of REEs by China, the world's leading producer and exporter of a group of strategic metals. Most of the deposits in China are located in East Mongolia. In recent years, Africa has emerged as a viable alternative source of rare earth minerals, although production has faced several challenges including political instability and insecurity.

Economic deposits of REE occur in diverse but generally uncommon geologic settings including (a) magmatic disseminations in carbonatites and peralkaline igneous systems, (b) magmatic magnetite-hematite bodies; (c) iron oxide-copper-gold (IOCG) deposits; (d) xenotime-monazite accumulations in mafic gneiss and (e) monazite-xenotime-bearing placer deposits. The minerals that contain REEs are numerous, diverse, and often complex in composition. They include silicates, oxides, carbonates, phosphates, arsenates, sulfates, and vanadates and halides. The main sources of REE in Africa are Madagascar, South Africa, Namibia, Burundi, Mozambique, Malawi, Tanzania Kenya and Ethiopia (Fig. 4.7). Many of the deposits are small and associated with carbonatites and peralkaline and alkaline igneous complexes. Some occur as alaskite veins and dykes in surrounding basement rocks.

### 4.6.2 Uranium and Thorium

Uranium (U) and thorium (Th) are naturally occurring, radioactive heavy metals with the ability to generate nuclear energy by the natural breakdown of its components that can be used in nuclear reactors. Both metals are silvery white-grey and malleable (can be pressed into shape) and ductile (can be beaten and drawn into a wire). Thorium is much more abundant than uranium in rocks of the Earth's crust; and found in small amounts in most rocks and soils. On the other hand, uranium is more soluble and found in low levels within all rock, soil, and water and relatively enriched in organic-rich rocks.

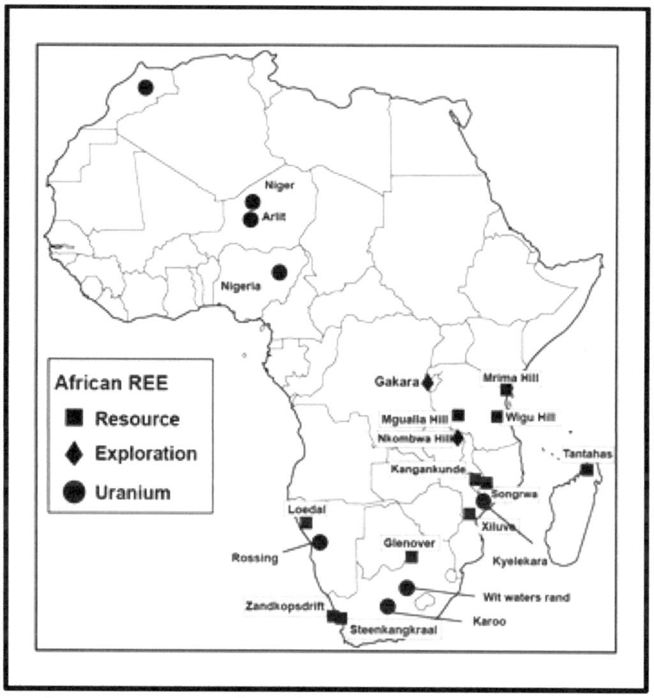


Fig. 4.7: Rare Earth and Uranium Deposits in Africa

Uranium is used in nuclear reactors to generate electrical power. Electricity generation in nuclear reactors has many similarities to more conventional electricity generation by converting thermal energy to mechanical energy that is used to drive turbo-generators (mostly by steam or superheated water). Thorium oxide ($ThO_2$) has one of the highest melting points of all oxides and has been used in light bulb elements, arc-light lamps and welding electrodes as well as in heat resistant ceramics. However, non-radioactive substitutes have been developed for many applications of thorium. Uranium is found mostly in the oxide form as uraninite or pitchblende. ($UO_2$ and coffinite ($USiO_4$)). Secondary alteration may produce several uranium hydroxide and silicate minerals. Uranium combines with most other metals to form complex minerals. Thorium is typically found in the minerals thorite, thorianite ($ThO_2$) and monazite, a rare earth phosphate mineral. However, monazite is the only mineral from which thorium is currently mined. These monazite deposits are found in placer deposits. India, Brazil, Australia and the United States currently have the most thorium deposits in the world. The leading producers of uranium in 2019 were Kazakhstan 42% (22,000 tons), Canada 13% (7000 tons), Australia 12% (6500 tons), Namibia 10% (5500 tons) and Niger 6% (3000 tons).

Economic concentrations of uranium occur in a variety of geologic settings including *magmatic* (felsic intrusives and pegmatites); contact metasomatic, hydrothermal veins, breccia deposits, sedimentary (paleoplacer, conglomeratic, sandstone-type, unconformity-related,); and surficial deposits. Sedimentary deposits account for over 75% of world uranium production. Ore grades can range from 0.03% up to 20% $U_3O_8$.

Uranium production is an important part of the African economy, with Namibia, Niger and South Africa accounting for about 16% of the world's annual production in 2019. Other African countries that have reported uranium occurrences include Nigeria and Gabon. Namibia's production is derived mostly from the Rossing uranium deposit which is localized within Precambrian pegmatites and alaskites. Niger Republic, the world's 5th largest, produces uranium from several Paleozoic-Mesozoic sedimentary (tabular and channel type) deposits around Arlit in central Niger Fig. 4.7). Uranium in South Africa is a byproduct of gold production in the Precambrian pebble conglomerates of the Witwatersrand Basin (e.g. Beatrix Mine, Free State Province), and from the sandstones of the Karoo Basin. Uranium which was discovered in the Karoo Basin in the 1960s is being mined from the Beaufort West district of the Karoo Province.

## 4.7 Review Questions

1. Define and classify metals into groups. What properties do metal have that make them more useful than other materials.
2. List the metals that constitute the ferro-alloys. What additional properties do they provide for iron and steel products?
3. What are precious metals? List them and briefly describe their uses and the geological environments in which they occur.
4. Nickel and platinum are metals that often occur together. Explain their geochemical affinity and the types of deposits in which they are formed. Give examples.
5. South Africa is one of the world's leading producers of several metals. Name two of these metals, their distribution mode of occurrence and geologic settings.
6. List the four base metals, describe their uses and explain why they are so important.
7. Copper and cobalt are two metals found together in Central Africa. List the countries and describe the ore nature of mineralization, principal ore minerals and origin of the deposits.
8. Niobium and tantalum are known as "twin metals". Explain this relationship and name the main ore and transition minerals. In what type of mineral deposits are both metals commonly found?
9. Aluminium is one of the light metals. Describe its uses, mineralogy, types of deposits. Why are the African deposits confined to certain elevations and equatorial latitudes?
10. Describe the rare earth elements and their uses. Why are they considered as strategic metals by the United States? In what type of rocks are the rare earth metals localized?

## References


Beukes, NJ, Swindell, EAJ and Wabo, H (2016) Manganese deposits of Africa. Episodes Vol. 39, no. 2, p. 285-317.

Evans, DM, Simmonds, J and Hunt JP (2016) An overview of nickel mineralization in Africa. Episodes Vol. 39, no. 2, p. 319-333.

Harmer, HA and Nex, PAM (2016) Rare earth deposits of Africa. Episodes 39, no. 2, p. 381-403

Kinnaird, JA, Bowden, P, Nex, PAM. and Milani, L (2016) Tin in Africa. Episodes 39, no. 2351- 360.

Kinnaird, JA and, Nex, PAM (2016) Uranium in Africa. Episodes 39, no. 2, 361-380.

Misra, K.C., (2000) Understanding mineral deposits: Kluwer Academic Publishers, 845 p.

Pohl WL (2011) *Economic geology: principles and practice.* In: Metals, minerals, coal and hydrocarbons – Introduction to formation and sustainable exploitation of min-eral deposits. Blackwell Publishing Ltd., Oxford, 663 p.

Revuelta, MB (2016) *Mineral Resources*: From Exploration to Sustainability Assessment. Springer Publishers, 641 p.

Taylor, CD et al (2005) Geology and Nonfuel Mineral Deposits of Africa and the Middle East. U.S. Geological Survey Open-File Report 2005–1294-E, 246 p.

# Chapter 4 Photographs

## Metallic Minerals

Plate 4.1: Gold Nuggets

Plate 4.2: Silver Ore

Plate 4.3: Platinum Ore

Plate 4.4: Iron Ore

Plate 4.5: Manganese Ore

Plate 4.6: Chrome Ore

Plate 4.7: Nickel Ore

Plate 4.8: Cobalt Ore

Plate 4.9: Titanium Ore

Plate 4.10: Vanadium Ore

Plate 4.11: Tungsten Ore

Plate 4.12: Molybdenum Ore

Plate 4.13: Tantalite Ore

Plate 4.14: Coltan Ore

Plate 4.15: Columbite

Plate 4.16: Forms of Copper Ore

Plate 4.17: Galena Lead Ore

Plate 4.18: Zinc Ore

Plate 4.19: Cassiterite Tin Ore

Plate 4.20: Alluvial Cassiterite

Plate 4.21: Aluminium Ore

Plate 4.22: Aluminium

Plate 4.23: Uranium Ore Uraninite

Plate 4.24: Yellow Uranium Ore

# Chapter 5
# Non-Metallic Mineral Deposits and Rocks

## 5.1 Introduction

Non-metallic minerals are earth materials that do not contain metals or mineral fuels. In economic geology, most non-metallic materials are considered as "*industrial minerals*" which was defined in the literature as "*a mineral, rock or other naturally occurring material*"...... The ambiguity of describing an industrial mineral as a rock or mineral led to the use of the more inclusive or acceptable term "*industrial minerals and rocks*". Industrial minerals and rocks are valued not for their metal content or energy source but mainly for their physical properties, and sometimes chemical composition. Most of these minerals (and rocks) are widely used for building and construction, and in various manufacturing industries to produce a variety of products. Examples include limestone, mica, barite, gypsum, marble, phosphate, salt, diamond, granite, talc and soapstone. Other industrial minerals are valued as gems for the jewelry industry. Compared to metallic minerals, non-metallic minerals show the following differences:

- Do not contain metals.
- Do not show shiny surfaces or luster typical of metals
- Valued for physical and chemical properties and not metal content
- Good insulators rather than good conductors of electricity and heat
- Lack malleability and ductility, and break or fracture easily when hit hard
- Found predominantly in sedimentary rocks
- More abundant than metallic minerals

### 5.1.1 Nomenclature of Industrial Minerals and Rocks

The current nomenclature of industrial minerals and rocks is a bit confusing and will require some definition of usage in this book. The term "*industrial minerals and rocks*" is widely used in the literature to describe collectively several types of earth materials that are different in composition, such as minerals being fundamentally different from rocks. As part of industrial minerals and rocks, the term "*construction materials*" is often used to describe the loose, unconsolidated aggregates that are utilized in the building and construction industry, such as sand and gravel or crushed stone. Sand and gravel should not be described as rocks, because a rock is a hard "solid mass" of indurated material (compare sand and sandstone). So, in this book, construction materials will be described as a separate group and not lumped with the "industrial rocks" as used by some authors. Industrial minerals are distinguished from industrial rocks because their composition, characteristics and uses are completely different. *Gemstones* are a group of non-metallic minerals that could be considered as industrial minerals because they are cut and polished for use in the jewelry industry. However, gemstones because of their unique identity and physical properties will be classified as a special class of nonmetallic minerals.

### 5.1.2 Types of Industrial Minerals and rocks

Based on the above analysis, industrial minerals and rocks are subdivided into four groups on the basis of their composition and uses as follows:

1. **Industrial minerals** (e.g. barite, feldspar, graphite, salt, talc)
2. **Industrial rocks** (e.g. phosphate rock, diatomite, limestone, nepheline syenite)
3. **Construction materials** (e.g. sand, gravel, crushed stone, laterite)
4. **Gemstones** (e.g. diamond, emerald, sapphire, aquamarine)

***Industrial minerals*** are those non-metallic, non-fuel minerals used in the manufacturing of various products. They are mostly 'mono-mineralic' (one mineral) substances.

***Industrial rocks*** are solid materials composed of one or more minerals that are indurated and used as raw materials in industrial production and as ornamental stone without any mineral separation,

***Construction materials*** are mostly loose minerals or rock aggregates that are used in building and construction.

***Gemstones*** are a unique class of minerals that are used for jewelry and other adornments. Table 5:1 provides a list of common industrial minerals and rocks.

The above classification scheme may not accurately 'pigeonhole' all industrial minerals and rocks or exclude the possibility that some minerals or materials may fall into more than one category. For example, clay is an industrial mineral, but consolidated clay as claystone or shale may be used as industrial rock in cement manufacture. The same rationale applies to diamond or corundum which are both used in industrial applications, but can be cut and polished as gems. Although limestone is an industrial rock used in producing cement and lime, crushed limestone is a construction material used as road aggregate and rail ballast in countries with temperate climates.

## 5.2 Sources and Origin of Industrial Minerals and Rocks

Industrial minerals and rocks originate from various sources as a result of their formation by different geological processes similar to those for metallic minerals. Consequently, their sources will be described in accordance with their origin.

### 5.2.1 Magmatic Sources

During the crystallization of mafic and felsic magmas, several industrial minerals are formed by magmatic dissemination and crystal fractionation. Early magmatic processes in ultramafic and mafic rocks are associated with the formation of unique industrial minerals derived from the mantle such as diamonds and olivine. Diamonds of gemstone and industrial quality occur almost exclusively in ultramafic kimberlite pipes and ultramafic (potassic) lamprophyre dykes. In the same phase of magmatic differentiates are found refractory olivine minerals (forsterite) found in dunite and peridotite. A popular gemstone called peridot is found as xenocrysts in basalt flows and peridotite xenoliths in lavas derived from the mantle.

Alkali basalts are the sources of most of the world's corundum gemstone deposits (ruby and sapphire). More of these minerals occur secondary deposits (mostly alluvial) derived from the basalts. Other gem minerals are zircon and spinel which all seem to occur mostly as xenocrysts in these basalts and are believed to have been formed in the mantle and transported to the surface by volcanic activity. Alkali basalts are common within ancient rift systems.

Carbonatites and associated alkaline igneous complexes are major sources of apatite-derived phosphate minerals for agricultural use. Carbonatites are also sources of phlogopite (insulators, filler), and vermiculite (insulation). There are other alkali minerals and rocks such as apatite nephelinites in alkali complexes; alkali feldspar in granite and albitite. Nepheline syenite is an industrial rock used for ceramics. Carbonatite bodies exposed at the surface are often used as calcite raw material for many industries and for fertilizer production.

Pegmatites are one of the major sources of industrial minerals and gemstones. They are the parent rocks of K-feldspar (for ceramics and glass), piezoelectric quartz crystals, topaz, garnets, emerald, tourmaline, beryl (for Be-alloys), and muscovite (insulation); lithium bearing lepidolite, amblygonite, spodumene and petalite, and various gemstones, and valuable crystals of many rare minerals. Granites are important ornamental stones.

Skarn deposits that form along the contact between igneous intrusions and sediments commonly involves the addition of fluids from igneous sources that produce industrial minerals and gems including corundum, beryl, emerald, and boron minerals used in fiberglass, ceramics, and glass, and wollastonite for ceramics and filler products.

Table 5.1: Types of Non-Metallic Minerals and Rocks

| Industrial Minerals | | Industrial Rocks | Construction Materials | Gemstones |
|---|---|---|---|---|
| Andalusite | Kyanite | Carbonatite | Aggregate | Amethyst |
| Asbestos | Mica | Diatomite | Building Stone | Aquamarine |
| Barite | Olivine | Limestone | Cement | Diamond |
| Bentonite | Magnesite | Nepheline Syenite | Brick Clay | Emerald |
| Borates | Potash | Phosphate Rock | Gravel | Garnet |
| Clays | Quartz /Quartz Sand | Basalt | Laterite | Ruby |
| Corundum | Sillimanite | Marble | Pumice & Perlite | Sapphire |
| Feldspar | Talc | Granite | Quartzite | Topaz |
| Fluorspar | Trona (Soda Ash) | Serpentine | Sand | Tourmaline |
| Graphite | Wollastonite | Soapstone | Sandstone | Zircon |
| Gypsum | Zeolites | | Slate | |
| Halite | Zircon/Zircon Sand | | Quartzite | |
| Kaolin | | | | |

## 5.2.2 Hydrothermal Sources

Hydrothermal solutions that emanate from magmatic sources are often enriched in alkalis, silica and volatiles which combine to form industrial minerals in the following deposit types.

- ***Hypothermal Veins*** originating at higher temperatures crystallize graphite in crystalline schists, and fluorite (fluorspar) in greisenized granites.
- ***Mesothermal Quartz-gold Veins*** are sources of massive quartz deposits.
- ***Epithermal Deposits*** include veins of barite and fluorspar in sediments and metasediments.
- ***Geothermal-Epithermal*** environment produces mineralized waters, and low temperature minerals such as adularia, zeolites and phlogopite produced by the alteration of ultramafic rocks.
- ***Volcanogenic Minerals and Rocks:*** Extensive barite deposits occur in association with volcanogenic massive sulfides (VMS). Volcanic fumaroles and hot springs produce sassolite (source

of boron) and native sulfur. Rhyolitic volcanic rocks such as perlite and pumice (very porous) are both excellent heat insulators.

- *Seder* deposits of barite.

### 5.2.3 Sedimentary Sources

Sedimentary deposits are the leading sources of industrial minerals and rocks through precipitation and sedimentation in marine and non-marine environments.

- ***Evaporite Deposits**:* Evaporite minerals and rocks including rock salt, potassium salt, gypsum/anhydrite, and borate minerals, trona and other salts in playa lakes are important industrial minerals.
- ***Biochemical/Biogenic Precipitates:*** Biochemical minerals and rocks - examples include limestone (cement, lime): sulfur in cap rocks of salt domes, concretionary phosphates in marine sediments, and nitratine (nitrate source for agriculture, explosives). Biogenic accumulation of marine skeletal remains can form the following industrial minerals and rocks - diatomite, phosphorite and most calcareous limestones.
- ***Clastic Sediments:*** Alluvial and coastal placers contain heavy minerals such as diamond, sapphire, ruby, topaz, zircon, andalusite, kyanite, and sillimanite. Paleo-placer (diamond, ruby) and buried gravels and sands; sandstone, siltstone and shale; clays and calcium bentonite.
- ***Diagenetic Deposits**:* Mississippi-Valley type barite-fluorite in marine carbonates; bentonite from altered volcanic piles and zeolites. Saline brine-related deposits: Emerald in black shales; dolomite and magnesite in marine carbonates.

### 5.2.4 Metamorphic Sources

Metamorphic sources of non-metallic minerals are those originating from regional metamorphism of sedimentary rocks and contact metamorphism involving emplacement of intrusive rocks into sedimentary or metasedimentary rocks.

- ***Regional metamorphic*** deposits of graphite formed by metamorphism of carbonaceous matter; asbestos, talc and serpentine from metamorphosed ultramafics; alumina-rich minerals sillimanite, kyanite and andalusite (all for refractories); garnets for (gemstone and abrasives).
- ***Thermal metamorphism*** along the contact of intrusive rocks (without metasomatism) has produced industrial minerals such as phlogopite, graphite and corundum. Other industrial minerals include andalusite, magnesite, and talc.

### 5.2.5 Surficial Sources

Processes that occur near the earth's surface such as weathering, supergene enrichment and fluvial deposition are important sources of industrial minerals and materials such as clay, laterite, gibbsite, sand and gravel.

- ***Residual Deposits:*** These deposits include residual clays, eluvial placers of diamond, kyanite, sillimanite; phosphate-rich soil ("phoscrete") above carbonatite; bauxite and kaolin blankets on aluminous silicate rocks; laterite soil and stone. Chemical weathering and supergene enrichment of copper ores produce gem minerals such as azurite and malachite.
- ***Fluvial Deposits**:* Recent fluvial deposits of diamond and gemstones; river sand and gravel are products of surficial processes

## 5.3 Industrial Minerals

Industrial minerals are a group of non-metallic minerals that are used as raw materials for industrial applications. Industrial minerals can also be defined more specifically as "geological materials that are not sources of metals, fuel, or gemstones, but are mined for their commercial value". Industrial minerals are valuable, not for the metals they contain, but for their physical and chemical properties. They are used mostly in their natural state or after minimal beneficiation, either as raw materials or as additives in a wide range of household and industrial products. Based on this definition, certain ore minerals such as chromite, bauxite, and rutile that have industrial applications are not considered as industrial minerals *sensu stricto*. A list of common industrial minerals is shown in Table 4.1.

The end uses of industrial minerals are varied. Some are used as sources of important chemicals (e.g. halite for sodium chloride *and* borax for borates). Some are used for the production of building materials (e.g. gypsum for plaster and kaolin for ceramics). Others are applied as agricultural fertilizers or animal feed (e.g. apatite for phosphate and sylvite for potassium). Still, others are utilized in many other ways, such as abrasives (e.g. diamond and corundum) or fillers (e.g. mica, kyanite). Many industrial minerals have a low unit value (low cost per ton of material) which makes them less valuable except close to consumers. It is often said that "*Without a market, an industrial mineral is merely a geological curiosity*"

In recent years, the interest in industrial minerals (and rocks) has increased considerably due to the substantial contribution of this class of mineral resources to the economy of many developed and developing countries. In the advanced countries, industrial minerals provide inputs in many industrial processes, while in some developing countries with little industrial infrastructure, significant portions of their foreign exchange derive from exports of industrial minerals. The rise in the exploitation of industrial minerals and rocks as a result of their widespread occurrence and uses has galvanized the efforts of geological exploration for these deposits, as part of a concerted strategy to enhance the solid minerals economy of many developing countries. A good example is Nigeria where over 80% of its annual solid mineral production is attributable to industrial minerals and rocks, particularly limestone which is used for cement production, a material in high demand for building construction, and granite which is used as aggregates for road construction.

### 5.3.1 Characteristics of Industrial Minerals

Industrial minerals possess some unique physical and chemical characteristics that distinguish them from other earth materials. Industrial minerals are:

1. Utilized because of their physical characteristics and/or chemical composition rather than their metal content.
2. Used in industrial applications with very little or no beneficiation.
3. Most industrial minerals have high *place value* but low *unit value*. (Place value reflects the importance of the value of the product in relation to its market while unit value indicates the real market value of a unit of the product e.g. one ton of sand)
4. Economic importance depends on the availability of markets, market location, and transportation costs. Without a market, an industrial mineral deposit is merely a geological curiosity.
5. Variable functions in industry. (Products derived from industrial minerals are interrelated in various industrial processes, e.g. use of barite and bentonite in drilling mud.)
6. A single material may form the basis for a wide range of applications in different industries, e.g. limestone is used for cement, production of lime, road aggregates etc.

## 5.3.2 Uses of Industrial Minerals

Industrial minerals are widely used in the following major areas of industrial activity:

a. ***Building Industry***: Several industrial minerals are used in the production of building materials. Gypsum is used in the production of drywall, cement plaster and additive for hardening of Portland cement. Andalusite, kyanite and sillimanite are used as additives in ceramic tiles for walls and floors, Asbestos is used in roofing and ceiling tiles and for insulation. Clays, feldspar and quartz are important raw materials in the manufacture of floor and wall tiles and structural clay products.

b. ***Manufacturing Industry*****:** Several products are manufactured from industrial minerals. In the manufacture of sheet glass and containers, quartz and silica sand, with additives of fluorspar, borates and soda ash are used. For ceramics, a variety of industrial clays is combined with feldspar as flux in clay bodies and glaze. For abrasives, minerals such as diamond, zircon, garnet, corundum are used. For filler manufacturing, ground calcite and zeolites are used as a filler in paint, paper and plastics. Kaolin is used in coating of paper and as paint fillers and medicinal products. Fire clay is used for refractories while other minerals such as halite, fluorite and apatite are used to produce chemicals while others are applied as metallurgical fluxes, soil stabilizers and neutralizers, and paper, and water treatment.

c. ***Agriculture Industry*****:** Most fertilizers are produced from natural minerals and rock products such as apatite (phosphates), nitrates, potash, salt, and sulfur. Calcite, and to a lesser degree dolomite, when pulverized (ground to a fine powder), may be used to condition clayey soils and neutralize acid soils in order to promote plant growth.

d. ***Chemical Industry***: Industrial minerals are used in the production of several chemicals. Such minerals include limestone (for lime), fluorite, sulfur, halite (salt), nitrates and borates.

e. ***Metallurgical and Refractory Industry:*** The metallurgical industry uses industrial minerals as fluxes in the processing of iron ores and other metals. Limestone (for lime), borax, silica and dolomite are metallurgical fluxes magnesite, fire clay, graphite, bauxite, zircon, silica, and dolomite, Minerals commonly used for refractories include magnesite, bauxite, graphite, kaolin, and zirconium.

## 5.3.3 Importance of Industrial Minerals

Industrial minerals constitute the most widely exploited mineral resource in both developed and less developed countries. In developed countries such as the United States, the annual dollar value of industrial mineral production has since surpassed that for metals, and continues to grow rapidly. In 2018, the value of non-metallic minerals and rocks produced in the United States was valued at $53 billion (US dollars) compared to the value of metals production worth $30 billion (USD). Industrial minerals are of great economic value as they serve as major raw materials in the manufacture of several industrial and daily-need products such as glass, abrasive, paper, chemicals, detergents, ceramics, fertilizers, and metallurgical and agricultural products. A wide range of everyday domestic and vital industrial products are made from industrial minerals. According to the Industrial Minerals Association of North America, the average American resident consumes about 24 tons of industrial minerals each year.

While in developed countries industrial minerals provide inputs in many industrial processes, it is a little different in some less developed countries with little industrial infrastructure, where significant portions of their foreign exchange derive from exports of industrial minerals and rocks. Also, the fact that industrial minerals (and rocks) are found almost everywhere (widespread occurrence) makes these materials good sources of income for local suppliers and communities. For example, sand and gravel are available in most local communities in Africa where they are excavated and used for building construction which generate income for local entrepreneurs and transporters. Industrial minerals are therefore vital to a modern economy – they underpin the manufacturing industry, construction and agriculture – they also have important environmental applications by conversion of mine wastes to beneficial use.

Table 5.2: Industrial Minerals and their Uses

| Industrial Minerals | Uses | Industrial Minerals | Uses |
|---|---|---|---|
| **Andalusite**, Kyanite, Sillimanite | Refractories, ceramics and alumina bricks | **Kaolin** | Used in paper coating: for ceramic whitewares and pigment in paint |
| **Asbestos** | Roof and floor tiles, fire and heat resistant products. | **Mica** | Use as insulators, lenses of optical instruments |
| **Barite** | Weighting material in drilling mu; and filler or extender. | **Quartz / Quartz Sand** | Glass, ceramics and chemical apparatus |
| **Bentonite** | Component of drilling mud for deep wells | **Salt (Halite)** | Table salt, food seasoning and chemicals |
| **Borates** | Insulators, fiberglass and specialty glass, textile | **Talc** | Cosmetic powder, personal care products; filler in paints & ceramics |
| **Clays** | Ceramics, refractories and structural clay products | **Trona (Soda Ash)** | Flux for glass making, water purification; removal of pollutants |
| **Corundum** | Used for abrasives and alumina refractories and bricks | **Wollastonite** | Use in ceramics and friction products; filler in paints |
| **Feldspar** | Flux in ceramics. Fillers and extenders in paints and plastics | **Zeolites** | Cement additive; use as adsorbent, ion exchangers. Water filtration. |
| **Fluorspar** | Flux for melting of glass and enamels. Used for chemicals. | **Zircon** | Refractory bricks and liners; foundry casting applications. |
| **Graphite** | Used in lead pencils, lubricants | | |
| **Gypsum** | Manufacture of drywall, cement and plaster of Paris. | | |

Owing to the scarcity of resources and supply issues, the European Commission on Raw Materials, recently designated some industrial minerals as *strategic minerals* that are "critical" to industries in Western Europe, e.g. graphite, fluorspar, phosphate, borates etc. Nigeria has designated limestone as a strategic mineral for national development. For developing countries such as Nigeria, the prospects for selling industrial minerals on the export market has been very encouraging in recent years, and this has led to increased mining activity by artisans. Moreover, the desire to locally manufacture products that use industrial minerals is seen as a mechanism to conserve foreign exchange reserves by reducing the importation of such products. The growth in the demand for industrial minerals can be attributed to the combination of high and increasing per capita consumption of goods produced from industrial minerals, as well as increasing income levels in developing countries. For example, the consumption of cement and ceramic products has grown exponentially due to the booming building construction in African countries. Several household goods and life conveniences, such as automobiles, and solar panels require the use of industrial minerals.

Another example is the Dangote Group, a Nigerian multinational industrial conglomerate, founded by Alike Dangote. It is the largest conglomerate in West Africa and has invested significantly in the production of cement by using locally available industrial minerals and rocks. Currently, the company owns Dangote Cement which is the largest cement production company in Africa, with a market capitalization of almost US$14 billion and has subsidiaries with cement factories in Benin, Cameroon, Ghana, Nigeria, South Africa and Zambia

The widespread exploitation of industrial minerals has produced positive economic impacts on many communities in developing countries, in large part due to the multiplier effect in which a single mineral in one production process often involves the use of several others. For example, the production of glass from quartz sands may require the use of soda ash, calcite, dolomite, feldspar, borax, gypsum and fluorspar. Above all, these minerals and rocks provide the raw materials for infrastructure development in which large volumes of sand, gravel, clay, crushed and dimension stone are consumed locally. Industrial minerals are therefore vital to a modern economy where they form the bedrock for a fast-growing manufacturing industry, construction and agricultural activities

## 5.3.4 Classification of Industrial Minerals

Industrial minerals can be classified according to three criteria: (1) geological occurrence; (2) uses (utility), and (3) economic value.

### 5.3.4.1 Geological (Genetic) Classification

Industrial minerals are formed by geological processes similar to those responsible for the formation of rocks and metallic minerals with which industrial minerals are closely associated. The three main geological processes are igneous, metamorphic and sedimentary. Grouping by geological association is as follows:

- ***Igneous and Hydrothermal Minerals**:* Industrial minerals that form through igneous processes (intrusive, extrusive, pegmatitic and hydrothermal) include many pegmatite minerals, such as muscovite, feldspar, spodumene, tourmaline, and beryl; hydrothermal vein minerals quartz crystal, fluorspar, barite, and magnesite;
- ***Sedimentary Minerals:*** Clastic, biogenic and chemical processes produce industrial minerals and rocks that include diatomite, potash minerals, sodium minerals, borate, nitrates, sulfur, salt and sedimentary rocks: sand and gravel, sandstone, clay, limestone and dolomite, phosphate rock, gypsum, rock salt from marine and non-marine environments.
- ***Metamorphic Minerals**:* Regional and contact metamorphism both produce industrial minerals classified as metamorphic in origin, including graphite, asbestos, talc, phlogopite, and zeolites.
- ***Residual Minerals:*** These are industrial minerals formed by surficial weathering and concentration. Examples include clay minerals gibbsite and kaolin.

### 5.3.4.2 End-Use or Utility Classification

This classification is based on the end-use (utility) of industrial minerals...Six groupings can be identified:

- ***Construction Minerals***- gypsum, asbestos, clays, feldspar and quartz
- ***Ceramic Minerals*** - clays, silica , feldspar, quartz, and bauxite
- ***Metallurgical and Refractory Minerals*** - magnesite, fire clay, graphite, bauxite, zircon.
- ***Abrasive Minerals*** - chalcedony, quartzite, chert, corundum, garnet
- ***General Manufacturing Materials*** - glass sand, borate, fluorite
- ***Chemical and Fertilizer Materials***.- salt, lime, potash

### 5.3.4.3 Economic Value Classification

This is a classification based on price and bulk or production volume of industrial minerals. There are three classes:

1. ***Low Price – Large Volume***: - These are materials that are used in large volume (bulky) but are relatively cheap e.g. quartz sand, and gypsum.
2. ***High Price – High Volume:*** - These are minerals that command high prices but are still used in large volume for industrial application e.g. borates, bentonite, potash and salt.
3. ***High Price – Low Volume***: - These are high-priced minerals that are needed in small quantities: e.g. barite, kyanite and beryl.

## 5.3.5 Description of Selected Industrial

Approximately one hundred minerals are known to have some economic value, but only about thirty are of real commercial value. A selected few will be described below with emphasis on their mode of occurrence and uses.

### 5.3.5.1 Andalusite (Kyanite and Sillimanite)

Andalusite, kyanite and sillimanite (Al2SiO5) are alumino-silicate polymorphs with the same chemical composition but different structural configuration. They are metamorphic minerals formed at varying temperature and pressure conditions (Fig. 5.1). Andalusite is the lower pressure and mid-temperature polymorph that converts to sillimanite at higher temperatures and pressure. It is produced during low-pressure amphibolite facies regional metamorphism or contact metamorphism of pelitic rocks. Andalusite is used as refractory material to withstand high temperatures in furnaces, kilns and other industrial processes because of its alumina and silica contents, Andalusite is also used in the manufacture of high-temperature porcelain of spark plugs. High-quality crystals are sometimes used as gemstone. World-leading producer of andalusite is South Africa, where it is derived mostly from the vast contact zone of the Bushveld Igneous Complex. Initially, the mineral was exploited from alluvial placers, but currently, it is produced from hard rock mining of primary deposits.

*Kyanite* is the high pressure polymorph of Al2SiO5, found in high grade metamorphic terrains and formed by the metamorphism of siliceous and pelitic sediments (gneiss, schists, and metamorphic pegmatites). It is used to manufacture a wide range of products, particularly refractories such as fire bricks, mortars, and kiln furniture that are used in high-temperature furnaces. The United States and India are the world's largest producers of kyanite which are found within Precambrian schists and gneisses.

*Sillimanite* is the highest temperature Al2SiO5 polymorph, found mostly within metamorphic gneisses, quartzites and schists (Plate 5.3). Mineable deposits of sillimanite occur as two main types: podiform deposits which are massive high-grade ores in schists, and stratiform type deposits as sillimanite-quartzite horizons with 20-50% of sillimanite minerals. Most of the sillimanite produced is used as raw material for the manufacturing of high-alumina refractories. India and South Africa are the leading producers of commercial sillimanite.

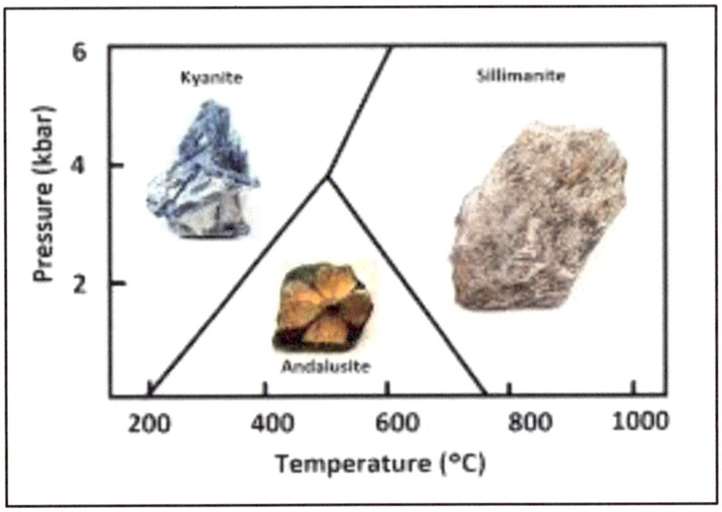

Fig. 5.1: Kyanite, Andalusite and Sillimanite Stability Fields

### 5.3.5.2 Asbestos

Asbestos is a generic name for six naturally occurring asbestiform or fibrous minerals that belong to either the serpentine or amphibole group, namely: chrysotile (serpentine) and the other five are amphiboles: crocidolite, amosite, tremolite, actinolite and anthophyllite. Chrysotile, crocidolite and amosite are the most common. Chrysotile (white asbestos) is by far the most abundant and widely distributed of all the asbestiform minerals. Asbestos with its fibrous texture, has certain industrial properties including being an excellent electrical insulator and highly heat-resistant.

For several decades, asbestos was widely used in the manufacture of building materials such as roofing and ceiling sheets, heat insulation, vehicle gaskets and brake linings, and in common household products. When it became known as a potential health and safety hazard in the 1980s, its use and production were later banned or discouraged in several countries in Europe, USA and South Africa, However, it continues to be used in large quantities around the world and become an important commodity in global trade.

An estimated 1.5 million tons of asbestos continue to be used around the world each year, although this is significantly less than peak annual consumption of nearly 5 million tons three decades ago, India, China, Russia, and some developing countries are the major consumers of asbestos. Chrysotile is the most widely used variety of asbestos (Fig. 5.2a). Crocidolite is considered the most hazardous type of asbestos because it is made up of extremely fine and sharp fibers that are brittle and particularly easy to inhale. However, about 50% of asbestos world production is of chrysotile (asbestiform serpentine) which is less dangerous than crocidolite. Production has declined considerably from about 4.3 million tons in 1977 to 1.3 million tons in 2017 when Russia was the largest producer with 53% of the world total, followed by Kazakhstan (16%), China (15%), and Brazil (11%). Asia consumes some 70% of the asbestos produced in the world with China, India and Indonesia being the largest consumers.

The mode of occurrence of chrysotile asbestos is mostly as small veins in serpentines, serpentinized ultramafic rocks, and in serpentinized dolomitic marbles. It is formed by the process of 'serpentinization' of ultramafic rocks containing olivine, Mg-rich pyroxenes, and amphiboles by either hydrothermal alteration and/or low-grade metamorphism. The chrysotile fibers grow at right angles to vein walls of massive serpentine formations to form chrysotile deposits. Most of the industrial chrysotile fibers are extracted from deposits where fiber lengths can reach several centimeters (cm), but most often do not exceed 1 cm. The asbestos fibers usually grow perpendicular to the vein walls and their fiber length and value is determined by the vein thickness. Asbestos is used in hundreds of products including cement building materials, boiler pipes, insulating mattresses, fire-resistant insulation boards, sprayed fire-proofing products, floor tiles, roof tiles, and friction materials for vehicle brakes and clutches.

Fig. 5.2: (a) Chrysotile Asbestos (green) (b) Crocidolite Asbestos (blue)

In Africa, the leading producers of asbestos were South Africa and Zimbabwe. Until 2009, when asbestos mining was banned, South Africa was a leading producer of amosite, crocidolite and chrysotile asbestos from mines in northeastern Transvaal and Northern Cape Province. At its peak in 1997, asbestos production was about 380,000 tons per annum.

### 5.3.5.3 Barite

Barite (or Baryte) ($BaSO_4$) is a sulfate mineral, and the main ore of barium. The mineral has a wide range of colors and crystal forms, but is commonly colorless, reddish-brown to white. Barite is heavy, with a specific gravity varying from 3.5 to 4.5. The largest proportion of barite production is used in the petroleum industry as a weighting material in drilling mud. Other uses of barite include as filler or extender, and additive to cement, plastic, rubber, and urethane foam; and plastic automobile paint primer; friction products (brake and clutch pads) and as a white pigment in paints. Global production of barite in 2019 was 9.5 million tons (Mt), and the leading producers were China (2.9 Mt), India (2.2 Mt), Kazakhstan (0.62, Iran (0.49), and Mexico (0.42 Mt).

The geological occurrence of barite is varied. Commercial deposits are of five types: (i) *contact metasomatic deposits,* (ii) *vein deposits,* (iii) *sedimentary (bedded)* deposits, (iv) *volcanic bedded deposits,* and (v) *residual deposits.* The contact metasomatic deposits are formed by replacement of host rocks along the contact with intrusive rocks; while vein deposits occur as fracture filling often associated with lead-zinc, silver and fluorite minerals. Thick beds of barite deposits occur as sediment-hosted (stratabound) deposits in association with Pb-Zn in sedimentary exhalative (Sedex) and Mississippi Valley-type (MVT) deposits. Volcanic bedded barite occurs along the flanks of volcanogenic massive sulfide deposits. Residual deposits are products of weathering of pre-existing barite ore which may end up as detrital ores if transported downslope.

South Africa and Nigeria are the main producers of barite in Africa. Other countries are Morocco, Algeria and Egypt. In South Africa, barite occurs as metasomatic replacement deposits in Namaqualand, and as stratabound barite associated with sedimentary exhalative (Sedex) type massive sulfides in the Northern Cape Province. Barite mining facilities exist in seven provinces, namely: KwaZulu Natal, Northern Cape, North West Province, Mpumalanga, Free State, Limpopo, and Gauteng.

In Nigeria, barite occurs as veins and cavity filling deposits closely associated with Pb -Zn hydrothermal veins within the Cretaceous sediments of the Benue Trough in Cross River. Benue, Nasarawa, Taraba and Plateau states. Barite veins have also been found in the basement complex adjacent to the sedimentary basins. The barite veins vary in width but may reach 5 m wide and 5000 m long. Azara in Nasarawa State has the largest deposit while Cross River State has the best quality and most quantity of barite, with occurrences in Obubra, Yala, Biase, and Ikom LGAs, Annual production of barite in Nigeria is estimated as 30, 000 tons.

### 5.3.5.4 Bentonite

Bentonite ($Al_2H_2Na_2O_{13}Si_4$) is a naturally occurring, very fine-grained and plastic clay that is composed 95% of the clay mineral, *montmorillonite* (Na,Ca) $Al_2Si_4O_{10}(OH)_2(H_2O)_{10}$. It has a high swelling capacity and viscosity which are essential properties for use as a drilling mud component. There is also an increasing demand for bentonite for use in other areas, such as binders for iron pellets, cosmetics, pharmaceuticals etc. Generally, bentonites that are required for these applications are pure white bentonite with high viscosity. For industrial purposes, two main varieties of bentonite exist: sodium and calcium bentonite. The sodium bentonite is the "true" bentonite which is very plastic with high swelling propensity. The calcium bentonite has more calcium than sodium and less swelling capacity. However, the Ca-bentonite can be beneficiated by adding Na from Na2CO3 which improves its swelling capacity but not as much as the natural Na-bentonite.

Bentonite is used in drilling fluids to lubricate and cool the cutting tools, to remove cuttings, and to help prevent oil well blowouts. Much of bentonite's usefulness in the drilling and geotechnical industry comes from its unique rheological properties. Relatively small quantities of bentonite suspended in water can form a viscous, shear-thinning material. At high enough concentrations (60 grams of bentonite per liter of suspension), bentonite suspensions begin to take on the characteristics of a gel that makes it suitable as part of drilling mud used to curtail drilling fluid invasion by its propensity for forming mud cake. Bentonite usually forms from the hydrothermal alteration or weathering of volcanic ash, or the devitrification and chemical alteration of some volcanic material, most often in the presence of alkaline solutions (Fig. 5.3a). This allows for the enrichment in sodium that enhances the swelling capacity of bentonite.

The world leading producers of bentonite are the USA, China, Greece, India and Turkey. The largest producer in Africa is Morocco. Most Nigerian bentonites are of the calcium variety and not suitable for use except when beneficiated which can raise the cost of production. There are also a group of clays and shales that are sometimes described as "bentonitic clays", because they have certain properties similar to bentonite but are mostly Ca-montmorillonite with a small amount of Na. Nigeria is a major oil producer and the drilling activities associated with the development of petroleum and groundwater resources require the use of bentonite for drilling mud. Nigeria imports most of its bentonite in quantities exceeding 70,000 tons per annum.

### 5.3.5.5 Borates

Borates are naturally-occurring minerals containing boron. The element boron does not exist by itself in nature but combines with oxygen and other elements to form boric acid, or inorganic salts called borates - mostly sodium and calcium borates. There are eleven important borate minerals, of which the most common are borax (tincal) (Na2B4O5(OH)48H2O), kernite (Na2B4O6(OH)23H2O), ulexite (NaCaB5O6(OH)65H2O, and colemanite (CaB3O4(OH)3), Borates possess some useful physical and chemical properties such ahs great hardness, low mass density, high thermal conductivity, high chemical inertness and excellent wear resistance. Borates have numerous uses including as insulators, additives in fiberglass and specialty glass, and in textiles, ceramics, wood treatment and agriculture (for plant growth). They are also used as a component of detergents and cosmetics.

Borates occur mainly as evaporites deposited in lacustrine and marine environments, mostly in lakes and playas in desert regions. Borates in Playa Lake sediments occur as massive beds or crystals dispersed in layers of salt and gypsum (Fig. 5.3b). Older borate deposits form thick sedimentary units within evaporite sequences. Rare metamorphic and skarn deposits are mined in China and USSR respectively. Borate deposits are of variable sizes, and several million tons are mined, processed, and distributed around the world every year. The sources of the boron in these deposits are attributed to the weathering of previous deposits, seawater precipitation, thermal springs and volcanic activity.

Global production of borates in 2015 was estimated as 3.8 million tons of which the leading producers were Turkey (30%), USA (28%), Russia (22%), Argentina (8%) and China (7%). The United States production comes from California in the Miocene playa lake sediments in Boron, Death Valley and Searles Lake in the Mojave Desert region. The Turkish production is from four districts about 200 km south of Istanbul. World reserves are estimated at 1.3 billion tons of which 85% are in Turkey. The global consumption of borates annually is about 3.5 million tons, mostly in Asia (60%) for agricultural use, while Europe consumes 18% mainly for additives in fiberglass and detergents.

Fig. 5.3: (a) Bentonite Mine, USA

(b) Borates Mine, California

### 5.3.5.6 Clays and Kaolin

Clays are naturally occurring, fine-grained materials containing one or more clay minerals which are layered silicates with sheets of SiO4 tetrahedra combined in various ways with sheets of Al(OH)6 and/or Mg(OH)6 octahedra. There are three main groups of clay minerals, each with its own particular properties: *kaolinite, montmorillonite* (smectite) and *illite*. Clay rocks may contain a mixture of these minerals (mixed layer clays), and may end up with very variable properties.

Structurally, here are two types of clays, described as 1:1 or 2:1 types which reflect the ratio of tetrahedral silicate sheets and octahedral hydroxide sheets. The 1:1 clay consists of one tetrahedral sheet and one octahedral sheet (e.g. kaolinite). The 2:1 clay consists of an octahedral sheet sandwiched between two tetrahedral sheets (e.g. montmorillonite and vermiculite). The 2:1 clays can be classified into expanding (smectite) and non-expanding clays (illite and sericite). In some clay minerals exchangeable cations (Na +, K +, H +, Ca2 +, or Mg2 +) are able to enter into the clay structure and give them important chemical properties with industrial applications.

The main industrial properties of clay minerals include plasticity, porosity, firing color, tensile strength and vitrification (fusibility). All these properties are important in evaluating the potential industrial applications of clays. *Plasticity* is the most outstanding property that makes clays easily workable. *Porosity* is a necessary property that clay must have so that clay products can dry without cracking. *Firing characteristics* refer to the behavior and appearance (color) of clay products after being "fired" (subjected to high temperatures) at above 900° C. They are very relevant to clay's aesthetic value. *Vitrification* is the process of becoming glass-like when fired, which is a property related to its ability to fuse at appropriate temperature (fusibility). Although clay products never become absolutely vitrified or glass like, it is necessary that the clay becomes hard (or almost vitrified) at a reasonable temperature, usually 1200 to1500°C.

Clay deposits of commercial value normally consist of one or more clay minerals with varying proportions of non-clay minerals, the most common being quartz, feldspar, carbonate minerals, gypsum and organic material. Clay deposits are formed by the primary process of hydrothermal alteration of granitic rocks, or secondary processes of chemical weathering and/or sedimentation. Primary clays are usually kaolin or china clay whereas secondary clays are varied. The various types of secondary clays: include kaolinitic clays (weathering), alluvial clays (river deposition), marine clays (sea deposition) and glacial clays (ice action).

Clays are important industrial minerals widely used in building construction and manufacturing of ceramics, bricks, hydraulic cement, structural clay products, fillers and extenders, and additives in drilling mud, fiberglass, paper coating, furnace lining etc. Industrial clays can be classified into the following groups based on their industrial properties and usage.

- *China Clay/Kaolin* (primarily kaolinite with minor quartz, mica)
- *Ball Clay* (primarily kaolinite with illite, chlorite, smectite minerals, quartz)
- *Fire Clay* (kaolinite, halloysite, diaspore)
- *Common Clay* (illite, kaolinite and chlorite, others)
- *Bentonitic Clay* (smectite with feldspars, biotite, quartz)

### *Kaolin (China Clay)*

China clay or Kaolin (derived from the Chinese word 'Gaoling') is a very fine, soft white clay made up of more than 80% kaolinite. Highest quality kaolin contains almost pure kaolinite with >90% of particles less than 0.002 mm in size and of low to moderate plasticity. Minor amounts of mica and quartz may be present. Kaolin is the "king" of clays, and the purest of all clays. It is formed mostly by the hydrothermal alteration of granitic rocks and pegmatites, and also by chemical weathering of several feldspathic rock types.

Kaolin has some unique physical properties that are useful in industrial applications. These properties include shape, particle size, color, gloss, viscosity, softness and non-abrasiveness. Kaolin has long been used in the ceramic industry, especially in the production of china wares, fine porcelain and other table wares. However, the primary use of kaolin is in the paper industry where it is applied as paper coating for brightness and glossy appearance. Other industrial applications include use as a pigment in white paint and a filler in fiberglass, rubber, plastics, and cable insulators. Kaolin deposits can be evaluated for their brightness, particle size distribution, viscosity (important for the valuable paper-coating clays), the iron and potash contents, fired brightness, casting behavior and other ceramic properties. The quality of kaolin can be improved by additional processing and purification.

Kaolin is a special clay with a high unit value and is a commodity traded around the world. In 2019, a little over 28 million tons were produced worldwide with a market value exceeding $4 billion US dollars Most of the demand is from paper, ceramics, fiberglass, paints, rubber, plastics end-use industries, The world-leading producers are the United States (7 Mt), Uzbekistan (5 Mt), Czech Republic (3.8 Mt), Germany (3.8 Mt), and Korea (2.9 Mt). Nigeria produces some kaolin and kaolinitic clays that are widespread within areas underlain by the crystalline rocks of the Precambrian basement complex in central, southwest and northwest regions. Relatively pure kaolin is found in a few places such as the well-known Kankara Clay in Katsina State and Potter Smith near Jos in Plateau State.

***Ball Clay***

Ball clay is a white burning, but usually buff to grey-colored, very plastic clay that burns to a light color. It is a member of the kaolinite family that is renowned for its plasticity and green strength, and as such, it forms a major part of many ceramic applications. Ball clay consists of varying proportions of kaolinite, mica and quartz, with small amounts of organic matter and other minerals. The very fine-grained texture of ball clay enhances its plasticity and commercial value because it is added to kaolin to increase the workability and strength of various ceramic bodies, as well as burn white or near white when fired. Ball clays are in general, different from kaolin by their higher plasticity and much finer-grained texture.

Ball clay is used in many different industries, but in particular, forms a vital component in ceramic manufacturing because it provides the plastic body for these products. Ball clay is used in the production of sanitary ware, tableware and floor/wall tiles. A 'ceramic body' for sanitary ware typically includes 30% ball clay to provide plasticity and workability, 20% kaolin, 30% feldspar and 20% quartz/silica. Ceramic tableware and sanitary ware utilize ball clay to provide high plasticity and good white-fired color, combined with kaolin, feldspar and quartz. In wall and floor tiles, ball clay is added for its plasticity and bonding properties is combined with talc, feldspar, quartz/silica and kaolin. Ball clay is also used in the production of glazes which are coatings for ceramic products to ensure a perfect finish.

The origin of ball clay is obvious by its common association with other rocks of sedimentary origin. They are usually formed in a marine environment (marine clay). As a result of its origin, raw ball clay may have a wide range of off-white colors but is most valued by the ceramics industry for its white-firing properties, Ball clays are not very common, but fortunately they are found abundantly in most of Nigeria's sedimentary basins, including in Ondo, Edo, Anambra and Delta States.

***Fire (Refractory) Clay***

Fire Clay is a clay capable of resisting high degrees of heat at temperatures of up to 1700$^{o}$C without **damage**. To be referred to as a "fire clay", the clay material must be refractory, i.e. being able to withstand a minimum temperature of 1515$^{o}$C without melting. Fire clays are also described as refractory clays. Fire clays may be plastic or non-plastic and composed of kaolinite with some mica or illite and some quartz and organic matter. Fire clays vary widely in physical properties as long as they are refractory. Some fire clays are very plastic (*plastic fire clay*), while some are stiff and lack plasticity (*flint fire clay*), with a range in between. The fired color may vary, but most fire clays burn white. Chemically, fire clays are characterized by relatively high alumina content, with a Si/Al ratio of about 1:2. Such refractoriness or resistance to heat means that the clay is relatively pure and free from iron. Fire clays are used for a variety of products, mainly in the manufacture of fire bricks and other refractory parts for kilns, furnaces, boilers, and melting pots. The origin of fire clays ranges from residual to sedimentary, but most are of sedimentary origin. In Nigeria, fire clays within the sedimentary basins; in the Anambra Basin where they occur beneath the coal beds of Enugu in Enugu State, and in the East Dahomey Basin at Onibode Clay in Ogun State.

***Common Clays***

The term common clays was devised by the US Geological Survey to describe a variety of earthy clay materials, including soil clays, lateritic clays, weathered shales and glacial clays that are plastic when wet and composed of some clay minerals, other silicates and organic matter. These clays are fine grained and typically exhibit plastic behavior when wet and can be molded into many desired shapes, dried and fired to produce clay products with a rock-like hardness. The common minerals are kaolinite, and/or illite and/or chlorite, sometimes with minor montmorillonite, quartz, feldspar, and calcite. In the raw form, common clays are red, brown, greenish, or gray, as a result of the presence of iron oxide and some organic matter. They are the principal raw materials used in the production of traditional pottery wares, construction bricks, roofing tiles, drain tiles and other structural clay products such as drainage or sewage pipes and decorative bricks.

Common clays are the most widely used ceramic materials in the world. They are quite common, and are usually found along the earth's surface, in soils, stream beds and within the weathered regolith. They contain iron and other mineral impurities in sufficient quantities to cause the clay to become tight and hard-fired at about 950 to 1000°C. When fired, the clay may vary in color from pink to buff to tan, red, brown, or black, depending on the clay, the impurities and the condition of the firing.

### 5.3.5.7 Corundum

Corundum is an aluminum oxide (Al2O3) that often appears as well-formed hexagonal crystals. It is an industrial mineral as well as a colorful gemstone - ruby and sapphire. Pure corundum is colorless and clear if transparent or pale white if opaque, but shows attractive colors of blue and purple-dark red gem varieties (sapphire and ruby) which arise from the presence of impurities of Fe, Ti and Cr.

Corundum is a very hard, tough and stable mineral, second only to diamond in its hardness. The extreme hardness of corundum makes it suitable as an abrasive for grinding and polishing. Crushed corundum is used to produce uniformly sized granules and powders for grinding media, polishing compounds, sand papers, grinding wheels, and other cutting applications. Corundum has many other uses. Because it is chemically inert and resistant to heat, corundum is seen as a perfect material for making refractory products such as fire brick, kiln liners, and kiln furniture.

Corundum occurs as a primary mineral in igneous rocks (magmatic) such as syenite, nepheline syenite, and pegmatite. Corundum is also found in metamorphic rocks in locations where aluminous shales or bauxites have been subjected to contact metamorphism and metasomatism. Schist, gneiss, and marble produced by regional metamorphism will sometimes contain corundum. Some alkali basalts contain corundum as gems. Secondary corundum deposits are found as eluvial or colluvial accumulations, and alluvial and marine placers. Some of the richest deposits of industrial corundum are found in India, Myanmar (Burma), Thailand, Sri Lanka, Russia, Madagascar, Zimbabwe, and South Africa.

### 5.3.5.8 Feldspar

Feldspar is an aluminosilicate of potassium, sodium and calcium, and is by far the most abundant group of minerals in the earth's crust. The three end members of the group are orthoclase ($KAlSi_3O_8$), albite ($NaAlSi_3O_8$.) and plagioclase (CaAl2Si2O8). Feldspar minerals are used in industrial application for their alumina and alkali content, particularly for glass and ceramics manufacture. In glass manufacture, alumina from feldspar improves product hardness, durability and resistance to chemical corrosion while in ceramics, feldspar is used as a flux, melting at an early stage in the firing process, thus forming a glassy matrix that bonds together the other components of the ceramic product. Feldspar is therefore the main vitrifying agent in ceramics and the second most important ingredient after clay. Fused feldspar is also the main ingredient in the glaze on porcelains and other white wares. Feldspars also are used as fillers and extenders in applications such as paints, plastics and rubber.

Among the three feldspar minerals, the alkali feldspars, K-feldspar (orthoclase or microcline) and Na-feldspar (albite) are the most desirable for industrial application because they contain both alumina and alkalis, and are readily available for exploitation. Feldspar with higher calcium (plagioclase) content is used less for its alumina content or in construction. Feldspars are one of the most common silicate minerals that occur in granitic rocks and pegmatites. For commercial production, the best feldspars for industrial applications are the alkali feldspars found in substantial quantities in pegmatite dikes and massive bodies (Fig. 5.4a). These are mostly K-feldspars, although albite-rich intrusive bodies (albitites) which are rare have been found in some localities including Quebec, Canada, and Andra Pradesh in India. Global production of feldspar in 2015 was 21 million tons. The leading producers were Italy (25%), Turkey (24%), China (12%), Thailand (5%), and the United States (2%).

In Nigeria, economic deposits of alkali feldspar are found in the Older Granite pegmatites and granitic intrusions in the basement complex. Important sources of feldspar rich pegmatites are in Kaduna, Ogun and Kogi States, and around Osogbo in Osun State. Near Osogbo, potash feldspars attain lengths of 0.9m, while around Ijero in Ekiti State, gigantic crystals are up to 1.2m long. Feldspar-rich pegmatites are also common in the Gwoza area of Borno State and parts of Adamawa and Taraba states.

Fig. 5.4: (a) Feldspar deposit in Pegmatite (b) Pile of Fluorspar Deposit

#### 5.3.5.9 Fluorite (Fluorspar)

Fluorite ($CaF_2$), known commercially as 'fluorspar' is a mineral with a vitreous luster, perfect octahedral cleavage, and relative softness (hardness of 4). Fluorite comes in a variety of colors: colorless, white, yellow, blue, green, rose, red, brown or nearly black, but the purple variety is the most common (Fig. 5.4b). Ideally, fluorite contains 51.1% calcium and 48.9% fluorine.

For industrial applications, fluorite is used for its chemical composition as a fluxing agent for smelting in the steel industry, and as additive in the production of decorative glassware and enamels. The purest grades of fluorite are a source of important industrial chemicals, hydrogen fluoride (HF) and hydrofluoric acid (70% HF), which are the intermediate source of most fluorine-containing fine chemicals, such as CFCs (chlorofluorocarbons), HFCs and HCFCs. These are used as industrial solvents, refrigerants, aerosol propellants and plastic foams. Industry-wide practice has established three grades of fluorspar: *acid grade* (>97% $CaF_2$); *ceramic grade* (85% to 95% $CaF_2$); and *metallurgical grade* (60% to 85% $CaF_2$). Over 70% of fluorspar used in developed countries is transformed into HF products, while in developing countries over 65% of fluorite is used in steel-making

Fluorite is found in a variety of geological settings. Fluorite mineralization is commonly associated with or is a part of carbonatites and alkaline complexes, Mississippi Valley-type Pb-Zn-F-Ba deposits, hydrothermal Pb-Zn veins (with fluorite and barite), fluorite-iron oxide veins, and disseminations, REE deposits, fluorite/metal-bearing skarns, Sn-polymetallic greisen-type deposits, zeolite rocks and uranium deposits. However, vein-type deposits (fluorite ore and gangue) are the most common source, and historically have supplied the majority of fluorspar in the world market.

Global production of fluorite in 2019 was 6 million tons (Mt). China was by far the largest producer at 4 Mt, followed by Mexico (1.2 Mt), and South Africa (240,000 tons). Other African countries with fluorite production are Namibia (132,000 tons), Kenya (130,000 tons) and Morocco (110,000 tons). The major consumers of fluorite for steel production and chemicals are China, Japan, and India. South Africa has the largest reserves of fluorspar in the world.

In Nigeria, fluorite and barite occur together or separately within hydrothermal Pb-Zn vein, in argillaceous sediments of the Benue Trough. Fluorite is generally formed at low temperatures as a gangue mineral. Particular locations of fluorite are in Ebonyi State (Onicha, Ohiazora); Benue State (Katsina-Ala), and Taraba State (Wukari). The deposit in Katsina-Ala is estimated to exceed 500,000 tons. Fluorite is extracted and sold by artisanal miners.

### 5.3.5.10 Graphite

Graphite is a crystalline form of pure carbon that normally occurs as black crystal flakes and masses (Fig. 5.5a). It has important properties that make it a vital raw material for many industrial applications, such as chemical inertness and non-toxicity, thermal stability, high electrical and heat conductivity, flexibility, softness, sectility (ease of cutting) and lubricity (slipperiness), Graphite has been used since ancient times and has a wide range of applications as writing material (lead in pencils), automobile brakes and clutch lining, electronics, lubricants, batteries, graphene sheets and refractories in metallurgy and steelmaking. For some of these uses, there are no suitable substitutes.

Steelmaking and refractory applications in metallurgy use the largest amount of produced graphite. However, other applications in emerging technology usage in large-scale fuel cells and lithium batteries of electric vehicles, and lightweight high-strength composite applications may increase future demand. Currently in the United States, the major uses of graphite are for refractories (30%), foundries (15%), lubricants (14%), brake linings (13%), pencils (7%), steel making (5%) and batteries (3%).

Graphite occurs in three forms based on the degree of crystallinity, grain-size, and morphology: (i) *amorphous* (microcrystalline), (ii) *flake* (disseminated-crystalline), and (iii) *lump* or chip (vein-crystalline). *Flake* graphite is the most desirable type of graphite, largely due to its applications in the auto industry as anodes for lithium batteries in electric vehicles. Vein graphite is mostly used in advanced, thermal and high-friction applications such as car brakes and clutches. The amorphous graphite is of the lowest quality, and used in the refractories industry to manufacture crucibles, ladles, molds, and electrodes for electrical metallurgical furnaces.

Almost all graphite deposits mined today were formed from the metamorphism of carbonaceous sedimentary rocks. Contact or thermally metamorphosed coal is the most common source of amorphous graphite. Flake graphite formed as flat, plate-like particles disseminated uniformly throughout the rock bodies of carbonaceous metamorphic rocks, mostly schist, marble and gneiss. Lump or chip graphite is found in hydrothermal veins that occupy fissures or fractures, in high-grade metamorphic regions. Vein deposits such as those in Sri Lanka were formed in granulite facies metamorphic terrains. Residual deposits derived from the weathering of graphite-bearing schists and gneisses are mined in small quantities in Madagascar and China.

Production of natural graphite in 2019 was dominated by China (780,000 tons), India (150,000 tons), and Brazil (95,000 tons). Other producers are Canada (30,000 tons) and Madagascar (23,000 tons). China is the world's dominant exporter providing about 80% of world export in 2019 of natural graphite. Mexico and Canada export graphite mainly to the United States, which has not produced natural graphite since the 1950s. The estimated world reserve base of natural graphite amounts to 380 million tons, of which 70% are in China. In Africa, apart from Madagascar, other countries, Mozambique, Namibia, and Tanzania have substantial deposits of graphite. As a matter of fact, Mozambique has one of the largest deposits of high-quality graphite in the world, and about 10 to 20% of total global reserves. Most of the country's graphite, yet to be exploited, are found within the Precambrian metamorphic terrain of Cabo Delgado, in the country's northern-most provinces.

#### 5.3.5.11 Gypsum

Gypsum is a soft mineral made up of hydrated calcium sulfate ($CaSo4.2H2O$). It is often white or colorless when pure, but can be tinted grayish, reddish, yellowish, or brownish due to impurities of $SiO2$, $Al2O3$, $MgO$, $K2O$ and $Na2O$. Texturally, gypsum can appear in three forms: *massive* (alabaster), *fibrous* (satin spar), or in *transparent crystals* (selenite). Gypsum is a very versatile mineral with varied applications in industry due to its softness (2 on Mohs' scale), and being moderately water soluble at low temperatures. Nearly all gypsum is used in three prime applications: (i) building construction (plaster, plasterboard, etc.) involving 33% of the market; (ii) cement (where it is used as a setting retarder): 61% of the market; and (iii) agriculture (mostly for soil conditioner and fertilizero.

Gypsum is one of the more common minerals in sedimentary evaporite basins (Plate 5.20). It is an evaporite mineral that occurs as massive or layered beds in sedimentary deposits in association with other evaporites such as halite, anhydrite, sylvite and other salts or carbonates. It originated from the evaporation of sea or lake water, and precipitation of the minerals. Gypsum can also be found as veins in argillaceous sediments.

In 2019, the world's largest producer of gypsum from mines was the United States, with a production of 20 million tons (Mt), followed by Iran and China, both producing 16 Mt. African countries that produce gypsum are Morocco, Mauritania, Egypt, South Africa, and Tanzania which is reported to have one the largest gypsum deposits in the world that are yet to be developed due to lack of infrastructure. The demand for gypsum has increased in Africa in recent years due to the increased manufacture of cement for building and road construction. Gypsum represents 5% of cement components. Nigeria alone has about thirteen cement factories with an annual production of over 40 million tons of cement. Although Nigeria still imports gypsum to satisfy demand, some gypsum occurrences are being mined especially in the northeastern region of Borno, Yobe, Gombe, and Adamawa States where they are intercalated within the Cretaceous to Cenozoic sediments of the ancient Lake Chad basin. The most promising deposits are located in the Fika-Fune area of Yobe State where over 10 million tons of good quality gypsum has been indicated. Traces of gypsum occur within the phosphate deposit near Ifo in Ogun State.

Fig. 5.5: (a) Graphite (b) Muscovite Mica

#### 5.3.5.12 Mica

Mica is a group of aluminum silicate minerals with sheet or layered structure (phyllosilicates), and are common rock-forming minerals in igneous, sedimentary, and metamorphic rocks. There are several varieties of mica, but the three primary ones are, *muscovite*, a hydrated silicate of aluminum and potassium

($KAl_2[AlSi_3O_{10}](OH)_2$); *phlogopite*, which is a magnesium bearing mica ($K_2Mg_6(Si_6Al_2O_{20})(OH)_4$); and *biotite*, a potassium ferromagnesian variety with a formula ($K(Mg,Fe)_3AlSi_3O_{10}(OH)_2$). *Lepidolite* is a rare variety of mica that is rich in lithium ($K(Li,Al)_3(Si,Al)_4O_{10}(F,OH)_2$).

Muscovite (potash mica) is the most common and most important mineral of the mica group (Fig. 5.5b). It has some unique properties including perfect cleavage and tendency to form very thin, shiny and transparent sheets that have a pearly to vitreous luster (Plate 5.21). The sheets are quite elastic and flexible and have been found to form in dimensions of up to 5m x 3m. Of all the types of mica, muscovite has the highest thermal resistance, close to $900^{o}$ C; low thermal and electrical conductivity, and high dielectric strength. It is also chemically inert, lightweight and resilient, making it suitable for use as insulators in industrial applications. Sheets of muscovite of particular thicknesses are used in optical instruments. Ground muscovite is used in many other ways, as in the manufacture of shiny wallpapers, and as a lubricant when mixed with oils;

Phlogopite (magnesian mica) is next to muscovite in industrial importance. It is brown in color, and like muscovite, it easily splits into thin sheets that are transparent, flexible but because of the color, it is not used where color of the particles is important (Plate 5.22). Phlogopite is used in plastics, rubber, asphalt roofing, and drilling mud. Ground phlogopite is useful as a substitute for asbestos in automobile brake linings and clutch plates. Phlogopite is also used as an additive to increase the strength of epoxies, nylons, and polyesters.

Lepidolite (lithium mica) is one of the important sources of lithium ore, and also provides rubidium and cesium as by-products. It is used only as a gemstone and for metaphysical healing.

Mica is a common mineral of various rocks, but mineable deposits of good quality (sheet mica) are found mostly in granitic pegmatites. Phlogopite occurs most frequently in metamorphic rocks where dolomitic limestones or ultramafic rocks are subjected to hydrothermal metamorphism or contact metasomatism,

Global production of mica in 2019 was estimated at 400,000 tons, in which the largest producer was China (100,000 tons), followed by Finland (64,000 tons), United States (38,000 tons), Madagascar (36,000 tons), Canada (23,000 tons), and India (22,000 tons).The mica industry in India has long been one of the world's largest in terms of mica production and mica exports. Most of the Indian deposits in mica pegmatites are of high quality and located in Jharkhand, Bihar, Andhra Pradesh and Rajasthan. Madagascar is the leading African producer of mica, muscovite and phlogopite from pegmatites of Pan-African age within the Precambrian basement complex. Mining is mostly artisanal using manual and crude equipment sometimes involving child labor. In Nigeria, muscovite is found associated with the Older Granite pegmatites, and approximately 1000 tons of crude flakes were exported yearly.

### 5.3.5.13 Quartz and Quartz Sand

Quartz is a compound of silicon and oxygen ($SiO_2$), and one of the most common minerals on the earth's surface. Pure quartz is a colorless, transparent, and very hard crystalline mineral that may appear in a variety of colors due to impurities (e.g. amethyst - a gem). Quartz is found in most rock types, including igneous, sedimentary and metamorphic, but economic deposits occur mostly in hydrothermal quartz veins. Quartz sand or silica sand represents the accumulation of detrital quartz grains released from primary sources by chemical weathering and concentrated in rivers, beaches, dunes and marine environments. River deposits can produce quartz sands if free of clay and other impurities. Best silica sands are beach sands, usually sharp with gritty texture and containing more than 98% $SiO_2$. While quartz deposits are common, and quartz is present in some form in nearly all mining sites, high purity and commercially viable deposits are not that common.

For industrial applications, quartz is valued highly for its combination of certain electrical or optical properties with great physical strength and chemical resistance. Pure quartz after fusion is used for producing glass, ceramics, and chemical apparatus. Quartz glass (fused quartz) has a number of interesting properties such as low thermal expansion, transparency to ultraviolet light, chemical inertness, and ability to form strong threads used in physical instruments. Pure quartz is also used as an oscillator in electric circuits in watches and computers. It is the main ore of silicon, which is used in the integrated circuits (chips) of computers.

Quartz sand deposits usually containing >95% quartz are a very important raw material for making different kinds of glass that require different levels of purity and grain size. Silica is hard, chemically inert, and has a high melting point, which requires the use of fluxing material to reduce the melting temperatures. High quality quartz sand is also used in applications like foundries and water filtration systems. In ceramics manufacture, quartz is added to the clay mix for toughness and resilience, and for use in ceramic glazes. Quartz is an important component in refractory (temperature resistant) bricks for lining furnaces,

Quartz crystals of economic value are mostly derived from hydrothermal quartz veins that vary in size and can be as wide as 100 meters and traceable for hundreds of kilometers (e.g. Taunus Mountains, near Frankfurt, Germany). They are almost all fracture fillings or bodies along bedding planes. Well-formed crystals of quartz are often found growing in miarolitic cavities within the vein bodies. Some quartz known as "gangue quartz" is also derived from the gangue of metallic ores, particularly gold and copper deposits. Silica sand deposits provide large quantities of quartz compared to primary quartz deposits. They represent the final products of rock weathering that yield detrital quartz grains that are transported mainly by water or wind and subsequently deposited as sand in river beds, beaches, and marine environments. In deserts, the sands are piled up as sand dunes.

Silica sand is produced in several countries worldwide, and because it is a high-volume and low-unit value industrial material, silica sand production is mostly used for domestic consumption. The world's largest producers are the United States, Italy, China, and Germany. In Nigeria, silica sand deposits of industrial quality are found in Lagos, Ondo, Delta, Rivers, Abia, Kano, and Jigawa States.

### 5.3.5.14 Salt (Halite)

Halite is the mineral of salt, which is composed of sodium chloride (NaCl). When halite occurs as beds of solid rock, it is commonly called "rock salt." Halite is found as crystals usually white but can appear in a variety of colors due to impurities. Halite is an evaporite mineral that is very common worldwide, that is deposited in solid underground masses, and as a dissolved solution in many arid-region inland lakes and oceans. Salt deposits are mined for common table salt (NaCl), which is in great demand and has many uses, including food seasoning, road safety in the winter months to melt snow and ice, water softeners, salt licks for horses and cattle, medicinal purposes and primary ore for both sodium and chlorine. Halite is also an important component in the production of detergents, bleach, soaps, dyes, and chemicals.

Salt is obtained from two sources; *rock salt,* which is crystallized salt produced by evaporation of seawater, and *salt brine,* which is water containing a high concentration of salt (>15% NaCl). Rock salt occurs as beds of sedimentary deposits formed by evaporation and precipitation of halite from seawater in ancient marine environments. Salt deposits are found in two environments; as subsurface or surface deposits. *Subsurface* deposits are buried evaporite sediments that occur as ancient bedrock where large extinct salt lakes and seas evaporated millions of years ago, leaving behind thick deposits of salt in beds that range from a few feet to more than 1,000 feet in thickness. Surface deposits of salt occur in many inland evaporative lakes in desert regions, such as the Great Salt Lake, Utah, and Searles Lake in Mojave Desert, California, USA.

Global production of salt in 2019 was estimated as 350 million tons (Mt), and the six largest producers are China (I60 Mt), United States (42 Mt), India (20 Mt), Germany (14 Mt), Australia (13 Mt), and Canada (12 Mt). Several African countries produce salt, notably Egypt (3 Mt), Namibia, South Africa, Mozambique, Senegal, and Kenya. In Nigeria, occurrences of rock salt deposits have been widely reported in Benue State, although the quantity and quality have not been fully determined. There are also salt brines that emanate from the Cretaceous sediments along a stretch of the Benue Trough, extending from Cross River/Akwa Ibom States in the southeast to as far north as Gombe/Adamawa/Taraba States in the northeast. Salt brines in springs and lakes that have been exploited for several centuries using traditional methods are found in Gombe (Gombe State), Awe (Plateau State), Okposi, Afikpo and Abakaliki-Ishiagu (Ebonyi State), Uburu (Imo State), Bende (Abia State}, Uyo (Akwa Ibom State), and Okpoma (Cross River State). Salt brines at Okposi Lake (Ebonyi State). Nigeria imports about 90% of its domestic needs for salt and spends about $700 million annually on salt importation.

### 5.3.5.15 Talc

Talc is a hydrated magnesium silicate ($Mg3Si4O10(OH)2$) which is widely used not for its chemical composition but for its unique physical properties. The mineral is usually white, platy and soft (talc is the softest known mineral), with a pearly luster and soapy feel when touched (Fig. 5.6a). Other properties include being hydrophobic (water repellant) and chemically inert, and does not generate a chemical reaction when applied on the skin or ingested, except if it contains asbestos impurities. Talc is also organophilic, which means it has affinity for certain organic compounds; a property, for example, that is useful in paper making and as an adsorbent. In its natural form, talc is usually white, but can be green, grey or brownish, depending on its impurities. A rock composed primarily of talc is called steatite, while with some impurities it is called *soapstone.*

Talc is an industrial raw material in a wide range of applications, particularly as a major ingredient in cosmetic powder and other personal care products. Other applications include its use as filler in ceramics, paints, paper, plastics, rubber, fertilizer and food, and in biological wastewater treatment. Talc powder is used as a carrier for insecticides and pesticides and as a lubricant in high temperature applications.

Talc deposits are formed by low grade metamorphism of pre-existing mafic and ultramafic rocks (including serpentinite), and magnesium carbonates, at low to medium temperature and pressure. It can also be produced by retrograde metamorphism of eclogites and blue schists, and hydrothermal alteration of existing minerals in mafic host rocks. Talc deposits appear as relatively small bodies of translucent masses that can be classified into two types according to source rocks and mineralogical association: (1) talc-chlorite deposits, and (2) talc-carbonate deposits. The talc-chlorite deposits usually contain talc (sometimes 100%) and minor chlorite. This type of deposit is characteristic of igneous host rocks. The talc-carbonate deposits are composed of talc and carbonate minerals (magnesite and dolomite), produced by metamorphism of magnesium carbonate or dolomitic rocks.

In Nigeria, talc occurrences of varying quality and extent have been reported from the Precambrian Schists Belts in Oyo, Kogi, Kaduna and Niger states. They are usually small bodies or lenses of impure talc schist, talc-tremolite schist or talc-carbonate rock. Talc is mined at two locations in Niger State where they are crushed and processed for sale.

Fig. 5.6: (a) Talc (b) Trona (c) Wollastonite

### 5.3.5.16 Trona (Soda Ash)

Trona ($Na_2CO_3.NaHCO_3'2H_2O$) is a naturally occurring hydrated sodium carbonate which is processed into soda ash or baking soda - a white, odorless and non-toxic alkali compound (Fig. 5.6b). Trona normally contains about 60% $Na_2CO_3$, and formed in inland lakes under arid conditions by the evaporation of saline water containing sodium, alkali and bicarbonate. Soda ash has a wide range of applications in the manufacturing of glass, chemicals, paper, detergents, and textiles. The largest application is in glass manufacturing where it is the main ingredient added to quartz to reduce melting temperatures, and its content of sodium helps support the shaping of the glass items such as glass containers, fiberglass and sheet glass for buildings and automobiles. Soda ash is also used in air purification and softening of (hard) water. In environmental protection, soda ash assists in the removal of sulfur dioxide and hydrochloric acid from stack gases (air pollutants) in chimneys emanating from coal-fired power plants. Producers of chemicals use soda ash to manufacture products that sweeten soft drinks (corn sweeteners), relieve physical discomfort (sodium bicarbonate) and improve foods and toiletries (phosphates). Household detergents and paper products are some of the identifiable products using soda ash.

Global production of natural soda ash in 2019 was 50 million tons. The largest producers were the United States (12 Mt), Turkey (3.5 Mt), Kenya (300,000 tons), Botswana (250,000 tons) and Ethiopia (8,000 tons). Wyoming in the United States has the world's largest deposit of trona, supplying about 90% of the US soda ash. The Wyoming deposits belong to the Eocene Green River Formation of lacustrine origin. The trona deposits contain over 40 beds of trona that vary in thickness from 1 to 12 m and interbedded with marl, oil shale, halite and clastic sediments that can be traced for over 250 km. The Kenyan deposit is in Lake Magadi, one of the salt lakes in the East African Rift Valley which floods and dries out every year to form trona beds up to 35 m thick. In Nigeria, trona, can be found in large quantities as an evaporite rock from the Lake Chad region of Borno, Yobe and Adamawa States. Chemical analysis of the trona from Borno State shows that it contains only 20% sodium carbonate, which is low grade compared to the average 50% in good trona deposits.

### 5.3.5.17 Vermiculite

Vermiculite is a clay mineral formed by the weathering or hydrothermal alteration of ferro-magnesian mica (biotite and phlogopite). Vermiculite has a layered structure that contains water in between the layers, which undergoes significant expansion when heated. Expanded vermiculite has unique and versatile characteristics; it is lightweight and suitable for insulation, it is non-combustible, highly absorbent, pH neutral, inert, non-reactive to all but very strong acids and compressible. These desirable characteristics allow expanded vermiculite to be widely and in industrial and commercial applications including: packaging, fire protection, refractory and high temperature insulation, lightweight concrete, plasters, and horticultural potting mixes.

World mine production of vermiculite was an estimated 500,000 metric tons in 2019.South Africa, the United States, Brazil and China were the leading producers. South Africa's production of 180,000 tons represented 36% of global production and 72% of production in Africa. Other African countries with some production are Zimbabwe 12%), Uganda (12%), Egypt (3%) and Kenya (5%). Vermiculite reserves in the western world are estimated as 50 million tons, 50% in the United States, 40% in South Africa and 7% in Brazil. In South Africa, vermiculite occurs within the Palabora carbonatite where it is produced by the hydration of phlogopite (and biotite) by the loss of alkali and the addition of water through infiltration of water during weathering processes. In contrast, the vermiculites in the United States are products of hydrothermal alteration of mafic rocks.

### 5.3.5.18 Wollastonite

Wollastonite is a calcium pyroxene with the chemical formula CaSiO3. Although most wollastonite is relatively pure, some may contain small amounts of Fe, Mg and Mn substituting for Ca. Pure wollastonite has a characteristic bright white color, but may appear cream, brown, pale green or even red due to impurities (Fig. 5.6c). Wollastonite has several physical properties that make it useful as an industrial mineral, including its chemical inertness, high brightness, white coloration, structural stability at high temperatures (>1,120° C), low moisture and oil absorption capacity, and low volatile content. Wollastonite is used primarily in ceramics, friction products (brakes and clutches), and additive in metal making (flux), filler in paints and plastics, and a substitute for asbestos in thermal insulation applications.

Wollastonite is formed by two processes: (i) thermal metamorphism of limestone, and (ii) magmatic crystallization in carbonatites. Thermal metamorphism and/or metasomatism of limestone can occur when silica and limestone are raised to a temperature of 400-450°C, either as a result of deep burial (regional metamorphism) or by proximity to an igneous intrusion (contact metamorphism). The chemical reaction forms wollastonite and gives off carbon dioxide. Carbonatites are products of crystallization of carbonate rich magmas in which some residual silica may combine with calcium carbonate to form wollastonite as a primary mineral.

Most of the global production of wollastonite is derived from contact metamorphism of impure limestones. Based on the latest available statistics, the estimated world production of crude wollastonite ore in 2016 was in the range of 700,000 to 720,000 tons of which the leading producers were China (425,000 tons), India (185,000 tons), United States (130,000 tons), Mexico (67,000 tons) and Finland (16,000 tons). Kenya and South Africa have some unexplored wollastonite deposits.

### 5.3.5.19 Zeolites

Zeolites are a group of hydrated aluminosilicate minerals made of interlinked tetrahedra of alumina (AlO4) and silica (SiO4) in an open framework (tectosilicate) structure (similar to feldspars). However, the zeolite structure is porous with intercellular space for water molecules and a wide variety of cations, such as Na+, K+, Ca2+, and Mg2+ to get in. More than 50 minerals are known to belong to the zeolite group, but the most common are analcime, chabazite, clinoptilolite, heulandite, mordenite, natrolite, and phillipsite. They often occur as well-formed crystals (Fig. 5.7a).

Zeolites possess some unique properties that make them useful for industrial processes. The molecular sieve properties (having pores of uniform size) of zeolites in particular are extensively employed in industry for liquid separation and filtration. They are also stable solids that can resist high temperatures and pressures without burning or deformation, and won't dissolve in water or other inorganic solvents, or oxidize in the air. The largest end use for natural zeolites is as a cement additive, to produce a slower-hardening and stronger product. Zeolites are found to be good adsorbents, ion exchangers and molecular sieves, and widely used as water softeners and water filters. Many everyday laundry and dishwasher detergents contain zeolites to remove calcium and magnesium and soften water so they can wash more effectively.

Zeolite deposits (containing >50% zeolite minerals) are formed mostly in volcanic environments by the alteration of volcanic glass and volcaniclastic rocks (volcanic ash) of varying composition and age through diagenetic and/or low-temperature hydrothermal alteration processes under alkaline conditions. Zeolites in sedimentary rocks are authigenic and formed during diagenesis by reaction of vitric (volcanic glass) material with interstitial water. The formation of zeolites is favored by a relatively high pH and high activities of alkali ions in the interstitial water.

The most common zeolite minerals that occur in mineable deposits are clinoptilolite-heulandite, mordenite, chabazite, analcime, and phillipsite. Annual world production of natural zeolites has stood at about 1 million tons, China is the largest producer of natural zeolites, with nearly 30% of global production in 2018, followed by South Korea (120,000 tons), New Zealand (100,000 tons), United States (95,000 tons), Turkey (70,000 tons), and Cuba (55,000 tons). Asia is the dominant market for zeolites.

### 5.3.5.20 Zircon

Zircon is a silicate mineral with a chemical formula of ZrSiO4, and is the main source of zirconium. Zircon is an industrial mineral as well as a gemstone. In its natural state, zircon forms well-shaped crystals usually white or colorless, but can show diverse colors due to traces of certain impurities (Fig. 5.7b). Zircon has some physical properties such as high bonding capacity, low acidity, low thermal expansion coefficient, high chemical resistance, low wettability by molten metal, and good recyclability, that makes it extremely suitable for use in foundry and refractories (furnace linings and foundry bricks), and for casting applications and fabrications where its resistance to shock, abrasion and high temperature makes it valuable. For ceramics, zircon dioxide (zirconia) is added to glazes for floor tiles and sanitary ware,

Fig. 5.7: (a) Zeolite Crystals (b) Zircon Crystals

Zircon is found as an accessory mineral in igneous, metamorphic, and sedimentary rocks. In igneous rocks, it is commonly associated with alkaline and peralkaline granites and pegmatites. In metamorphic rocks, zircon crystallizes at moderate to high temperature and pressure. In sedimentary rocks, zircon is a heavy mineral resistant to weathering and accumulates as zircon "sands" (Plate 5.33) in alluvial and coastal environments where it has been concentrated by fluvial and wave action, respectively.

World production of zircon in 2019 was estimated as 1.6 million tons, of which the leading producers were Australia (550,000 tons), South Africa (350,000 tons), United States (100,000 tons), China (80,000 tons) and Mozambique (50,000 tons). In Nigeria, zircon occurs as an accessory mineral in the alkaline and peralkaline granites of the Jurassic Younger Granite Complex in Plateau, Kano, and the Nasarawa States, and also as xenocrysts in the Cenozoic alkali basalts in the Mambilla and Jos Plateau. A large deposit of zircon sand occurs in the Azara-Lafia area in Nasarawa State. The deposit contains up to 96% zircon, with minor kyanite (0.5%) and rutile (1.5%) and other impurities (2%). The Nigerian export of zircon and zircon sands has increased considerably in recent years with an annual export of over 2000 tons per annum.

## 5.4 Industrial Rocks

Industrial rocks are rocks (indurated aggregates of minerals) that are used as raw materials for the manufacture of industrial products or for ornamental applications without any mineral separation. This definition distinguishes *industrial rocks* from *industrial minerals*. Apart from the metal mining industry, the ornamental stone industry is the oldest known mineral industry that used earth materials. Many geologists have sometimes described *construction materials* such as gravel, sand and clay as industrial rocks. However, because these materials are loose and not indurated, or compact, they are not rocks *in sensu stricto*. The best known industrial rocks are limestone, phosphate, coal, diatomite, carbonatite and nepheline syenite, and ornamental stones -marble, granite, serpentinite and soapstone.

### 5.4.1 Diatomite

Diatomite is a friable, light-colored, siliceous sedimentary rock that is made up of skeletal remains of marine organisms called diatoms which are unicellular aquatic algae that form deposits consisting almost entirely of silica. Diatomite occurs as a very porous rock or powdery earthy material (diatomaceous earth) with a fine particle size and low specific gravity. The properties which make diatomite valuable include low density, high porosity, high surface area, abrasiveness (due to silica content), insulating properties, inertness, absorptive capacity, brightness, and high silica content. Diatomite is used as a filtration aid, additive in paints and plastics, mild abrasive in metal polishes, facial scrubs and toothpaste and absorbent for liquids. Diatomite is also used for soil amendment and valued as mechanical insecticide due to its abrasive absorptive capacity.

Diatomite is a soft, very fine grained sedimentary rock formed in both marine (oceans) and freshwater environments (lakes and streams) (Fig. 5.8a). Diatomite may be intercalated with other shallow marine sediments such as shale and limestone. The origin or depositional environment is an important factor when a diatomite source is being considered for personal use because diatomite from saltwater sources can contain salts that are objectionable or toxic for human consumption.

In 2017, a total of 29 countries produced commercial amounts of diatomite. The United States was the leader, producing an estimated 700,000 metric tons. Czech Republic, Denmark and China each produced over 400,000 metric tons. Argentina, Peru and Japan each produced 100,000 metric tons or more. Other countries that produced at least 50,000 metric tons include Mexico, France, Russia, South Korea, Turkey, and Spain.

In Nigeria, diatomite is found in commercial quantities in the Borno/Chad Basin, at Bularafa, Alangafe, Abakire, Gujba and Maluri in Yobe State; Shani, and Mirinja in Borno State; Kwani in Gombe State and Udubo area of Bauchi State. The deposits occur as beds several meters thick within the Chad Formation of Pliocene to Pleistocene age and composed of fluviatile and lacustrine thick bodies of sand and clay with lenses of diatomite.

### 5.4.2 Limestone

Limestone is a common sedimentary rock composed mainly of one mineral, calcite (or aragonite), a carbonate of calcium ($CaCO_3$). Although limestones are mostly mono-minerallic, composed mostly of one mineral, they are not industrial minerals, but best described as industrial rocks. Calcite which is an industrial mineral, occurs as gangue in hydrothermal veins but not in economic quantities. Limestones are usually white in color, but may range from buff to grey due to impurities. In their mode of occurrence, exposed limestones are commonly bedded or massive with varying thicknesses that range from less than a meter to several hundred meters. They are usually intercalated with other shallow water marine sediments such as shales, siltstones and sandstones, with occasional evaporites. Limestones are formed in marine environments by two processes: (1) biochemical or biogenic processes that involve accumulation of skeletal remains of marine organisms; and (2) chemical or inorganic precipitation from sea water due to evaporation.

Limestone is a versatile industrial rock used in a variety of industrial applications, including the following: as a major component in cement manufacture; pulverized limestone powder used as soil conditioner; production of quicklime (calcium oxide) and slaked lime (calcium hydroxide); for use as fillers in paints, paper, plastics, rubber and textiles, and for production of cosmetic products, baking soda, pharmaceuticals, toothpaste and as an additive in glass manufacture and binder for slag in blast furnaces for iron purification. Limestone is mostly extracted from surface quarries.

Limestone is present in most Phanerozoic sedimentary basins around the world. The global limestone market is estimated at US $58.5 billion in 2020, mostly for lime production. According to the best available statistics, the three leading lime producing countries in the world as of 2019 were China (300 Mt), the United States (18 Mt), and India (16 Mt).

In Nigeria, over 90% of exploited limestones are used primarily in the manufacture of cement at several locations in the country. The remaining 10% are crushed into powder for various industrial applications. Limestone is classified as a "strategic mineral" (or rock) in Nigeria, which means it is of critical importance to the country's economic development. Geologically, limestone deposits are widespread within Nigeria's Cretaceous sedimentary basins namely; the Dahomey basin, Anambra basin, Benue Trough (lower, middle and upper), Calabar basin (embayment), Sokoto and Borno basins.

Limestone deposits in Nigeria belong to two age groups; those formed either in the Cretaceous or Tertiary periods. The Cretaceous limestones occur along the Benue Trough extending from Calabar in the southeast, through Igunmale and Yandev in the central region to Bauchi and Ashaka in Gombe State, and the Sokoto basin. They are interbedded with shales, siltstones and sandstones. The Tertiary limestones are thick and well defined units confined to the southwest region with occurrences in Sagamu, Ewekoro and Ibeshe. The resource base of the known limestone deposits is estimated as 2.3 trillion metric tons with 568 million tons of proven reserves .Most of the deposits being exploited for cement manufacture are quite pure and range from 80% calcite in Sokoto limestone to 96% calcite in the Mfamosing limestone in Calabar.

Fig. 5.8: (a) Diatomite

(b) Phosphate

## 5.4.3 Phosphorite (Phosphate Rock)

Phosphorite or phosphate rock is a chemical sedimentary rock composed mainly of phosphate minerals that occur as compact or nodular masses (Plate 5.39). The dominant phosphate minerals are varieties of the mineral apatite: fluorapatite, ($Ca_5(PO_4)_3F$,) chlorapatite ($Ca_5(PO_4)_3Cl$), and hydroxylapatite, ($Ca_5(PO_4)_3OH$), usually admixed with other detrital materials. Variety of textures include; oolitic, nodular, pelletal, or micritic, and closely mixed with calcite, quartz, and clay minerals. The phosphates originate from sources including shells of marine invertebrates containing calcium phosphate, and the bones and excrement of vertebrates such as birds. Phosphorites are deposited mostly in shallow marine or low-energy environments. Thick deposits occur as beds intercalated within limestone, shale and carbonaceous mudstone.

The phosphate content of phosphorite varies greatly, from 4% to 35% phosphorus pentoxide (P2O5), but typical phosphorites contain about 30% P2O5. The grades of phosphorite are classified as high-grade (>30% P2O5), medium-grade (20%–30% P2O5), and low-grade (15%–20% P2O5). The very high-grade ore blends with medium-grade ore for direct sale. The low-grade ore is beneficiated by flotation techniques to remove impurities. For use in the chemical fertilizer industry, beneficiated rock phosphate must be concentrated to levels of at least 28% ($P_2O_5$), although most marketed grades of phosphate rock are 30% or more

Phosphorite constitutes the main source (75%) of phosphate used worldwide while the other sources are 24% from carbonatites and other igneous rock complexes, and 1% from Guano. About 90% of rock phosphate production is used for fertilizer and animal feed supplements while the rest 10% for industrial chemicals and other applications including manufacture of soap and detergents, food preservatives, baking flour, pharmaceuticals, anti-corrosion agents, cosmetics, fungicides, insecticides, ceramics, water treatment and metallurgy. Elemental phosphorus and phosphoric acid are produced in liquid form and shipped to fertilizer plants worldwide in tankers. Sedimentary rock phosphate is found as bedded deposits in shallow marine environments where they are composed of phosphate material mixed with carbonates, detrital quartz and mica (Plate 5.40).

Global production of phosphate in 2019 totaled 240 Mt, with China as the largest producer (110 Mt), followed by Morocco and Western Sahara (36 Mt), United States (23 Mt), Russia (18 Mt) and Jordan (8 Mt). Togo is the leading producer of phosphate in West Africa, with extraction from large open pit mines (Plate 5.41). In Nigeria, sedimentary phosphate deposits (nodular) are found within the Cretaceous-Tertiary Sokoto, and Eastern Dahomey Basins. In Sokoto State, phosphate of Paleocene sedimentary deposits occur in Dange, Gidan Bauchi, Illela, Gada, and Kalambaina. In Ogun State, phosphate occurrences are located in Oja Odan and Ifo. These deposits although of adequate quality are small, not exceeding a few tens of thousand tons.

## 5.4.4 Nepheline Syenite (for Alumina)

Nepheline syenite is a light-colored alkaline igneous rock composed mainly of alkali feldspar (albite and microcline) and nepheline in appreciable amounts. It is similar in appearance to coarse-grained granite but characterized by the absence of quartz and less mafic minerals. (Fig. 5.9a). In chemical composition, nepheline syenite contains on the average substantial alumina (23% Al2O3), sodium (19%) and potassium (5%) and can be used in a variety of industrial applications. Nepheline syenite is an indispensable raw material in glass and ceramic industries.

In the ceramics industry, the low fusibility temperature and high fluxing capacity of nepheline allow it to act as vitrifying agent by contributing an early glassy phase that binds other constituents of the ceramic mix. This really helps in reducing the flux content in the ceramic body, increases alumina content in the mix, lowers the firing temperature, and saves money by shortening firing times. Furthermore, nepheline syenite finds uses in the manufacture of paints and pigments, and as filler in, paper, rubber, and plastics. It is also used in the manufacture of alumina, alkali carbonates, and as additives in glass making by achieving the lowest melting temperature while acting as a source of alumina in glass batches. Nepheline syenite is better than feldspar as a ceramic flux because of lower silica and higher alumina and alkalis .It is more valued for its whiteness as a flux in tile, sanitary ware, porcelain, and vitreous and semi-vitreous bodies.

Fig. 5.9: Nepheline Syenite

(b) Carbonatite rock

Nepheline syenite is not a common rock... The main producing countries of commercial nepheline syenite are Brazil, Canada, China, Norway, Russia, and Turkey. Most of the Russian nepheline syenite production is consumed locally for the production of alumina. Available trade statistics show that in 2016, about 3,440 tons of the material was exported to a number of countries, including Mexico (56%), Canada (28%), and Australia (7%) and also to nine other countries (about 1% each). The global market for nepheline syenite in 2019 was $80 million.

## 5.4.5 Coal

Coal is a combustible black sedimentary rock containing high amounts of carbon and hydrocarbons. Coal is a major source of energy used worldwide to generate electricity. It is also a major *industrial rock* used in the production of chemical fertilizers and several chemical products, and as coking coal to extract iron from iron ore during steelmaking. There are four types or rank of coal based on their degree of coalification (carbon content): anthracite (86-97% C), bituminous (45-65% C), semi-bituminous (35-45% C), and lignite (brown coal) (25-35% C).

Industrial applications of coal include its use as a feedstock in the production of a wide range of chemical fertilizers and other chemical products. The main pathway to these products was coal gasification to produce *syngas*. The process converts coal to methanol, carbon monoxide, and hydrogen which are primary chemicals that form the chemical building blocks from which a whole spectrum of derivative chemicals are manufactured, including olefins, acetic acid, formaldehyde, ammonia, urea, and others. Coal gas is also a clean fuel to generate electricity.

Metallurgical (met) coal, which includes coking coal and pulverized coal for injection (PCI) is a primary ingredient in steelmaking. Coke is also used to produce carbides, ferroalloys and other compounds. Worldwide coal production in 2019 was about 7.4 billion tons (Bt). China is the largest producer with 47% of global production and approximately 13% of the world's total reserves, with the USA, India, Australia, Indonesia, Russia, South Africa, Kazakhstan, Columbia, and Ukraine being the top ten producers with significant resources coal, Based on current extraction rates, coal with an estimated 1.1 trillion tons of proven coal reserves worldwide. Most of the reserves are anthracite and bituminous coal (70%). In 2019, global met coal consumption was1 080 Mt. By far, China is the largest met coal consumer accounting for 64% (691 Mt) of the global total in 2019. Other significant met coal consumers were the Russian Federation (hereafter "Russia") (7%), European Union (5%) and India (6%).

Coal production in Africa accounts for about 5% of world production compared n to Russia (30%), North America (25%) and Asia/Australia (29%). South Africa is the leading producer with an annual output of 278 Mt in 2019 representing over 95% of coal produced in the continent. Other countries with substantial production are Mozambique, 6.8 Mt (2.5%), Zimbabwe 3 Mt (1.2%), Botswana 2 Mt (.07%), Tanzania (304,000 tons), Niger (272,000 tons), Zambia (139,000 tons), and Nigeria (51,000 tons).

In 2019, reserves of coal in South Africa amounted to approximately 9.9 billion metric tons, and almost 80% of the energy needs are met by the use of coal. South Africa has no petroleum resources. Also, in addition to the extensive domestic use of coal, South Africa exports about 28% percent of its production making it the fourth-largest coal exporting country in the world. The largest coal deposits in South Africa occur within the sediments of the Ecca Group in the Paleozoic Karoo Basin. In recent years, Africa's energy needs are growing and this has led more countries, such as Botswana, Mozambique, Tanzania, Niger and Nigeria to turn to coal-powered plants. Zimbabwe has become the only other country actively coal mining in Africa on a large scale, with around 0.1% of the global reserve. Botswana has over 200 billion tons of coal reserves and the development of the coal industry has become a major priority. Mozambique is also expected to experience a sharp increase in coal production. Despite its oil wealth, Nigeria is using coal to generate power for cement and other industrial production.

## 5.4.6 Carbonatite (for Quick Lime)

Carbonatites are unique rocks that consist mostly of more than 59% calcite and/or dolomite of magmatic origin (Fig. 5.9b). They are similar to limestones in mineralogy. Where carbonatites are exposed at the surface and relatively fresh, they have been used as sources of calcite for several industries. Quicklime obtained from the calcination of carbonatites has been used in synthesizing precipitated calcium carbonate (PCC) through a lime-soda process. Hydration of quicklime made from carbonatite yields much higher Ca2+ in solution than from marble.

## 5.4.7 Basalt (for Mineral Wool)

Stone wool or mineral wool is an insulation material formed by spinning molten rock and minerals with steel slag to create a cotton-candy-like wool product. Pressed into rolls and sheets, stone wool creates incredibly effective insulation with sound-absorbing and fire-resistant properties. Basalt with added dolomite is the raw materials that are melted like lava in a cupola furnace heated to over 1,500 °C with a carbon-containing energy source through which a stream of air or steam is blown. More advanced production techniques utilize the spinning of molten rock in high-speed spinning heads to create a cotton-candy-like wool product. Stone wool can be used for ceiling tiles and all types of sound or thermal insulation and fire protection in many applications. It is particularly suitable for industrial use given its high-temperature resistance and non-combustibility. Several countries produce mineral wool including Nigeria.

Fig. 5.10: Rock Wool (b) Wool Insulation Sheets (c) Raw Materials for Rock Wool

## 5.4.8 Ornamental Rocks

Ornamental rocks are those rocks used for decorative purposes in construction, architectural purposes and monumental sculptures. These include external and internal decorations, such as columns, house trimmings, flooring, countertops and monumental work. The ornamental stone industry has a large potential market worldwide and is an important industrial sector for many European countries. The most common ornamental rocks are *marble, granite, serpentine* and *soapstone*. It is pertinent to note that ornamental stones used for decorations are distinguished from "building stones" that are cut into certain dimensions and used in building construction (walling, roofing and cladding).

### 5.4.8.1 Marble

Marble is a metamorphic rock composed of calcite and/or dolomite, and formed by regional metamorphism of limestone, and other carbonate rocks. Marble has some unique properties as a decorative stone. It is hard and compact, and occurs in a variety of attractive colors (including white, red, black, gray, pink and green), and can take a good polish. Mottled marble and banded marble are also used as ornamental stone. The specific gravity of marble is 2.89 and compressive strength of 70 Mpa. Other properties of marble are durability, heat resistance and fire resistance. Marble is one of the most prestigious stones that provides elegance and beauty to buildings, and has been used since ancient times for interior and exterior decorations including flooring applications, table countertops, facing works, steps, fountains, etc. It can also be carved into sculptures, and decorative and ornamental structures.

Good quality marble for decorative stones is not as common as granite, and is involved in global trade that is estimated at US$50 billion annually. Annual production of marble in 2016 was estimated as 816 Million square meters. Until 2016, countries in the European Union (e.g. Italy, Spain, Greece) were overall the dominant producer and exporter of marble, but strong growth has occurred in emerging markets, especially in China, Pakistan and India.as both leading producers and exporters.. The world's top ten natural marble producers include China, India, Iran, Turkey, Italy, Spain, Brazil, Egypt, Portugal and Greece. In Nigeria, high quality marble that can be cut into slabs are not common, but crushed marble is obtained from several occurrences in the Schist Belts of SW and Central Nigeria such as Igbeti, Jakura, Ukpilla, Igara and Kwakuti.

### 5.4.8.2 Granite

Granite is an igneous rock formed at relatively high temperature and pressure. They show varying textures from fine-grained to coarse-grained or porphyritic. Granite is made up of mostly quartz and feldspar with minor mica and amphibole. Depending on the type of feldspar, the color of granite may be brown, pink, red, black or grey/green which makes granite one of the most beautiful natural stones. Texture varies from very fine grained to coarse grained or porphyritic. It also takes good polishing. Granite offers high compressive strength and durability, and is commonly used as structural and ornamental stone. Large pieces of granite are carved for monuments such as ornamental columns, statues and headstones.

Granite slabs are polished and used in fabricating granite countertops for kitchens and granite flooring and facing for buildings. Some of the positive qualities of granite stone products are the beauty, heat resistance, stain and scratch resistance and easy maintenance. Granite is a fairly common igneous rock that is found in most countries. Nigeria has abundant granite sources, particularly in the porphyritic Older Granite series and the Younger Granite series. A granite slab manufacturing facility was recently established at Afo-Akoko in Ondo State.

Fig. 5.11: (a) Ornamental Granite Slabs (b) Marble Flooring and Wall Facing

### 5.4.8.3 Serpentinite

Serpentinite is a greenish to dark fine grained rock consisting of a variety of serpentine minerals formed by low-temperature metamorphism or hydrothermal alteration of low-silica, mafic and ultramafic rocks, including peridotite and dunite. Chrysotile, antigorite, and lizardite are three of the primary serpentine minerals, but there are non-serpentine minerals including calcite and dolomite, carbonates that typically form the tiny white veins that often cut randomly through the stone. Minerals such as chlorite, talc, magnesite and brucite are also often present. Serpentinites are found in relics of oceanic crust that occur in ophiolite complexes or as small intrusive bodies within orogenic belts.

Serpentinite has historically been used as decorative or architectural stone for their marble-like qualities. Some of their attributes include the distinctive and wide variety of green and greenish colors, often has an attractive pattern, works easily, and polishes to a nice luster. It has a Mohs hardness of 3 to 6 which is softer than granite, and usually harder than most marble. Serpentinite is currently used in internal decorations and small sculptures, but is easily weathered or becomes scaly when exposed to tropical or subtropical climate or used for external decoration. Serpentinites are found in different places in the world, with global reserves estimated as 200 million tons. The most important serpentinite natural deposits occur in Australia, Armenia, Italy, Russia, USA, Canada, Paraguay, New Zealand, and Brazil. California became the first state in the USA to have an officially designated state rock was bestowed on its serpentinite

### 5.4.8.4 Soapstone (Talc Rock/Schist)

Soapstone is a metamorphic rock consisting predominantly of the mineral talc, a magnesium silicate which may appear massive (in the rock *steatite*) or schistose (in *talc schist*). Apart from talc, soapstone is composed of varying amounts of chlorite and amphiboles (tremolite and anthophyllite) formed by low-grade metamorphism (and metasomatism) of ultramafic rocks and siliceous dolomites. Soapstone has some useful properties of softness and smoothness due to high talc content as well as the ability to hold heat and retain it for long periods of time. Also, soapstone is a beautiful rock when polished with varying colors ranging from lime green, dark green to black. Despite its softness, the stone is very dense and used for indoor decoration of buildings, including kitchen countertops, laboratory countertops, sinks, stoves, fireplace and ceramic tiles, and for cookware (bowls, pots, slabs and utensils). Soapstone is produced worldwide, but usually in small deposits particularly in the regions with ophiolite complexes near plate margins. World's major producers are China, India, the United States and Brazil.

## 5.5 Construction Materials

Construction materials are a group of unconsolidated or loose earth materials and crushed or cut rocks that are used directly in building and road construction. Examples include sand and gravel, crushed rock aggregates, laterite, brick clay, building stone such as slate and sandstone, and cement materials.

### 5.5.1 Aggregates

The term "aggregate" refers to any medium- to coarse-grained granular material formed from a natural rock substance and used in construction. Crushed stone of various sizes, gravel and sand are usually described as aggregates. They are the most widely used industrial materials. Aggregates are used in producing cement concrete, asphalt concrete and in the construction of roads, sidewalks, railroads, subway tunnels and airport runways, as well as our homes, offices, hospitals, schools and shopping centers. Aggregates are also important components in some manufactured products such as glass, cement, paint and other building materials.

In geotechnical parlance, aggregates are divided into two distinct categories--fine and coarse. *Fine* aggregates generally consist of natural sand or crushed stone with most particles less than 5 mm (0.19 inch) in diameter. *Coarse* aggregates are any particles greater than 5 mm but generally range between 9.5 and 30 mm in diameter. The two groups can be separated by screening. The quality of aggregates is determined by several factors including; (i) the particle size and shape of the component minerals, (ii) the type of minerals they are composed of, and (iii) physical and chemical properties. These physical properties determine their strength and durability in construction. Examples of important physical properties are: particle size distribution (gradation), toughness/hardness, abrasion resistance, durability (strength) and soundness, particle shape, surface texture, cleanliness and absence of deleterious materials. Aggregates are usually tested for their physical properties, using such geotechnical tests as particle size distribution, abrasion test, angularity test, and soundness test. Production of aggregates mainly of crushed stone is often one of the largest extractive industries in many developing countries including Nigeria which produces more than 40 million tons annually.

*Concrete* is a composite material consisting of cement, water and aggregates. It is the most important material used in building and pavement construction. For a good concrete mix, aggregates need to be clean, hard, strong particles free of absorbed chemicals or coatings of clay and other fine materials that could cause the deterioration of concrete. Aggregates constitute 60 to 75% of the total volume of concrete, Gravels are the majority of coarse aggregate used in concrete, with crushed stone making up most of the remainder.

### 5.5.2 Sand and Gravel

Sand and gravel are naturally occurring materials that are found in stream beds or terraces. Sand is a fine- to medium-grained water-borne or wind-borne granular material made of fine-grained rock and minerals of a size less than gravel (<2mm) but coarser than silt. Their properties usually depend on the composition of parent rocks. Sand grains can be sharp ('sharp sand') or rounded ('smooth sand') depending on the angularity of the grains. Sand is usually excavated from buried or current river beds, as well as from coastal areas where beach sands are available. After excavation, sands may be washed to remove undesirable material or 'fines', or screened to separate or divide into desired size groupings. Sand is a major ingredient of concrete, plaster, mortar and road pavements.

Gravel is a naturally occurring, water-borne or water-worn pieces of rock that are >2mm in size and found in old or current stream beds. They are normally rounded with smooth surfaces, and their properties may be dependent on parent rock. Gravels of varying sizes are mixed depending on the level of sorting. River gravels are usually made up of minerals or rocks that are resistant to weathering, such as quartz, quartzite and some fine-grained mafic rocks. Prior to use, gravels may be washed to remove undesirable material or may be screened to separate into desired size groupings.

Sand-and-gravel deposits are products of erosion of bedrock and surface materials and the subsequent transport, abrasion and deposition of the particles. The principal geologic agent that affects the distribution of deposits of sand and gravel is water, although wind is another agent in transportation of sand deposits in deserts (sand dunes). Sand and gravel deposits are excavated or mined through an open pit (sand or gravel pit) in old river beds and stream terraces, or mined from beaches and inland dunes or dredged from ocean and river beds. Construction sand and gravel is a *high-volume, low-value* commodity, and for economic reasons, should not be transported long distances - not exceeding 30 miles. Consequently, sand and gravel pits are usually located near the market, and that is why they are common in large, expanding urban areas.

Most countries of the world have their own sand and gravel supplies. For example, in the United States, all fifty states produce sand and gravel, where more than a billion tons are produced every year. The USA has extensive, quality deposits of sand and gravel, along with the technology that allows it to process the material into nearly any quality for nearly any application. Worldwide, the leading nations in the sand and gravel processing industry include the US, Australia, Austria, Belgium, Brazil, Canada, France, Germany, India, Spain, Sweden, and South Africa.

In Nigeria, sand and gravel are extracted from stream or river beds or from abandoned stream terraces all over the country. In areas around Lagos, yellow sand is dredged from the surrounding lagoons. River gravels of various sizes are widespread in Nigeria, especially in Akwa lbom and Cross River States. Sometimes, weathered regoliths overlying quartzite ridges are processed as gravel, although it is obvious that the quartzite fragments are not rounded for lack of transportation by running water. Manufactured or recycled aggregates of gravel size do exist, and they are often the byproduct of other manufacturing industries, such as slag, and recycled concrete.

Fig. 5.12: (a) Rock Aggregates (b) Sand and Gravel Quarry

### 5.5.3 Crushed Stone

Crushed stone is an artificial aggregate composed of rock fragments produced from hard rocks by mechanical means. Geologically, they can be derived from igneous rocks such as granite or basalt; sedimentary rocks such as limestone, dolomite or sandstone; or metamorphic rocks such as quartzite, marble and gneiss. Crushed stones are produced in a quarry or mine. After being extracted from a rock face by blasting, crushed stones, usually in large fragments or boulder size are processed through several stages of crushing and screening to create the final product. Crushed stone is the most widely used subgrade material for road construction, and is a major ingredient in various types of concrete including cement concrete, asphalt concrete, and as ballast for railroads. Crushed stone is also a high volume, low value commodity usually located near the point of consumption so as to avoid excessive transportation costs.

In Nigeria, crushed stone is widely distributed because over 50% of the country is underlain by crystalline rocks of the Precambrian basement complex. Production is from open pits and quarries, in outcrops or inselbergs of granites, gneisses, quartzite, marble, dolerite limestone, sandstone and basalt where they are transported short distances for use as subgrade material for roads and asphalt concrete as well as cement concrete for high-rise buildings. Rocks of the Precambrian Basement Complex (gneisses, quartzites, granites) and the Younger Granite complex provide rich sources of crushed rock. Basalts from the Mambilla plateau and dolerite dykes in the Benue Trough, SE Nigeria are also used as crushed stones. Limestone and marble (from Schist Belts) are crushed mostly for use in cement manufacture.

### 5.5.4 Building Stones

Building stones are naturally occurring rocks that can be quarried and cut or shaped into blocks or slabs for use in building walls, cladding, paving, roofing or other construction activities. Natural building stone is the oldest construction material, and has been used by human beings since prehistoric times. Many ancient structures built with limestone have survived as archeological remains for thousands of years. In modern times, rocks are still used as building and structural materials where they are locally available, although more are used for decorative purposes. The most common rocks used as building stones are gneiss, diabase/dolerite, sandstone, quartzite, slate, marble, limestone and granite.

The main attributes of a good building stone are appearance and color, weight, resistance to fire, hardness and toughness, strength and durability. Also, a building stone should be readily available, easy to quarry and worked into desired shapes. The most common types of building stones and their applications are summarized as follows.

#### 5.5.4.1 Slate

Slate is a dark or grey, usually fine grained, foliated metamorphic rock formed by low-grade metamorphism of shale, and has been used as a building material since the ancient Roman times. Natural slate is composed mainly of fine micas or clay minerals, and the relative concentrations depend on the intensity of metamorphism that the rock has been exposed to. Some of its striking properties include its fissility (ability to be split into thin slabs), attractive dark color, water resistance, fire resistance, environmental friendliness, and durability which make slate a very good choice for building construction. Available in a variety of shades, designs, and colors, slate is extensively used for roofing, wall cladding, flooring, flagging, and partitions.

Slate of the right quality and quantity for building construction is most common in Europe, particularly in parts of Great Britain, such as Wales and Scotland. Ninety percent of Europe's natural slate used for roofing originates from Spain. Other countries where slate is used are Portugal, Germany, Italy and Brazil,

Fig. 5.13: External Wall Cladding

(b) Slate Roofing Tiles

### 5.5.4.2 Gneiss

Gneiss is a foliated metamorphic rock that ranges in texture from near equigranular (granite gneiss) to banded (banded gneiss). It is formed by the regional metamorphism of granite and pelitic rocks. Granite gneiss which is very similar in composition to granite, and composed chiefly of quartz and feldspar, is more widely used as building material because it looks homogeneous, and normally does not split along foliation planes.

Gneiss is a natural stone of considerable variety, color, texture and beauty. It is also very strong and durable because it has been subjected to high temperature and pressure. Some gneisses, including "augen gneiss" are durable enough to perform well as a dimension stone. Gneiss can be cut or sheared into blocks and slabs, and used for interior and exterior spaces in buildings, as paving stone, facades and walls.

Granite gneiss and banded gneiss are common rock types in Precambrian metamorphic terrains worldwide. In Nigeria, there are localities with extensive ridges and inselbergs where gneisses are exposed in southwestern and northeastern Nigeria. They can be quite fresh and exploited for dimension stone.

### 5.5.4.3 Sandstone

Sandstone is a sedimentary rock composed of cemented quartz and feldspar grains. It is hard and tough, and can be cut into stone blocks and slabs for walling, columns and flooring. Sandstone was widely used as building material in North America and Europe in the olden days, but was later replaced with marble as popular building material. In Nigeria, the Nupe Sandstone in the Bida Basin of central Nigeria and the Bima Sandstone in the Upper Benue Trough are cut into small blocks and slabs, and used by the local people for walling and flooring.

### 5.5.4.4 Quartzite

Quartzite is a non-foliated metamorphic rock that is composed of almost 100% of interlocking crystals of quartz formed by metamorphism of quartz-rich sandstone. Due to the hardness of quartz and the interlocked nature of the crystals, quartzite is a hard, tough and durable rock that is relatively difficult to break or crush. Some of the properties of quartzite that make it popular in the construction industry include; resistance to wear and tear (durability), high abrasion resistance, resistance to chemical attack, smooth and sometimes glossy surface, and appearance in a variety of colors. Quartzite is a very versatile building material used as bricks for building walls, and quartzite tiles are used in applications ranging from roofing to flooring, and for fireplace surroundings because it is heat resistant.

In Nigeria, thick quartzite bands are intercalated within the Archean gneisses and migmatites (Older Metasediments) and within the schist belts (Younger Metasediments). However, the quartzites are rather impure and schistose, and the outcrops that form prominent ridges are often deeply weathered.

Fig. 5.14: (a) Gneiss Stone Blocks (b) Sandstone Building Stone (c) Quartzite Building Stone

### 5.5.5 Cement

Cement is one of the most widely used materials in building and construction, Cement is a binder that sets, hardens, and sticks to other materials to bind them together. Cement mixed with fine sand produces mortar for plastering and masonry, or with sand and gravel, produces concrete. Concrete is the most widely used material in existence and second only to water as the world's most-consumed material. There are various types of cement but the most commonly used is the Ordinary Portland cement (OPC) which is the basic ingredient of concrete. Portland cement is a mixture of argillaceous and calcareous materials, sintered and then pulverized into a fine powder to which gypsum (5%) is added to prevent rapid setting.

The main geological raw materials for cement manufacture are; (1) *lime* (limestone/marble/marl); (2) *alumina/silica* (clay-shale); (3) *iron (*iron ore or lateritic soil); and (4) *gypsum.* (Silica sand may be added if needed). Lime (CaO) and silica ($SiO_2$) make up about 85% of cement. The mix ratios of cement raw materials vary, depending on their chemical composition, usually 4 to 1 ratio (limestone to clay). However it is important to have the correct proportions of *calcium, silicon, aluminum, and iron.* Cement raw materials are mixed in proportions formulated to satisfy the required chemical composition of Portland cement as follows: Lime (CaO) 60-67%; silica ($SiO_2$) 17-25%; alumina ($Al_2O_3$) 3-8%; iron oxide ($Fe_2O_3$) 0.5-6.0%. Any deficiencies in silica and iron may be rectified by addition of silica and iron oxide respectively as quartz and laterite respectively. Deleterious constituents such as magnesium, sulfur and phosphorus must be less than the maximum allowable concentrations of 5% MgO, 1% $SO_3$ and 1% $P_2O_5$. Portland cement is manufactured in a four step process: 1) Quarrying; 2) Raw Material Preparation; 3) Clinkering; and 4) Cement Milling.

In Nigeria, there are more than a dozen cement factories using limestone or marble as source of lime, and shale or clay for source of alumina/silica, with the addition of laterite and silica sand if needed. Because the limestone raw material is bulky, most cement factories are located near the source of lime. Fortunately, in Nigeria, all the needed raw materials except gypsum, are often found in the same quarry or in close proximity. Cement is in high demand in Nigeria, and the cement factories which are quite evenly distributed in the country produce over 40 million tons annually.

### 5.5.6 Brick Clay

Brick clay is an earthy material that is plastic when wet and composed of some clay minerals, with minor quartz and iron oxide. Brick clay is used as raw materials in the manufacture of construction bricks, roofing tiles and other structural clay products used by the building industry, such as floor/wall tiles, drainage pipes. Brick clays are quite 'common', and are usually found as surface deposits and won from unconsolidated or soft sediments found along stream beds and within the weathered lateritic regolith. In the raw form, brick clay is red, brown, or gray, as a result of the presence of iron oxide and some organic matter. The main minerals are quartz, and clay minerals which are mixtures of kaolinite, illite, smectite and chlorite. They contain iron (as iron oxides) and other mineral impurities in sufficient quantities to cause the clay to become tight and hard-fired at about 950 to 1000°C when clay particles fuse together. When fired, the clay may vary in color from pink to buff to tan, red or brown, depending on the clay, the impurities and the condition of the firing. For brick making, the red or tan colors are preferred for their aesthetic value.

Burnt or fired brick clay, usually reddish in color, is one of the oldest of all building materials, and found in many of the ancient structures of the world. It is also one of the most durable materials, since there are brick walls, foundations, pillars, and road surfaces constructed thousands of years ago that are still intact. Today, bricks are still widely used for wall construction in homes and fences, especially as an outer wall facing.

In Africa, bricks have been recommended as sustainable building material since common clays are found easily within the lateritic soils and alluvium. However, due to several factors including the scarcity of brick-making factories, power shortage to fire the kilns, the advent of cheaper sandcrete cement blocks and shortage of red brick laying masons, the usage of red bricks for buildings has virtually disappeared in countries like Nigeria and Ghana.

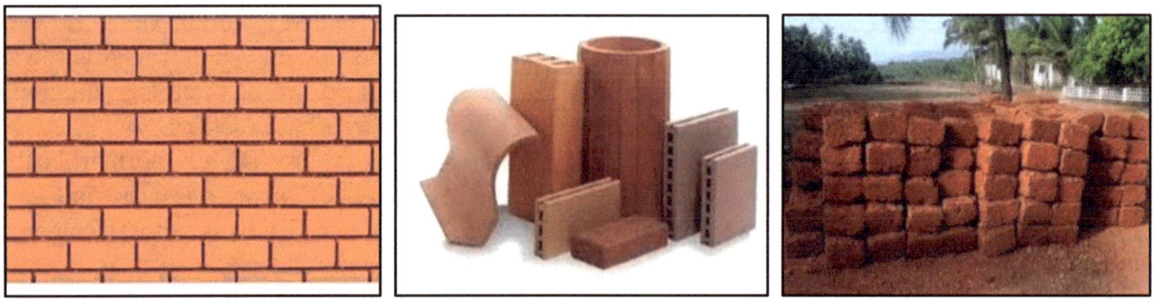

Fig. 5.15: (a) Fired Brick Clay (b) Structural Clay Products (c) Bricks from Laterite Stone

## 5.5.7 Laterite

Laterite is a soil or indurated material (stone or rock) rich in iron and alumina and formed by intense chemical weathering under hot and wet tropical and subtropical conditions. Laterites can form thick residual soil profiles that are tens of meters thick sometimes with distinct horizons, particularly in Africa, India, SE Asia and Australia. The pervasive reddish color of laterite is due to iron content. Laterites can be soft and easily broken into smaller pieces, or firm (indurated) and physically resistant. Laterite 'stone' is developed where the laterite is indurated or cemented by higher concentrations of iron oxide. Laterites when crushed are usually found to be well-graded, comprising both cohesive (silt and clay) and loose sand and gravel, with colors varying from ochre through red, brown, violet to black, depending largely on the concentration of iron oxides. Mineralogically, they are composed of iron- aluminum oxides and sesquioxides, mainly goethite, limonite, and clay minerals, sometimes with minor gibbsite (Al-hydroxide).

Laterite has been used as building material in Africa and Asia since time immemorial. In modern times, laterite is still used as building material by moistening plastic lateritic soil and making bricks that are allowed to dry and then used for building walls. Where the laterite is well indurated, it is hard as a rock, and the laterite stone can be excavated and cut into bricklike shapes for use directly as walling units, particularly in India and South Africa where such material is common. Recently, there has been advancement in the technology of using laterite mixed with about 5% cement to make interlocking bricks to construct walls without the use of cement mortar.

Geotechnical studies have shown that laterite bricks are of adequate strength and durability that make them sustainable under different kinds of severe weather and seasonal variations. For road construction, laterites have been used as stable material for surfacing rural roads and as base and sub-base materials when compacted for asphalt road construction. Laterite is also suitable as fill materials in embankment and dam construction.

### 5.5.8 Pumice and Perlite

Pumice and perlite are volcanic rocks made up of rhyolitic volcanic glass. Pumice is a light-colored volcanic rock that consists of highly vesicular rough textured rock glass that may or may not contain crystals. Pumice is formed when over-pressured volcanic gases form bubbles that escape from a rapidly cooling magma as it forms glass. It is a common product of explosive volcanic eruptions. Pumice from silica-rich lavas is white, those from lavas with intermediate silica content are often yellow or brown, and rarer silica-poor varieties are black. The hollows created in the glassy lava can be rounded, elongated, or tubular, while the glassy material can be in threads, fibers, or thin partitions between the hollows. Pumice has a low density due to its numerous vesicles, and with a porosity of 64-85%, it floats on water for a very long time. Because of its light weight and low density, pumice is widely used as a wall insulator. It is commonly mixed with cement to make lightweight concrete and decorative perforated blocks for fence walls.

Perlite is a silica-rich volcanic rock composed entirely of amorphous volcanic glass (obsidian) that has a relatively high water content and the unusual property of high rate of expansion (7-16 times) when heated to about 900 °C. The expanded material is brilliant white and lightweight and after processing, it can be used in industrial applications. Perlite is an amorphous volcanic glass that has a relatively high water content, typically formed by the hydration of obsidian. It occurs naturally and has the unusual property of greatly expanding when heated sufficiently. It is an industrial rock and a commercial product useful for its low density. In building construction, perlite is used for insulation as loose fill material in hollow concrete blocks or brick walls, and as an ingredient in lightweight concrete and plasters, and ceiling tiles.

Fig. 5.16: (a) Pumice (with vesicles) (b) Perlite Crystals (lightweight)

## 5.6 Gemstones

Gemstones (or gems) are a special class of industrial minerals used in the jewelry industry. A gemstone is defined as a mineral or rock, which can be used in jewelry after cutting or faceting and polishing. Gemstones often occur as well-formed crystals of colorful minerals that are spectacular objects of beauty when cut and polished for use as jewelry or personal adornments. Gems are so valuable in their rarity, and sufficiently durable to give lasting pleasure to their owners. Although almost all gemstones are minerals, there are a few rocks (such as lapis lazuli) and organic materials that are not minerals (such as amber, jet, and pearl) but are also used for jewelry, and are therefore often described as gemstones.

The world trade in gemstones is very substantial. In 2018, the total value of the global production of rough diamonds amounted to 17.4 billion U.S. dollars. In 2019 the value of rough diamond sales worldwide amounted to 13.9 billion U.S. dollars. It has been calculated that the world colored-gem trade is worth over $6 billion per year. Although synthetic forms of many gems now exist, they have not had a serious impact on the international gemstone market.

*Gemology* is the science of gemstone that deals with both natural and artificial gem materials. It is a geoscience, and a growing branch of mineralogy. Some jewelers (and many non-jewelers) are academically trained gemologists who are qualified to identify and evaluate gemstones.

## 5.6.1 Characteristics of Gemstones

For a mineral to qualify as a gemstone, it must possess at least these three qualities (i) *beauty* - an aesthetic value determined by appearance and color; (ii) *durability* - the ability to last for a long time and show resistance to wear and tear; and (iii) *rarity* - not common. Other useful characteristics that distinguish gemstones include extraordinary colors, crystallinity, purity, transparency, luster and clarity (no inclusions), specific gravity, hardness and toughness to resist scratching. These attributes combined in various ways, make gemstones unique and valuable. Gemstones have managed to endure through time because of their certain characteristics which are highlighted below. Also, we must consider other extraneous factors such as fashion, superstition (magical powers) and social customs that can play a part in determining the importance of a gemstone. The three most valuable and prized gemstones in the world today are: tanzanite, black opal and red beryl.

### 5.6.1.1 Beauty and Appearance

Gemstones are very much valued especially for their great beauty or perfection without any blemish. Consequently, appearance becomes the most important attribute of gemstones. A gemstone's beauty normally depends on several features, mostly its color, size, shape and sparkle. Color is the most important gem characteristic whereas size and shape also play a vital role in enhancing the beauty of gemstone. A gemstone's luster or sparkle definitely enhances its aesthetic beauty in a great manner.

### 5.6.1.2 Durability and Hardness

Durability and hardness are both related attributes that are considered as important for a gemstone. No matter how beautiful a gemstone may appear, for it to be valuable as a precious gemstone, it will need to be reasonably durable which partly depends on its hardness and resistance to abrasion or scratching. Using the well-known Mohs Scale of Hardness, most valuable gemstones range from 7 to 10 in hardness (Table 5.4).

### 5.6.1.3 Rarity

Rarity which means 'not common' is a desirable attribute for a gemstone. The rarity of a stone imparts a sense of exclusiveness and worth that increases our desire to possess and pay the high price for a gemstone. Rarity most commonly determines the prices placed on gems and famous jewelry worldwide. Gems may be rare for a variety of reasons. Many gems are varieties of common stones, but their exceptional color or clarity determines them to be rare gemstones.

## 5.6.2 Classification of Gemstones

There are over 200 varieties of natural gemstones that are recognized, and some are so rare that only a few specimens are known to exist and are extremely expensive. Gemstones have been classified by a variety of criteria, but the three most common are:

- Chemical Classification (inorganic/organic/rock)
- Natural Classification (natural/synthetic/treated)
- Commercial Classification (Precious/Semi-Precious)
- Mineralogical Classification ( Species and Varieties)

***Chemical classification*** is based on the chemical constituent of the gemstone; whether it is inorganic and crystalline, or organic and non-crystalline. A mineral is defined as inorganic substances with a crystalline structure and a defined chemical composition. The only exception is an inorganic substance that is not crystalline, such as opal which is made of rapidly cooled volcanic glass that is non-crystalline. It is called a *mineraloid.*

***Natural Classification*** is based on whether the gemstone is natural or artificial (synthetic). From a gem collector's point of view, this is the most basic classification of gemstones. With the eye of a professional gemologist or mineralogist, natural gemstones may not be that difficult to differentiate from artificial ones, on the basis of good knowledge of their physical properties Treated or refined gemstones have also been considered as artificial, if the observer understands the physical properties well. Owing to the improvement in technology and chemical products, it could be a difficult task to distinguish between an artificial and natural stone by a first time buyer.

**Commercial Classification** has been used since the mid-1800s by jewel marketers, and adopted by gemologists and other professionals to classify gemstones based on their apparent quality and value into two types: - *precious* and *semi-precious* stones,

### 5.6.2.1 Inorganic Gemstones

Inorganic gemstones are naturally occurring, solid crystalline substances (minerals) that are made up of one or more elements. This group includes over 95% of known gemstones. Opal, which is widely used as a gem is an inorganic substance made up of amorphous silica (obsidian) but non-crystalline, and is therefore not considered as a mineral, but a "mineraloid".

### 5.6.2.2 Organic Gemstones

Organic gemstones are non-crystalline organic materials produced by organisms (plants and animals), that are solid and can be cut into gems and other ornamental objects. Examples include amber, jet and pearl. There are also organic materials that have been mineralized (replaced and infilled by mineral or inorganic material, such as chalcedony, but are still classified as organic because they do retain a relict organic structure. Examples include petrified wood, fossil coral, turritella agate, and dinosaur bone. Although diamond is made up of organic carbon, it is in a recrystallized form of carbon (allotrope) but is no longer organic because it has a crystal structure. Organic gemstones are one of the oldest materials used in jewelry, and they continue to be popular.

### 5.6.2.3 Rock Gemstones

Rock gemstones are aggregates of minerals -rocks that have been accepted as gemstones, due to their unique color or distinctive features. When these rocks are polished, they are attractive enough to be used as gems, but are not minerals. Examples include lapis lazuli, a rock composed of three different minerals; lazurite, calcite and pyrite, jasper (form of silica), 2K granite and obsidian.

### 5.6.2.4 Synthetic Gemstones

Synthetic gemstones are manufactured or 'man-made' materials, grown in laboratories with essentially the same chemical composition, crystal structure and optical and physical properties as the natural gemstone. Some of the popular artificial/ synthetic gemstone varieties include diamond, corundum, and spinel, sapphires, garnets, alexandrite, rubies and peridot. They can be distinguished based on UV radiation tests

Table 5.3: List and Classification of Most Common Gemstones

| **Precious Stones** | **Semi-Precious Stones** | | **Rock Gemstones and Mineraloids** | **Organic Gemstones** |
|---|---|---|---|---|
| Diamond | Amazonite | Kyanite | **Rocks** | Amber |
| Emerald | Amethyst | Labradorite | Agate | Ammolite |
| Ruby | Ametrine | Lapis Lazuli | Jasper | Ammonite |
| Sapphire | Ammolite | Lepidolite | Lapis Lazuli | Coral |
| | Andalusite | Malachite | K2 Granite | Crinoids |
| | Anyolite | Moldavite | Obsidian | Dinosaur Bone |
| | Apatite | Moonstone | Unakite | Fossilized Coral |
| | Aquamarine | Morganite | Wonderstone | Ivory |
| | Aventurine | Nephrite | | Jet |
| | Azurite | Peridot | **Mineraloids** | Mary Ellen |
| | Azurmalachite | Perthite | Opal | Jasper |
| | Beryl | Prasiolite | | Mother of earl |
| | Benitoite | Quartz | | Opalized Wood |
| | Birthstones | Red Beryl | | Peanut Wood |
| | Bloodstone | Rhodochrosite | | Pearls |
| | Charoite | Rhodonite | | Petrified Palm |
| | Chrysoberyl | Rose Quartz | | Petrified Wood |
| | Chrysocolla | Ruby | | Turritella Agate |
| | Chrysoprase | Sapphire | | |
| | Citrine | Scapolite | | |
| | Cordierite | Serpentine | | |
| | Diopside | Smoky Quartz | | |
| | Emerald | Sodalite | | |
| | Fluorite | Sphene | | |
| | Gaspeite | Spinel | | |
| | Garnet | Spodumene | | |
| | Gem Silica | Strontium | | |
| | Geodes | Titanate | | |
| | Goldstone | Sugilite | | |
| | Goshenite | Sunstone | | |
| | Green Beryl | Tanzanite | | |
| | Helenite | Titanite | | |
| | Heliodor | Topaz | | |
| | Hematite | Tourmaline | | |
| | Iolite | Turquoise | | |
| | Iris Agate | Variscite | | |
| | Jade | Zircon | | |
| | Jadeite | Zoisite | | |

Table 5.4: Mineralogical Groups and Varieties of Gemstones and Their Properties

| Gemstone Mineral Group | Gemstone Varieties | Hardness | Specific Gravity | Color Description |
|---|---|---|---|---|
| **Beryl** ($Be_3Al_2Si_6O_{18}$) | | 7.5-7.8 | 2.63-2.91 | |
| | Emerald | | | Intense green or bluish green |
| | Aquamarine | | | Greenish blue or light blue |
| | Goshenite | | | Colorless, greenish yellow, yellow green, |
| | Heliodore | | | Golden yellow to golden green |
| | Morganite | | | Pink, purple pink, or peach |
| | Red Beryl | | | Raspberry red |
| **Chrysoberyl** ($BeAl_2O_4$) | | 9.0 | 3.63-3.78 | |
| | Chrysoberyl | | | Yellowish green to greenish yellow |
| | Alexandrite | | | Green in daylight |
| | Cat's eye: | | | Usually yellowish or greenish |
| **Corundum (** $Al_2O_3$) | | 9.0 | 3.96-4.05 | |
| | Ruby | | | Intense red |
| | Sapphire | | | Blue |
| **Diamond (C)** | | 10.0 | | |
| | Diamond | | | Colorless to faint yellowish tinge, & variable colors |
| **Feldspar (**Ca)/$KAlSi_3O_8$) | | 6-6.5 | 2.55-2.76 | |
| | Labradorite | | | Colorful, iridescent, also transparent stones in yellow, orange, red, and green |
| | Amazonite | | | Yellow green to greenish blue |
| **Garnet** Complex formula) | | 6.5-7.5 | 3.5-4.3 | |
| | Almandine | | | Orange red to purplish red |
| | Andradite | | | Yellowish green to orange or to black |
| | Grossular | | | Colorless; also orange, yellow, & brown |
| | Tsavorite | | | Green to yellowish green |
| | Pyrope | | | Colorless; also pink to red |
| | Chrome Pyrope | | | Orange red |
| | Rhodonite | | | Purplish red to red purple |
| | Spessartine | | | Yellowish orange |
| | Uvarovite | | | Emerald green |
| **Jade** (Na,$AlSi_2O_6$) | | 6.0 | 2.9.3.5 | |
| | Nephrite | | | White, deep green, creamy brown |
| | Jadeite | | | Blue green, emerald green, lavender, dark blue green and greenish black |
| **Lapis Lazuli** | (Rock) | | | |
| | Lazurite | | 2.7-2.9 | Bluish color with flecks of white & gold) |
| **Olivine** ((Mg,Fe($2SiO_4$) | | 7.0 | 3.22-3.45 | |
| | Peridot | | | Olive to lime green |
| **Quartz** ($SiO_2$) | | 7.0 | 2.65 | |
| | Amethyst | | | Purple |
| | Citrine | | | Yellow to amber |
| | Morion | | | Black |
| | Crystal Quartz | | | Colorless |
| | Smoky Quartz | | | Smoky grey o brown |
| | Rose Quartz | | | Translucent pink |
| | Green Quartz | | | Green |
| **Spinel** ($MgAl_2O_4$**)** | | 8.0 | 3.58-4.06 | |
| | Balas Ruby | | | Red |
| | Almandine Spinel | | | Purple red |
| | Rubicelle | | | Orange |
| | Sapphire spinel | | | Blue |
| | Chlorspinel | | | Green |
| **Topaz** ($Al_2SiO_4(F,OH)_2$) | | 8.0 | 3.5-3.6 | |
| | Topaz | | | Wine yellow, pale blue, green, violet, |
| **Tourmaline** (Complex formula) | | 7-7.5 | 3.03-3.25 | |
| | Achorite | | | Colorless |
| | Dravite | | | Brown |
| | Indicolite | | | Dark blue |
| | Paraba | | | Multiple colors |
| | Rubellite | | | Pink to red |
| | Siberlite | | | Violet |
| **Turquoise** ($CuAl_5(PO_4)_4(OH)_4H_2O$**)** | | 5-6 | 2.6-2.8 | |
| | Turquoise | | | Sky blue; greenish blue |
| **Zircon** ($ZrSiO_4$) | | 7.5 | 4.6-4.7 | |
| | Zircon Crystal | | | Colorless |
| | Jargon | | | Variable color |
| | Matura diamond | | | Colorless |
| | Hyacinth | | | Yellow, orange, red brown |

#### 5.6.2.5 Precious Stones

Precious stones are natural gems that are visually appealing, highly valuable and attractive. This group comprises four specific stones; *diamonds, rubies, emeralds, and sapphires.* All precious stones are translucent and are valued by the richness of their color, except for diamond, which has a higher value based on being colorless but sparkling. Their rarity, beauty, and method in which they are produced, all add to the allure of precious stones. Diamond which is made of recrystallized carbon is the most precious and alluring of all gemstones; often sparkling and colorless or faintly colored. Emerald is a green variety of the mineral beryl ($Be_3Al_2(SiO_3)_6$. Ruby is a pink to blood- red colored variety of the mineral corundum ($Al_2O_3$) while sapphire is a typically blue variety of the same mineral corundum.

#### 5.6.2.6 Semi-Precious Stones

Semi-precious stones refer to all gemstones other than the four precious stones. Examples of semi-precious stones are shown in Table 5.3.Semi-precious stones are considered as less valuable or less appealing, although attitudes have changed regarding what gemstones are precious today. Some semi-precious stones have been prized higher than some of the so-classified precious stones. For example, the three most valuable and prized gemstones today are tanzanite, black opal and red beryl.

### 5.6.3 Mineralogical Groups

Gemstones can be classified into mineral groups called "species" as shown Table 5.4 There are fifteen groups, and each group has one or more members termed "varieties". The most important groups in terms of variety and color are Beryl, Corundum, Diamond, Garnet, Tourmaline, Quartz and Zircon. Most of the gemstones have a Mohs hardness of about 7 and above.

### 5.6.4 Geological Environments of Gemstones

Gemstones are minerals formed mostly by geological processes that occur at great depths under conditions of relatively elevated temperatures and pressures. Surficial processes only form a few gemstones but assist in recirculating already formed primary minerals through erosion and sedimentation. Gemstones are formed in four different geological environments:

1. Igneous (Plutonic ,Volcanic & Magmatic-Hydrothermal)
2. Metamorphic (Regional and Contact Metasomatic)
3. Sedimentary Rocks (Placer)
4. Surficial Materials I Weathering)

Gem deposits are rare because of the unique conditions that promote their formation are unusual. They generally occur in small quantities, and do not form "ore deposits" in the real sense as we know in metallic minerals, but when they occur, tend to be disseminated or scattered sparsely throughout a large body of rock or to have crystallized as small aggregates or as small veins and cavity fillings or as weathered fragments in regoliths. When stream gravels contain concentrations of gem minerals, they also tend to be small in size, appearing as small pockets or small gravel lenses in a stream bed. So, the term "gem deposits" will be used in this context while describing the occurrence of gemstones where single mineral crystals can be so valuable to be worth millions of dollars. Based on the origin and ore forming processes, gem deposits can be classified into two major types: *primary* deposits and *secondary* deposits.

### 5.6.4.1 Primary Gem Deposits

Primary deposits refer to the concentration of gems due to primary crystallization at high temperature and pressure as a result of magmatic and metamorphic processes.

### (1) Igneous Rocks

Magmatic processes are of two settings: *plutonic* and *volcanic*. Plutonic rocks are formed at considerable depth by crystallization of magma and associated hydrothermal solutions, while volcanic rocks are formed at the surface by the discharge of lava (molten rock) or hot water, steam, and gases (hot springs and fumaroles). Gemstone deposits associated with igneous rocks are found in kimberlites, lamprophyres, alkali volcanic rocks, pegmatites and magmatic hydrothermal veins.

a. ***Kimberlites***: Kimberlites are ultramafic magmatic rocks that form deep in Earth's interior and are brought to the surface by explosive volcanic eruptions. They are often brecciated, intensely altered and contain xenoliths of other rocks. They are believed to have originated from the mantle. Diamond deposits of economic significance are found exclusively within kimberlite dykes that form diatremes or breccia pipes. However, only about 1% of kimberlites contain diamonds of economic value. Diamonds in kimberlites are believed to be xenocrysts brought up from deer-seated sources in the mantle, and are commonly associated with other gem minerals such as garnet, spinel and diopside.

b. ***Lamprophyres***: Lamprophyres are a group of unusual alkaline ultramafic igneous rocks that occur mainly as dykes or small intrusions, and somewhat similar to kimberlites. Lamprophyres contain corundum (sapphire) and some diamonds that are usually not in economic concentrations.

c. ***Alkaline Volcanic Rocks***: Gem minerals found in these rocks are corundum (sapphire), peridot (olivine), zircon and spinel that occur mostly as xenocrysts carried up from depth or phenocrysts that crystallized in the magma as early crystals before rapid extrusion of the volcanic rock.

d. ***Pegmatites and Syenites:*** Pegmatites are coarsely-crystalline igneous rocks formed from the residual fluids in fractionated magmas. Most pegmatites are granitic in composition, but less common are gabbroic pegmatites. Gemstones are common in pegmatites, including apatite, beryls (emerald, aquamarine), corundum (sapphires), feldspars (including perthite and amazonite), quartz, fluorite, garnet, lepidolite, quartz varieties (crystal, rose, and smoky), spodumene, topaz, zircon, and tourmaline. Some albitized alkali granites (e.g. Nigeria) contain beryl and emerald. Syenite and nepheline syenite also contain gem corundum.

e. ***Hydrothermal Veins:*** Hydrothermal veins are fracture-filling minerals formed by hot solutions derived primarily from magmatic sources, although some are derived from sedimentary and metamorphic processes. Gem minerals include quartz of all varieties (see Table 5.4), apatite, fluorite, beryl, emerald, topaz, zircon, geodes (small-crystal filled cavities), and opal. Porphyry copper deposits of hydrothermal origin contain gemstones including azurite, chrysocolla, malachite, turquoise, and others.

f. ***Igneous Rocks:*** Some igneous rocks with beautiful color, texture and crystallinity are polished and used as rock gems. Examples include K2 Granite, Obsidian, Unakite (a granite) and Wonderstone (tuff).

### (2) Metamorphic Rocks

Metamorphic processes including regional metamorphism, contact metamorphism/metasomatism and hydrothermal metamorphism, all contribute to the formation of gem minerals.

a. ***Regional Metamorphic Rocks:*** These are rocks (igneous and sedimentary) that have been changed over a wide area by increased heat and pressure associated with deep burial and deformation. The rate of metamorphism ranges from very low to high grade. Gemstones, like other minerals, are commonly formed by the crystallization of new minerals or recrystallization of existing minerals during prograde metamorphism. High grade granulites and gneisses, and lesser grade schists and amphibolites are known sources of corundum deposits. Gemstones associated with regional metamorphic rocks include variety of garnets (pyrope, almandine, rhodonite, and spessartine), cordierite, turquoise, beryl, jade (jadeite), nephrite (a green fibrous mineral) spinel, ruby, sapphire, emerald, kyanite and zircon. Gem corundum is found in high grade metamorphic terrain produced by the transformation of Al-rich and/or Si-poor rocks into ruby and sapphire bearing gneisses, marbles and granulites Emerald is a unique example of gems formed by the metamorphism of shales. Metamorphic rocks used as gems include lapis lazuli, metachert and aventurine (quartzite).

b. ***Contact Metamorphic - Metasomatic Rocks:*** Contact metamorphic/metasomatic rocks are formed when magma is intruded into pre-existing rocks in which heat and fluids (metasomatism) change or lead to the formation of new minerals, particularly around carbonate and aluminous rocks. Gemstones formed by this process include andalusite, scapolite, diopside, corundum (including ruby and sapphire) especially those associated with coarse-grained oligoclase-corundum rock called *plumasite*. These gem corundum deposits result from small-scale metasomatism involving reactions between silico-aluminous rocks (pegmatites, gneisses, etc) and silica-poor rocks (ultramafites, meta-carbonates). Emerald is also found in some veins of hydrothermal origin in association with contact metamorphism.

c. ***Hydrothermal Metamorphism:*** Hydrothermal metamorphism is the process in which hot, volatile solutions circulate and react with pre-existing rocks, both volcanic and sedimentary in basins. The heat and volatile solutions produce extensive alteration and formation of new minerals including some gemstones. Examples are benitoite, serpentine, zoisite (epidote) and tanzanite - a blue variety of zoisite.

### 5.6.4.2 Secondary Gem Deposits

Secondary deposits are formed by surface weathering, erosion of primary deposits and concentration of the heavy minerals as eluvial, colluvial and alluvial deposits.

- ***Sedimentary Deposits***

Some primary gemstones are of high specific gravity and resistant to erosion (durability). Consequently, when they are eroded and sometimes transported by rivers (or wind), they can accumulate as sedimentary placer deposits. Examples of gems in this category include diamonds, corundum and chrysoberyl, which are minerals with hardness of 9 or 10 on Mohs' scale. Garnets, spinels and beryl are sometimes found in both eluvial and alluvial deposits not too distant from source. The majority of diamond and corundum gem (ruby and sapphire) deposits are found as eluvial and alluvial placers. Paleo-placer deposits containing corundum and diamond in consolidated sands and gravels occur in Madagascar and South Africa respectively. Black shales are one of the richest sources of emerald in Columbia. Colombia-type emerald deposits consist mainly of carbonate-pyrite-albite quartz veins forming "en échelon" systems and cementing breccias. Emeralds are disseminated in the veins as clusters, single crystals or crystal fragments; however, the best gemstones are found in cavities. Country rocks are black carbonaceous and calcareous shales. The deposits are considered diagenetic to epigenetic.

- ***Supergene Materials***

Supergene alteration associated with chemical weathering of copper deposits has produced some materials that are used as gems, including azurite (blue), malachite (green), azurmalachite (blue and green), chrysocolla and turquoise.

## 5.6.5 Description of Precious Gemstones

The precious gemstones are diamond, emerald, ruby and sapphire. Their physical properties and mode of occurrence are described below. Photographs of typical gemstones are presented at the end of this chapter.

### 5.6.5.1 Diamond

Diamond is the most highly valued and most popular gemstone. It is a naturally occurring crystalline form of carbon formed at high pressures, and transported from great depths to the surface by a unique group of mantle-derived ultramafic dykes (usually pipes) called kimberlites. . Diamond has some unique properties including being the hardest mineral, a high specific gravity that ranges between 3.4 to 3.6 and a very bright luster - the highest non-metallic luster - known as "adamantine." which makes diamonds “sparkle”. There are industrial and gemstone varieties of diamond. Industrial diamonds are mostly used in cutting, grinding, drilling, and polishing procedures. Gem diamonds are of higher quality and are valued for their size, color, clarity, and sparkle. They are rare and make up a minor portion of world production of diamonds. Most gem -quality diamonds are mostly colorless, but there are yellow, brown, or gray.as well as rare blue, brown or red varieties. Geological occurrence of gem diamonds is confined to kimberlite pipes that are believed to be of mantle origin. They occur mostly within the ancient cratonic shields of Africa, Canada and Russia. Approximately 5000 kimberlite pipes are known worldwide, but only a few (less than 100) contain diamonds of gem or industrial quality.

Global production of natural diamonds from mines in 2019 was estimated as 142 million carats with worldwide reserves of 1.2 billion carats. Major producing countries in the order of production are Russia, Botswana, Canada, South Africa, Angola, the Democratic Republic of Congo, Namibia and Lesotho. The United States is the leading consumer of diamonds accounting for 40% of world diamond jewelry trade. Other countries are India, China, the European Union, Japan, Hong Kong, and the Middle East.

### 5.6.5.2 Emerald

Emerald is a bright green gem variety of the mineral beryl ($Be_3Al_2Si_6O_{18}$.). It is found in pegmatites and some alkali granites, regional metamorphism of shales and ultramafic lavas, contact metamorphism, and in hydrothermal veins, black shales, and rarely in alluvial deposits. Emerald is a very valuable gem, third after diamond and ruby, and is only found as gem deposits in a small number of locations worldwide. Emerald is defined by its bright green color. Whereas pure beryl is colorless, the presence of trace amounts of *chromium* or *vanadium* in the mineral causes it to develop a green color. Also, trace amounts of iron will tint emerald a bluish green or a yellowish green color depending upon its oxidation state. Columbia is the world’s leading producer of emerald, with an estimated 60% of world production. Other major producers are Brazil, Zimbabwe, South Africa, Afghanistan, and the USA.

#### 5.6.4.3 Ruby and Sapphire

Ruby and sapphire are gem varieties in red and blue colors of the mineral corundum ($Al_2O_3$. Both of these gemstones have the same chemical composition and the same mineral structure, but the trace amounts of impurities determine their different colors of red ruby and blue sapphire. The red color of ruby is produced by the presence of chromium while trace amounts of titanium and iron produce the blue color of sapphire. Ruby and sapphire are perhaps the world's most widely sold colored gemstones, accounting for approximately one-third of sales by value.

Corundum gems (ruby and sapphire) are found in primary and secondary deposits. They are associated with alkaline basalts and lamprophyres, and in metamorphic rocks within marbles meta-ultramafites and metasediments, Contact metamorphic-metasomatic skarn deposits and plumasite contain rich corundum deposits. Because of the hardness and specific gravity of corundum, secondary deposits in eluvial and alluvial placers are common. Most of the world's sapphires are produced from gem placers related to alkali basalts, as in eastern Australia or southern Vietnam, while placers in metamorphic environments, such as in Sri Lanka and Madagascar produce the highest-quality sapphires. World leading producers of ruby and sapphire are Myanmar, Sri Lanka, Madagascar, Vietnam and Kashmir.

## 5.6.6 Gemstones from Africa

Africa is arguably the most richly mineralized continent in the world. It is endowed with abundant metallic minerals, such as gold, copper, iron ore, bauxite, etc., and beautiful gemstones of which diamond is the best known. Six of the seven leading producers of gem quality diamonds in the world are located in Africa, namely: Botswana, Angola, South Africa, Democratic Republic of the Congo, Namibia and Lesotho. They collectively produce over fifty million carats of diamond annually, and own some of the largest diamond mines in the world. Apart from diamonds, Africa is also a major source of colored gemstones that are found in more than half a dozen different countries in Southern and Eastern Africa, stretching from Namibia in the southwest through Zimbabwe, Zambia, Malawi and Tanzania, to Kenya and Mozambique in Eastern Africa, and in the nearby island of Madagascar. They are all part of the same mineral-rich region known as the "Mozambique Belt" underlain by metamorphic and igneous rocks of Pan African age (Neo-proterozoic). Outside the Mozambique Belt, Nigeria in West Africa is the only notable producer of gemstones that are derived from the Pan African granitic pegmatites and Cenozoic alkali basalts. The colored gemstones commonly found in Africa are ruby, sapphire, garnets of several varieties Paraiba tourmaline, tanzanite (epidote), amethyst, topaz and zircon.

#### 5.6.6.1 African Diamond

Africa is the world's largest producer of gem quality diamond, the "mother of all gems". Most of the diamonds are derived from kimberlite pipes and alluvial deposits derived from them. Botswana stands out as the world's second-leading producer of diamonds on the basis of carat weight, and the leading producer on the basis of value, assessed on the larger size and higher quality of its diamonds compared to world leader, Russia. Botswana's Jwaneng mine is often referred to as "the richest diamond mine in the world.", and the mine in recent years has been producing about 15 million carats of high-quality diamonds per annum. South Africa which was once the world's leading producer of diamonds but now faced with dwindling resources, is still a major player in the gem trade with production of a few million carats annually. Angola, Democratic Republic of the Congo, Namibia, Lesotho, and Sierra Leone are other African producers (Fig. 5.17).

### 5.6.6.2 African Ruby

Ruby, a "bloody-red" variety of corundum which is the second most valuable precious stone next to diamond. Rubies can be valued at up to $1,500 a carat depending on its clarity, origin, and color. Myanmar (Burma) has for a long time dominated the world trade in rubies, until discoveries were made in Africa. The so-called "African Ruby" originating mainly from Mozambique is of the highest quality and compensated for the demand for clear ruby specimens. More recently, there have been more discoveries of top-grade ruby found in eastern Mozambique, including some material of excellent color that does not require heat treatment. Other sources of African ruby are the neighbouring Kenya, Tanzania, and Madagascar which is known to also contain high quality ruby. The Mozambique Belt of Eastern Africa is therefore a unique area rich in corundum gems.

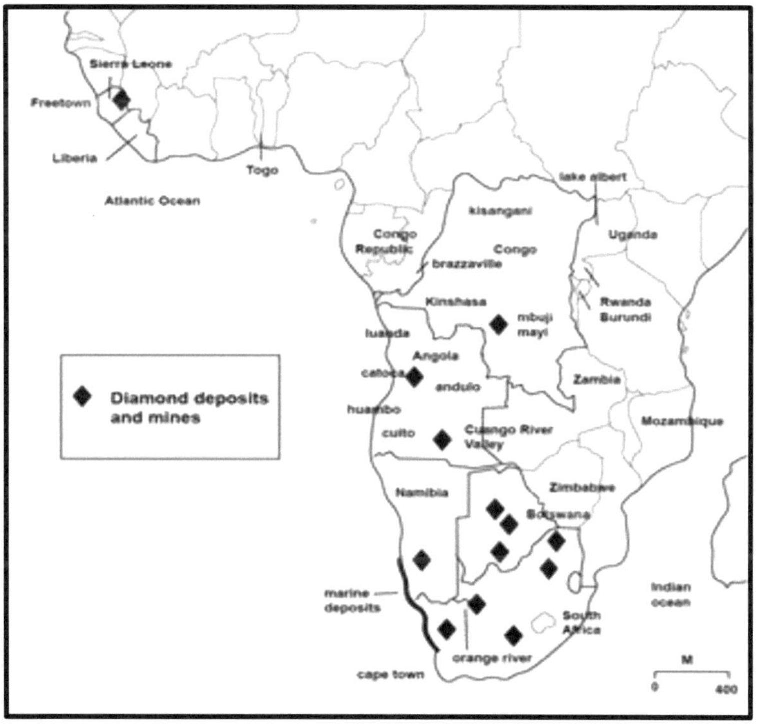


Fig. 5.17 Diamond deposits in Africa

### 5.6.6.3 African Sapphire

Sapphire is the royal-blue variety of corundum. In 1998 large deposits of high quality sapphire were discovered mostly as xenocrysts in alkali basalts in the Limagne Rift; in northern, central, and eastern Madagascar. A combination of size and quality of the gems have made the Madagascar sapphire popular worldwide. Good quality sapphire is also found in the alkali basalts of the Mambilla Plateau of Nigeria and Cameroon, and new discoveries were made recently at Aksum in Ethiopia. Sapphire is a beautiful gem that is in high demand worldwide.

### 5.6.6.4 African Garnets

Africa has become a major source of excellent and beautiful garnets with great color and clarity. Garnets found in Africa include: *red* almandine, *orange* spessartine, *pink-red* rhodonite, *green* tsavorite, *yellow* Mali, and *blue* garnet. They are found mostly within the metamorphosed Precambrian gneisses and pegmatites. The first blue garnets known in the world were found at Bekily, in Madagascar in 1998. It is a garnet that resembles sapphire and has also been found in Kenya.

*Rhodonite* is probably the world's finest and most valued of the garnets because of its reddish-purple color, rarity and luster. *Tsavorite* with its bright green color is sometimes used in place of emerald. Tanzania and Kenya are the leading producers of high quality tsavorite garnet and rhodolite gems. *Spessartine*, the orange garnet is beautiful but scarce. Nigeria is known to contain some of the finest specimens of spessartine from Oyo State, and a little bit of rhodolite, pyrope and almandine garnet. These gem deposits come from colluvial and alluvial deposits derived from mostly granitic pegmatites. Some rhodolite and spessartine also occur in Mozambique. Madagascar is a producer of red almandine garnet. Yellow garnet named "Mali" was first discovered in 1994 in the West African country of Mali which is still its only known source. This rare gemstone has great brilliance and a high dispersion and is in high demand but with a limited supply of just a few crystals at a time. .

### 5.6.6.5 African Paraiba Tourmaline

Tourmaline is a boron aluminium silicate that occurs in a variety of colors (pink, red, green, blue or multicolor). Paraiba tourmaline is a variety that has an electric blue-green colour that is both unusual and stunningly beautiful. The term "Paraiba tourmaline" is generally used as a trade name for this copper-bearing elbaite tourmaline that is found in only three countries in the world (Brazil, Mozambique and Nigeria). Paraiba tourmalines of this color and composition were first reported from Brazil in 1988, and in Nigeria around 2000 and Mozambique in 2005. They are valued for their "glow" that is not seen in other gemstones due to the presence of copper. Mozambique has recently become famous for its high quality Paraiba tourmaline, and most of the world supply now originates from Mozambique rather than the original source in Brazil. Nigeria produces large quantities of Paraiba tourmaline. It is the most popular gemstone in Nigeria and substantial quantities are produced in Oyo, Kaduna and Nasarawa states.

### 5.6.6.6 Tanzanite

Tanzanite is a trade name for gem-quality specimens of the mineral zoisite (epidote group) with a blue color (blue zoisite). It is named after the country of its original discovery, Tanzania where the only known deposit of commercial importance is mined near the base of Mount Kilimanjaro and the city of Arusha in northern Tanzania. Tanzanite is a relatively new gemstone discovered in the 1960s that has become the second most popular and probably the most valued blue gem in the market. Apart from its strong presence, tanzanite also stands apart among other gemstones because it exhibits trichroism - showing three different colors of blue, burgundy and violet when viewed from three different directions.

### 5.6.6.7 Topaz

Topaz is a silicate mineral with a chemical composition of $Al_2SiO_4(F,OH)_22$. As a gemstone it is well known for its hardness (8 on MOhs' scale), which is why it is a popular choice for jewelry. There are different colors of topaz, but the varieties from Africa are typically yellow, golden brown, and varying shades of blue. .Nigeria has good supplies of white and yellow topaz, while other varieties, while the blue variety is more common in Namibia and Zimbabwe.

### 5.6.6.8 Zircon

Zircon occurs in a wide range of colors including colorless, yellow, red and brown. It has a brightness and 'sparkle' that rivals those of diamond, and the colorless version of this gemstone is often used as a substitute for diamonds, South Africa is the leading producer of zircon in Africa, with main deposits located around the Limpopo Belt where it is referred to locally as "Matura diamond". Tanzania also produces good quality zircon of various colors. In Nigeria, zircon in well-formed crystals is common but not in commercial quantity in the tin mining areas of Jos Plateau (Jos-Bukuru, Kigom, Ropp) and in Nasarawa State (Rafin Gabas)), where it is associated with the Younger Granites. Zircon mostly brown in color is found in decomposed alkali basalts near Jemaa where it accompanies sapphire as xenocrysts.

### 5.6.6.9 Other Gem Minerals

Other gem minerals identified or produced in small quantities are, emerald and aquamarine which are both members of the beryl family found in Zambia, Mozambique, and Nigeria.

### 5.6.6.10 Origin of African Gemstones

The origin and sources of gemstones in Africa can be attributed to three geotectonic domains and associated igneous and metamorphic processes.

1. Cratonic shields which are the ancient stable platforms are the areas intruded by kimberlite pipes that are the source of primary and secondary diamond deposits in Southern, Central and Western Africa.
2. The Pan African orogenic belts where metasediments and metavolcanic rocks and reworked high grade gneisses contain corundum, garnets and emeralds. In particular, the Mozambique Belt of Eastern Africa, the Limpopo Belt of Southern Africa, and the Pan-African belts of Western Africa are the best sources of gemstones.
3. Precambrian granitic pegmatites of mostly Pan-African age and contact metasomatism associated with granitic intrusions are good sources of gemstones in Mozambique, Tanzania and Nigeria.
4. Cenozoic to Recent alkali basalts that are connected with rifting and magmatism contain sapphire and zircons that are believed to be xenocrysts from the mantle.

Table 5.5: Producing Countries and Properties of African Gemstones

| | Gemstone | Mineral Species | Color | Hardness (Mohs') | Producing Countries |
|---|---|---|---|---|---|
| 1 | Amethyst | Quartz | Purple | 7 | Madagascar, Namibia, South Africa, Zambia, Nigeria |
| 2 | Aquamarine | Beryl | Faint/Light Blue | 7.5-8 Brittle | Mozambique, Nigeria, Zambia, Madagascar, Malawi |
| 3 | Diamond | Carbon | Colorless | 10 | Botswana, South Africa, Angola, Democratic Republic of the Congo, Namibia, Lesotho, |
| 4 | Emerald | Beryl | Green | 7.5-8 | Zambia, Ethiopia, Nigeria, Madagascar |
| 5 | Ruby | Corundum | Red | 9 | Mozambique, Kenya, Tanzania, Madagascar |
| 6 | Sapphire | Corundum | Blue | 9 | Madagascar, Nigeria, Cameroon, Kenya, Ethiopia |
| 7 | Rhodolite | Garnet | Pink-Red | 6-6.5 | Tanzania, Kenya |
| 8 | Tsavorite | Garnet | Green | 6-6.5 | Tanzania |
| 9 | Mali | Garnet | Yellow | 6-6.5 | Mali |
| 10 | Tanzanite | Zoisite | Blue | 6.5 | Tanzania |
| 11 | Paraiba | Tourmaline | Blue-Green | 7.5 | Mozambique, Nigeria |
| 12 | Topaz | Topaz | Yellow-Brown | 8 | Nigeria, Namibia, Zimbabwe |
| 13 | Zircon | Zircon | White/Brown | 6-7.5 | South Africa, Mozambique, |

## 5.7 Review Questions

1. Briefly describe the types of non-metallic minerals and rocks. In what geologic settings are they commonly found? Give examples.
2. Define industrial minerals and list their main characteristics. Briefly explain their importance in the national economy of developing countries.
3. Attempt a classification of industrial minerals and rocks based on (a) geological setting (b) end use and (c) economics. Give examples.
4. Describe how the physical or chemical characteristics of the following industrial minerals allow for the commodity to be used by society: (a) Barite (b) Diamond (c) Asbestos (d) Ball Clay (e) Feldspar.
5. What are industrial rocks? How are they distinguished from construction materials? Select one industrial rock and describe its uses and origin.
6. Clay minerals are a group of important industrial minerals. Describe the various types of industrial clays, their industrial properties and industrial applications.
7. Describe the characteristics and highlight the differences between ornamental and building stones. Select one example from each group and explain their uses and geological occurrence.
8. Gemstones are valuable substances of beauty when polished and used in jewelry. Briefly describe the unique attributes of gemstones. List seven mineral groups (species) and at least one variety in each group.
9. Describe the geologic settings and origin of gemstones. List some gemstones found in the following African countries: (a) South Africa (b) Nigeria (c) Madagascar
10. Write short notes on the following highlighting their importance as geological raw materials: (a) borates (b) soapstone (c)pumice (d) limestone (e) granite

## References


Bates R.L. (1969): Geology of the industrial minerals and rocks. 2nd edition (translated into Russian). Dover Publications Inc., New York, 459 pp.

Carr o.o. (1994) Industrial minerals and rocks, 6th edition. Soc. of Mining, Metallurgy and Exploration, Littleton, Colorado, 1196 p.

Chang, LLY (2002) Industrial Mineralogy: Materials, Processes and Uses. Upper Saddle River, NJ: Prentice-Hall.

Christidis, GE (2011) Industrial minerals: Significance and important characteristics. European Mineralogical Union Notes in Mineralogy 9(1):1-12

Ciullo, PA (1996). Industrial Minerals and Their Uses. Westwood, NJ: Noyes Publications.

Evans, A. (1993): Ore geology and industrial minerals - an introduction (3rd edition). Blaekwell Scientific Publication, Oxford, 390 p.

Foster, RP. (2016) Mineral resources of Africa, 2008-2016. Society of Economic Geologists, Paper, 8 p.

Giuliani, G. et al. (2014) The geology and genesis of gem corundum deposits. Mineralogical Association of Canada Short Course 44, Tucson AZ, February 2014, p. 29-112

Groat, LA. (ed). (2007) The Geology of Gem Deposits. Mineralogical Association of Canada, Short Course Volume 37.

Groat, LA (2012) Gemstones. American Scientist, 100: 2, March-April 2012.

Harben, PW. And Kuzvart, M. (1997) Industrial minerals - Global geology. 462 pp., Industrial Minerals Information Ltd. (IMIL), Worcester Park, Surrey, UK.

Kogel, JE, Tricedi, NC. et. al. (2006) Industrial minerals and rocks – commodities, markets and prices. Soc. Min. Metall. Explor. Publication, 1353 p.

Kuzvart M. (1984): Industrial minerals and rocks. Elsevier, Amsterdam, 454 pp. Annex 3

Kuzvart, M. (2006) Industrial Minerals and Rocks in the 21st Century

Neukirchen, F. (2020) The World of Mineral Deposits: A Beginner's Guide to Economic Geology, Springer, 353 p.

Prentice J. E. (1990): Geology of construction materials. Chapman and Hall, London, 201

Pohl WL (2011) Economic geology: principles and practice. In: Metals, minerals, coal and hydrocarbons. Blackwell Publishing Ltd., Oxford, 663 p.

Revuelta, MB (2016) Mineral Resources: From Exploration to Sustainability Assessment. Springer Publishers, 641 p.

Simonet, C. Frisch, E. and Lasnier, B. (2008) A classification of gem corundum deposits aimed towards gem exploration. Ore Geology Reviews, 34. 1-2, p. 127-133.

# Chapter 5: Photographs of Industrial Minerals and Rocks

## Industrial Minerals

Plate 5.1: Andalusite

Plate 5.2: Sillimanite

Plate 5.3: Kyanite

Plate 5.4: Chrysotile Asbestos

Plate 5.5: Crocidolite Asbestos

Plate 5.6: Amosite Asbestos

Plate 5.7: Barite

Plate 5.8: Barite Vein

Plate 5.9: Bentonitic Clay Powder

Plate 5.10: Bentonite Mine

Plate 5.11: Borax

Plate 5.12: Borates Mine, California

Plate 5.13: Kaolin Deposit

Plate 5.14: Ball Clay

Plate 5.15: Refractory Clay Bricks

Plate 5.16: Corundum Crystals

Plate 5.17: Feldspar in Pegmatite

Plate 5.18: Fluorspar

Plate 5.19: Graphite

Plate 5.20: Gypsum

Plate 5.21: Muscovite Mica

Plate 5.22: Phlogopite Mica

Plate 5.23: Quartz Crystals

Plate 5.24: Quartz (Silica) Sand

Plate 5.25: Talc

Plate 5.26: Trona

Plate 5.27: Wollastonite

Plate 5. 28: Zeolites

Plate 5.29: Zircon

Plate 5.30: Zircon Sand

## Construction Materials

Plate 5.31: Aggregates

Plate 5.32: Sand and Gravel

Plate 5.33: Sand -Gravel Quarry

Plate 5.34: Granite Slabs

Plate 5.35: Marble Floor & Walls

Plate 5.36: Slate Roofing

Plate 5.37: Gneiss Building Stone

Plate 5.38: Sandstone Blocks

Plate 5.39: Quartzite Building Stone

Plate 5.40: Fired Brick Clay

Plate 5.41: Structural Clay Products

Plate 5.42: Laterite Soil & Stone

Plate 5.43: Laterite Bricks

Plate 5.44: Pumice Rock

Plate 5.45: Perlite Fragments

## Industrial Rocks

Plate 5.46: Diatomite

Plate 5.47: Diatomaceous Earth

Plate 5.48: Diatomite Mining

Plate 5.49: Limestone

Plate 5.50: Limestone Quarry

Plate 5.51: Calcium Oxide (Lime)

Plate 5.52: Phosphate

Plate 5.53: Phosphorite (P-Rock)

Plate 5.54: Phosphorite Mining

Plate 5.55: Nepheline Syenite

Plate 5.56: Nepheline Syenite Mine

Plate 5.57: Glaze of Nepheline Syenite

# Photographs Of Best Known Gemstones (From Geology.Com)

Agate

Alexandrite

Almandine

Amazonite

Amber

Amethyst

Ametrine

Andalusite

Andesine

Andradite

Aquamarine

Aventurine

Benitoite

Beryl

Bloodstone

Carnelian

Cat's Eye

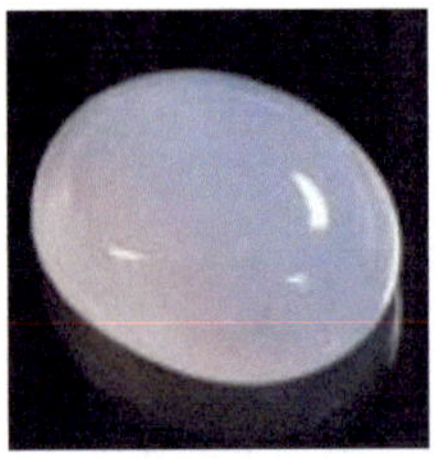
Chalcedony

Chrysoberyl

Chrysocolla

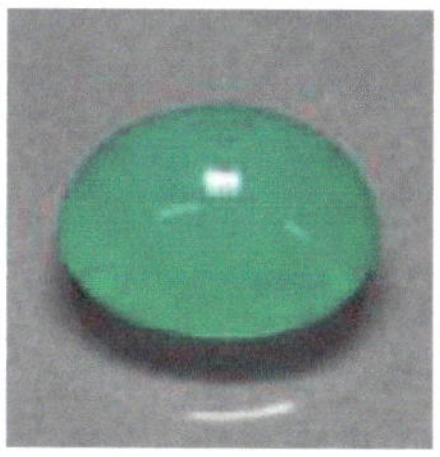
Chrysoprase

Citrine

Copper

Coral

Danburite

Diamond

Diaspore

Diopside

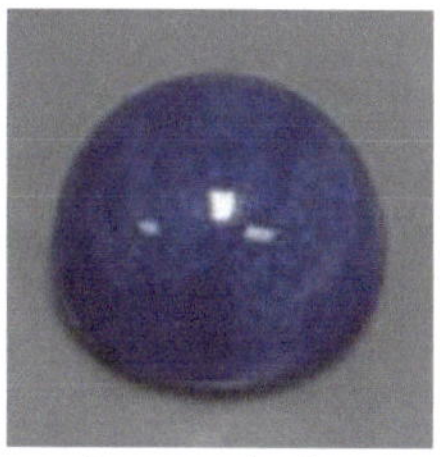
Dumortierite

Emerald

Fluorite

Garnet

Gold

Golden Beryl

Goshenite

Grossular

Hiddenite

Iolite

Jade

Jasper

Kunzite

Kyanite

Labradorite

Lapis Lazuli

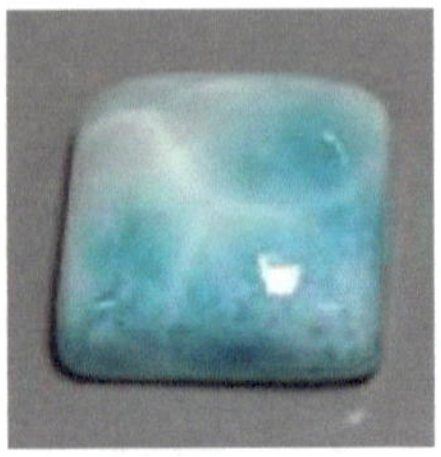
Larimar

Malachite

Moonstone

Morganite

Onyx

Opal

Orthoclase

Pearl

Peridot

Platinum

Prehnite

Pyrite

Pyrope

Quartz

Rhodochrosite

Rhodolite

Rhodonite

Rose Quartz

Ruby

Rutilated Quartz

Sapphire

Scapolite

Silver

Smoky Quartz

Sodalite

Spessartite

Sphene

Spinel

Spodumene

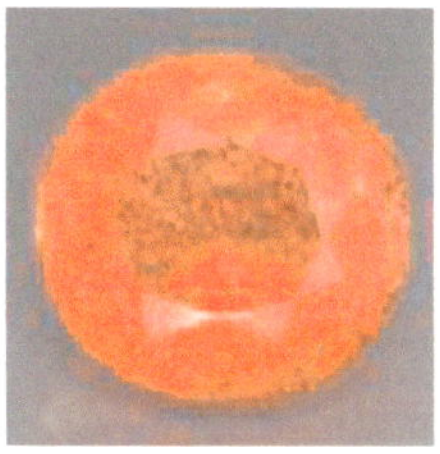
Sunstone

Tanzanite

Thulite

Tiger's Eye

Topaz

Tourmaline

Tsavorite

Turquoise

Uvarovite

Variscite

Zircon

Zoisite

# Chapter 6
# Mineral Exploration

## 6.1 Introduction

**Mineral exploration** is simply the systematic search for mineral deposits, and is essential for ensuring that the world has a stable and sustainable supply of minerals to meet human needs now and in the near future. Mineral exploration has become a very important aspect of economic geology because it provides an organized process for locating new mineral deposits, and also serves as a link between mineral deposits geology and mining (extraction of minerals). Mineral exploration is undertaken by all types of private organizations (companies, partnerships, corporations), and public (government) agencies, and some programs are funded by international organizations, such as the United Nations and the World Bank,

The search for useful minerals is not new. It has been practiced since ancient times when minerals became prominent commodities for the use of humankind. Ancient Mayan Indians in South America and ancient Egyptians in Africa mined copper and gold respectively, and developed methods of finding such deposits based on visual inspection and prospecting. What has changed in modern times is the use of modern technology in aiding the search for mineral deposits. Mineral resources are the foundation of human civilization and are essential to human existence because most of the household and industrial products that we need or desire for the good quality of life are produced from metallic and non-metallic minerals.

Since mineral resources are non-renewable, there is a continuing need to find more minerals to satisfy human consumption. However, the discovery of new minerals has become more difficult in recent times because the easy to find and near-surface minerals have already been discovered. So, for economic geologists involved in the search for mineral deposits, methodology has become more sophisticated, and the use of advanced technology has become necessary to aid in the search for new mineral discoveries.

*Mineral exploration* can be defined in more detail as "an organized and systematic process of searching and finding mineral deposits, and evaluating their economic potential". Another relevant definition is: "the process of finding commercially viable concentrations of minerals for profitable extraction". In practice, mineral exploration often involves a complex and sequential process of information gathering and analysis, and using such information to locate and evaluate potential concentrations of mineral deposits.

*Mineral prospecting* is a term that is sometimes used as synonymous with mineral exploration, although they are not exactly the same. *Mineral prospecting* is a process of conducting a field search for minerals of economic value by examining outcrops and field exposures. It is a small-scale activity by one or few individuals (prospectors) looking for minerals using traditional methods. Prospecting that may be integrated with the reconnaissance phase of a mineral exploration program. Compared to mineral prospecting, mineral exploration is a much more intensive, systematic, organized and professional form of ore search.

In many parts of the world, such as Canada, Norway, Australia, Finland or the United States, prospecting for minerals used to be carried out by "laymen" or non-professionals called "mineral prospectors". They generally don't have significant training in geology, but are usually field-oriented individuals with interest and some knowledge in minerals who go out into the field, using visual techniques to find useful minerals. Sometimes they work with or accompany field geologists as field assistants, (The author had the opportunity of working with prospectors in the search for uranium in northwest Canada). The traditional prospectors searched for valuable minerals by direct observation of mineralization in outcrops, panning of sediments, and soil whereas modern prospectors of today are better equipped by acquiring more knowledge about rocks and mineral identification, and adopting more technology that employs a variety of portable field equipment which can help identify economic minerals, such as the Geiger Counter for radioactive minerals, metal detectors for iron-bearing minerals etc. In developing countries, such as in Africa, there are local people in the villages who have "a good eye" for minerals, and sometimes called "pathfinders". They assist prospective miners in locating minerals such as gold, gemstones, cassiterite in soils and alluvial gravels. A lot of artisanal miners depend on such help in finding extractable minerals.

During the mid-19$^{th}$ century in North America, prospectors were very common, and they were the individuals who made. The State of California renowned for its "gold rush" by finding massive specks of gold through visual scanning and panning stream beds and rock outcrops (Fig. 6.1).

Fig. 6.1: California Gold Rush Prospectors, 1848

## 6.2 Approaches to Mineral Exploration

Mineral exploration is a systematic process that involves sequential activities ranging from searching for a new mineral occurrence, making a discovery and conducting an evaluation of the discovery. Most mineral exploration in developed or advanced countries is undertaken by private companies, whereas in Mineral exploration often begins with the application of a concept or conceptual model in economic geology that helps identify prospective areas worthy of detailed evaluation. For already known or discovered mineral occurrences, mineral exploration is more focused on delineating the extent and quality of the ore deposit, and the feasibility of extracting such minerals at a profit. A mineral exploration project is usually an organized and large-scale operation with the following objectives:

- Identifying what minerals and what areas will be the focus of ore search
- Finding the mineral resource

- Evaluating the ore potential for mining
- Extracting minerals at a profit

There are two main approaches to mineral exploration; 'Greenfield' exploration, and 'Brownfield' exploration.

### 6.2.1 Greenfield Exploration

Greenfield exploration is a grassroot exploration undertaken where there has never been prior work. It involves exploration in the unknown, relying on professional expertise and the use of conceptual ideas of ore genesis to find mineral deposits in previously unexplored areas or in areas where they are not already known to exist. Greenfield exploration projects are considered very risky due to lack of prior information, and some statistics indicate a high rate of failure, with only about 1 in 5,000 greenfield exploration projects ever reaching the mining stage. In many advanced countries, the risk associated with greenfield projects has pushed investors and exploration companies into brownfield projects where risk of failure is reduced.

### 6.2.2 Brownfield Exploration

Brownfield exploration, unlike greenfield exploration, takes place near or adjacent to an already operating mine or known mineral prospects. Mineral deposits tend to form in clusters within metallogenic belts, and exploration typically occurs outward from known mineralization. The availability and use of existing data lower the risk of failure. Consequently, the reduced risk and capital savings in brownfield exploration has made it more attractive to exploration companies, and in recent years became a focus of exploration projects. A good example is the success of Thor Exploration Ltd, a Canadian company involved in the exploration for gold in southwest Nigeria that acquired a brownfield property at Iperindo, near Ilesha. After advanced exploration involving subsurface drilling, the company discovered extensive gold deposits tagged the "Segilola" gold mine, currently the largest gold deposit in Nigeria.

## 6.3 Stages of Mineral Exploration

Mineral exploration is a complex set of activities that involves systematic phases similar to a "life cycle" that starts with developing an idea of what mineral to look for and ending in a mineral discovery, evaluation and mine development. The following stages have been recognized in the mineral exploration cycle (Fig. 6.2):

1. Program Planning and Desk Study
2. Reconnaissance Exploration and Prospecting
3. Intermediate Exploration
4. Advanced Exploration
5. Feasibility Study and Mine Planning

The activities at every stage are summarized in Table 6.1. As a rule, positive results in any stage of exploration will normally result in an advance to the next stage and an increase of the exploration effort and expenditure. Some authors have used a different scheme to describe the stages of the mineral exploration cycle as follows: (1) area selection, (2) target generation, (3) resource evaluation, and (4) reserve definition. This scheme is almost equivalent to the scheme proposed above for the mineral exploration cycle. The difference is just a matter of emphasis, and not of scope. The latter scheme is used commonly by mining engineers, and lacks some of the detailed activities important to economic geologists.

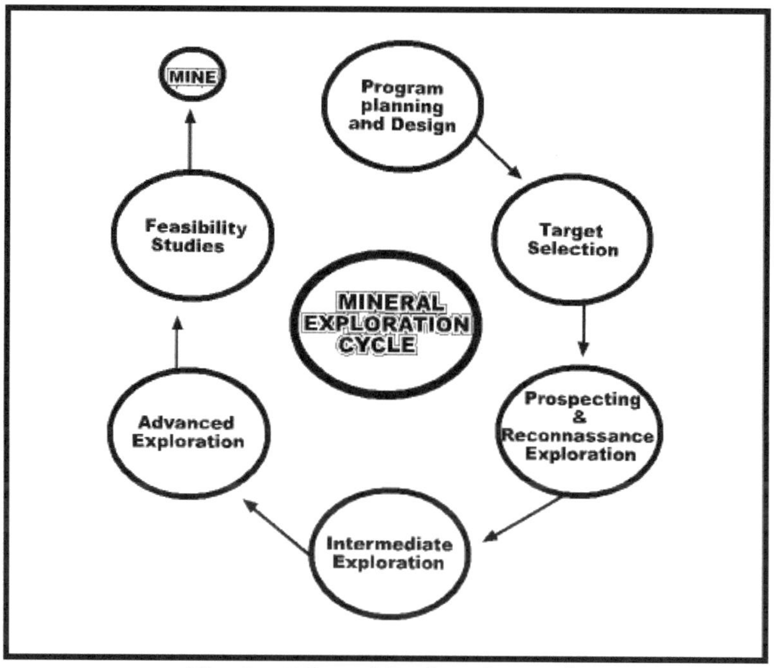

Fig. 6.2: Mineral Exploration Cycle

### 6.3.1 Program Planning and Desk Study

Most mineral exploration in developed or advanced economies are undertaken by private companies or corporations of various sizes, that are sometimes classified into *senior* and *junior* companies depending on the size of their venture capital. Exploration by individual prospectors is still common in countries with large unexplored areas such as Canada and Australia. However in most developing countries and controlled (socialist) economies, most exploration is carried out by state-owned agencies or public companies such as geological surveys or mines departments.

Notwithstanding the nature of company organization, most mineral exploration typically begins with *project planning* which is undertaken by the management and professional staff, along with the gathering and review of publicly available information on various aspects of the proposed exploration project. During the initial planning effort, geological and economic issues of critical importance are decided. Geological issues include identifying what areas will be the focus of the exploration (target area), and selecting mineral(s) and ore deposit type to be targeted. Economic considerations will vary widely depending on the type of mineral deposit sought, the nature of the target area (e.g. remoteness, access, and infrastructure), market conditions and a proposed budget. Most of the time, the basic economic goal of mineral exploration projects is to find ore deposits at minimum costs and extract them at maximum profit.

The choice of what minerals or ore deposit type to look for may be determined by several factors, including market economics (market demand, metal prices) and a company's preferred or strategic interests that are more or less dictated by existing policy. For example, a government-funded exploration agency may decide to explore for iron ores because of the need for raw materials for a proposed iron and steel industry (e.g. Nigeria), or to locate uranium ore for nuclear energy reactors to supply power... The choice of the target area(s) is often left to the technical team of experienced geologists and geophysicists who are knowledgeable and capable of making such decisions, as well as choosing the appropriate exploration methods and the applicable field technologies.

#### 6.3.1.1 Target Minerals

It is pertinent to note that the minerals or deposit types of prime interest in mineral exploration are usually the *metallic* minerals from which one or more metals can be extracted (e.g., gold, iron, copper, nickel). Of lesser importance are the industrial minerals and rocks (e.g., limestone, clays, barite, gypsum, salts and fertilizers), although they have attracted a lot of attention in recent years, particularly in developing countries where they are mined to their industrial applications. For example, there has been increasing need to explore for easily accessible limestone deposits in Nigeria for the manufacture of cement which is in high demand in a booming construction industry. Also, gemstones, such as diamonds, rubies, sapphires and opals, are a unique class of non-metallic minerals that are sought after because of their value and beauty. Thus, apart from economic factors such as market demand and commodity prices, there are special circumstances in which certain minerals are targeted for exploration because they belong to a group of minerals known as *'strategic'* or *'critical'* minerals that are prioritized for exploration due to their relevance in national and international development.

Table 6.1: Stages and Summary of Activities in Mineral Exploration

| | Exploration Stage | Activities and Techniques |
|---|---|---|
| 1 | **Planning** | Database search<br>Economic modelling<br>Economic forecasting<br>Exploration model review<br>Target mineral/metal selection<br>Target area selection<br>Field Activities planning |
| 2 | **Desk study**<br>*(Target Selection)* | Computer database research<br>Literature search and review<br>Remote sensing analysis<br>Review of geology and mineral deposits reposts<br>Review of topographic and geology maps<br>Compilation of data and reports<br>Land acquisition / Exploration permit |
| 3 | **Reconnaissance Exploration and Prospecting**<br>*(Target Generation)* | Photogeology<br>Remote sensing<br>Regional geological mapping<br>Aerial geophysical surveys<br>Stream sediment geochemical survey<br>Prospecting |
| 4 | **Intermediate Exploration**<br>*(Target Appraisal)* | Detailed geological mapping<br>Photogeology<br>Rock, and Soil geochemical surveys<br>Stream sediment surveys (small scale)<br>Ground magnetic, electromagnetic, electrical surveys<br>Ground radiometric survey |
| 5 | **Advanced Exploration**<br>*(Target Testing)* | Drilling<br>Pitting and Trenching<br>Core logging<br>Sampling and Assaying<br>Geostatistics and Geological model studies<br>Mineral resource evaluation<br>Ore reserve definition |
| 6 | **Feasibility Study**<br>*(Target Evaluation)* | Scoping Study<br>Pre-Feasibility report<br>Full Feasibility report |
| 7 | **Mining**<br>*(Target Extraction)* | Mine design<br>Mine development<br>Mine production |

*Strategic minerals* are those minerals and rocks that are of vital importance to a country's national defense, energy and industry sectors, and are prioritized for exploration and development. For example, the United States has identified certain minerals that are used in military equipment, jet engines, nuclear plants and space research as strategic minerals, e.g., manganese, titanium, zirconium and yttrium. Nigeria has identified seven mineral commodities and rocks as strategic minerals; iron ore, lead-zinc, gold, coal, bitumen, barite and limestone.

*Critical minerals* are those minerals for which a threat to supply could create harm to the economy. These are minerals that occur only in small quantities and the supply chain is considered fragile. For example, the United States has essentially no reserves of rare earths, columbium (niobium), strontium, manganese, tantalum, or cobalt; sheet mica; or bauxite ore. To the extent that these minerals are important in various industrial processes, they are regarded as critical and are given priority in geological investigations and exploration.

### 6.3.1.2 Target Area Selection

One of the primary objectives of the planning phase is to identify exploration target areas for the reconnaissance phase. Target area selection (or targeting) is the process of selecting the area(s) with a good prospect of containing mineral deposits of interest. This is a critical step in mineral exploration decision making, because it will obviously impact the ability of the exploration program to succeed and find ore deposits easily and quickly. Several factors affect the choice of target areas, most of which are *geological*, but some are *economic* (e.g. taxation, trade practices, stability), *political* (e.g. instability, leadership quality), and *social* (e.g. insecurity, unrest, civil wars). Other minor but related concerns are metal prices which may be too low or fluctuate widely, and the issue of remoteness and lack of infrastructures (e.g. power, transportation). Most of the above considerations are straightforward business-related decisions that are usually made by companies or agencies in accordance with their philosophy and policy.

#### *Geological Factors in Target Area Selection*

*Geological factors* are of utmost importance in selecting target areas for exploration. They are affected by the company's approach to exploration and exploration strategies. First of all, a choice may be made in the selection of either the greenfield or brownfield approach. In *brownfield* exploration, target selection is relatively easier because the potential targets will be located in areas near or adjacent to known ore deposit or an already operating mine. Such targets are able to use existing information to make less risky decisions on the area for detailed exploration. In contrast, greenfield exploration starts from the grassroots with no previous data to rely on. It depends on the potential predictive power of conceptual ore genetic models to select target areas in previously unexplored territory or in areas where deposits are not already known to exist. Some statistics indicate that only 1 in 5,000 to 1 in 10,000 grassroot exploration projects ever reach the production stage. Also, in most cases, the choice of what commodity (e.g. gold or copper) and ore deposit type will dictate the area selection strategy. For example if the chosen metal is gold and the ore deposit type is "orogenic gold", the area selection will focus on orogenic belts in Precambrian greenstone terrains.

Several mineral exploration companies are multinational corporations that explore for minerals worldwide as a result of investor interests. Many such companies from advanced countries such as Australia, Canada and the United States are exploring for minerals all over the world, particularly in mineral-rich developing countries of Africa, South America and Asia. The global exploration budget for nonferrous metals was an estimated US$9.8 billion in 2019 of which Africa took the second largest share. Exploration has become a global enterprise, and exploration targeting has become a matter of scale, ranging from global to regional, district and local levels. In the process of selecting an area with the best mineralization potential, economic geologists have developed models for exploration targeting based on the application of some concepts of economic geology, in a process so-called "*conceptual targeting.*" What is a model? A *model* is defined as a "functional idealization of a real world situation used to aid in the analysis of a problem." These conceptual models include two types which are often combined: *genetic* and *deposit* models which are based respectively on (a) theories of ore forming processes, and (b) the characteristics and geological controls of ore deposit type. These models may be integrated with other datasets derived from geophysics, geochemistry, and spatial analysis.

In addition to conceptual targeting, another approach to area selection is called "*empirical targeting*" which uses datasets from regional exploration methods that involve widespread systematic collection of geophysical or geochemical measurements and typically result in the production of large numbers of anomalies. Since little will be known about any of these anomalies other than the fact of their existence, but any one anomaly could reflect an ore body and must be regarded as a prospect to be followed up with a preliminary assessment. This exploration strategy is a bit speculative and unpredictable, but can yield surprise discoveries of world class deposits. The required data sets derived from regional exploration may not be easily available in developing countries, and this approach will not be further discussed. Conceptual targeting will be addressed in more detail below.

### *Conceptual Targeting*

*Conceptual targeting* focuses on understanding the processes controlling the distribution of the commodity of interest and predicting how and where these processes would combine to create an economic deposit within an Earth system. Conceptual targeting is applicable without the requirement of extensive data on known mineralization, and can deal with incomplete or partial datasets. The application of target parameters generally depends on the scale of the target areas, ranging from global through regional (province) to district and local or deposit scale (Fig. 6.3). What is notable is that in the final analysis, the mineral deposit which is the ultimate target of the exploration effort lies within a relatively small locality within a designated 'local' area where the final stage of mineral exploration will be carried out to define the ore deposit. Table 6.2is a summary of some of the geological parameters that can be applied in targeting at various scales. These parameters involve geodynamic, tectonic and litho-structural concepts that evolve progressively from global to local scale targeting. The different targeting parameters are important at different scales and should be used appropriately to scale of the target area.

An important constraint on the quality of exploration targeting decisions is the availability of high-quality geoscientific data, particularly reliable regional geological maps, and high-resolution geophysical and remotely sensed datasets at the district to local scale. In developed countries, such as Australia, Canada and the USA, targeting is more effective than in developing countries, particularly in Africa and parts of Asia, where data availability is poor or are completely absent. It is definitely more expensive to generate regional geophysical maps for exploration purposes.

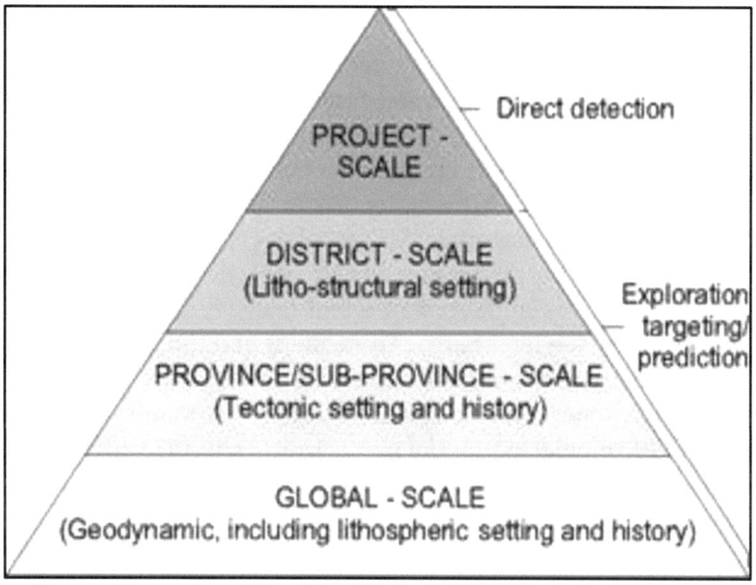


Fig. 6.3: Hierarchy of Scale-dependent Targeting Processes and Parameters (after Hronsky and Grove, 2008)

Table 6.2: Some Targeting Parameters According to Scale of Target Areas

| **Target Scale** | **Target Characteristics** | **Potential Target Parameters** |
|---|---|---|
| *Global Scale Targeting* | Global regions favorable to some ore deposits; Major sedimentary basins; Precambrian cratonic shields; Volcanic belts; Mountain belts | Major deep-seated structures that are probably trans-lithospheric as well as trans-crustal in scale, as many also lie on, or near, lithospheric boundaries or sutures |
| *Regional Scale Targeting* | What basin/belt/arc/district has the potential to host a substantial mineral system? Activity at this scale is relatively inexpensive and tends to rely on framework geoscience information. | Metallogenic belts, Orogenic belts, Plate margins and Subduction environments. Intracratonic rift basins. Deep-seated lineaments, Major tectonic breaks between lithostratigraphic groups. Crustal-scale shear zones, dominant greenschist-facies domains and complex lithostratigraphic sequences, |
| *District Scale Targeting* | Camp/Cluster-scale targeting – where within the basin/belt/arc/district could a number of deposits be clustered? At this scale exploration starts to deploy expensive detection technology. | Lithostratigraphic, metamorphic and igneous controls. Greenstone belts. Synorogenic granites. Major fault systems. Mineral zoning. Contacts between sedimentary and volcanic sequences. |
| *Deposit Scale Targeting* | Ore body of sufficient quality within the camp or cluster of deposits? This scale can be costly and time consuming | Structural controls - faults, folds, shear zones, fractures. Quartz veining. Lithologic association and control. Lineaments, Hydrothermal alteration zones. Geophysical and Geochemical parameters. |

### 6.3.1.3 Desk Study

The next stage after target area selection is to conduct a desk study. This is a process of carrying out research and acquiring information on the selected target area(s). Besides background information on geology, other relevant data on the occurrence of producing or closed mines and prospects in the area (if available) are essential. This information will normally be obtained through extensive literature review of published information, including scientific articles, reports by government agencies e.g. Geological Survey and Mines Department, and research materials (dissertations, theses, publications) on existing mines, mineral occurrences and geological characteristics of the potential target area(s). Other items such as topographic maps, geological maps, aerial photographs and Landsat imagery should be reviewed and used in developing a good knowledge of the target area. These maps and imagery may not be easily available in many developing countries where geoscientific information is inadequate.

In Nigeria, for example, topographic maps on a scale of 1:100,000 or 1:25,000 are available, but may be old and date back to the colonial era. Aerial photographic coverage and Landsat Imagery are rare or difficult to obtain. Other relevant information such as aerial geophysical (magnetic, radiometrics) and geochemical maps (if available) should be examined so as to shed more light on prospective areas and help in developing a comprehensive strategy for reconnaissance exploration. At the end of the desk study and planning, the geological staff will recommend the appropriate methodologies for field exploration to the management for approval, and a time schedule and general budget are established for the project.

As shown in Fig 6.4, a schematic diagram of the exploration pyramid, portrays how after the initial desk study, each subsequent stage of exploration becomes less and less likely to happen because project outcomes are evaluated at the end of each stage, and many are terminated due to lack of viability.

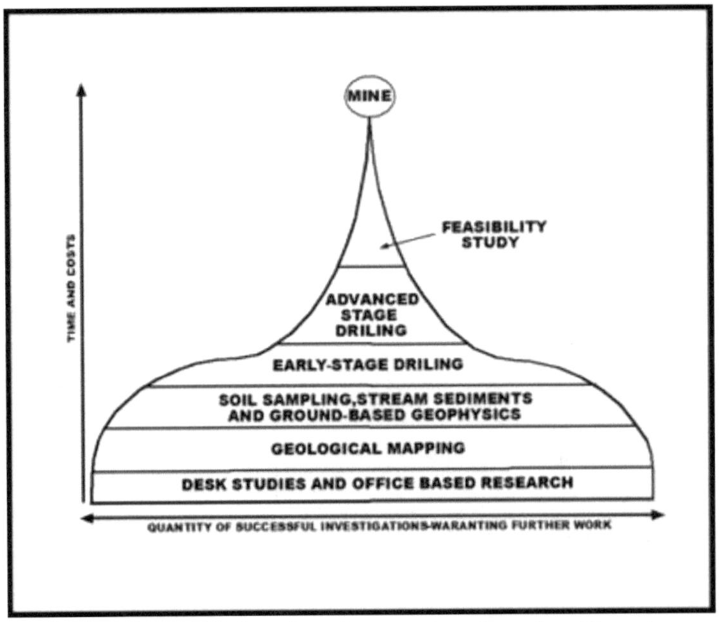


Fig. 6.4: Pyramid showing time, cost and exploration stage

## 6.3.2 Reconnaissance Exploration and Prospecting

After the initial process of area selection and desk study is complete, the reconnaissance stage is initiated with the main purpose of evaluating the target areas of interest rapidly, and identifying potential anomalies that may become the focus of further investigation. Finding potential targets or anomalies in the reconnaissance stage often involves systematic geological investigations through field work and remote sensing analysis; aerial geophysical surveys, and if applicable, conducting wide-spaced (regional) geochemical surveys using stream sediment sampling which allows coverage of a large catchment area.

Reconnaissance exploration aims at rapid and low-cost sorting out of prospective parts of an area. It may also incorporate "mineral prospecting" which involves "putting the boots on the ground" by traversing the target area searching for valuable minerals, primarily by direct observation of outcrops, stream sediments, and soils. In areas of reasonable outcrop the first stage will be to check the geology by driving along roads or from the air if ground access is impossible. Apart from the geological hammer, portable geophysical equipment such as the Geiger Counter for radiation detectors (for uranium) and metal detectors for magnetic ores are used in spot testing for mineral showings. For heavy minerals such as gold, tantalite and corundum, panning of stream beds may apply (Fig. 6.5).

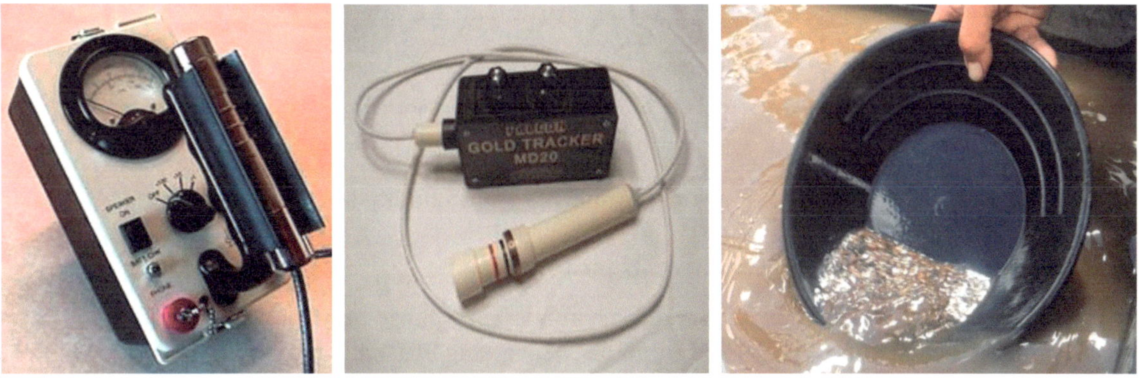


Fig. 6.5: (a) Geiger Counter (b) Metal Detector (c) Panning for Gold

### 6.3.2.1 Remote Sensing Analysis

Remote sensing involves gathering information about the Earth's surface by measuring electromagnetic radiation, and field signals that emanate from objects. Remote sensing allows us to collect information about an object on the Earth's surface without making physical contact with that object, and allows data acquisition before extensive field observation. Using aerial photographs and satellite and spectral imagery (e.g. Landsat), remote sensing is *a* sophisticated but valuable tool that can be used during reconnaissance exploration to "look down" on potential geological and structural features or targets that may indicate potential mineralization.

Remote sensing images are used for reconnaissance mineral exploration in two key ways: (1) mapping and analysis of the geological and structural features (e.g. Faults and fractures) that may be associated with an ore deposit. (2) recognizing hydrothermally altered rocks by their spectral signature. It is therefore plausible that mineralized zones exposed on the earth's surface may be detected by identified anomalous alteration zones or structural features that localize mineralization that appear as lineaments on satellite imagery (Fig. 6.6).

Applications of remote sensing can also provide value by reducing the risk of a project and helping prioritize which sites to explore first. Expensive operations like drilling and intensive fieldwork can come after reliable information is gathered. The benefits of using remote sensing in exploration of remote areas where accessibility is minimal cannot be overemphasized, particularly in many parts of tropical Africa with dense vegetation and no accessible roads. Also, remote sensing data can be used to prepare preliminary geological maps that identify main lithologies.

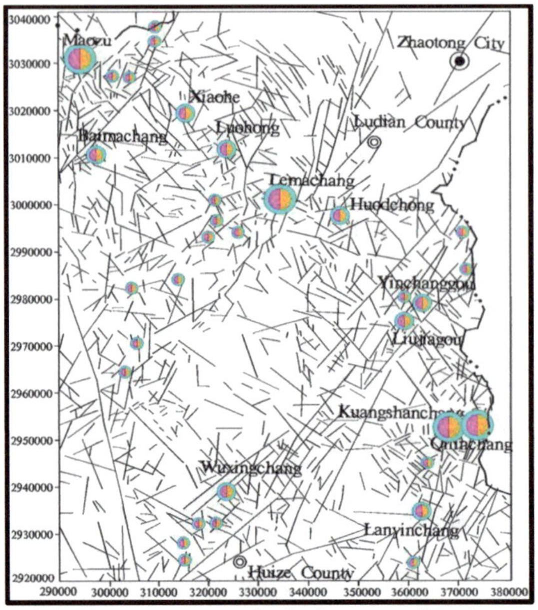

Fig. 6.6: Lineaments analysis from satellite imagery

### 6.3.2.2 Reconnaissance Geological Investigations

Systematic geological investigations during the reconnaissance phase may include some outcrop mapping, outcrop sampling, and testing of rocks and minerals. Traversing is usually undertaken by foot, field vehicles or by helicopter in areas that are not easily accessible. The preliminary field work helps to provide basic knowledge of the rocks and familiarization with the target area. It may also be carried out on a broader scale to follow up objects of interest that were identified from remote sensing analysis or aerial photographs, such as lineament traces, zones of alteration and depth of weathering. Existing geological maps provided by government geological surveys at a scale of 1:25,000 or 1:100,000 can be useful in reconnaissance geological investigation.

Rock and mineral samples collected during field visits to outcrops may be sent to the laboratory for thin sectioning and preliminary petrographic and ore microscopic studies. Systematic sampling of fresh rock samples can be submitted for geochemical analysis as part of a lithogeochemical survey. Mineral prospecting should be combined with geological activities.

### 6.3.2.3 Aerial Geophysical Surveys

Airborne geophysical surveys are the most effective and widespread method of gathering reconnaissance data at this stage of exploration. Low flying winged aircraft, helicopters and drones are used to carry geophysical equipment to collect data on the magnetic, electromagnetic, and radiometric properties of rocks that can generate anomaly maps. Airborne geophysical survey is an obvious choice for rapidly scanning regional target areas in a cost-effective manner during initial exploration particularly in remote terrains. Several ore deposit types have unique geophysical attributes that make them detectable from airspace.

The techniques which find application from an airborne platform include magnetic, radiometric, electromagnetic and infrared. Gravity survey application is currently under advanced development and will be available for commercial use in the future. Depending on the metal or mineral being explored, each method has its own advantages because each method measures different physical properties, and a combination of methods may be diagnostic of mineral deposit associations. The real outcome from these surveys are broad anomalies that could be attributed to several geological factors that could be filtered to identify potential exploration targets. Such targets can be followed up by ground-based surveys.

In addition to detecting potential anomalies, airborne methods are useful in other exploration scenarios such as helping to differentiate various rock types, major faults/folds and many other geological features, which cannot be detected by conventional geological mapping due to extensive soil cover. Airborne geophysics can be very expensive, and the choice of which geophysical tool will be used dependent on many factors, including the size of the prospect area, local geology, the physical characteristics of the target, and the surrounding rock types, the size and orientation of the target, location and access to the survey area, terrain features, weather conditions and the exploration budget, etc. Also, data resolution will be controlled by the height above the terrain, line spacing, flight speed, and sampling rate. Different survey techniques can be used for different types of mineral deposits as summarized below:

- ***Aerial Electromagnetic*** - most suited for conductive ore bodies, including massive sulfide deposits, graphitic horizons, hydrothermal alteration zones because of clay content; some lithologic types; basement-sedimentary rock contacts; and major fault zones
- ***Aerial Magnetic*** - most suited for magnetic bodies; iron ores, manganese ore bands; basement/sediment boundary; ultramafic rocks
- ***Aerial Radiometric*** - Uranium/thorium deposits; potassium alteration zones and granitic intrusives; lithologic mapping.
- ***Aerial Infrared*** - surface thermal patterns; geothermal resources; fault zones with thermal water, hot-spring related deposits (sulfur, mercury, silver etc.)

Airborne magnetic, radiometric, and electromagnetic datasets and maps that cover large areas are like geological maps, available in most advanced countries, but scarce or non-existent in most developing countries of Africa and Asia. Lack of such useful geoscientific information may increase the cost of exploration projects. For example in Nigeria, the Geological Survey Agency is trying to cover large tracts of the country with aerial radiometric, electromagnetic and magnetic survey maps to encourage explanation by outside investors.

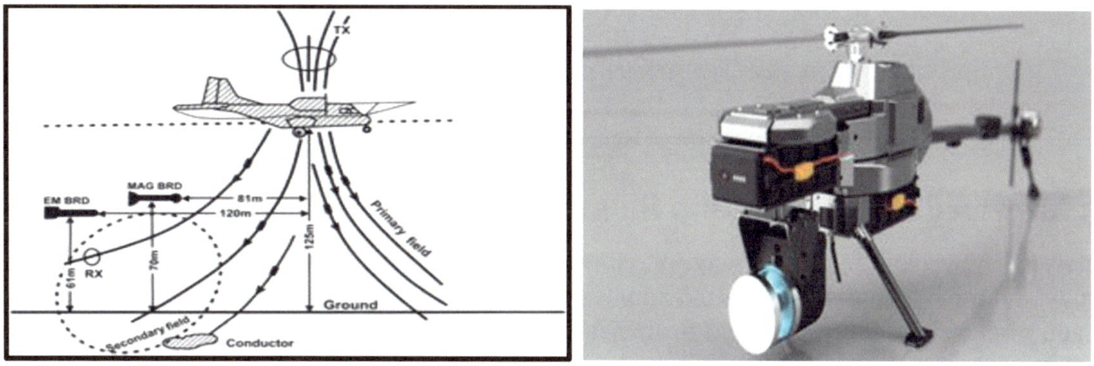


Fig. 6.7: (a) Aerial geophysical detection of orebody (b) Drone (UAV) Geophysical Surveyor

### 6.3.2.4 Reconnaissance Geochemical Surveys

Geochemical surveys using widely-spaced stream sediments are commonly used in reconnaissance exploration particularly in areas with adequate drainage systems and annual precipitation that allow stream flow. Stream sediments represent the best possible composite of weathered and primary rocks over a large area. They are composed of clastic and hydromorphic materials that include clays, detrital fine-grained rock and mineral particles, organic matter, and iron and manganese coatings on clasts derived from upstream catchment areas. If mineralized areas exist within the catchment area, it is likely that some components are incorporated in the sediments. Sampling and chemical analysis of the clay/silt fraction (minus 75-80 mesh) may show secondary dispersion of ore and pathfinder elements, and geochemical anomalies following the "path" of increasing values upstream may lead to mineralization in the catchment area of drainage basins. Low sampling density such as one sample per 10 or more square kilometers are considered appropriate for reconnaissance exploration. The outcome of the reconnaissance stage may lead to a follow-up and detailed geochemical survey (Fig. 6.8).

Rock geochemical survey on a reconnaissance level has been successful in identifying mineralization potential of rocks particularly in temperate regions where fresh rock exposures are abundant. Examples include regional sampling in Nigeria of alkali granites in order to locate intrusions that may contain tin-niobium deposits, and identification of rare metal pegmatites in the Older Granite suite with ore-bearing potential.

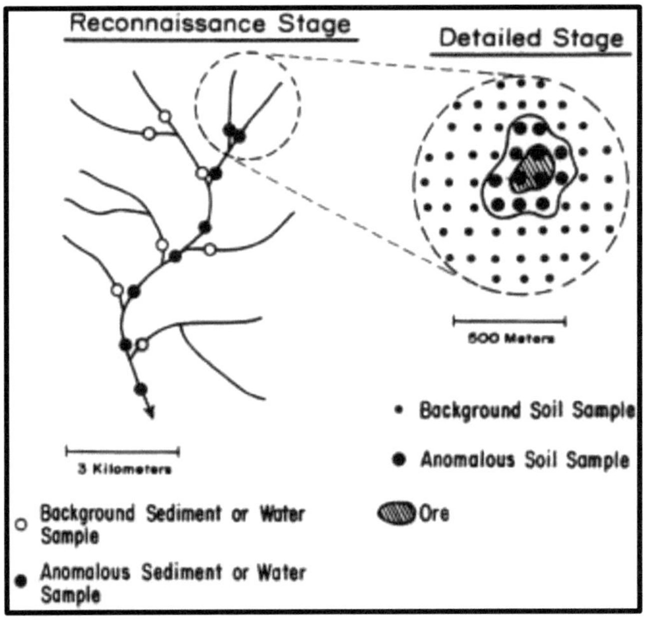


Fig. 6.8: Diagram of Reconnaissance Geochemical Survey

### 6.3.3 Intermediate Exploration Stage

The intermediate exploration phase may sometimes be combined with the reconnaissance stage to permit continuity and rapid follow-up of detected anomalies in the field. However, this stage is usually focused on target appraisal to help investigate and further define any anomalies obtained during the reconnaissance stage. Some mineral showings identified by prospecting are further investigated before embarking on expensive drilling programs. The intermediate exploration process may generally begin with establishing a regular grid on interesting areas serving as a base for further investigations involving more detailed ground geophysical and geochemical surveys using a variety of techniques as well as systematic and detailed geological mapping undertaken typically at 1:5,000 to 1:500 scales. Mineralogical studies are also involved, as well as pitting or trenching to expose mineralization beneath overburden.

#### 6.3.3.1 Geological Mapping

Geological maps are graphical representations of geological observations and interpretations on a plane surface. Geological mapping is an essential and valuable tool in mineral exploration, and has been in use for hundreds of years. While geological maps published by government Geological Surveys are sometimes available, they are usually at a "small" scale of about 1:25000 to 1:100,000 which is not sufficiently detailed to help delineate mineral deposits that are typically 1 to 2 km2 in outcrop area even for world-class deposits. Consequently, exploration projects must conduct the geological mapping at this stage at a more detailed and larger scale ranging from about 1:5000 to 1:500.

The objectives are to provide, observe and record information on major lithological units, age relationships, alteration zones, structural features, and mineralization types present on or around an exploration property, along with the nature and location of major physiographic features (Fig. 6.9). Geological mapping should also provide useful information on mineral deposits, their morphology and other characteristics, including mineralogy of ores and hydrothermal alteration. The information is collected and stored on paper and/or in digital format. In addition, geological maps are important control documents for all subsequent stages of detailed exploration such as geophysics, rock sampling and geochemistry, geostatistics, drilling and mining. They continue to be updated as more geological information becomes available.

The main equipment used in geological mapping include a base map, global positioning system (GPS) unit, compass, geological hammer and chisel, hand lens, stationery and field notebooks. Beyond the use of traditional paper-based mapping tools, recent technological advances incorporate global positioning systems, pen tablet computers, and laser ranging devices that all support direct (paperless) field-based digital geological mapping. In this sense, geographic information systems (GIS) has revolutionized exploration practices by providing tools for digital display and storage of information. If mineralized zones are found during the initial exploration, petrographic and mineralogical studies of the host rocks, ore minerals of and alteration minerals, will be helpful in understanding the nature of mineralization.

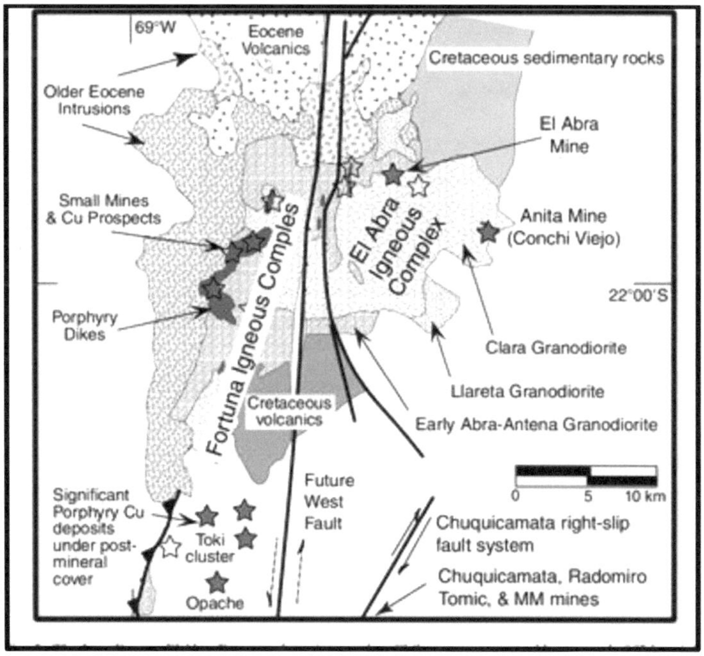


Fig. 6.9: Example of Geological Map for Mineral Exploration

### 6.3.3.2 Ground Geophysical Surveys

Geophysical techniques are diverse, but they all measure the physical properties of rocks or mineralized zones, including gravity, magnetism, electromagnetism, radioactivity, and many other different variables that help in identifying anomalies related to mineralization. They are most suited to searching for buried and deep targets since there are few large ore bodies, and play a large role in generating data which are widely used in mineral exploration. . Certain geophysical methods can detect buried anomalous orebodies 500 meters below the surface. At the intermediate stage of exploration, only ground geophysical surveys are most commonly used, or aerial surveys at very close intervals. . They are designed to detect conductive minerals such as sulfides or magnetic and radioactive minerals within a variety of host rocks. Ground-penetrating radar (GPR) is becoming a widely used technique in exploring for shallow residual deposits such as bauxite and clays that occur near the surface. As shown in Fig. 6.10, some buried ore deposits can be detected by multiple geophysical techniques.

When developing and designing a geophysical survey, the geoscientist needs a clear understanding of the geophysical characteristics of the type of mineralization the exploration program is seeking, and its possible location and orientation. This understanding must govern the choice of geophysical survey method and the design of the survey specifications. A qualified geophysicist should plan, supervise, and interpret the geophysical survey

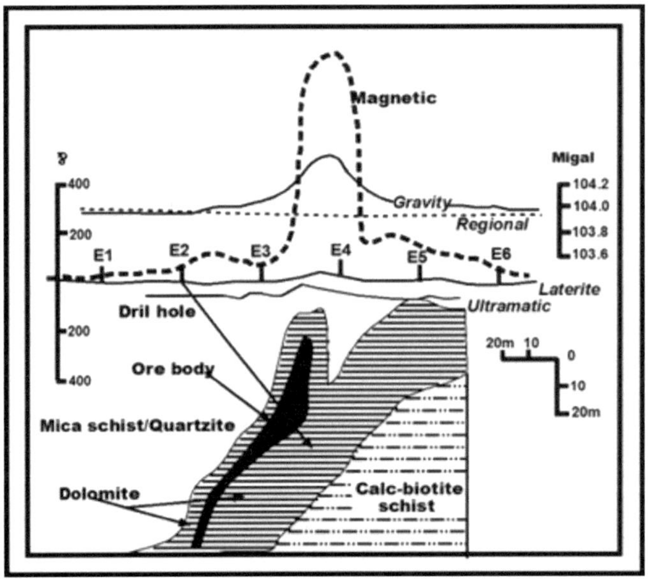


Fig. 6.10: Geophysical anomalies from buried orebody

The following is a summary of the characteristics and type of deposits that can be detected by geophysical techniques.

- ***Magnetic*: -** Testing for differences in magnetic susceptibility and indicates the amount of magnetite/pyrrhotite in a given rock. Typically useful for finding iron ores, volcanogenic massive sulfides, manganese ores, titano-magnetite and kimberlites.
- ***Electromagnetic* (EM): -** Measures subsurface conductivity contrasts. Generally used to find deposits hosting conductive sulfide minerals of interest (e.g. Cu, Ni), and shallow buried ferrous and non-ferrous metallic bodies.

- ***Induced Polarization*** **(IP):** - Measures the response of the ground after charging with an electrical current. Can be useful in identifying certain porphyry copper, volcanogenic massive sulfide (VMS) and precious metal vein deposits.
- ***Radiometric*****:** - Measures natural radiation or radioactivity emitted by rocks on the surface. It is used to detect uranium, thorium or potassium that can lead to the discovery of uranium, iron oxide -copper-gold (IOCG), porphyry copper, VMS and epithermal gold deposits.
- ***Gravity*****: -** Measures density contrasts in subsurface rocks. It is used in identifying lithologies and structural features that may be linked to mineralization, and sometimes detecting orebodies themselves.

### 6.3.3.3 Detailed Geochemical Surveys

Geochemical methods of mineral exploration are based on the systematic measurement of chemical elements or parameters that can be used to identify anomalies related to mineralization. Several types of materials or sampling media are used including stream sediments, soils, rocks, and sometimes water and vegetation. However, not all these materials are available for sampling, so the decision to use detailed geochemical survey is dictated by the environment. For example, in tropical and sub-tropical environments, rock geochemistry may not be practicable because of intense weathering while soils and stream sediments are ubiquitous. In desert environments, where surface flow is minimal, stream sediments are not available for sampling.

Understanding the different survey methodology options, including their limitations, is important for choosing an appropriate sample medium, sample spacing, preparation protocol, and analytical method. A useful approach at this early stage is to conduct an orientation survey over an area of known mineralization in a similar area and under conditions similar to those anticipated in the target survey area. The goal of an orientation survey is to test different media and methodologies in order to identify which set of survey parameters offers the highest probability of detecting the target mineralization type under the conditions present in the proposed search area.

#### *Detailed Stream Sediment Geochemical Survey*

*Stream sediment* geochemical surveys are applicable wherever active erosion, surface drainage and precipitation are adequate to allow stream flow. The geochemical analysis of stream sediments collected at moderate to high sample density, such as one sample per 1km2 can be used to identify geochemical anomalies as a follow-up of reconnaissance sampling to further define areas with potential mineralization (see Fig. 6.8). Detailed stream sediment survey is particularly useful in detecting metals that are soluble and mobile and generate extensive secondary dispersion patterns. Such metals include Cu, Pb, Zn, Ni, and pathfinder elements such as As, Cd, Some metals such as uranium that are extremely soluble may remain dissolved in water and not precipitate in stream sediments due to oxidizing and acidic conditions. Resistant minerals that occur as heavy mineral particles in stream sediments, such as Au, Cr, and Pt may remain in clastic form with short distance dispersion and may not show enhanced values in stream sediment surveys. In stream sediment sampling, care must be taken/silt to ensure that the samples are representative and handled to avoid contamination. The clay fraction of samples is the best material for chemical analysis and can be obtained by sieving during sample preparation.

#### *Soil Geochemical Survey*

Soil geochemical *surveys* are based on the chemical analysis of soil samples usually collected from the top one foot of the soil profile. Soil sample geochemistry is a powerful tool in the exploration of anomalies particularly as a follow up of those found by stream sediment investigations (Fig. 6.11). The method works very well since weathering and leaching of buried deposits can discharge anomalous concentrations of trace elements to soil and groundwater. The released heavy metals spread outward and originate a dispersion halo in the soil that is much larger than the mineral deposit itself. As a rule, the dispersion halos in soils are smaller than those in stream sediments but larger than those in primary halos in rocks. Similar to stream sediment sampling, samples collected are commonly the fine silty-clay fraction that may contain some organic matter.

Soils are composed of several horizons produced by chemical weathering, and some of these horizons may accumulate metals derived from underlying mineralization. The top horizon is usually preferred for sampling because it is a zone of metal enrichment while the deeper horizons are often associated with leaching. Soil samples are usually collected in a grid at specific points on a rectangular pattern along sampling lines with constant intervals which may depend on the purpose of the survey and the expected size of the dispersion halo to be detected For detailed anomaly investigation, soil samples are typically taken at 100 m intervals or less, and are usually collected at a depth of less than a foot below the surface, using a trowel, spade or a hand auger. Equipment should be cleaned between samples to avoid cross contamination.

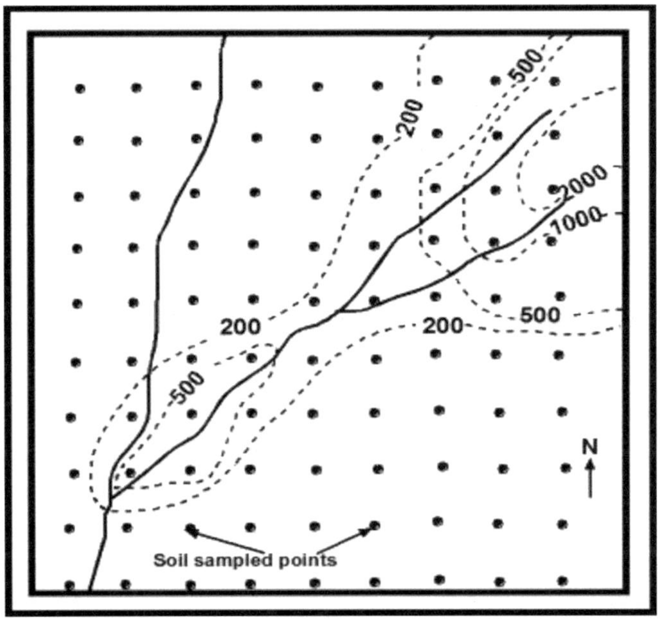


Fig. 6.11: Zinc Distribution in Follow-up Soil Geochemical Survey

Soil samples are usually analyzed using sensitive instrumental techniques with low detection limits and high precision which can help anomalous concentrations that represent the surface "footprint" of the ore body. Significant anomalies may be detected of the target metal and elements that may be associated with the type of mineralization being explored. As shown in Figure 6.11, soil samples collected at fixed intervals of 100 m are used to define anomalous zones associated with a stream sediment anomaly.

### *Rock Geochemical Survey*

Rock geochemical surveys (lithogeochemistry) involve the collection of rock-chip samples from outcrops for chemical analysis to determine if they contain elevated concentrations of elements that may indicate proximity to mineralization. Rock geochemical surveys search for the primary dispersion halo around mineral deposits, but because this type of halos is restricted to a small area immediately surrounding any prospective min-eral deposit, rock surveys are mainly applied to evaluate specific targets outlined by regional surveys. Such surveys are only practicable where there are fresh exposures of rocks around the potential mineralized zone. As shown in Fig. 6.12, the sampling of sandstone host rocks for copper across a mineralized zone shows distinct anomalies across fault zones that contain the hydrothermal deposit. Such definition can be very helpful in targeting areas for detailed exploration by drilling.

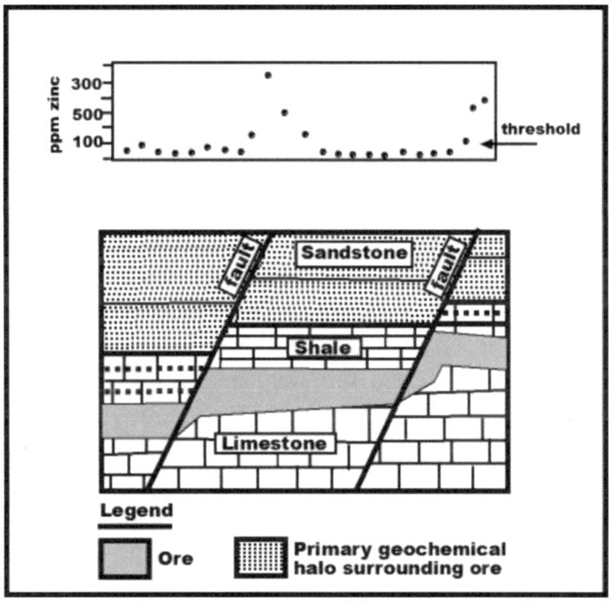


Fig. 6.12: Rock Geochemical Survey Profile

### 6.3.3.4 Pitting and Trenching

After anomalies have been confirmed by detailed follow up surveys, it is often necessary to probe the target area further if under soil cover at shallow depth by excavation, pitting or trenching, and sometimes by shallow drilling to access the potential orebody if near the surface. Pitting is best applicable where the mineralized zone is flat lying, such as placer deposits or sedimentary deposit type whereas trenching is preferred for zones or rock bodies that are elongated, such as structurally-controlled mineralization (e.g. veins, dykes). Stripping the overburden and trenching or pitting are activities that can involve heavy machinery such as a bulldozer or backhoe to remove the overlying soils, and may require the use of explosives to blast a trench in the subsurface rock. Hand-dug pits and trenches may be used in places where power excavation equipment cannot be brought to a remote locations,

Despite their relatively shallow depth, pitting and trenching provide several benefits in creating better exposure of the ore zone for petrological and mineralogical examination, and to obtain a larger volume of material for bulk sampling and assays. In tropical regions, thick lateritic overburden provides optimal conditions for pitting, and if the soil is dry, pits to 30 m in depth can be safely extracted by manual digging with hand tools. The sinking of 1-2 m diameter pits through the overburden into weathered bedrock has been a standard practice in many parts of Africa, where exposure is poor due to the depth of weathering. The use of surface excavations can help minimize costs and provide better insights into the nature of mineralization. On the other hand, these techniques are obviously limited to the cutting depth of the excavation equipment involved.

In very rare situations, special drifts or tunnels may be driven in order to better expose and explore a deposit, but this is a very expensive and time-consuming practice. In general, the purpose of driving such drifts is to provide drilling sites from which a large volume can be explored and a three-dimensional model of the potential ore body developed. Old shafts and drifts often provide a valuable and convenient way of exploring extensions of mineralization particularly in brownfield exploration projects.

## 6.3.4 Advanced (Detailed) Exploration

On the basis of field investigations during the intermediate exploration stage, a number of prospects may be identified, and the most promising of these become the focus of the advanced or detailed exploration phase. Detailed exploration is restricted to relatively small areas and can be intensive and expensive, particularly where drilling is involved. The principal aim of detailed exploration is to further define and evaluate the mineral prospect(s) in detail. The objectives are to: (a) obtain information on the subsurface characteristics of the prospects, including the lithology, structural features and mineralization characteristics; (b) define the vertical and horizontal dimensions of the ore body; (c) provide samples for assay and mineralogical studies; (d) conduct mineral resource and ore reserve estimates; and determine the worth of the ore deposit.
.

Several exploration techniques are used, depending on the type of deposit and its proximity to the surface but drilling is foremost and accompanied by core logging, sampling, assaying, and eventually, ore reserve estimation. If results are encouraging, the project will proceed to pre-feasibility and feasibility studies, where other non-geological factors are considered to establish if a mining project will be feasible or doable based on all the data that have been accumulated.

Advanced exploration g is the most important and resource-consuming phase of mineral explorations, and on the average, accounts for over 50% of exploration costs. The following activities are associated with advanced exploration.

1. Drilling
2. Core logging and Sampling
3. Assaying
4. Ore reserve estimation.

### 6.3.4.1 Drilling

Drilling is the process of mechanically putting a hole in the ground that extends into the subsurface to obtain fresh and representative samples and geological information. Drilling is the most widely used and expensive tool of mineral exploration, and on the average, accounts for more than 40% of exploration costs. Drilling is used in relatively small areas that have been selected as "targets" based on geological, geophysical, and/or geochemical evaluation, and used to obtain detailed information about rock types, mineral content, rock fabric, and the relationship between the host rocks and mineralization at depth. Over 95% of mineral sampling is done by drilling.

Drilling helps locate and define ore bodies and provides most of the subsurface information required for the economic evaluation of a mineral prospect. In setting up a drilling program, the geologist in charge of the project will make several decisions including (a) where to drill, (b) the type of drilling required, (c) the drill hole spacing (d) the timing of drilling, and (e) the contractor to be used. The pattern of drilling used is usually based on the assumed outline attitude and thickness of the ore body. This depends on the information obtained from geophysical data and geological extrapolations based on the deposit model which may not be accurate. Vertical drill holes are the easiest and cheapest to drill and widely used for mineral deposits that are flat lying, disseminated or gently dipping (Fig. 6.13). On the other hand, inclined holes are usually preferred for steeply-dipping mineralization. Drill holes are usually installed systematically and at close intervals along the strike and dip of the target deposit). A surveyed grid system (probably established during the intermediate stage) is usually used to accurately determine drillhole location, the shape, size, and disposition of the orebody, and to facilitate mathematical calculations in ore reserve estimate (Fig. 6.14). The logistics of drilling should be considered carefully as the drill will need drill crews, consumables, and spare parts; this will require helicopter support in remote areas and vehicle access in more populated areas.

There are several methods of drilling but the three most common methods employed in exploration drilling are; (a) *diamond core drilling*, (b) *percussion air drilling,* and *(c) rotary drilling*, Selection of the right drilling method depends on several factors including the type of rocks, depths of the drill holes, required sample volume and quality, logistics, speed, cost, and environ-mental considerations.

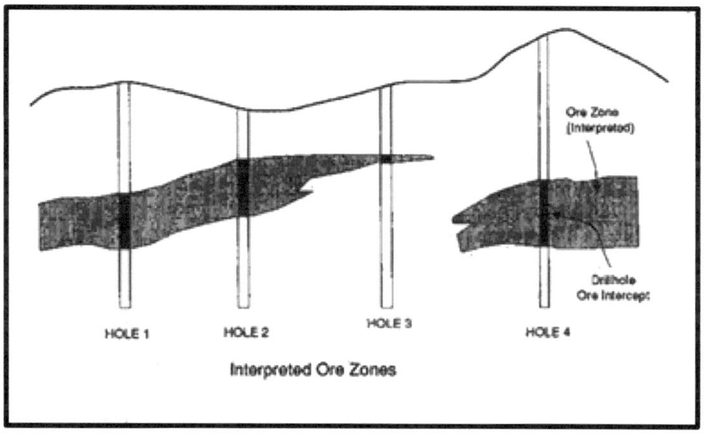


Fig 6.13: Schematic diagram of drill holes intersecting orebodies

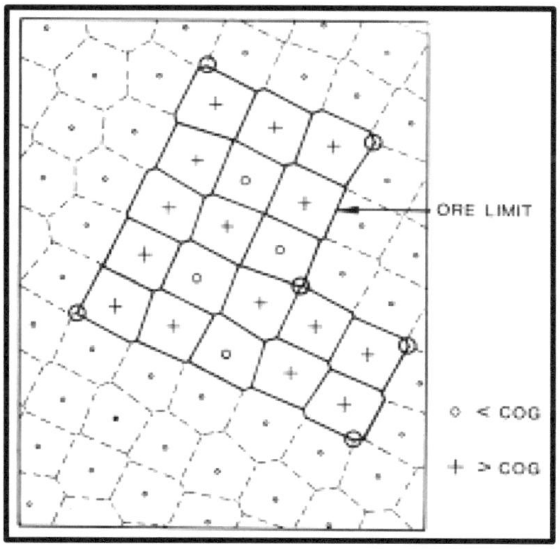


Fig. 6.14: Borehole locations and drilling patterns

### *Diamond Core Drilling*

Diamond core drilling is the technique most widely used in the exploration of hard-rock minerals. It differs from other geological drilling methods in that it produces a solid core extracted from depth for examination at the surface More than any other exploration techniques, drilling provides the exploration geologist with the most concrete and accurate material upon which an economic evaluation of a mineral prospect can be made. It also provides a detailed, continuous look at the subsurface geology that may be displayed in 3D. Diamond core drilling is best used in consolidated or hard-rock formations because of better core recovery.

Compared to rotary drills, diamond core drills are small in size, and are well-suited for remote exploration sites.

The core holes in diamond drilling may be hundreds or even thousands of meters in depth, and samples can be obtained throughout the length of the holes (Fig. 6.14). There are several core diameters but the most common (standard) diameter is about 50 mm (2 inches). In the drilling process, a mounted drill usually on a truck (Fig. 6.15), uses a diamond-tipped bit to cut a narrow kerf of rock, extracting intact a cylindrical core of rock in the center, The key technology of the diamond drill is the actual diamond bit itself. It is composed of industrial diamonds set into a soft metallic matrix. The diamonds are scattered throughout the matrix, and the action relies on the matrix to slowly wear during the drilling, to expose more diamonds. The cylindrical drill cores extracted from the subsurface are usually preserved for future storage by being arranged in sequential order in core boxes, according to the depth from which they were obtained (Fig. 6.16). The drill cores may be split longitudinally, and half of the split core is assayed while the other half is permanently stored for future use and re-assaying if necessary.

Fig. 6.15 (a) Diamond drilling (b) Drill Cores (c) Diamond Drilling bits

***Percussion Rotary Drilling***

Percussion air drilling or rotary air blasting (RAB) is a method in which the rock is broken by means of a piston that delivers rapid impacts to the drill stem, thereby transferring energy to the drill bit (Fig. 6.16). The 'blows' to the rock downhole are delivered by the bit while a rotational device ensures that the bit impacts a new rock surface with each blow. A feed force is applied to maintain rock/bit contact, and compressed air is used to remove or flush the drill cutting from the hole, advancing the hole depth efficiently. The piston can be mounted out of the hole or down the hole for quiet and efficient drilling. Generally, down the hole (DTH) hammers with tungsten carbide bits are used in hard rocks while a tricone bit may be used in softer material, or where the hammer will no longer function due to a head of water entering the hole. Percussion drilling can reach depths of a few hundred meters.

*Reverse circulation* (RC) *drilling* is a type of percussion drilling that uses a double-wall drill pipe. Compressed air is forced between the inner and outer tube down to the bottom of the hole. The air cleans the face of the bit and brings the sample cuttings through the center hole of the inner tube up to the surface Fig. 6.17). Drilling can be performed with either a tricone bit or DTH hammer depending on the formation. RC drilling allows for uncontaminated chip samples and can be used in all formation types. Although it is more expensive and requires more equipment than percussion rotary air blasting, its accuracy makes it one of the most popular mineral exploration drilling techniques.

Percussion drilling has advantages compared to core drilling because it is a simpler technique that is less expensive, fast and provides useful information about the underlying rocks. However, it is not possible to obtain a representative sample because the chips are mixed with overburden. Reverse air drilling provides uncontaminated chip samples and performs in all formation types.

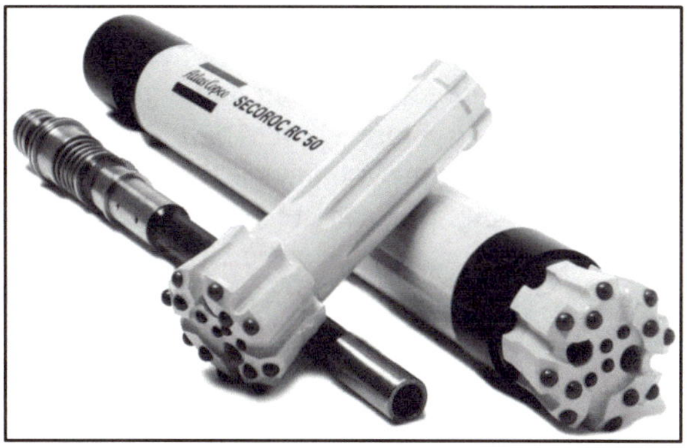


Fig. 6.16 (a) Percussion Air Drilling (b) Percussion Down the Hole Hammer and Bit

***Rotary Drilling***

Rotary drilling as opposed to percussion drilling cuts the rock by rotating a bit at the bottom of the hole. In addition to rotation, downward pressure is exerted on the bit as it goes deeper and cuts through the rock or soil. Rotary methods are generally of two types: *auger drilling* and conventional *mud-air rotary drilling* (Fig. 6.18). ***Auger drilling*** uses a drill bit that looks a lot like a helical screw made of steel casing, with curved flights that are rotated while the rig drill head applies pressure to move the bit further into the ground. The rotation of the flights is important, because it mechanically moves material to the surface. Augers can be mounted on a variety of machines or trucks, and come in different lengths and diameters ranging from 2 to 18 inches. As the drill advances, extra rod sections are added to the top of the drill string. Large augers are used in drilling unconsolidated to weakly consolidated earth materials such as alluvium, soils and residual deposits. An auger rig can drill 200-400' a day and allows visual examination and sampling of the subsurface "in-place" (while not a true core, it is similar) compared to cuttings from percussion drilling. In addition, no fluids or air are required to lift the cuttings to the surface, and the drilling equipment is relatively lightweight. Auger drilling is a relatively cheap method that is used in exploration for alluvial and beach placer deposits, as well as lateritic ores and bauxites.

There are two types of augers: *solid stem* and *hollow stem* (Fig. 6.19a). In the hollow stem auger, drilling uses large diameter (up to 14 inch outside diameter) continuous flight augers which mechanically excavate and continuously transport cuttings to the surface. A center bit, which is attached to the drill rod and bolted to the auger drive cap, is inserted through the cutter head to excavate the center of the boring. The cuttings are collected in a small barrel behind the bit which, when full, is simply pulled from the ground to be emptied. As the borehole is advanced by adding sections of auger, sections of drill rod are added, maintaining the center bit at the face of the cutter bit. Hollow stem augering has a number of advantages. The augers act as a temporary casing during and at the completion of drilling. This facilitates reliable and not much disturbed sampling. It is relatively rapid and little or no fluid is required in the drilling process. Additionally, this drilling method readily supports thin wall and split barrel sampling. Hollow stem auger drilling is most suitable for unconsolidated or swampy deposits.

Solid stem augers use continuous flight augers with no central hole which mechanically excavate and continuously transport cuttings to the surface. Solid stem augering has a number of advantages. It produces a moderate amount of easily contained cuttings and little or no fluid is required in the drilling process. However, this drilling method also has a number of disadvantages. In order to obtain soil samples, the solid stem auger must be removed from the hole. Therefore, this technique is limited to stable soils and sediments which will not collapse when the augers are removed.

***Mud-air rotary drilling*** is executed usually with a roller-cone or blade bit, in which the bit breaks or grinds the rock with energy supplied to it by a rotating drill stem (Fig. 6.19b). The drill stem is rotated while a thrust is applied to it by a pull-down mechanism using the weight of the machine. The bit breaks and removes the rock by either a ploughing-scraping action in soft rock, or a crushing and chipping action in hard rock, or a combination of both. Water (with drilling mud) or compressed air is supplied to the bit via the drill stem. The air/water cools the bit and provides a flushing medium for the cuttings.

Rotary drilling methods are relatively simple, fast and inexpensive but they are not a preferred method of drilling in most exploration programs because it is not possible to obtain a representative sample. However, mud-air rotary is very convenient in the exploration of residual and placer deposits.

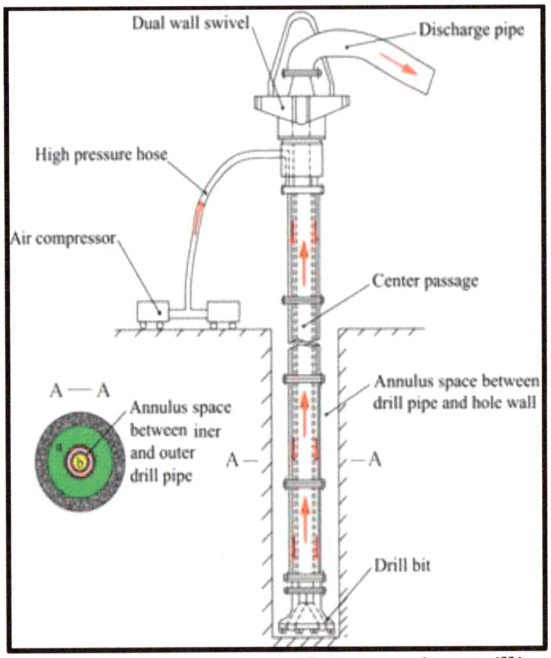


Fig. 6.17: Reverse Rotary Air Percussion Drilling

Fig. 6.18: (a) Auger Drilling (b) Mud Rotary Drilling

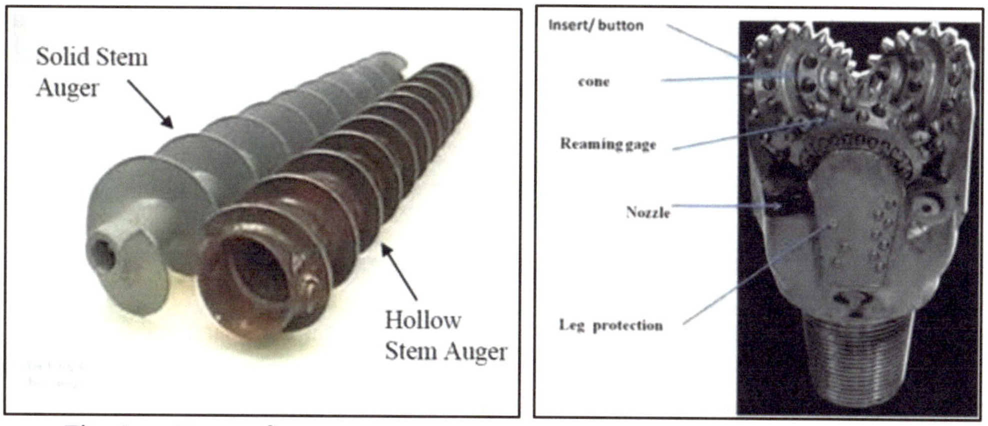

Fig. 6.19: Types of Augers (b) Tricone Rock Roller Bit bits

#### 6.3.4.2 Core Logging

After diamond drilling is complete, a series of drill cores (Fig. 6.20) set in core boxes become available for geological examination. The geological logging of other non-core materials such as rock chips from percussion drilling and auger drill samples also take place but in a different format because of the low reliability of such samples. Drill cores provide the best and most reliable samples of the mineral deposit. Core logging is the geological evaluation of drill cores involving a systematic recording and measuring of as much information as possible regarding the lithology (rock types), mineralogy, potential geological history, structure and alteration zones that occur underground through a tiny piece of cylindrical rock. Drill cores provide a complete record of the rock and ores encountered in the subsurface during detailed exploration.

Core samples most commonly appear as continuous cylindrical bodies of rock (if recovery is 100%) usually of the standard 2 inch diameter but may be composed of broken interlocking pieces of rock/ore if recovery is poor. Poor core recovery often reflects zones of tectonic activity, such as faulting or shearing. Diamond drilling produces rock samples that can be examined and used to determine the mineral concentration, thickness, and depth of mineralized zones. Geochemical analyses of the drill core samples provide the basis for determining the average grade and value of the ore deposit. Core logging is usually undertaken by the project geologist, and requires highly specialized skills.

Fig. 6.20: (a) Drill core s in storage (b) Cure logging:

In practice, the basic information required during core logging should include the following: lithological description (e.g., color, texture, mineralogy, rock alteration, and rock name), fracture spacing and orientation, core recovery (including the location of excessive core loss, >5%),, photographic documentation, description of the geological structures visible in the core, preliminary geological profile, rock properties for calculating geotechnical parameters and content and distribution of mineral and ore components, including as possible in situ assaying of ore. The description must be quantitative as much as possible and systematic.

The angles of intersection of given geological structures, measured relative to the axis of, or the perpendicular to, the drill core, are also recorded as part of the logging program. For deposits that are structurally controlled, oriented core provides better spatial description of structural features and improves the predictive values of models based on structural measurements. Since structural features must be captured before splitting of the core, the most useful way is to take photographs of the wet core to produce a permanent photographic record. In non-core drilling, descriptions must also be systematic and quantitative.

### *Core Logging Methods*

The data from drill core observations are usually presented in a concise format in prose or summary forms or graphs to be utilized in the interpretation of the subsurface geology in the holes drilled or those to be drilled. The way such data are recorded influences in a significant way the amount and type of data that are observed. Using the most appropriate system for recording geological observations of drill cores is therefore very important in the understanding of the mineralization. Although a huge number of different logging forms have been devised and used by companies during exploration, there are only three basic methods for recording observations on drill core and cuttings: (a) prose logging, (b) graphical scale logging, and (c) analytical spreadsheet logging.

***Prose logging*** uses description of observations in words for a selected interval of drill cores defined by depth limits, e.g. depth of 145 to 150 feet (Fig. 6.21). Words are a powerful means of summarizing conclusions and prose. However, long passages of prose are a laborious and ineffective way of recording the complex spatial relationships that can exist between the observable features of a rock. It is also unlikely that any two geologists would ever describe a rock using quite the same words. This means that extracting precise and objective information from a prose log, so as to construct a drill section or to understand the relationships seen, is difficult and time consuming. Nevertheless, when used, such descriptions should be as precise as possible and short in length.

| DEPTH | DESCRIPTION |
|---|---|
| 123.45 - 136.9 | *Sandstone with calcareous cement.Pale khaki colour with a medium,somewhat gritty grainsize.Bedding defined by thin shale partings at 123,34m 126.58 & 132.12 also pebbly band at128.4m. Bedding at generally 45° to LCA but becomes a bit flatter to end section. Numerous thin wispy qtz veins assoc. with pyrite speckling along really distinct margins(looks a lot like the last hole).You can see a weakish cleavage in places but mostly its not there..* |

Fig. 6.21: Example of prose logging

***Graphical scale logging*** uses standard forms that can include several mapping columns along with extra columns for recording digital data, sketches, verbal comments, etc. (.Fig. 6.22). For example, columns can be available for lithologic types while another column is used for structures that are simply drawn in the attitude that they appear in the core. For example a 50 cm vein, which cuts the core at 45°, will be drawn (at 1:100 scale) on the core map as a 5 mm wide vein at 45° to the core trace. The important feature of graphic logs is that they assemble many different types of geological observations on one form. All important relationships, especially spatial relationships, can thus be seen at a glance. Graphical scale logs are a powerful tool in helping to understand geology and to correlate structures between holes. This is probably the best method of logging.

***Analytical spreadsheet logging*** is less commonly used, but recommended in second-phase drilling programmes (resource evaluation and definition) where the main geological problems associated with the ore body have been solved, and the aim of the logging is the routine recording of masses of reproducible data. In analytical spread-sheet logging, the characteristics of the rock are described under a number of precise, prior-defined categories such as color, grain size, mineral content, number and type of veins, type of alteration, intensity of alteration etc. These descriptive parameters form the headings for the columns of a spreadsheet. To keep the log compact and precise, symbols, standard abbreviations and numbers (geocoding) are used whenever possible to record this information. (Fig. 6.23).

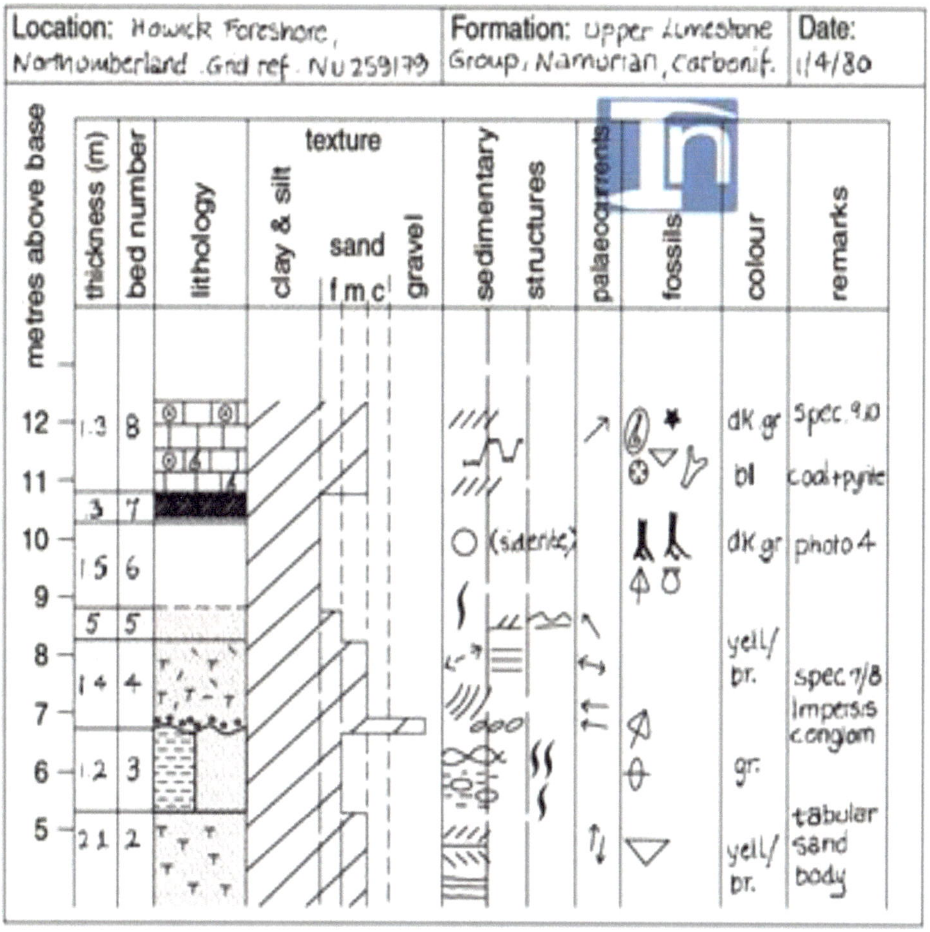

Fig. 6.22: A simple example of graphical scale logging

| XYZ EXPLORATIONS | Logged by : *Steve McKenzie* | Date : 10/14/96 | Sheet: 1 Of: 1 | Hole No : HID 228 |
|---|---|---|---|---|

| LITHOLOGY | | | | | | | TEXT | GRAIN SIZE | | ALTRN. | OXIDN. | STRUCTURE SYMBOLS (Graphic Log) | |
|---|---|---|---|---|---|---|---|---|---|---|---|---|---|
| Ab: metabasalt | Gb: meta gabbro | Pf: fel. por. | Pg: peg/tite | H: hornblende | A: amphibole | Q: quartz | mv: massive | p: pegmatitic | f: fine | 1: weak | ox: complete | : lithological contact | : schistosity |
| Am: Hi-Mg basalt | Pq: quartz porphyry | L: Lode | D: dolerite | B: biotite | Ac: actinolite | Gn: garnet | s: schistose | c: coarse | a: aphanitic | 2: moderate | tr: partial | : ductile shear | : brittle fracture |
| Ag: garnet Ab | Ph: horn. porphyry | Gr: granite | Sp: saprolite | S: sericite | Fd: feldspar | C: cordierite | b: banded | m: medium | | 3: strong | fr: fresh | : breccia | : broken core |

| Metres | | Recov ery | Oxid ation | Lith ology | Texture | Grain size | Alteration | | Mineralisation | | Structure | | Comments | Summ ary |
|---|---|---|---|---|---|---|---|---|---|---|---|---|---|---|
| From | To | | | | | | biotite | sericite | % sulph. | % quartz | graphic | digital | | |
| *0* | *65* | | | | | | | | | | | | *PRE COLLAR* | |
| *65* | *74* | *100* | *tr* | *Ab* | *m* | *f-m* | | | | | \ | *35* | | *Ab* |
| *74* | *78* | *100* | *fr* | *BS-SB* | *s* | *f-m* | *3* | *3* | | | \ | *35* | | *Ab(alt)* |
| *78* | *80* | *100* | *fr* | *Ab* | *s,b* | *f* | *2* | *2* | *0* | *0* | \ | *30* | *black "spots" to 10mm-cordierite?* | *Ab(alt)* |
| *80* | *90* | *100* | *fr* | *Ab* | *m-s* | *f* | *1-2* | *0* | *0* | *0* | \ | *30* | | *Ab* |
| *90* | *92* | *100* | *fr* | *L* | *s,b* | *f* | *3* | *3* | *30* | *10* | \ | *30* | *dominant po, minor sp, asp* | *L* |
| *92* | *102* | *100* | *fr* | *Ab* | *s,b* | *f* | *3* | *3* | *5* | *0* | \ | *40* | | *Ab(alt)* |
| *102* | *118* | *100* | *fr* | *L* | *s,b* | *f* | *3* | *3* | *5-30* | *10* | \ | *40* | *dominant po, minor sp, asp* | *L* |
| *118* | *125* | *100* | *fr* | *Ab* | *s,b* | *f* | *1-3* | *1* | *0* | *0* | \ | *35* | | *Ab* |
| *125* | *148* | *100* | *fr* | *Ab* | *m* | *f* | *0-1* | *0* | *0* | *0* | \ | *35* | | *Ab* |
| *148* | *153* | *100* | *fr* | *L* | *b,s* | *m* | *1* | *3* | *20* | *5* | \ | *40* | *dominant po, minor sp, asp* | *L* |
| *153* | *156* | *100* | *fr* | *SB* | *s* | *f* | *0* | *3* | *0* | *5* | \ | *45* | | *SB* |
| *156* | *180* | *100* | *fr* | *Ab* | *m-s* | *f* | *1-2* | *0-2* | *0* | *0* | \ | *35* | | *Ab* |
| | | | | | | | | | | | | | | |
| | | | | | | | | | | | | | | |
| | | | | | | | | | | | | | | |
| | | | | | | | | | | | | | | |
| | | | | | | | | | | | | | | |

| Prospect/Mine : Mount Impossible | Collar: N9800 E5025 | Inclined: 60 To: Grid W | General: CORE STORED WEST YARD |
|---|---|---|---|

Fig. 6.23: Sample of a simple analytical logging spreadsheet

### 6.3.4.3 Sampling

Sampling is the collection of a representative part of a mineral deposit, sometimes including the host rock. In any exploration program involving the delineation or definition of an ore deposit, sampling is an essential step to acquiring information that can help establish the limits, volume, mass, and grade of the ore deposit. Thus, the main goal of sampling is to generate values about the mineralization (e.g., assays of metal grades) that can be used in ore reserve estimates. Sampling is also important in providing materials to study the mineralogy of the ore body and the geotechnical properties of the overburden and the host rock that will be utilized in mine design and development.

During advanced exploration drilling, sampling of drill cores (or non-cores) is usually carried out during or immediately after logging. The standard practice is to cut or split the cylindrical core along its length into two identical halves with respect to mineral distribution as observed during logging. One half is grinded, reduced, and sent to the laboratory for chemical analysis. The other half is preserved in the core boxes as a primary record for future check studies including mineralogical and metallurgical evaluation. All sampling should be carried out carefully and diligently, using scientifically established sampling practices designed and tested to ensure that the results are representative and reliable. A geoscientist should supervise the sample collection and ensure that a chain of custody of the samples is established and recorded.

Geological samples are of various types: (a) *Grab* samples which are random samples taken from various source points; (b) *Channel samples* which are continuous sampling (usually chips) over an interval in a line such as along a trench or drill hole samples, and depth profile sampling of overburden; (c) *Bulk samples*: consisting of a large mass of material that could be crushed and split into fractions; (d) *Chip samples* made up of multiple pieces (chips) collected from a surface or subsurface location: and finally, (e) *Composite samples* **c**onsisting of a collection of numerous individual discrete samples or chips. Composite samples are the most representative and recommended medium for geochemical analysis. Each type of sample may be more appropriate for a particular phase of exploration; for example, grab sampling during reconnaissance exploration; channel sample during trenching; chip or composite samples are good during detailed exploration, and bulk sampling during feasibility study involving metallurgical testing.

Accurate reserve estimates depend on the collection of reliable samples during mineral exploration. Sampling errors should be avoided. The major cause of sampling errors in exploration and mining is the heterogeneity of ore samples which depends on the shape and size of the particles that constitute the sampled material. Care should be taken to collect representative and large enough samples in accordance with sampling protocol. Sampling programs must therefore be carefully designed to reduce the chances of gathering biased, unrepresentative, or contaminated material. Whether it is by a grab sample from trenches, or chip sample from drill cores, sampling must be designed to fit the target mineralization and interpretation requirements, so that the objectives of the exploration can be achieved. Stringent efforts must be made to avoid cross contamination of samples or "salting" which is the process of adding a valuable metal, especially gold or silver, to a sample to change the value of the sample with intent to deceive potential buyers of a mineral property. .A sampling and analysis plan (SAP) should be developed for each exploration project that must be followed during sampling, sample preparation and analysis.

***Sample Preparation***

Prior to sending samples to the laboratory for chemical analysis, samples need to be crushed, pulverized and packaged appropriately. The sample preparation procedures used should be appropriate for the objectives of the exploration project. Sample preparation should avoid cross-contamination or mis-labeling of samples. Where the volume of individual field samples is reduced prior to shipping to a laboratory for analysis, unbiased splitting procedures to obtain representative subsamples should be applied. Sample preparation procedures should be appropriate to the material being tested and the elements being analyzed. Representative fractions of the material sufficiently large to be analyzed or tested should be retained for a period of time, in case there is a need to re-analyze samples.

#### 6.3.4.4 Assay

An assay is the chemical analysis or testing of a rock, mineral, or ore sample to determine its metallic contents or grade. Assays are very critical in evaluating whether a newly discovered mineral deposit has economic potential. The assay value of an ore body is the same as the ore grade or tenor. It measures the metal content as a percentage of the total weight and is expressed as a percentage, e.g., 4% iron by weight (kilograms), but for precious metals that occur in very low concentrations, different units of measurement are used, such as ounces/ton or grams per kilogram. Assays are carried out in designated laboratories using instrumental or classical methods of analysis.

Considerable variability may exist in assays because of sampling errors or the difficulty of obtaining a representative sample from drill cores. Usually, the larger the core sample, the better the chances of obtaining accurate results. During sampling and analysis, quality assurance and quality control (QA/QC) guidelines should be followed to ensure reliability of results. QA/QC procedures include replicate analysis and inclusion of blanks and standards with known values in each batch of samples forwarded to a laboratory for analysis. The *precision* of analytical data is estimated by analyzing the same sample twice utilizing the same methodology (duplicates). The variance between the two data is an estimation of their precision. Precision is often affected by mineralogical factors such as grain size and distribution as well as by errors in the sample preparation and analysis processes. Reference standard samples (or reference materials) which are samples with a known concentration or grade of the metal(s) sought, are usually included as "blind" samples among samples submitted for assays. They are commonly used to assess analytical *accuracy* and bias by comparing the assay results against the expected metal concentration of the standard or reference sample. Accuracy is how close the chemical concentration is to the true value.

When all the samples collected from a mineralized zone are assayed, the results can be utilized in the estimation of ore reserves. Various statistical methods of averaging are used to obtain a single ore grade average for an orebody. Details will not be discussed here.

### 6.3.4.5 Statistical Analysis of Exploratory Data

Statistical analysis of exploratory data is a major requirement in ore reserve estimation so as to understand the statistical and spatial character of variables on which the estimate depends, to detect possible errors, and also recognize any information that is useful for validating ore deposit models. Such data analysis normally includes interrelationships among variables of interest, recognition of systematic spatial variation of the variables (e.g. grade, thickness, bulk density, etc.), and identification and understanding of results that lie outside expected distributions, known as "outliers". Apart from the statistical summaries (mean, median, standard deviation, etc.), graphical summaries such as frequency distributions, probability plots, scatter plots for correlations, and various multivariate procedures such as trend analysis, multiple regression analysis (e.g. bulk density–metal relationships) and multiple variable plots (e.g. ternary diagrams) may also be computed to provide a better understanding of the exploratory data.

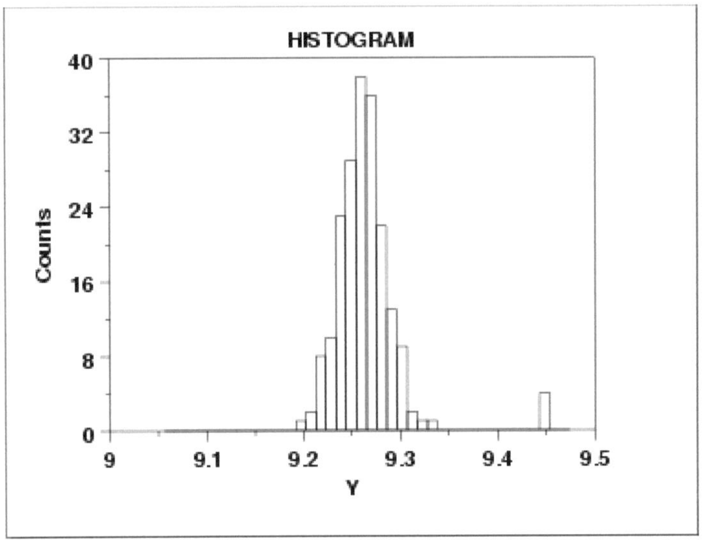


Fig. 6.24: Frequency distribution of exploratory data showing outlier

*Outliers* are anomalously high values that lie outside the main population (based on frequency distribution) which often result in grade bias or observations that appear to be inconsistent with the vast majority of data values. How to handle these errant high values is one of the essential problems in ore evaluation. There are a couple of possible reasons including sampling and assay errors or genuinely reflecting the way in which valuable minerals are distributed throughout the ore body. A decision on how to handle such high grade values include

A number of options: (a) reanalyzing the samples, if possible, (b) cutting the values to some predetermined upper limit based on either empirical cutting method or arbitrary percentile value derived from frequency distribution (e.g 95th percentile of data). However, the most common method to resolve the problem of outliers is to cut the grade to the average of the adjacent samples, use an arbitrary percentile value, of which the 95th or 97th percentiles are most common.

## 6.4 Ore Reserve Estimation

Ore reserves are, simply, the estimates of the quantity (tonnage) and quality (ore grade) of a mineral deposit that has been tested. After the completion of exploration activities involving drilling into the subsurface, and detailed examination, sampling and assay of the drill cores, ore reserve estimates can be calculated. A geological model or ore deposit model can be developed that provides adequate representation of the information on the nature and extent of mineralization. The calculated tonnage and grade of the ore deposit may be integrated into such models to show the vertical and lateral disposition of mineable ore bodies. Ore reserve estimation and grade determination are significant aspects in the exploration and subsequent mining of a mineral resource, and has become of crucial importance due to the large investments involved in the development of a mining project.

The accuracy of information gathered during exploration is very important in reserve estimates. The more accurate the exploration data, the more precision is achieved during estimation of mineral reserves. With the increase in the level of geological knowledge and confidence in exploration data gathered, a mineral reserve estimate can be reported as a mineable asset. Reserve estimates are made for various reasons, including creating a basis for placing a value on a mining property for the benefit of the owners or investors, and also for tax purposes. To ensure that the basis for making estimates are adequate, there are proposed national and international standards or regulations regarding the process and reporting of results? The density of drilling (number of borings per unit area) should be sufficient to satisfy the economic and statutory standards of an ore reserve definition. However, there are exceptions depending on the financial situation and size of the deposit, and the structure of the company. The level of detail required to generate these reserve estimates may vary. For small companies or non-corporate entities, very low level of detail may be accepted compared to big corporations that depend on raising money from capital markets and investors where greater accountability is required for success.

The term "*mineral reserve*" is often used interchangeably with "*ore reserve*", but they both refer to the same thing - the value of minerals, although some authors prefer to restrict mineral reserves to non-metallic minerals and rocks (e.g. limestone, coal, feldspar, gypsum). The term "ore reserve" is more applicable to deposits with metallic content, e,g. iron, copper. Results are expressed as million tons of the obtained percentage of the metal in the deposit; for example, 50 million tons of 72% Fe. Tonnage may also be expressed at a specific cut-off grade. A *cut-off grade* is the minimum grade required in order for a mineral or metal to be economically mined. Material found to be above this grade is considered to be ore, while material below this grade is considered to be waste. The concept of a cut-off grade or value is a fundamental component in the preparation of ore reserve estimates, and should be clearly indicated.

### 6.4.1 Classification of Ore Reserves

Ore reserves are the parts of a Mineral Resource that can be economically mined. A Mineral Resource on the other hand, is just the concentration of earth materials of economic interest in or on the earth's crust. The relationship between ore reserve and mineral resource is shown in Figure 6.25. Resources become reserves as more data is required and confidence improves. Only proven or indicated resources count as ore reserves. While the same exploration data can be used for resource and reserve estimates, calculation of reserves is more stringent, and extrapolation of mineralized zones is not acceptable, whereas such procedure is acceptable for resource estimate. In addition, ore reserves involve some technical (engineering pre-feasibility) evaluation aimed at statistically quantifying the ore grade continuity and the size of the orebody

The purpose of classifying ore reserves is to help indicate the level of reliability and confidence that is placed on the numeric ore reserve estimates provided by companies. This is especially important when results of exploration programs are being reported to the public domain, and the need to protect private individuals or corporations that have invested in exploration and mining ventures. There have been several instances in the past when investors were duped by fraudulent activities in reporting the true values of ore deposits. For this purpose, ore reserves have been classified into three categories: *proven* or *measured, probable* or *indicated* and *inferred* or *possible.*

### 6.4.1.1 Proven or Measured Reserves

Proven reserves are quantitative estimates for which the tonnage, grade, and orebody shape have been computed from dimensions revealed in outcrops, trenches, underground workings or drill holes. The ore grade is then calculated from assays based on adequate sampling to a high degree of confidence. This category is most commonly applied when there is a high degree of certainty, usually only in mines where the ore is blocked out and thoroughly sampled. It is not recommended for use when exploration data are from drill holes only, except where the ore material is of uniform mineral content.

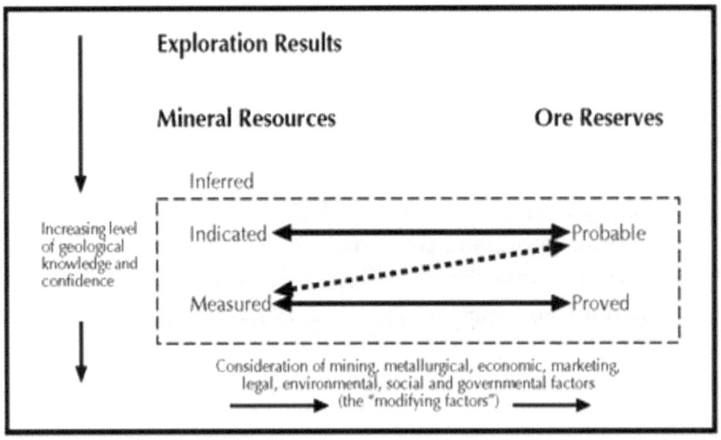


Fig. 6.25: Relationship between mineral resources and Ore reserves

### 6.4.1.2 Probable or Indicated Reserves

Probable reserves are quantitative estimates for which tonnage and grade are calculated partly from specific measurements, and samples from well-spaced drill holes or production data, or partly from the projection for a reasonable distance on geological evidence. A fairly high degree of confidence is implied by the indicated or probable category, and should only be used for ore that has been clearly outlined in three dimensions by drilling, pitting, trenching, etc. Most reserve estimates from drillhole data should be described as probable or indicated reserves. Further improvement in the calculations can be obtained as the orebody is developed.

### 6.4.1.3 Inferred or Possible Reserves

Inferred reserves are estimates based largely on inferences or extrapolations derived from a broad knowledge of the geological character of the area and for which there are few, if any, samples or measurements. The inferred or possible category implies a degree of uncertainty and is used in cases where there is good geological evidence for continuity, but only a limited amount of sample data such as a few widely spaced boreholes are used.

In quoting ore reserves, most companies combine only the proven and *probable* categories together, while excluding the inferred reserves because they are considered speculative.

## 6.4.2 Evaluation of Exploration Data

Raw data obtained from exploration needs to be carefully examined to ensure that decisions to proceed with the exploitation of a new deposit are based on objective, reliable, accurate, relevant and transparent evaluation. There are a number of steps that the project evaluation process needs to go through before major decisions are made.

1. Collection of geological information
2. Development of geological model
3. Integration of mineral deposit model
4. Determine mineral resource
5. Calculate ore reserves
6. Conduct Pre-feasibility study

Deposit models are simulations created from the combinations of diverse geoscience information on the geological and mineral deposit characteristics of an exploration target in resource evaluation (Fig. 6.26).

### 6.4.2.1 Geological Model

There are relevant geological data that are collected during the long process of exploration, including results of geological mapping, rock chip samples, trench data, geophysical and geochemical data, and drill log data. Which are used directly or indirectly in developing the geological model (Fig. 6.26a). The geological model reflects the geologists' understanding of the mineralization; its host rocks and its structural deformation. The geological model is built using 3-D modeling software. Its main objective is to define three-dimensional envelopes that reflect the geological characteristics that may impact ore localization, such as host-rock distribution, the erosion surfaces, geological contacts or faults. These features are likely to have some direct influence on either the limits of the mineralization or on mining or processing costs; such as rock density, hardness or the degree of weathering.

### 6.4.2.2 Ore Deposit Model

Once the geological model is complete, an ore deposit model will be developed that shows the ore deposit boundaries and the distribution of ore grades. The first step in constructing the grade model is to superimpose a digital 3-D block model or framework onto the geological model. The ore deposit may be subdivided into blocks, and the dimensions of the blocks depend largely on the size and the shape of the orebody (6.26b). The mineralization envelope which defines the outer limits of the mineralization may have *hard* boundaries where the mineralization is known to be cut off sharply, for example by a fault, or *soft* boundaries which define the outermost portions of the gradational transition of the mineralization. In the case of soft boundaries the ultimate limits of mineralization will be ultimately defined by the application of a cutoff grade.

Having established the framework, available data such as assay data, thickness of ore intersections and mineralogical information are built into the framework to assign, for example an estimated grade to every block in the model. Only reliable data from drill cores or other related sources are recommended for use. The development of realistic geological and ore deposit models will culminate in the estimation of mineral resources and calculation of ore reserves, usually block by block.

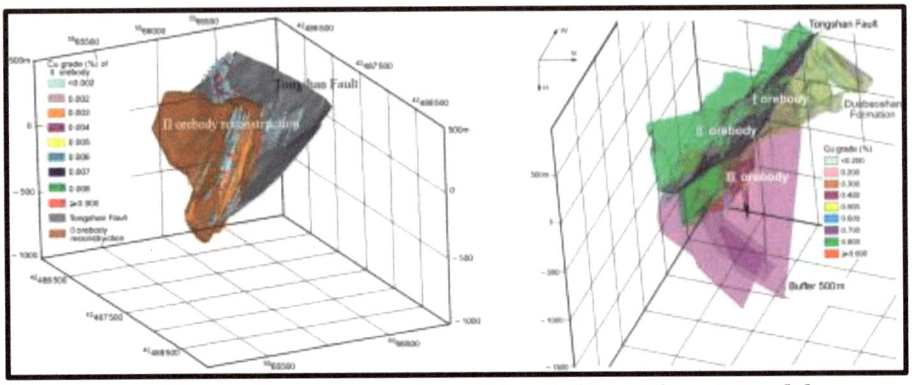


Fig. 6.26 (a) Geological block model and (b) Ore deposit model

## 6.4.3 Calculation of Ore Reserves

Ore reserve calculations are based on the results of detailed exploration activities involving drilling of several holes and chemical analyses of the samples obtained from them (Fig. 6.23). Since the sample points (drill holes) are spaced at specific intervals, certain assumptions have to be made regarding the geological continuity and grade of ore between drill holes. The level of confidence in the data increases with the density of sampling. Calculations can vary from a very simple manual effort to highly complex mathematical, geostatistical and computer applications.

In the simple calculation of ore reserves, the main objective is to determine the *quantity* (tonnage) and *value* (grade) of mineable ore within the deposit. The first step is to calculate a volume and establish the average grade of the deposits or parts of it. The tonnage is derived from the volume by multiplying by the tonnage factor. If the volume is determined in cubic meters, the tonnage factor is simply the specific gravity or bulk density of the ore. If the volume is obtained in cubic feet, then a different tonnage factor expressed as "cubic feet per ton of ore" is used. The volume is commonly determined by calculating an area in two of the dimensions and then multiplying by thickness - the third dimension in order to determine the final volume. To accurately determine total area, it is usually possible to divide the area under consideration into a number of regular geometric figures such as squares, rectangles, triangles, trapezia, etc. Calculations using the traditional techniques may look simple, but ore reserve calculations can be complex depending on the level of detail required and the estimation method used which may involve geostatistics and computer modeling.

### 6.4.3.1 Calculation Parameters

There are four important parameters that are relevant to ore reserve calculation, for which reliable data must be obtained from exploratory drilling and laboratory testing. Input parameters may vary depending on the estimation method selected, but usually includes most of the following:

1. *Area* of the ore body (calculated from geometric shape used)
2. *Thickness*; (average thickness obtained from drill hole data)
3. *Average Ore Grade* (average of assay values for ore body or segments thereof);
4. *Tonnage Factor* (determined from measurements)

***Area***

The area is calculated based on geometric shape. For a rectangle or block shape, the surface area is obtained by the multiplication of length x width of the orebody

***Average Thickness***

This is the third dimension required for volume calculation. It represents the vertical thickness of an ore body as measured in drill holes. Average thickness is based on the average of values obtained in ore intersections.

***Tonnage Factor***

The tonnage factor provides the mechanism for the conversion from volume of ore to weight of ore. In the *metric* system, the tonnage factor is the specific gravity or bulk density of the ore which is obtained by laboratory measurements. If the ore volume has been computed in cubic meters, the volume multiplied by the specific gravity produces the tonnage in *metric tons*. In the *English imperial* system where feet is the unit of measurement, the tonnage factor is normally expressed as "*cubic feet per ton of ore*" which is obtained by calculation or measurement. The tonnage is expressed in *short tons*.

***Specific Gravity***

Specific gravity or bulk density is measured in the laboratory. The most common way to determine specific gravity of an ore in the laboratory is to weigh a sample in the air and then weighing the same sample suspended in water, and later apply the formula:
*Specific gravity = weight in air/ weight in air - weight in water*

***Average Grade***

This is calculated as "weighted" grade. The (weighted) average grade is obtained by dividing the sum of the products of the thickness of ore in each hole and the corresponding assay value by the sum of the total thickness of the ore in the hole expressed as follows:
Average Grade= $\sum Thickness*Grade/\sum Thickness.$
Average Grade of Entire Deposit = $\sum Tons*Grade/\sum Tons$

***Tonnage Formula***

The simple formula or variant of it that is typically used for calculation of tonnage is: *Tonnage = Area* x *Thickness* x *Specific Gravity*: $T= A\times Th \times BD$ or T = V× BD where V is volume and BD is the specific gravity or bulk density obtained by laboratory measurements.
If the ore deposit is partitioned into blocks, *Tonnage* will be expressed as the sum of all blocks where total tonnage is expressed as T $\sum Ti$ = Vb × BD where $\sum Ti$ is the total tonnage of one block in percentage, Vb is the volume in m3 of one block and BD is the density.
For metallic ores, the final Tonnage of Ore is obtained by multiplying Tonnage by average grade. *Tonnage of Ore = Tonnage* x *Average Grade (*T × AG) where AG = Average Grade.
For *metallic ores*, tonnage obtained per grade of metal, is converted to metal ore, e.g. 500 million tons of 2% copper = (500 million x .02) = 10 million tons of copper ore.
(For *non-metallic minerals* and rocks, e.g gypsum which is utilized for its physical properties there is no need for grade in calculation.

### 6.4.3.2 Calculation Methods

There are two broad groups of techniques used in ore reserve estimation; *Classical* and *Geostatistical* methods that are computer aided. All methods used in ore reserve estimates make use of two basic assumptions: (1) samples taken close to each other are likely to have the same value, and (2) average grades are calculated using some form of a weighted average.

**(1) Classical Methods**

- Tabular Method
- Block/Rectangular Method
- Polygon Method
- Triangle Method
- Cross-section Method
- Inverse Distance Method

**(2) Geostatistical Methods**

- Variograms
- Kriging
- Artificial neural network Method

### 6.4.3.3 Classical Methods of Ore Reserve Estimation

Classical or traditional reserve estimation methods use simple geometric criteria (shape, area and volume) to estimate the tonnage of a mineral deposit. Several different geometric shapes, e.g. rectangular, block, polygonal, triangular, or cross sectional methods are used in calculating the volume. Both the tonnage and average ore grade are calculated based on the supposed geometry of the ore body. The various geometric shapes could be divided into blocks of smaller sizes. The volume of each block or shape is its surface area times the average thickness of the ore in the block. The volume multiplied by the tonnage factor gives the tons of ore in the block. Multiplying the tonnage in each block by its average assay value, adding the products, and dividing this sum by the sum of the tonnages gives the average assay value of all the blocks.

The classical methods have the advantages of being simple, quick, cheap, with low technological and sparse dataset requirements, and be used with minimal knowledge of the deposit. On the other hand, these methods are not particularly reliable because of the subjective assignment of the areas of influence of drill holes, and the tendency to give too much weight to occasional high grade results. Nevertheless, they do provide fairly good resource estimates that can be used promptly, and to check the results obtained using more complex geostatistical estimation methods. As a matter of fact, comparative studies of classical and geostatistical have shown differences of usually less than 10%, in tonnage estimates with classical methods providing lower results which are not false positives. The classical methods have stood the test of time, but, because of the uncertainties and subjectivities involved in assigning areas of influence, they are now largely superseded by geo-statistical techniques for the past three decades Comparatively, geostatistical techniques provide powerful tools for obtaining very accurate reserve estimates and ore deposit modelling, using the spatial distribution of ore grades.

***Tabular Method***

This is the simplest method involving a stratified or horizontal tabular body of uniform thickness, such as beds of industrial minerals and rocks (e.g. limestone) (Fig. 6.27), or stratiform metallic ore deposits with uniform thickness and ore grade. The tonnage is obtained simply by calculating the volume multiplied by the specific gravity.

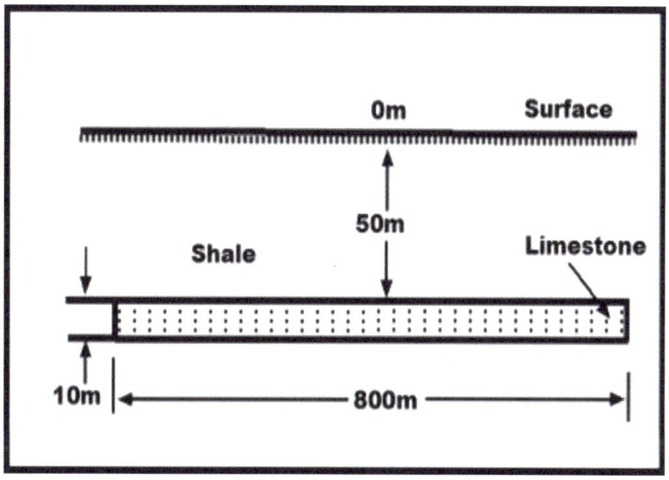

Fig. 6.27: Schematic diagram of a stratified deposit

Example: The Arimogija Limestone deposit in the Osse LGA of Ondo State, Nigeria was explored by drilling of several boreholes to delineate the extent of the deposit. Several samples were also analyzed to determine the CaCO3 content. The outline of the deposit was delineated with the following dimensions: Length is 8000 m and width is 1300m. Thicknesses of the limestone in four drill holes are averaged as 10 m. Specific gravity of the limestone is 2.5. Mineral Reserves (not Ore Reserves) will be calculated as Tonnage x Specific Gravity

*Volume* = 8000 x 1300 x 10 = 104,000,000 m3 x 2.5 = **260, 000, 000 metric tons or (2.6 x10$^8$ mT).**

### *Block (Rectangular) Method*

In the block method, the ore body is divided into rectangles or blocks, or blocks within a rectangle, with defined length, width and thickness, with drill holes located within the blocks or rectangle (Fig. 6.28). The volume is obtained by multiplying the area of each block by the average thickness of the ore in that block. Summation of the volumes of all blocks gives the total volume of the ore. This value is multiplied by the tonnage factor (specific gravity) to obtain the extent of the deposit.

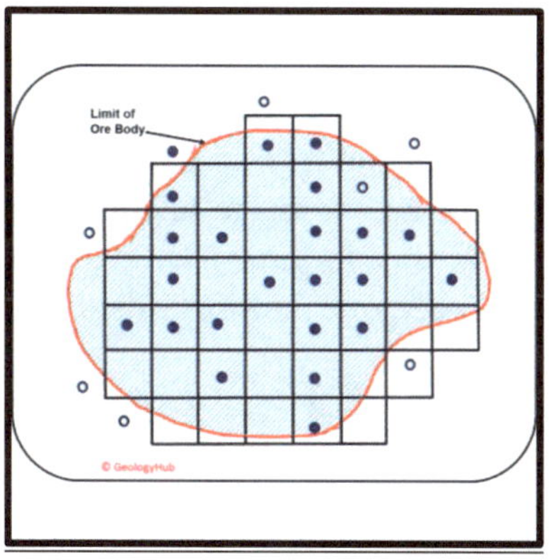

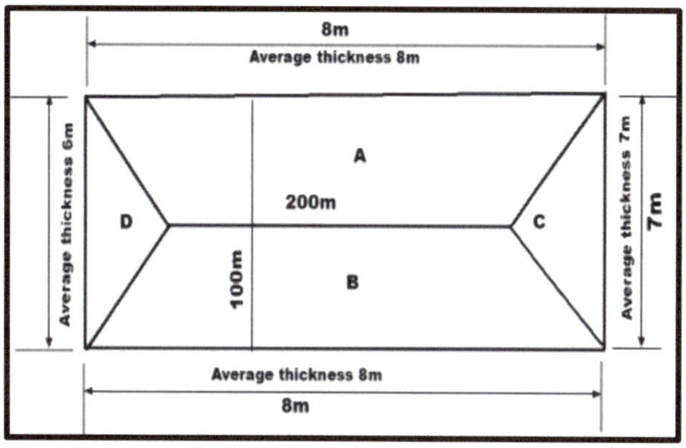

Fig. 6.28 (a) Ore body divided into Blocks (b) Rectangular ore dimensions in resource estimate

Example

An underground mine has an orebody marked out as a rectangular block divided into trapezoids and triangles with systematic assays of the ore block at 10m intervals along horizontal drifts and vertical raises (Fig. 6.28b). The sampling and assays were used in computing average thickness and weighted average grade, for each side of the rectangle in accordance with the length, area, or volume it represents. The data obtained is summarized in Table 6.2.

Table: 6.2: Summary data for ore reserve calculation

| | Length (m) | Width (m) | Thickness (m) | Average Grade |
|---|---|---|---|---|
| Area A (Trapezoid) | 300 + 200/2 (= 250) | 50 | 6 | 2% Cu |
| Area B (Trapezoid) | 300 = 200/2 (= 250) | 50 | 8 | 2% Cu |
| Area C (Triangle) | 100/2 (= 50) | 50 | 7 | 3% Cu |
| Area D (Triangle) | 100/2 (= 50) | 50 | 6 | 3% Cu |
| Specific Gravity 3.0 | | | | |

Tonnage in each block = length x width x thickness x specific gravity
Block a Tonnage = 250 x 50 x 6 x 3 = 225,000 tons
Block B: Tonnage = 250 x 50 x 8 x 3 = 300,000 tons
Block C: Tonnage: = 50 x 50 x 7 x 3 = 52,500 tons
Block D: Tonnage = 50 x 50 x 6 x 3 = 45,000 tons

**Total Tonnage = 225,000 +300,000+52,500+45,000 = 622,500 tons**
**Weighted Average Grade = 2.24% Cu (** $\Sigma$ Tonnage*Avg. Grade/Tonnage)
**Ore Tonnage = 622,500 x 2.23% Cu = 13,881 tons of Copper Ore**

### *Polygon Method*

The polygon method is based on a simple geometric algorithm, in which a 5-sided polygon is constructed around each hole to determine an area of influence for that hole; and then the total volume directly beneath the polygon is assigned the same values as the drill hole from which it was constructed (Fig. 6.29). The polygon method is the most widely used method, especially before the wide availability of computers. The weight for overall estimate is determined by the tonnage of the polygon that surrounds a particular sample or drill hole.

The surface area of the polygon is calculated by using the geometric function in which the *Area =1/2 x perimeter x apothem* (*1/2pa*). Apothem = a segment that joins the polygon's center to the midpoint of any side that is perpendicular to that side. The total volume is determined by multiplying the surface area with the average thickness of the ore in each polygon, and adding them up. Tonnage is then obtained by multiplying the volume with the tonnage factor (specific gravity) in metric tons. The average grade is calculated from taking the weighted average of all grades of individual polygons that were identified.

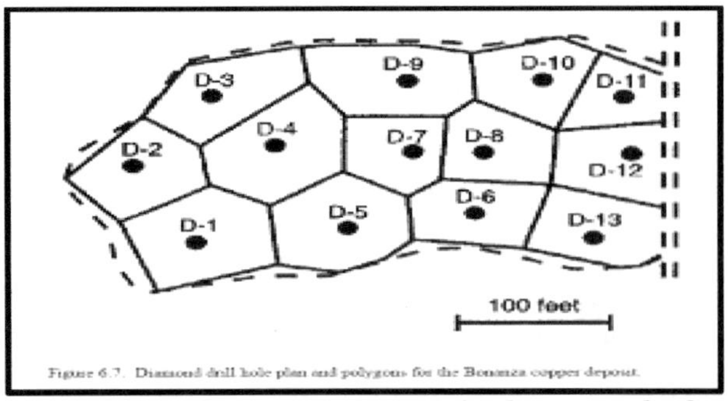


Fig. 6.29: Schematic diagram of polygon method

Example

A copper deposit in the middle Benue Trough, Nigeria was explored with 12 boreholes drilled in a grid pattern to determine the extent and grade of the deposit. The polygon method was used in the reserve estimate of which 5 polygons were developed with the relevant information on Table summarized in Table 6.3.The specific gravity of the copper ore is 2.7:

Table 6.3: Summary data for calculation using the polygon method

| Polygon | Area (m2) | Thickness (m) | Volume (m3) | Av. Ore Grade (% Cu) |
|---|---|---|---|---|
| P1 | 500 | 8 | 4000 | 7,5 |
| P2 | 500 | 10 | 5000 | 7.5 |
| P3 | 500 | 12 | 6000 | 8.5 |
| P4 | 500 | 6 | 3000 | 4.5 |
| P5 | 500 | 9 | 4500 | 6.0 |
| **TOTAL** | **2000** | **9 (Ave)** | **22,500** | **6.8% (Average)** |
| Specific Gravity = 2.7 | | | | |

As summarized in Table 6.3, the total volume of the five polygons is 22,500 m3.

**Tonnage: 22,500 x 2.7 = 60.750 metric tons of 6.8% Cu Ore Tonnage = 60,750 x 6.8% = 4,131 metric tons of Copper ore**

### *Triangle Method*

This is a modification of the polygon method but differs from it by constructing a group of triangles in such a way that each hole is situated at the corner of the triangle with the drill holes forming the three apices of the triangle (Fig. 6.30). The included area of each triangle is assigned the characteristic grade not of a single hole, but of the weighted average of the three holes forming the triangle. The weighting of the three holes is based on the length of the drill holes. The advantage of this procedure is that the data in three point holes enter into the calculations, which then proceed the same way as in the polygon method. To determine the extent of the deposit, the area of triangles are calculated and added up. The volume and tonnage of the deposit is then determined using simple geometrical equations.

In the triangular method, the surface area is calculated from base (b) and height (h) measurements using the formula $[(b \times h) \div 2]$. The thickness *(Th)* measurements of the ore section in the three holes adjoining the triangle are averaged and used in calculating the volume of each triangle is calculated by multiplying the area by the average thickness. Tonnage is calculated by adding the volumes of the triangles and multiplying by the specific gravity.

Example

The Akiri copper deposit was explored in which 10 holes were drilled (D1-D10) as shown in Fig. 6.28. The mineralized area was divided into 10 triangles T1 to T10 in which a drill hole is located at the corner of each triangle. Calculate the ore reserves and average grade. Example of average grade calculation for triangle T1 is shown in Table 6,4 and for the other triangles in Table 6.5.

Table 6.4: Tonnage Calculation for Triangle T1

| **Drill Hole** | **Thickness (m1 m)** | **Grade (% Cu)** | **Grade * Thickness** |
|---|---|---|---|
| D-1<br>D-4<br>D-5 | 50<br>100<br>75 | 0.83<br>1.05<br>0.72 | 46.5<br>105.0<br>54.0 |
| Total | 225 | | 205.5 |

From Table 6.4, Weighted Average Grade for T1 is ∑Grade*Thickness/∑Thickness =: 205.5/223 = 0.92

Table 6.5: Ore reserves parameters for the Akiru Copper

| Triangle | Drill Holes | Tons | Average Grade (% Cu) | Tons * Grade |
|---|---|---|---|---|
| T-1 | D-1, D-4, D-5 | 26,400 | 0.91 | 24,042 |
| T-2 | D-1, D-2, D-4 | 26,400 | 0.94 | 24,818 |
| T-3 | D-2, D-3, D-4 | 27,500 | 0.91 | 20,475 |
| T-4 | D-3, D-4, D-7 | 22,260 | 1.23 | 27,380 |
| T-5 | D-4, D-6, D-7 | 19,550 | 1.15 | 21,332 |
| T-6 | D-4, D-5, D-6 | 27,260 | 0.79 | 21,535 |
| T-7 | D-4, D-7, D-10 | 26,240 | 1.07 | 28,076 |
| T-8 | D-7, D-9, D-10 | 44,500 | 1.15 | 46,575 |
| T-9 | D-3, D-7, D-8 | 24,418 | 1.20 | 29,307 |
| T-10 | D-7, D-8, D-9 | 28,917 | 1.19 | 34,411 |
| | | 263,445 | | 277,927 |

Specific Gravity = 3.5
Given this data, the tonnage and grade calculation for triangle Tl would be as follows:
Area of triangle T1 = 4,400 square feet
Avg. Grade = 0.92
Average Thickness = 225/3 =75.1 meters
Average Grade of Deposit from Table 6.5: = 277,027/263,445 = 1.05% Cu
Volume = Area x Thickness = 4400 x 75 = 330,000 m3
**Tonnage = Volume x Specific Gravity= 330,000 x 3.5 = 1,155,000 metric tons**
**Tonnage of Ore = 1,155,000 x 1.05% = 12,127 tons of Copper**

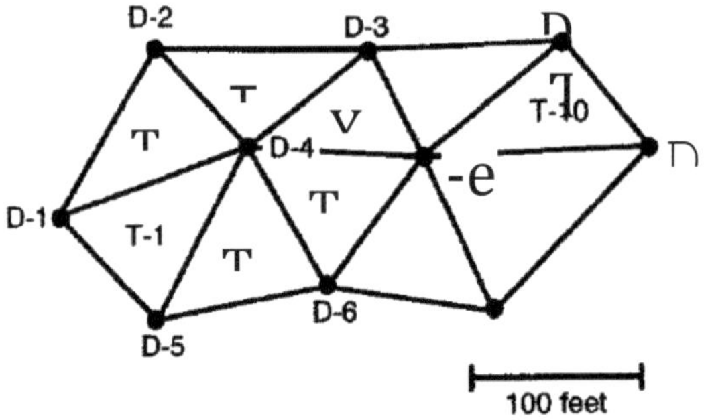


Fig. 6.30: Triangles showing drillhole locations and numbers

## *Cross Section Method*

In cross section method, the ore body is divided vertically into blocks or slices of ore along regularly or evenly spaced cross-sections, and each block is defined by a drillhole intersection (Fig. 6.31) the cross sections may be vertical, horizontal, or at right angles to the dip, usually parallel to each other, and often are spaced equal distances apart. Estimates are commonly made from sections normal to the strike of the ore body. Areas of influence are calculated in terms of the area surrounding each block of ore. The limits of the blocks are set exactly halfway between the drill holes (Fig. 6.31). If the sections are quite similar in outline the volume (V), in cubic meters, is obtained by multiplying the average area (A), in square meters, by the distance between them (H) in meters, as shown by the formula: $V = H [(A1+A2) \div 2]$ where A1 and A2 are the end areas of the block, H is the perpendicular distance between them, and V is the volume of the block. If there is a series of sections spaced equidistantly, this formula becomes $V = ½ H (A1+2A2+2A3+\text{--} An)$.

The area of the cross-section is multiplied by the thickness of intersected ore to obtain the volume. Total volume is obtained by adding the volume of each block. Average grade is calculated from the average assays of the drillhole section. The tonnage of the deposit is thereafter calculated by the volume times the overall specific gravity.

To improve the accuracy of computations, the number of blocks should be as many as possible, although care should be taken to avoid arbitrary locations and construction of sections not supported by exploration data. .The distance between sections should be based on the character of the ore body and the distribution of mineral values. The cross section method.

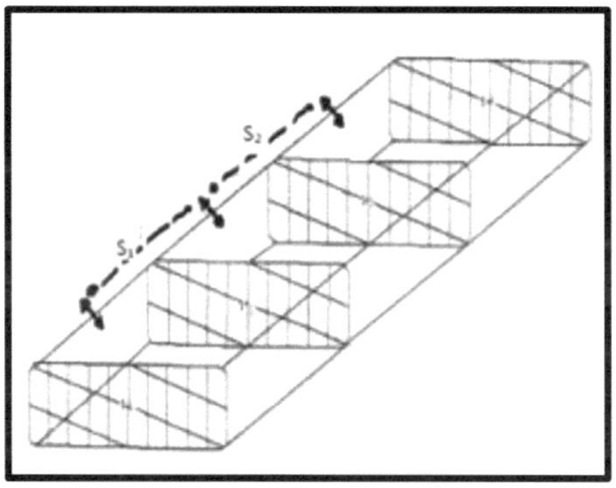

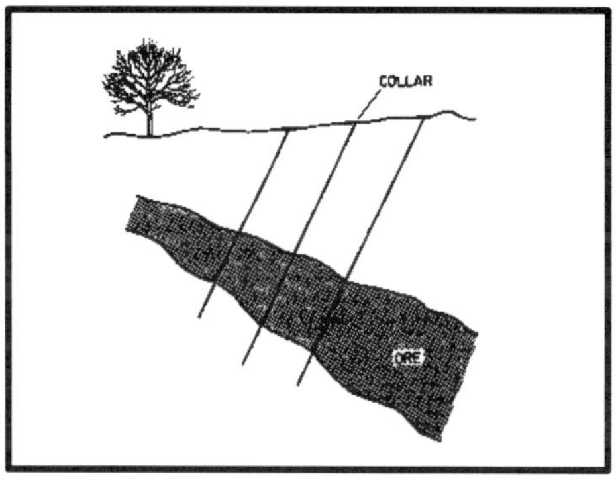


Fig. 6.31: (a) Blocks of ore in cross sections (b) Drill intersections of dipping orebody

<u>Example</u>

A tabular dipping copper vein as shown in Fig. 6.32 is sampled by two drill holes per section and one crosscut per section. The cross section blocks B and C are spaced 100 ft apart. The tonnage factor for the copper vein is 9.5 cubic feet per ton. The following data is provided in Table 6.6.

Table 6.6: Summary data for sections

| **Section Block** | **Area (Sq Ft)** | **Average Grade (% Cu)** | **% Cu/ sq ft** |
|---|---|---|---|
| B<br>C | 3260<br>4896 | 1.81<br>1.75 | 5901<br>8563 |
| Total | 8156 | | 14,463 |

From Table 6.6, Average Block Grade =∑ Area x Average Grade/ Average Grade
= 14463/8156 = 1.77% Cu

Formula: for Volume: V = (At + A°)/2 * H, (where H is spacing between blocks)
Volume of Ore Blocks = (Area B +Area C)/2 x Spacing (Spacing = 100)
Volume = = 3260 + 4895 x 100 /2 = 407,800 cu. ft. Tonnage Factor =1/ 9.5 (English Imperial System)
**Tonnage of Blocks = Volume /9.5 = 407,800 /9.5 - 42,926 tons of 1.77% Cu**

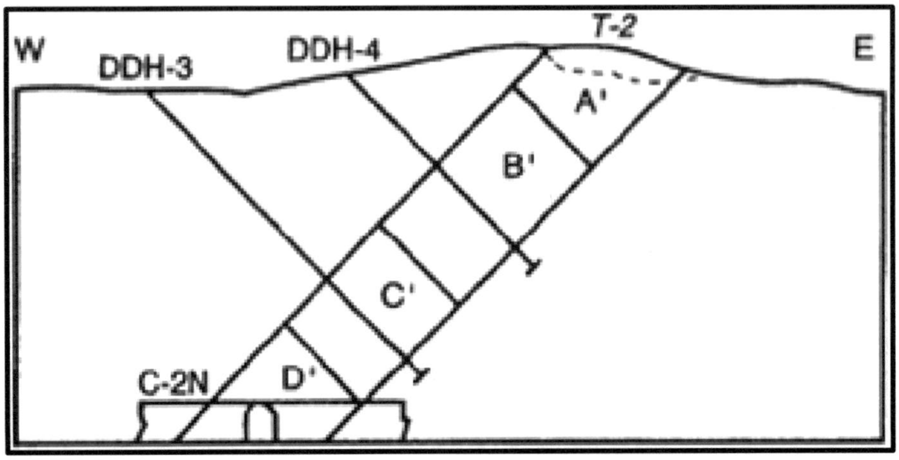

Fig. 6.32: Diagram of cross section blocks and reserve estimate

### *Inverse Distance Method*

The inverse distance weighting (IDW) method is the most advanced technique among the non-geostatistical methods. It is a deterministic spatial interpolation approach to estimate an unknown value at a location using some known values with corresponding weighted values. This method uses weighted averages for estimation between two data points. Thus, it provides for a gradual change in values between multiple sample points rather than an abrupt change at the boundary between adjacent blocks. The technique applies a weighting factor to each sample surrounding the central point of an ore block. This weighting factor is the inverse of the distance between each sample, and the block center is raised to the power «n, » where «n» usually varies between 1 and 3. For example if sample A is half the distance from the ore block than sample B then sample A will have twice the weight of sample B.

The inverse distance is taken into consideration by assuming that the influence of a borehole over a point varies inversely as the distance. The method begins to take the spatial distribution of data points into account in the calculations, a characteristic that will be repeated with geo-statistical methods, such as kriging. Although subjective, IDW estimation procedures are widely used. They have been found to generate results that are somewhat similar to geostatistical estimates produced using ordinary kriging methods. There is computer software available for the calculations after correct input of required datasets.

### 6.4.3.4 Geostatistical Methods in Ore Reserve Estimate

The basic theory of geostatistics in ore reserve estimation is the application of regional variability parameters in the analysis of ore reserve data using tools such as semi-variograms and spatial interpolation procedures. In geostatistical techniques, the knowledge of the geological characteristics of the deposit is used to improve the estimate of grade, at points where not sampled. They are widely recognized as superior and more robust techniques than classical methods. Classical statistics requires the use of a particular distribution model, e.g., the data are normally distributed or of lognormal distribution, and that the samples are independent of one another. However, in practice, there are usually insufficient samples of the ore body collected during exploration to assign a statistical distribution, and moreover, the samples are often correlated, i.e., they do not satisfy the independence requirement of classical statistics. Given that the samples, in the drill holes, are limited in number because they are expensive to acquire, often biased, and nearly always smaller in number than is desired, geostatistics has become a powerful tool for improving the quality of the estimation, using three-dimensional spatial statistics.

Geostatistical applications in ore reserve estimates have two basic steps: *variogram* model, and *kriging*. *Variogram* model is a main tool of geostatistics. It is the first step that involves a statistical treatment of the directional interdependencies which is the quantification of the spatial variability of the ore deposit and sample grades. *Kriging* is a mathematical procedure of interpolation by assigning weights to samples so that the point estimate error is minimized. *Artificial neural network* method is one of the most recent techniques used to extend spatial distribution of sample data. The main advantage of this approach is that it requires no complicated mathematical modeling and makes no assumptions about the spatial distribution of ore grade. Details on the applications of geostatistical methods are available in the literature.

Prior to using geostatistics, the exploratory data should be statistically evaluated to determine the frequency distributions whether they are normal or lognormal (skewed), and to determine the inter-relationships (statistical correlations) among the parameters to be used in the reserve estimation. In geostatistical application, it is generally assumed that the variables in the raw data are normally distributed, and if not, a normal transformation must be applied to the data. In general, the size of the database required for ore reserve modelling and calculations is too bulky to manage manually Consequently, numerical algorithms and mathematical approaches necessitate computer applications which are currently available in many computer aided systems and commercially available software. The accuracy and speed of computers enable evaluation of various scenarios within reasonably short times.

Geostatistical reserve estimation requires inputs that are provided in the following steps;

- Database of mineralogical characteristics, assays and geologic controls of mineralization and any other features of the deposit.
- Computation of variograms (graphical correlation of mineral characteristic) for each geologic zone.
- Division of orebody into matrices of blocks for kriging (calculation of block grade related to adjacent samples); blocks are then classed as measured, indicated and inferred.
- Estimate of tonnage and grade of each block at given cut-off-grade.
- Printing of recoverable grade distribution plans by level or bench in the proposed mine.

An example of comparative analysis of ore reserve estimates obtained by classical and geostatistical methods is summarized Table 6.7. Comparison of estimates of tonnage and grade by hand computed and geostatistical methods differ by less than 10%, with the manual results being generally lower in tonnage but higher in ore grade.

Table 6.7: Comparison of ore reserve estimates

| | No. of Blocks | | Tons of Ore | | Average Grade | |
|---|---|---|---|---|---|---|
| Cut-off Grade, % | 0.60 | 0.55 | 0.60 | 0.55 | 0.60 | 0.55 |
| Hand Polygon | ---- | ---- | 1,804,836 | 1,804,836 | 0.93 | 0.93 |
| Computer Polygon | 66 | 66 | 1,844,637 | 1,844,637 | 0.92 | 0.92 |
| Inverse Distance | 65 | 69 | 1,816,687 | 1,928,484 | 0.91 | 0.89 |
| Kriging | 68 | 71 | 1,900,534 | 1,984,382 | 0.86 | 0.85 |
| Neural Network | 61 | 72 | 1,679,407 | 2,012,331 | 0.87 | 0.83 |

*All data based on Hughes, Davis, and Davey (1979) except for neural network.

CAGEO 19/4—G

## 6.5 Feasibility Studies

Feasibility studies represent the final stage of a mineral exploration project. following the completion of ore reserve estimates and the development of the orebody model, As the term "feasibility" suggests, it is an attempt to answer the question of whether a mining project will be feasible or doable, based on all the data that have been accumulated. It is simply a technical and economic evaluation of the possibility of developing a mine.

There are three types of feasibility studies in mining:

1. Scoping Study or Order of Magnitude
2. Preliminary Feasibility Study or Pre-feasibility
3. Detailed Feasibility or Full Feasibility

The *scoping study* is a summary appraisal undertaken early in the final resource assessment. The *pre-feasibility study* is a preliminary evaluation of all the parameters relevant to the decision making process of moving a mining project forward to the next level of mine planning and design. A *full feasibility* provides a lot of detailed information on mine design, mine infrastructures, financial aspects of raising funds for capital and operational expenses, market estimates and environmental and risk management.

### 6.5.1 Scoping Study

Scoping study is an initial appraisal carried out early in the process of the final resource project evaluation, and normally involves a preliminary mine development plan, that forms the basis for determining whether to terminate a mineral exploration program, or determine how much further pre-development efforts are warranted. . Depending on the size of the project, a scoping study may be undertaken by a single individual. They are not accurate enough to carry out a meaningful assessment of the economic viability of a project, but should determine key project risks establish the probable type of mining operation, assess the potential for the delineation of resources capable of supporting a mine, and, in the case of coal and industrial minerals, confirm likely markets.

### 6.5.2 Preliminary Feasibility Study (Pre-Feasibility)

Pre-feasibility study follows a scoping study with the intent to determine whether a mineral resource is likely to support a viable mining project. It is more detailed than a scoping study and is the key intermediate step in the assessment of a mining project. Pre-feasibility study also contains preliminary mine planning and engineering evaluations based on the likely conversion of the mineral resources delineated during exploration into possible mining reserves. The report is generally sufficiently accurate to allow a comparison of alternative project configurations and include preliminary capital and operating cost estimates.
Pre-feasibility study should include, but not limited to the following:

- assessing reserves and saleable product from the delineated resources • generic mine design
- non-detailed, staged life of mine planning and production scheduling assessing the mining methods, treatment routes and identifying cut-off factors, recoveries, dilution and losses in both mining and treatment
- outlining probable plant, infrastructure, services and other facilities • producing a summary development structure and timetable
- determining capital and operating costs
- evaluating the specification and marketability of the commodity

### 6.5.3 Final (Detailed) Feasibility Study

The objective of the detailed feasibility study is to determine whether to proceed with a mining project into the detailed engineering and construction stage. This study is significantly more expensive than the preceding studies, largely because of the associated engineering work. Feasibility study deliverables will typically include, but not limited to:

- establishing the proven and probable reserves
- assessing project alternatives and selecting the desired development route
- proving the technical viability of the mine and extraction methods
- defining the mine's plant, equipment and infrastructure requirements and capacities
- establishing resource consent and other legal/governmental conditions and requirements for approvals to proceed
- defining the management control and timing of the project development and construction
- defining the commissioning of the project, recruiting and training of the management team and crew and hand over at the conclusion of construction
- estimating the development, capital and operating costs of the mine over the economic life
- identifying the market for the commodity
- assessing project risk areas, economic sensitivity to those areas and action that may be used if the risk becomes a problem
- completing economic assessments of the selected project configurations
- developing financial models
- setting a framework for the implementation of the capital investment in the mine development phase.

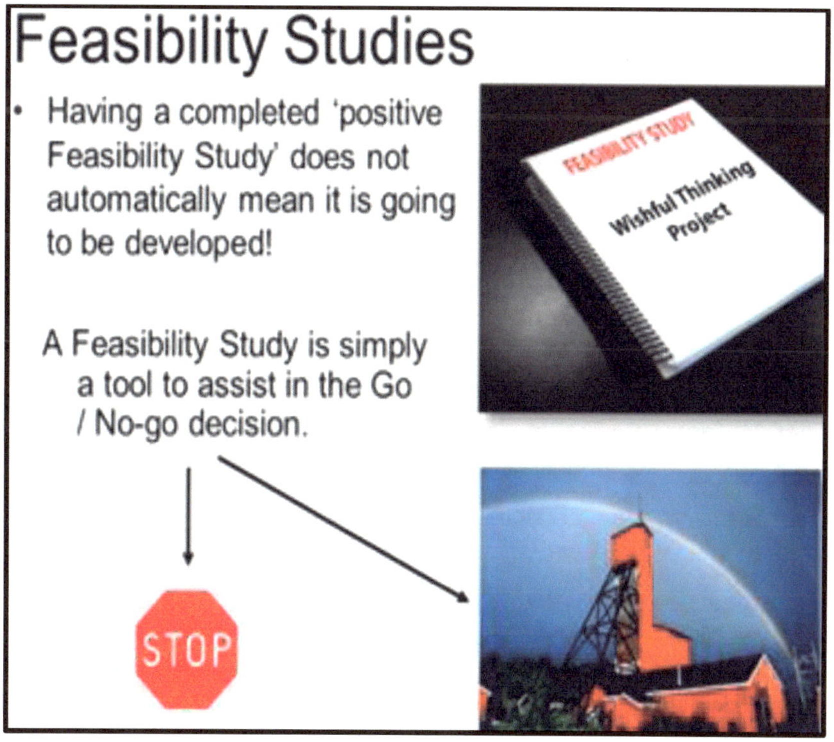

## 6.6 Review Questions

1. Define mineral prospecting and mineral exploration. What are the similarities and differences, if any?
2. What is "greenfield" and "brownfield" exploration? Explain the advantages and disadvantages.
3. Briefly describe the main stages of the mineral exploration cycle. What stage is most significant?
4. What is targeting? Explain the main parameters that are useful in targeting and their relationship to the scale of operations.
5. Describe the main types of drilling in mineral exploration. Explain what mineral deposit types are suitable for each method.
6. What is sampling? Describe the types of sampling applicable to exploration and mining, and explain the applicable QA/QC procedures.
7. Describe geological and ore deposit modeling and expatiate their importance in resource evaluation.
8. Define and differentiate mineral resource and mineral reserves. What is the relationship between the two?
9. Define the concept of geostatistics. What is the most important difference between classical and geostatistical methods regarding error estimation?
10. Write short notes on the following: (i) Target minerals (ii) Assay (iii) Muck sampling (iv) Outliers (v) Cut-off grade.

## References


Annels AE (1991) Mineral deposit evaluation: a practical approach. Chapman Hall, 450 p.

Brimhall GH, Dilles JH, Proffett JM (2006) The role of geologic mapping in mineral exploration. Soc Econ Geol Spec Pub 12:221–241

Carras SN (1987) Ore reserve estimation methods and their relationship to recoverable reserves. Management and Economics of Mining Operations (Symposium Proceedings, Western Australian School of Mines, Kalgoorlie, October 1986), pp 57–67

David M (1977) Geostatistical ore reserve estimation. Elsevier Amsterdam, 364 p.

Dentith M, and Mudge ST (2014) Geophysics for the mineral exploration geoscientist. Cambridge University Press, New York, 454 p.

Diehl PY, David M (1982) Classification of ore reserves and resources based on geostatistical methods. CIM Bull, 75, 127-135.

Erickson AJ, Padgett JT (2011) Geological data collection. In: Darling P (ed) SME mining engineering handbook. Society for Mining Metallurgy & Exploration, Englewood, pp 145–159

Evans AM and Moon CJ (2006) Mineral deposit geology and models. In: Moon CJ, Whateley MKG, Evans AM (eds) Introduction to mineral exploration, 2nd edn Blackwell Publishing, Ltd., Oxford, pp 33–51

Levinson AA (1974) Introduction to exploration geochemistry. Applied Publishing, 671 p.

Marjoribanks R (2010) Geological methods in mineral exploration and mining. Springer, Berlin, 238 p

Milson J (2006) Geophysical methods. In: Moon Whateley MKG, Evans AM (eds) Introduction to min-eral exploration, 2nd edn. Blackwell Publishing, Ltd., Oxford, pp 127–154.

Moon CJ, Whateley MKG, Evans AM (eds) Introduction to mineral exploration, 2nd edn. Blackwell Publishing, Ltd, Oxford, pp 253–278

Moon CJ, Whateley MKG (2006) Reconnaissance exploration and from Prospect to prefeasibility. In: Moon CJ, Whateley MKG, Evans AM (eds) Introduction to min-eral exploration, 2nd edn. Blackwell Publishing, Ltd., Oxford, pp 52–103.

Peters WC (1978) Exploration and mining geology. Wiley New York, 685 p

Pohl WL (2011) Economic geology: principles and practice. In: Metals, minerals, coal and hydrocarbons. Blackwell Publishing Ltd., Oxford, 663 p.

Revuelta, MB (2016) Mineral Resources: From Exploration to Sustainability Assessment. Springer Publishers, 641 p.

# Chapter 7
# Elements of Mining Geology

## 7.1 Introduction

**Mining** is the process of extracting valuable minerals or other geological materials from the Earth's surface or subsurface. Mineral extraction is the culmination of a process that probably started with mineral prospecting, discovery, exploration, evaluation, feasibility study and mine planning and development. Mining of minerals is a human activity that commenced since ancient times, and has served as a source of essential metal and mineral products used by human society to improve quality of life. The nature of mining creates a potential negative impact on the environment both during the mining operations and for years after the mine is closed. Geologists perform important roles in the successful and profitable extraction of ore minerals by advising mining engineers, and miners on the nature of the ores and the best approaches to mining at a profit. Mine geologists also provide supporting documents and vital services such as geological and geotechnical mapping, ore sampling, assays, structural analysis, guide on blasting points, environmental monitoring and ore grade control. There is a group of professionals known as *geological engineers* (not engineering geologists) who are a 'mixed breed' of geologists and mining engineers. They are better trained in geology than mining engineers and in engineering than most mine geologists.

**Mining geology** is a specialized branch of applied economic geology that deals with the application of geology in the extraction (mining) of economic minerals. Mining Geology, has also been rightly described as an "applied science which combines the principles of economic geology and mining engineering in the development of a defined mineral resource". After ore deposits are discovered, evaluated and approved for mine development, geological support is often required in evaluating the mining project and providing support for subsequent mine operations. Mining geologists and engineers work together to develop an identified ore deposit and economically extract the ore. Since mineral deposits are found in rocks, and many factors that affect their localization are controlled by the geological setting, an adequate understanding of geology will always help facilitate the extraction of economic minerals from the surface or subsurface environment. Consequently, the major goal of mining geology is to provide useful and detailed geological information to mining operations during the opening, preparation and extraction of ore deposits. Also, mine geologists ensure the cost effectiveness of extraction of ores by effective ore grade control and efficient separation of valuable minerals from waste rock and minimizing dilution.

Until the last few decades, most mine geologists did not obtain formal education and training in mining geology because such courses were not easily available. Consequently, the mine geologists were geologists or exploration geologists who obtained their knowledge and experience on the job. However, modern mining geology has become more sophisticated and requires apart from knowledge of economic geology, advanced stratigraphy, petrology, mineralogy and structural geology, other multidisciplinary knowledge in applied geochemistry, mining geophysics, sampling theory, geostatistics, geological modelling, mining engineering, rock mechanics, mineral economics, and environmental and computer sciences.

## 7.2 Brief History of Mining

Archaeological records indicate that mining commenced in prehistoric times as a result of the need of ancient people to use stones as tools for hunting. Primary stone particularly *flint*, a hard siliceous rock, was the first earth material mined because of its conchoidal fracturing which allowed it to be broken into sharp-edged pieces that were useful as scrapers, knives, and arrowheads. However, the earliest known underground mine for a specific mineral in the world was from the Ngwenya mountains in present Swaziland, Southern Africa where ancient (Paleolithic) humans mined hematite to make the red pigment called ochre used in burial ceremonies and as body coloring, and was dated as 43,000 years ago by radiocarbon method. Between 40,000 and 20,000 years ago, there was some evidence of coal mining in South Africa.

Mining did not become a significant industry until more advanced civilizations developed around 10,000 to 7,000 years ago when metals of copper, gold and mercury were found and valued. The first societies to use metals on a large scale emerged in Mesopotamia around 6500 years ago and Egypt around 6000 years ago. Gold was one of the first to be utilized, being mined from streambeds of sand and gravel (placer). Copper which occurs in native form was probably the second metal discovered and used. Silver was also found in a pure state and at one time was valued more highly than gold. The Egyptians and Sumerians smelted gold and silver from the gold mines of Nubia (now southern Egypt and Sudan) which were among the largest and most extensive of any in Ancient Egypt. As a result, these metals began to have a value that was transferable between people and between cultures.

Approximately 5,500 years ago in this history of mining, came the discovery of tin. Tin, mixed with copper, made bronze, the first alloy harder than its constituent metals. Metal working advanced over the next few thousand years, as did the use of other mined materials. With the development of the Mediterranean civilization, mining became one of the most critical industries in the world. Athens in Greece grew wealthy, due to the extractions from its silver mines until the fall of Athens when Sparta wrested away control of the silver mines. People mined through systems of shafts and galleries supported by stone columns as timber became scarce in the area. Additionally, it was the death penalty for extracting ore from one of the stone columns. However, it was during the Roman Empire that mining boomed, and the Romans introduced large-scale mining operations that included surface, underground and hydraulic mining. The growth of civilization expanded rapidly, and required more and more money to finance its operations, as well as more metals to create military equipment for its ongoing wars and, and build infrastructure.

In medieval Europe, mining as an industry underwent dramatic changes mainly focused on the extraction of copper and iron. Other precious metals were also used, mainly for making coins. Initially, many metals were obtained through open-pit mining, and ore was primarily extracted from shallow depths, rather than through deep mine shafts. Around the 14th century, the growing use of weapons, armour, and horseshoes greatly increased the demand for iron. One of the most complete early records of mining methods in Europe is by the German scholar Georgius Agricola considered as the "father of economic geology" in his book titled *De re metallica* (1556). He described detailed methods of driving shafts and tunnels. Soft ore and rock were laboriously mined with a pick and harder ore with a pick and hammer, wedges, or heat (fire setting). Great progress in mining was made when the secret of black powder explosives reached the West, probably from China in the late Middle Ages. This was replaced as an explosive in the mid-19th century with dynamite. During the rise of Europe in the $17^{th}$ to $18^{th}$ centuries, governments gave miners extensive rights to take land and profit from the minerals they found, and pay a portion of the revenue from mining as payments for granting the mining rights. Civilization progressed requiring an increasing amount of mined material to manufacture the necessities of everyday life and fund the economies of nations.

The influx of this mineral wealth from the New World comprising the United States, Canada, South America, Africa and Australia financed the Renaissance. , period of political and economic rebirth in Europe. Eventually, the creation and expansion of countries in the New World, along with the development of the Industrial Age, resulted in the generation of the mining industry we have today is just as vibrant as it has been throughout history. Coal and other metals are still essential to our day-to-day lives, and their extraction and production remains one of the most important industries in the world. A truly international pursuit, modern mining techniques are providing emerging economies with an opportunity to join the world market in a major way. A focus of modern mining practices is the use of the most efficient, safe, and environmentally friendly methods possible.

## 7.3 Mining Geologist - Role and Responsibilities

The terms "mining geologist" and "mine geologist" are used interchangeably in this book, although in my opinion, there may be a subtle difference: a *mine geologist* is defined as a geologist employed in mines, while a *mining geologist* is a geologist with formal training, preferably post-graduate in mining geology or mining engineering. The role of a mine geologist may vary from mine to mine, but the main roles are to provide, to advice on the geological aspects of development and production processes in mines and quarries, and provide geological support to the mining personnel to ensure that economic minerals are extracted in a way that maximizes profit with as few problems as possible.

The responsibilities of a mine geologist are varied (see list below) but are primarily focused on the acquisition and processing of geological information including geological mapping and logging, in support of mining operations, and the supervision of drilling, sampling and ore grade control processes. The mine geologist must work closely with the mining engineers and miners during production, as well as the drill and blast engineers and mine surveyors to determine the most appropriate way to blast the rock and ore exposed on the pit floors and mine faces. The role of mining geologists in mines is different from that of mining engineers who are employed to ensure the safe and efficient development of mines and other surface and underground operations

The responsibilities of mine geologists do not end when the extractions begin, but must continue with monitoring of mine surfaces and the petrographic study of ores and materials as they are extracted, so as to assess their quality, characteristics and continuity. During mine operations, mine geologists must continue to reevaluate the orebody, revise ore reserve estimates, and expand their knowledge and understanding of the relationships between the geology and ore formation, so as to be able to locate new mineral resources or extension of existing ore bodies. Mine geologists should always be ready to provide valuable advice on short and long term production plans, minimize dilution as well as advice on geotechnical stability of mine rocks, groundwater seepage and environmental contamination. Mine geologists must have computer skills to store and retrieve data, and use software for geological, ore reserve and ore grade modelling.

Fig. 7.1 (a) Mine geologists in underground opening (b) Geologist mapping of open pit wall

### 7.3.1 Duties or Job Description of a Mine Geologist

The duties or job description of mine geologists employed in a geological or mining department of a mine, either in open-pit or underground operations vary considerably but may include but not limited to the following:

- Conduct geological mapping and prepare geological mine maps, cross-sectional diagrams, charts, or reports.
- Plan, implement and supervise drilling and core sampling programs.
- Conduct core logging, petrographic study and mineralogical examination of ore minerals and mine rocks.
- Analyze and interpret geological, geochemical, geotechnical or geophysical information from sources such as survey data, well logs, bore holes, or aerial photos.
- Design and supervise grade control programs and interpretation of collected data.
- Assess ground or surface water movement to provide advice regarding issues such as mine drainage, waste management, and environmental remediation.
- Work closely with the drill and blast engineers to determine the best way to blast mine rocks.
- Assist the mine surveyor as needed and stand in as his substitute when quick surveying tasks are required.
- Conduct sampling of mine rocks and ores for assaying and mine development
- Use computer software for data entry, storage and modelling of orebody, ore grade and update of ore reserves.
- Calculate and update ore reserve estimates.
- Assist in mine development, including mine design and planning
- Assist in developing in-mine exploration and delineation drilling programs to locate new ore deposits or extensions.
- Prepare geological maps, charts and reports on mineral extraction for management review.
- Conduct geotechnical mapping and identification of risks for rockslides and mine failure
- Communicate geological findings by writing reports, research papers, participating in conferences and meetings.
- Collect data for compliance with environmental regulations and remediation.

### 7.3.2 Mining Engineer - Role and Responsibilities

**Mining engineers** are different from mining geologists in their role and responsibilities at mining operations. Mining engineers design mines and manage operations to ensure that mineral extraction is carried out safely and efficiently. Mining geologists assist mining engineers by providing advice on the quantity, quality, distribution, behavior and other properties of geological materials in mines, including host rocks and ores. Some of the duties and responsibilities of a mining engineer include but not limited to the following:

- Collaborate with geologists or geological engineers in geological exploration, and review maps and drilling logs to determine location, size, accessibility, and estimated value of mineral deposit.
- Determine methods to extract minerals, considering factors such as safety, operational costs, deposit characteristics, overburden depth, and surrounding strata.
- Design mines that allow workers to safely mine materials from the ground.
- Direct mine construction operations, such as location and development of shafts, tunnels, chambers, position of excavation benches, and access roads.
- Inspect new and existing mining sites to determine the best ways to extract specific materials from the ground.
- Manage mining operations and supervise mining projects and staff to ensure projects are moving forward on schedule and within budget parameters.
- Maintain, upgrade, and purchase mining equipment, and make decisions about the type of equipment that should be purchased to complete specific mining projects.
- Follow safety regulations set forth by the company and local, state, and federal laws.
- May apply knowledge of mining engineering to solve problems concerned with environmental risk and mitigation.
- Competent in computer applications to mining design and resources optimization.

## 7.4 Life Cycle of a Mine

A mine is an opening excavated in the ground either at the surface or subsurface for the purpose of extracting valuable minerals and other earth materials, usually from an ore body. The life cycle of a mine begins with discovery and ends with closure, and consist of five stages (Fig. 5.2):

1. Mineral Exploration
2. Mine Planning
3. Mine Development
4. Mine Production
5. Mine Reclamation

### 7.4.1 Mineral Exploration

Prospecting and exploration which can occur simultaneously are the precursors to mineral discovery and mineral resource development. Through these activities, an orebody is discovered, delineated and evaluated for its tonnage and grade. The ore reserve estimation is used to conduct a pre-feasibility study to formally determine whether it is economically worth developing the mineral deposit into a mine. A final feasibility report is generated that takes into consideration all the geological, technical, economic, financial and environmental needs of the project. Recommendations are made to the mining organization to help them make a final decision on whether the mining of the ore deposit will be a profitable venture for its investors.

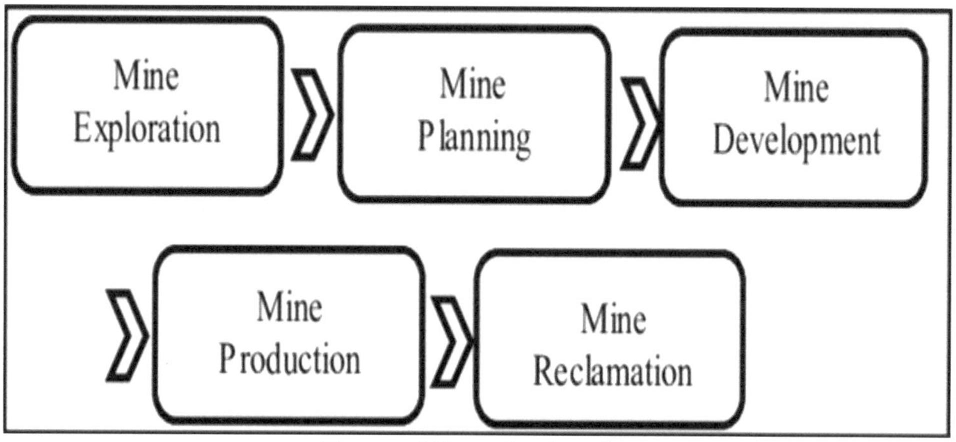

Fig. 7.2: Life cycle of a mine

## 7.4.2 Mine Planning

Mine planning is the process of developing a sort of 'business plan' to optimize the exploitation of the ore deposit with the objective of maximizing profit through appropriate capital investment in machinery, adopting the best possible mine design and production scheduling, and promoting sound financial management of mining operations. This phase is very important in developing a blueprint for how to proceed with the needed engineering plans and construction of mine infrastructure, and future production plans both in the short term and long term. All mine planning activities are implemented in the subsequent stages of mine development and production, and may continue throughout the life of a mine.

## 7.4.3 Mine Development

Mine development is concerned mainly with the construction of all the needed infrastructure at the mining site and production facilities which usually takes several years and may cost tens to hundreds of millions of dollars to complete, depending on the type of mine. Development involves extensive pre-development planning and paperwork to obtain approval from government agencies and municipal authorities. Plans are assessed regarding the: (1) the mining method and process technology that will be used, (2) building of access roads for mining and transportation, (3) identification of resources such as power and water sources, and (4) construction of ore processing facilities and disposal areas for waste. All the above-listed requirements are usually in the process of implementation, including mine access facilities, office buildings, storage areas, tailings areas, and mineral processing and haulage facilities.

## 7.4.4 Mine Production

The mineral production is the final and real phase of a mine project to extract ore and involves (1) the recovery of minerals which is the process of extracting the ore from rock using a variety of tools and machinery; (2) processing the recovered minerals through huge crushers or mills to separate commercially valuable minerals from their ores; and (3) smelting of the processed ore to extract the metal from its ore. The production processes may be integrated so that they take place within the mine site, but sometimes may require transportation to outside facilities, particularly for smelting. The operations of a mine to recover ores continues as long as the company operating the mine finds it economical to do so. Some exploration and development may continue at this stage, as well. Mine production may last several years depending on the ore reserves, although the average lifespan of a mine ranges from 25 to 30 years.

### 7.4.5 Closure and Reclamation

After the useful minerals in a mine are exhausted, the operations will cease and the mine will be formally closed. This is followed by reclamation; a process in which the land that has been mined is restored to a more natural or economically usable state. Restoration can include removing the buildings and roads, as well as covering up the open pits or re-vegetating the site. Remediation of contaminated soil and surface and/or groundwater may be involved as required by the environmental regulatory agencies.

## 7.5 Types of Deposits in Relation to Mining

Mining is the process of extracting useful commodities from ore deposits. The form and mode of occurrence of the orebodies often determine how they are extracted at a profit. Ore deposits are of various types, but in terms of mining, a simplified classification focuses on the geological controls, such as mode of occurrence and/or shape of the orebody which are very important parameters in mine design and mineral extraction. Using these criteria, five main types of deposits can be identified: (1) Vein deposits; (2) Porphyry-type deposits; (3) Stratiform deposits; (4) Placer deposits; and (5) Residual deposits.

### 7.5.1 Veins and Lodes

Veins are the most common forms of ore bodies: they are tabular and usually slightly inclined to vertical in attitude, and form in fissures or cracks in existing rocks that are filled by hydrothermal minerals. Vein deposits may have length and vertical extent of several hundred meters, and shorter width of a few meters. Mine design and extraction methods must take these physical features into consideration. Lodes are a series of closely separated veins that, if possible, are mined together in a single operation, usually by underground mining methods.

### 7.5.2 Porphyry Deposits

Porphyry deposits are unique in their occurrence as large, low-grade and mostly disseminated ores in felsic igneous rocks. Because of the proximity to the surface, large extent, low grade and large tonnage characteristics, this type of deposit is usually exploited by surface mining method (open pit) that uses huge and specialized equipment operated at relatively lower costs. This type of deposits account for most of world production of copper and molybdenum deposits

### 7.5.3 Stratiform Deposits

Stratiform deposits are horizontal or layered deposits that occur most commonly within sedimentary rocks or layered intrusive complexes. Their mode of occurrence as layers and their concordance with the host rocks help determine the best approach to mining which is normally by open pit or strip mining operations if they are close to the surface (e.g., coal deposits).

### 7.5.4 Placer Deposits

Placer deposits are concentrations of ore minerals in alluvial deposits transported by water. They are usually unconsolidated and the economic minerals are heavy and loosely held within the interstitial spaces of the granular materials. Their proximity to the surface and mode of occurrence as unconsolidated materials provide a unique opportunity to use surface and hydraulic methods of extraction (e.g. gold deposits).

### 7.5.5 Residual Deposits

Residual deposits are products of weathering that occur as blanket deposits or regolith on the surface, although they can reach depths of hundreds of meters. Their proximity to the surface allows direct excavation and the use of large, dedicated mining equipment (e.g. bauxite deposits).

## 7.6 Mining Methodology

After a mineral deposit has been discovered, explored, and evaluated, and sufficient information has been generated to warrant further consideration, the important process of selecting the most appropriate method or methods of mining usually begins. Initially, at the pre-feasibility phase, the selection is preliminary, serving only as the basis for a project layout and feasibility study. Later it may be found necessary to revise details, but the preliminary plan for ore extraction would remain a part of the final proposition.

### 7.6.1 Criteria for Selection of Mining Method

Mineral extraction is a very crucial phase in the life cycle of mining projects. Mining methods can be grouped into two main types: *surface* methods and *underground* (subsurface) methods. The selection of an appropriate mining method will require contributions from project personnel, including mining engineers, mining geologists, mine surveyors and miners/consultants who are expected to communicate and work together for the success of mining operations. Several factors are taken into consideration during the selection of either surface or underground mining, including the following:

1. type of ore deposit
2. location, size, shape, and depth of the deposit;
3. geological structure and geotechnical conditions;
4. capital requirements and operating costs;
5. ore recoveries and revenues;
6. productivity and machinery capacities;;
7. safety and potential injuries;
8. environmental impacts, pre- and post-mining;
9. reclamation and restoration requirements and costs;
10. availability of experienced workforce; and
11. Societal and cultural (community) expectations.

The above-listed factors can also be condensed under the following categories:

- *Geological* – depth to ore, thickness of overburden, shape of orebody, structural features and other geological factors.
- *Technical* – nature of equipment, size and efficiency; safety of use, and production rates.
- *Economic* –type and value of mineral resource, cost of investment, and economic viability.
- *Environmental* – toxicity of waste, proximity to water resources, impact on surrounding landscape and environment.
- *Human and Community* – work force availability, social and cultural responsibility

#### 7.6.1.1 Geological Considerations

In selecting an appropriate mining method, it is obvious that the geological characteristics and distribution of the ore body, and the depth from the surface are of utmost importance.

### *Depth of Ore Body and Surface Conditions*

Some ore deposits such as residual deposits of bauxite, clays and lateritic ores or sedimentary deposits that occur near the surface or at shallow depth are more suited to surface mining. Also, large, low-grade disseminated ores such as porphyry copper-molybdenum deposits are often mined at first, by the open-pit (surface) method, then, if there is continuity at depth (>1000 m),mining can be followed by block caving (underground) method. In contrast, ore deposits that are too deep to be mined from the surface are mined underground. Some vein deposits exposed at the surface but continue at greater depths may be mined first using surface methods and then changing to underground mining.

### *Morphology of Ore Deposit*

The mode of occurrence or attitude of an ore deposit matters in the selection of a mining method. Deposits that are horizontal or flat-lying can be mined by surface or underground methods, but not by both, whereas steeply-dipping or massive ore bodies are preferably extracted by subsurface methods, where selective mining can be implemented. Some can be mined by both methods, where surface mining transitions into underground mining. Orebodies of regular shape can generally be mined more cheaply than those of irregular shape, particularly when they include barren zones. For an open pit mine the shape and attitude of an orebody will also determine how much waste has to be removed during mining.

### *Distribution of Mineralization within Ore Body*

The variation of ore grade through the volume of an orebody may influence the mining strategy. The critical parameters are average grade, cut-off grades, and grade distribution. General grade distribution in an orebody may be uniform, uniformly variable, or irregular. Uniform ore grades are found in massive ore deposits, uniformly variable grades exhibit a spatial trend in ore grade, and irregular ore grades are found in deposits with high local concentrations of ore minerals, including vein, lens, and nugget deposits.

### *Geotechnical Characteristics of Wall Rocks*

Geotechnical characteristics of the ores and wall rocks, such as the strength and nature of discontinuities (e.g. joints, faults) are important in deciding the applicability of some mining methods particularly in underground mining where mine walls and stope stability and safety can be affected. Also, in underground mining, the decision to use supported or unsupported mine workings depends on the rock mass geotechnical properties.

### *Hybrid Mining*

While most ore deposits can be mined by either surface or underground methods, some may require mining by both methods (hybrid), starting with open-pit mining and ending up with underground mining. Such deposits are unique, and have considerable vertical continuity but with narrow width at depth that requires a multiple approach that saves costs.

Currently, about 85% of producing mines worldwide are operated by surface methods. Surface mining is often preferred to underground mining because of higher productivity due to large scale production, reduced costs and reduced development time. Other advantages of surface mining include flexibility in operations, better ore grade control, safety and working environment. However, proximity to populated areas and environmental impacts may limit the use of surface mining methods. For example surface quarries have been rejected in some urban areas, thereby resorting to underground quarries. Mining geologists contribute to the decision of selecting a mining method particularly because they are knowledgeable about the geological and geotechnical characteristics of the ore bodies and potential environmental factors are very important.

### 7.6.2 Mine Production Cycle

Mine production is the final and most expensive and durable phase of the mining cycle. During the extraction of ore from mines, significantly similar unit operations are commonly employed, independent of whichever mining method is selected. These steps in the mining process are called "*production cycle*", which include: (a) access to ore; (b) drilling and blasting (excavation); (c) loading; (d) hauling or transportation; and (e) dumping or delivery to crushers and mills where ores are processed. Getting access to the ore body is different in surface and underground mining. In surface mining, this process involves stripping the overburden, which is rock, soil, and ecosystem that lie above the ore body near the surface, whereas in underground mining, tunnels are dug and shafts are sunk to reach the ore body. The next step is the fragmentation of the ore and associated wall rocks through drilling of blast holes followed by blasting using explosives. After the ore is loose, this is followed by excavation which involves the use of heavy machinery to remove unconsolidated material or blasted ore-rock for loading and haulage from the mine by dump trucks, conveyor belts or rail system. In addition to the physical mining operations, certain auxiliary services are commonly performed, including production of compressed air for drilling, electrical power, mine ventilation, mine dewatering, or pumping and backfill distribution. Every mining method must accomplish the production cycle with minor variations related to the site conditions and equipment specifications.

## 7.7 Surface Mining Methods

Surface mining is the method of extracting ore minerals (and overburden) near the surface of the Earth. It is the most preferred mining method, and is responsible for the bulk (estimated 70%) of annual mineral production worldwide. Surface mining is usually a two-step operation: (1) removal of overburden (stripping), and (2) mining of the orebody. Surface mining is typically used for surface or near-surface and usually less valuable deposits. Surface mining can be classified into two groups on the basis of the method of extraction; (a) mechanical extraction, and (b) aqueous extraction. Also, the various methods differ in terms of the mine geometries, the techniques used, and the minerals produced.

*Mechanical extraction* methods employ mechanical processes in a dry environment to recover minerals from the ground. There are four main mechanical extraction methods: (1) quarry mining; (2) open-pit mining, (3) strip (opencast) mining; and (4) auger mining. *Aqueous extraction* methods use the access to water or an aqueous mixture during mining and processing. They recover the valuable mineral by jetting, slurring, dissolving or melting. These methods can be further subdivided into placer mining (panning, sluicing, hydraulicking, dredging) and in situ (solution) mining and leaching (Table 7.1).

Table 7.1 Classification of Surface Mining Methods

| | **Mechanical Extraction (Hard Rock Deposits)** | **Aqueous Extraction (Placer & Salt Deposits)** |
|---|---|---|
| 1 | Quarry Mining | Panning and Sluicing |
| 2 | Open-pit Mining | Hydraulic Mining |
| 3 | Strip Mining | Dredge Mining |
| 4 | ************** | Solution Mining |
| 5 | **************** | In situ Leaching |

### 7.7.1 Stripping, Cut-Off Grade and Dilution

In surface mining, the first step in the production cycle is providing access to the mineralized zones by removal of the overburden (stripping) which may include rock, soil, and ecosystem that lie above the surface. *Stripping* generally involves the use of mechanized devices for excavation and transportation of the waste material to a dump site outside the area of operation. To determine how much waste material is removed relative to the ore mined, a ratio has been formulated, known as the *stripping ratio* which is defined as the tonnage or volume of overburden or waste rock to be removed, divided by the tonnage of ore to be mined from the pit (Fig. 7.3). For example, a stripping ratio of 2:1 means that 2 tons of waste rock needs to be mined to mine 1 ton of ore. Generally, an open pit mine has a higher stripping ratio than an underground mine.

#### 7.7.1.1 Stripping Ratio

Stripping ratio is an important parameter in surface mining because it impacts operating costs and profit margin. The stripping ratio for metal mines is usually between 1:1 and 3:1 but can exceed 10:1 in mines with high-grade ore of considerable value. Stripping ratios greater than 20:1 occur in some coal mines, but are only workable if the operation is profitable. One important factor that influences the stripping ratio is the shape of the ore body which may vary considerably during the life of the mine. Consequently, as mining progresses, the unit cost of operations may increase over the life of an open pit due to increased stripping ratio and the longer haul for trucks out of the pit. Thus, an increase in the stripping ratio can render the deeper ore, more costly to mine by surface methods which may lead to the consideration of using underground methods. Stripping ratio can be calculated using the following formula.

For Metal: Stripping Ratio (Sr) = Waste rock (Volume) / Ore (Weight)

For Coal: Stripping Ratio (Sr) = Overburden Volume / Coal Weight (tons),

where volume is typically expressed as $m^3$ or cubic yard and weight is typically expressed as ton.

#### 7.7.1.2 Cut-Off Grade

Cut-off grade (COG) is generally defined as the minimum grade required in order for a mineral or metal to be economically mined (or processed). Cut-off grades are usually calculated and used at various stages during exploration and mine development, starting with the ore reserve estimates, mine design and mine operations. During calculations of ore reserves, the cut-off grade helps determine what part of an ore body can be included as ore reserves. During mine design; the cut-off grade helps determine the shape and size of open pit excavations and underground workings. Material found to be above the cut-off grade is considered to be ore, while material below this grade is considered to be waste. Cut-off grades are usually selected to achieve a certain objective, such as resource utilization or economic benefit. Sometimes blending of low-grade and high-grade mineralization may be carried out so as to maximize usage of the mineral resources.

Cut-off grades are usually determined by a combination of the geological characteristics of the deposit and the technological limits of extraction and processing, as well as by economic and social factors. The geological factors include mineralogy, grain size, and presence of deleterious constituents, shape and size of the deposit, structural complexity, or water infiltration problems. Economic factors (e.g., accessibility to markets, labor availability, current metal prices, cost of waste disposal and reclamation, capital costs and interest rates, and mining methods It is important to understand that the cut-off grade is not simply just a number but a strategic variable that has major implications on mine design and successful operations of a mine.. The cut-off grade changes with time and is usually adjusted as the economic environment changes with regard to metal prices and mining costs and is therefore a constantly changing variable consistent with the goals of mine development and profit maximization. For example, if metal prices fall or the cost of production increases, it will indirectly raise the cut-off grade.

### 7.7.1.3 Dilution

Dilution is the mixing of waste material with the ore, during production. Cut-off grades are usually strictly enforced to reduce dilution. However, dilution happens regularly during both surface and underground mining, and may be caused by the geological mix of ore and host rock as well as the nature of equipment used in mining. Dilution can be mathematically defined as the ratio (in percentage) of the tonnage of waste mined and sent to the mill to the total tonnage of ore and waste combined that are milled. The range of acceptable dilution varies between 5% and 30%. For example, if 100 tons of waste rock are mined with 900 tons of ore and all being sent to the mill, dilution is calculated as 100/100 +900 = 10%. Dilution can be classified as internal or external dilution. Internal dilution occurs within a mining block in which pockets of waste are unable to be separated and are mined with the block, whereas external dilution comes from the wall rocks surrounding the ore.

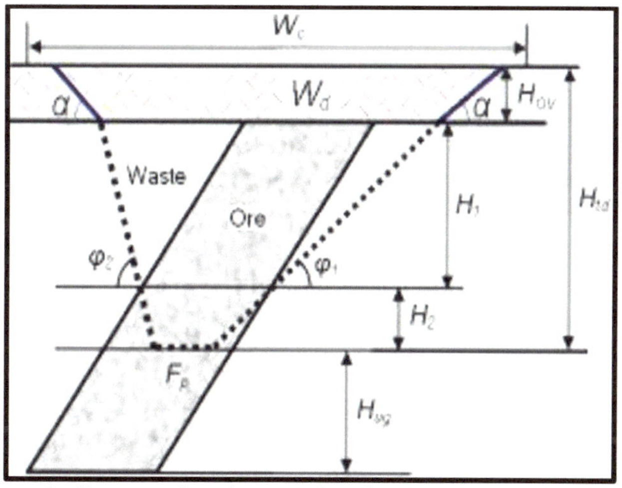


Fig. 7.3: Schematic diagram showing overburden, waste rock and ore relationship

## 7.7.2 Mechanical Extraction Methods

Mechanical extraction methods are used in the mining of mineral deposits in consolidated and competent rocks ('hard rocks') that require some force to break or crush. These include most mineral deposits (metallic and non-metallic) and industrial rocks.

### 7.7.2.1 Quarry Mining

A quarry is an open excavation or pit from which rock or minerals are extracted by digging, excavating, cutting or blasting. Quarrying is the simplest form of surface mining that is used commonly for extracting industrial minerals and rocks. Surface quarries resemble open pits described in the next section, but their benches are usually smaller. Quarrying is used when the rock is close to the surface of the land or crops out as an inselberg or on a hillside. Quarries are generally used for extracting mostly industrial minerals and rocks (e.g., feldspar and limestone), and construction and building materials, such as dimension stone, construction aggregate, riprap, sand, and gravel. Based on the nature of materials extracted, quarrying is of two types. The first type is the extraction of large ornamental stone blocks of specific colour, size, shape, and quality - an operation requiring special and expensive production procedures (Fig. 7.3a). For ornamental stone quarries, the common rocks extracted are marbles and granites. The second type of quarries applies to the recovery of sand, gravel, and crushed stone for the production of aggregates for road base, cement, and concrete (7.3b). Quarries of the second type are usually located close to urban areas where the products are in demand. For crushed stone aggregates, the most commonly used rocks are granites, gneisses, basalt and dolerite. The production process includes excavation which may be preceded by drilling and blasting, followed by crushing and screening into various sizes. Sand and gravel are usually excavated from pits in river beds and terraces using excavators. For ornamental stone quarries, rocks of high strength and beauty are preferred,

In the geological investigations for new quarry sites, focus should be placed on locating outcrops of the appropriate rock types that are exposed above ground and not weathered. Other factors for consideration include thickness of overburden, water table conditions, intensity of joints and fractures, nature of exposure, size of reserve, accessibility and absence of forest cover. Water flows into the quarry from the surface and from the ground may require dewatering. Many quarries are dug in hillsides so that water can drain better. For sand and gravel quarries, riverbeds or terraces provide the best locations.

Laboratory tests for determining suitability as aggregates include, abrasion test, crushing test and angularity test, for quarries close to urban areas, special measures are taken to minimize environmental impacts such as noise from drilling and blasting, and dust control from crushing and screening. Quarries for dimension stones are less common but require some specific requirements. The characteristics that make a deposit commercially viable for dimension stone include color, appearance, competence, uniformity, and freedom from defects. The rocks are usually cut into huge blocks along benches, and hauled to the factory where the blocks are further cut or sliced. Quarries are also used for mining industrial minerals such as feldspar, and talc, and industrial rocks e.g. limestone, marble and shale that are used in cement and lime production.

Many quarries fill with water to become ponds or small lakes after mine closure while others have been converted to landfills.

Fig. 7.3: (a) Quarry for marble dimension stone (b) Quarry for crushed stone

#### 7.7.2.2 Open-Pit Mining

Open-pit mining is a surface mining technique of extracting economic minerals and rocks from the earth's surface by their removal from an open pit. Other terms that are used to describe this technique are *open-cut mining* or *opencast mining*. Open-pit mines are used when deposits of commercially useful minerals or rock are found near the surface. Typical ore deposits mined in open pits if exposed at the surface are vein-type, pipe-shaped (e.g., diamond in kimberlite pipes), stratiform deposits, residual deposits and widely-disseminated deposits (e.g. porphyry type Cu-Mo or Au deposits). Copper, gold and iron deposits account for over 90% by volume of open-pit excavations worldwide.

Open-pit mines are excavated in horizontal layers called benches. Each bench is a horizontal surface of defined height and width (Fig. 7.4). The deposit is mined in vertical steps by adding more benches, although mining can be carried out on a number of benches at the same time, and a pit of any depth can be excavated. The top of each bench is equivalent to a working level, and access to different levels is gained through a system of ramps. The width of a ramp depends on the equipment being used, but typical widths are from 20 to 40 meters. Mining on a new level is begun by extending a ramp downward and a new cut is then progressively widened to form the new pit bottom. As excavation proceeds, successive benches which are arranged in circular form around the nine become smaller in diameter than the last one above, and narrower to maintain safety and stability inside the mine. With time, a cone-shaped geometry is produced in which the bench area becomes progressively smaller as the pit deepens with the walls formed by cascading benches that may sometimes be steeply dipping if the rocks are very competent and stable.

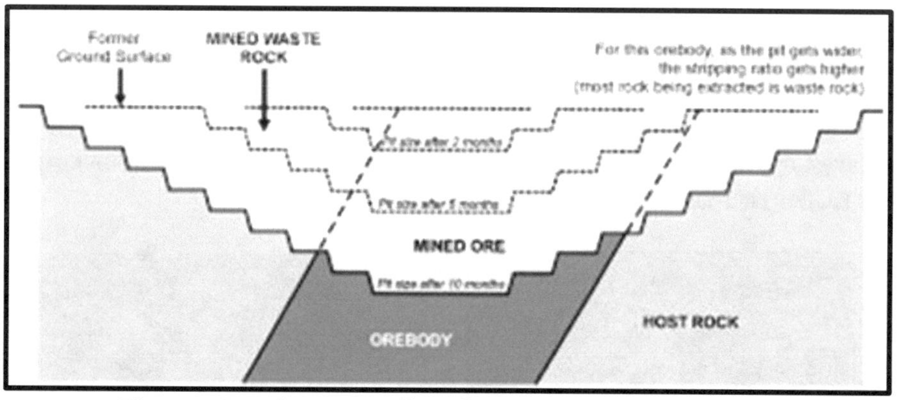


Fig. 7.4: Development of benches in open pit mine design

***Pit Geometry***

Open pit geometry refers to the dimensions of the pit and the benches. The *height* of the benches is the vertical distance between the highest point of the bench (crest) and the bottom (toe) (Fig. 7.5a). The height of these benches depends on the type of deposit, mineral being mined and the equipment being used. Typically, it varies from 2 to 30 m, but in general, for large mines it ranges between 12 to 15 meters, and less than 10 meters for smaller mines. The *width* of the benches is generally about 15 to 40 meters, and varies according to size of machinery being used and the type of bench. There are three main types of benches: (a) working benches where the ore is blasted and mined; (b) inactive benches which are remnants of working benches left in place for slope stability, and (c) safety benches to catch loose materials falling from benches above. Some benches are widened to serve as haul roads with downward ramps in which trucks can drive to another bench or, carrying ore and waste rock out of the mine. The walls of a pit have a certain *slope* that is determined by the strength of the rock mass and other factors. The pit slope is a very important parameter with respect to slope stability and mine safety. The pit slope is defined by two angles: the angle of the overall pit, and the angle of individual bench faces. Generally, bench faces are mined as steeply as possible but less than vertical, usually from about 60 to 80 degrees depending on the geotechnical characteristics of the rocks and equipment used. Increasing the pit slope angle by only a few degrees can decrease stripping costs tremendously or increase revenues through increased ore recovery, but it can also result in a number of slope failures on a small or large scale that can be costly in losses and operational delays.

The *depth* of the modern open pits varies from several tens of meters to more than a kilometer. Pit depth depends on the geometry of the ore body and the overall cost of production for which the overall pit wall angle remains one of the most important controlling factors. The pit wall angle must be maintained in a balanced position that will ensure overall slope stability while maximizing ore extraction. As a pit is deepened, more and more waste rock must also be stripped or moved away in order to uncover the ore. This may lead to unfavorable and high stripping ratios. Open-pit mines continue to be enlarged until either the mineral resource is exhausted, or an increasing stripping ratio that makes further mining uneconomic. When any of these happens, mining ceases, or if the orebody such as a lode or pipe still continues underground, mining may transition to using subsurface methods such as block caving or shaft mining. However, where the ore is of high value or precious gemstone such as diamond that occur as pipe-like bodies in very stable mafic host rocks, mining to extract more ore can continue to greater depth with relatively steep pit angles. Open-pit diamond mines (e.g. South Africa, East Siberia) are some of the world's deepest mines with steep pit walls (Fig. 7.5b). In the case of exhausted mines, the open pits may be closed and converted to landfills for disposal of solid wastes. Where the ore is a precious stone such as diamond, mining can continue to greater depth.

The size of open pits varies considerably, and includes some of the largest mines in the world. The world's largest open pit diamond mine near Mirny in Siberia is 540 m (1600 ft) deep, 120 km (75 miles) across and yields approx. 2 million carats in diamonds per year. The area above the mine is a no-fly zone since it creates its own downdrafts (pulldown) which can endanger low flying aircraft. The Bingham Canyon Mine, near Salt Lake City, is the world's deepest man-made open pit excavation for porphyry copper. The mine is 2.75 miles (4,5km) across and 0.75 mile (1,2km) deep.

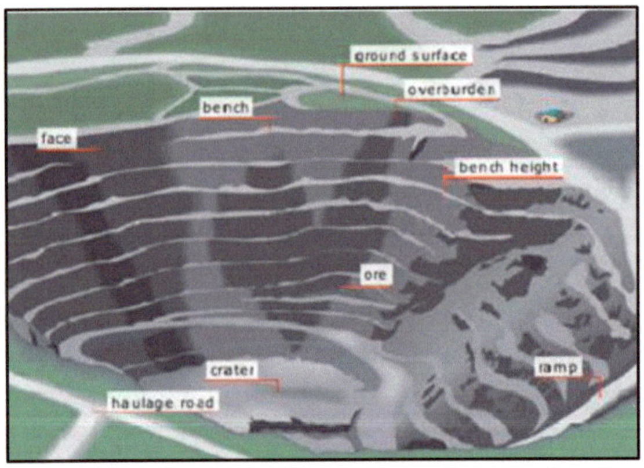

Fig. 7.5: (a) Schematic illustration of open pit mine, (b) Steep walls of open pit diamond mine

***Open-Pit Bench Terminology***

The following are the terms used to describe parts of benches (Fig. 7.6).

*Bench:* Ledge that forms a single level of operation above which mineral or waste materials are mined from the bench face

*Bench top:* Horizontal part of the bench limiting the bench height at the top

*Bench floor*: Horizontal or near horizontal part of the bench limiting the bench height at the bottom.

*Bench crest*: Edge formed at the intersection of bench top and bench face.

*Bench toe*: Edge formed at the intersection of bench face and bench floor.

*Bench height*: Vertical distance between bench floor and bench top.

*Bench width* Horizontal distance between bench toe of the upper bench and bench crest of the next lower bench.

*Bench slope angle (working slope)*: is the angle formed between the slanting edges of the bench face with the horizontal.
*Pit slope angle (ultimate slope):* is the average slope angle for the pit formed by an imaginary line drawn connecting bench toes and the horizontal.
*Berm*: Ledge or waste rock piled alongside a haulage road or along the edge of a bench or dump, intended as a safety measure.
Ramp: Moderately inclined access ways from the surface to mining levels for haulage of ore, materials, waste, workers, and equipment.

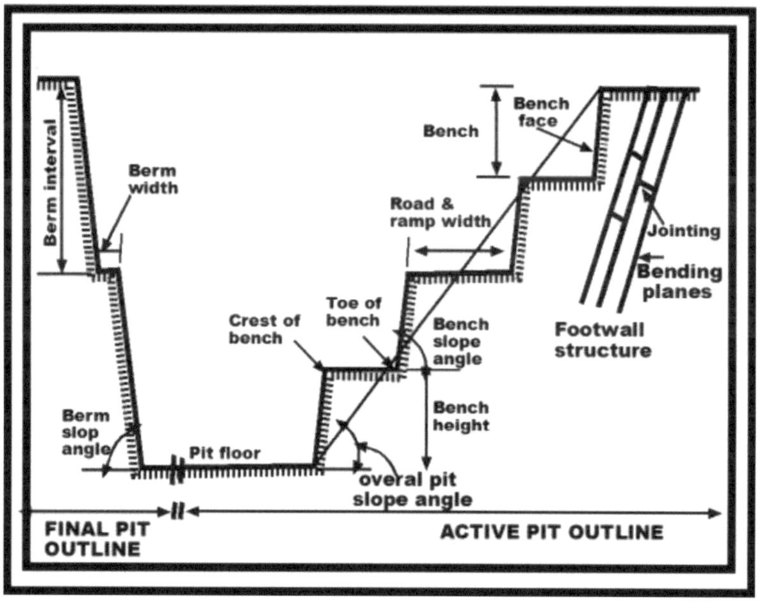


Fig. 7.6: Diagram showing the configuration of benches in open-pit mines

***Open Pit Design***

Open pit mines are excavations or cuts made at the ground surface to extract ore throughout the operational life of the pits. The *design* of open pits often require substantial geological and geotechnical information about the ore and host rocks, nature of structural features (folds, faults, fractures and joints), rock mass properties and groundwater flow. All these factors affect the efficient extraction of minerals and related processes such as rock blasting, ore grade control, slope stability and safety, and dewatering of mines. Most of the needed geological and geotechnical information are collected during the detailed exploration, site investigation and feasibility studies, but continuous monitoring and changes are implemented throughout the life span of the open-pit mine. The most economic final pit design will ultimately depend on a combination of factors including the geometric outline of the ore body, the distribution of ore grades within the ore body, topography, maximum allowable slope angles, etc. The ultimate pit outline specifies the areas in which ore bodies could be mined economically.

The first step in the evaluation and design of an open-pit mine is the delineation of the orebody to be mined and calculation of ore reserve estimates based on vertical holes drilled throughout the mineralized zone to provide a good idea of the ore body's vertical and lateral extent. From the vertical sections of the orebody, the tentative locations of the benches are selected. However, since the deposit is to be mined in horizontal benches, it is also convenient to evaluate the ore body in horizontal sections, with the thickness of each section equal to the height of the proposed bench. These horizontal sections are divided along coordinate lines into a series of blocks, with the plan dimensions (i.e., the length and width) of each block generally being one to three times the bench height. After the ore grade of each block has been determined, the blocks are assembled into a *geological block model* representation of the ore body (Fig. 7.7). After creating the grade (geological) block model, an *economic block model* is prepared based on some assumed economic parameters e.g., cost, revenue, and the total cash flow of mining operations. Such models can be used in the ultimate open-pit design and production planning with the main objective of maximizing the overall profit. Computer programs and algorithms are used for optimization of the open pit design and production plan.

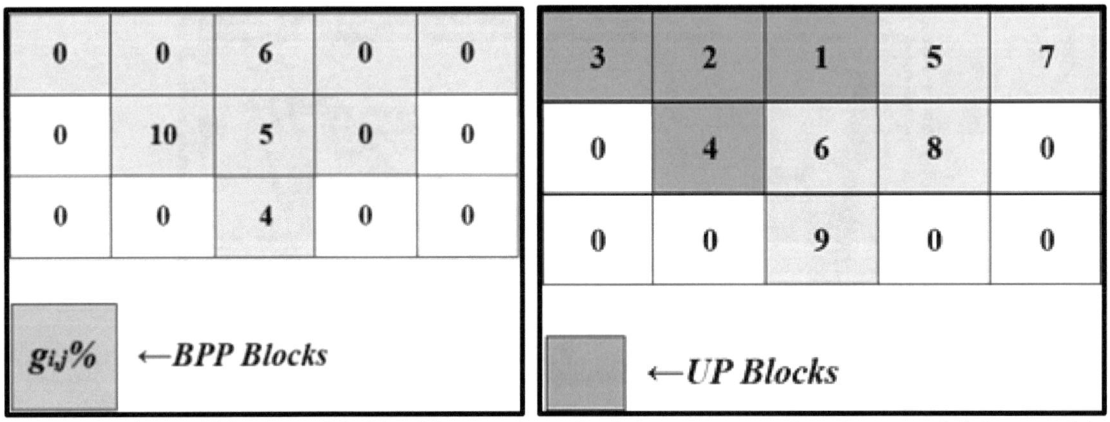


Fig. 7.7:(a) Geological block model (b) Ultimate pit and mining sequence for Fig 7a model (BPP -= biggest possible pit; UP = ultimate pit.

Figure 7.7 (a) is an example of a simple geological block model design based on ore grade and net value of each block. The higher the numbers the better the block value. The geological model is further optimized by considering the positional weight (PW) of each block which is the sum of the grades of ore blocks within the downward cone of a block without considering its grade. The meaning of a block PW is the metal amount that can be obtained from the underneath blocks after this block in the extraction path is mined. The higher the PWs of a block, the more lucrative is the block. The numbers in Fig. 7.7b are the recommended mining sequence.

### *Surface Mine Production Cycle*

Prior to the production of ore, mine development work is required in open-pit mines. This is a form of capital investment that involves the building of access roads and stripping the overlying waste material in order to expose the ore and establish the initial bench features. The production cycle in most open-pit mines consists of four unit operations; *drilling, blasting, loading* and *hauling*. These operations are accomplished by using systems of machinery for drilling, excavation, loading and haulage. The ore and waste rock are loaded and transported to a location outside the mine for storage or processing. Usually the waste rock goes separately to a dump while the ore goes to the processing facility. It is very cost effective to separate the ore from waste rock as much as possible to minimize dilution. The largest open-pit operations can move almost one million tons of material (both ore and waste) per day. In smaller operations the rate may be only a couple of thousand tons per day.

The types of equipment or machinery used in open pit mining are listed as follows:

a. drilling rigs: rotary drills (for large holes), percussion DTH and reverse circulation drills.
b. excavators: power shovel, hydraulic shovel, excavators, front end loader and others.
c. loaders: wheel loaders and power shovels

d. haulers, trucks, conveyor belt and rail.
e. other equipment: bulldozers and graders

The excavators are equipment used to extract loose materials, such as overburden, and ore and rocks after blasting. In most open-pit mines of moderate to large size, the dominant excavators are hydraulic with face shovel or backhoe configuration. The face shovel is preferred in harder rocks, whereas the face backhoe allows more digging of loose material. The bucket wheel excavators are rarely used except in strip mining of soft ores or rocks such as phosphate. Loaders lift the ore or waste material onto the trucks for removal from the mine. Loader types can include electric rope, hydraulic excavators including backhoe excavators, and front-end loaders (also called wheel loaders). Haulers used in transporting ore and waste rock from the mines are mostly trucks (off road trucks), and rarely railways. Rail haulage is restricted in surface mines due to the difficulty of attaining horizontal track placement. Mining trucks can be huge, varying from less than 40 tons to 400 tons. The size and cost of operating mining trucks is directly proportional to its tray capacity, while the speed at which the truck can travel is inversely proportional to its capacity.

### 7.7.2.3 Strip Mining

Strip mining or open-cast mining is a form of surface mining that involves stripping the surface away from the mineral that's being excavated (usually coal), and the overburden is backfilled into the mined area. The technique is suitable in the mining of shallow, large, tabular or stratified residual and sedimentary deposits such as bauxite, phosphate, evaporites, tar sands, manganese, but is most commonly in the mining of coal and lignite deposits.

The mining process often involves first removing a long strip of overlying soil and rock (overburden) to expose the underlying mineralization or seam which is excavated en masse. The overburden is dumped directly into previously mined areas rather than outside the pit, compared to open-pit mining. The process of excavation, loading and transportation to the conveyor belt is handled by a single machine which are draglines or continuous excavators, notably the "bucket wheel excavator". These are some of the largest and most powerful machines for mining soft rock and unconsolidated deposits, and can move as much as 12,000 cubic meters of earth per hour. (Fig. 7.8a).

A bucket wheel excavator is made up of a large, round wheel with a configuration of scoops which is fixed to a boom and is capable of rotating. Material picked up by the cutting wheel is transferred back along the boom into a stationary chute through which all of the buckets discharge. (Fig. 7.8a). A dragline bucket system consists of a large bucket which is suspended from a boom (a large truss-like structure) with wire ropes (Fig. 7.8b). The bucket is maneuvered by means of a number of ropes and chains. The hoist rope, powered by large diesel or electric motors, supports the bucket and hoist-coupler assembly from the boom. Drilling and blasting continues in advance with the movements of the excavators. There are three forms of strip mining; area stripping, contour stripping and mountaintop removal.

#### *Area Strip Mining*

This is the most common type of strip mining which is used on fairly flat terrain, to extract deposits over a large area. After removal of overburden, the mines excavate large rectangular pits, developed in a series of parallel strips or cuts which may extend several hundred meters in width and more than a mile in length. As each long strip is excavated, the overburden is placed in the excavation produced by the previous strip. This process continues along parallel strips of land so long as the stripping ratio makes it economically feasible to recover coal. Mining may cease in a particular area, for example, where the coal seam becomes thinner or where the seam pitches further below the surface. Mining equipment are the large excavators and draglines (Fig. 7.9a).

### *Contour Strip Mining*

The second method called contour *stripping* is used almost exclusively in mountainous regions such as the Appala-chian region of the United States (Fig. 7.9b), where coal seams outcrop from the sides of hills or mountains. Contour mining makes cuts on the slope where the coal seam that follows the contour of the land is located. The overburden is first removed followed by the coal itself. Overburden from adjacent cuts is used to fill previous cuts. The operator continues making cuts until the ratio of overburden to coal becomes uneconomical. Contour mining uses small earth-moving equipment such as power shovels, backhoes and bulldozers - similar to equipment used for many other kinds of construction activities. It is therefore a relatively cheaper method to operate.

### *Mountaintop Removal*

Mountaintop removal mining (MTR) is a radical form of strip mining that uses explosives to blast "overburden" off the top of mountains to reach stratiform ore deposits or coal seams at depth. The method is used to extract mineral beds in mountainous terrains. The removed waste material is usually deposited in a nearby valley, effectively filling it in and completely reworking the landscape. Mountaintop removal requires more capital and engineering skill than the contour mining method, but it allows the operator to extract virtually the entire coal seam.

Fig. 7.8: (a) Wheel excavator (b) Dragline excavator

Fig. 7.9 (a): Strip mining of phosphate, Togo (b) Contour strip mining of coal, Appalachia

### 7.7.2.4 Auger Mining

Auger mining is a surface mining technique that involves the drilling of horizontal large diameter holes to recover additional coal from a seam located behind a highwall produced either by stripping or open-pit mining. Augering is associated with strip mining of coal and conducted to maximize recoverability of mineral reserves remaining after the mining operations are completed. This method is specialized and of low cost, and limited to horizontal or slightly dipping coal seams that have been exposed by geologic erosion or along a pit face.

## 7.7.3 Aqueous Extraction Methods

These are methods that involve the use of liquids including water to assist in the extraction of minerals. They can be subdivided into two major groups: placer mining and solution mining. *Placer* mining is used to mine unconsolidated mineral deposits of detrital material such as sand and gravel that contain valuable heavy minerals such as diamonds, native gold, cassiterite, gemstones, platinum, and titaniferous mineral sands. These are minerals of high specific gravity and physical toughness. The two most economically important types of placer deposits are stream and beach placers. The placer mining techniques all involve the use of water to separate the valuable minerals from waste sands, but the choice of any particular technique depends on the extent, thickness, and orientation of the deposit; as well as the source and quantity of water available for mineral extraction.

In situ mining or solution mining is the extraction of the useful elements or minerals by dissolving them in a liquid without physical removal of the solid material, and the liquid is later removed for recovery of the valuable constituent. Extraction takes place "*in situ*", which means "in place". In-situ mining includes *in-situ leaching*, and *solution mining*. In situ mining techniques are most applicable to evaporite deposits that can easily be dissolved by aqueous solutions as well as sedimentary uranium and copper deposits that contain soluble primary and secondary minerals.

### 7.7.3.1 Panning and Sluicing

Panning is the simplest and most labor-intensive technique of placer mining. It involves filling a pan with placer dirt or sand and then submerging in still water. While underwater the contents of the pan are shaken with both hands until all the clay has dissolved and the lumps of dirt are thoroughly broken. Stones and pebbles are also picked out. Then the pan is held flat and shaken under water to permit the valuable mineral to settle to the bottom, and, in a series of quick motions, the pan is tilted and raised repeatedly until the lighter top material is washed off and the valuable mineral is left for harvesting (Fig. 7.19a).

In sluicing, a slightly sloping wooden trough called a *sluice box* is used as a channel along which mineral-bearing placer dirt or gravel is carried by a stream of water. As the water flows downward, specially shaped riffles made of transversely mounted wooden bars or poles agitate the current, preventing lighter material from settling while retaining the valuable heavy mineral (Fig. 7.10b).

Fig. 7.10: (a) Panning for Alluvial Gold (b) Sluicing for Heavy Minerals

#### 7.7.3.2 Hydraulic Mining

Hydraulic mining uses high pressure jets of water to dislodge sediment or rock material that contain valuable minerals (Fig. 7.11a). It is a low-cost method used in the mining of large quantities of unconsolidated material such as alluvial, eluvial and beach placer deposits, particularly of gold and tin, as well as residual deposits of clay and lateritic iron oxides. Hydraulic mining involves three steps: *hydraulicking, sluicing*, and *educing*. Hydraulicking is the process of breaking breaks up material and suspends it in a slurry usually by using a high pressure, large water cannon, followed by a process of diverting the sediment slurry through sluice boxes known as sluicing. Finally, the slurry is channeled into an enclosed circuit by a process called educing. Also, the method is applied in the mining of mineral sands along shoreline or coastal areas.

#### 7.7.3.3 Dredging

Dredging is the underwater excavation of a placer deposit or beach sand using floating equipment called a dredge. Dredging is the most common method of large-scale extraction of placer deposits, particularly around islands where the water is shallow. There are two types of dredges; *mechanical* or *hydraulic*, depending on the mode in which the material is transported to the surface. The mechanical dredges are excavators that utilize a mechanical method using steel "buckets" on a circular, continuous bucket line at the front end of the dredge (Fig. 7.11b). The material is then sorted/sifted using water and the material is dumped into containers for further separation. Hydraulic dredges use suction to suck placer material through a giant hose from the seabed or bottom of a river using water as the transport medium. The power used in suctioning comes from a jet-power intake. After material is suctioned up to the surface, it usually gets processed through a floating sluice at the surface. Any waste material from the box is then usually discharged over the side of the dredge. The valuable minerals mined by this method include gold, diamonds, cassiterite, precious stone (corundum) and heavy mineral sands. Sand or construction are also mined along the coastal areas using mechanical dredges.

Fig. 7.11: (a) Hydraulic Mining (b) Dredging of Placer Gold in Sediments

#### 7.7.3.4 In-Situ Leach and Solution Mining

In situ leach mining, also known as solution mining, or in situ recovery (ISR) in North America, involves leaving the ore where it is in the ground, and recovering the minerals from it by dissolving them and pumping the pregnant solution to the surface where the minerals can be recovered. Consequently there is little surface disturbance and no tailings or waste rock are generated.

***In-Situ Leaching (ISL)***

In-situ leaching (ISL) is a mining process that involves leaving the ore where it is in the ground, and recovering the minerals from it by dissolving them and pumping the pregnant solution to the surface where the minerals can be recovered (Fig. 7.11a). Consequently there is little surface disturbance and no tailings or waste rock is generated. However, the orebody needs to be permeable to the liquids used, and located so that they do not contaminate groundwater away from the orebody. This method is commonly used to recover minerals that are leachable such as uranium and copper through boreholes drilled into the deposit, from which the solutions are administered. Sometimes waste rock or tailings may be piled into heaps, and leaching solutions are passed through them to recover any leftover ore. Such a process is called "heap leaching", and it is used to recover gold and copper from tailings.

***Solution Mining***

Solution mining is slightly different from ISL in that the whole water soluble rock materials or minerals underground are converted to liquid using aqueous solutions (Fig. 7.11b). It is most applicable to water-soluble deposits of sodium and potassium evaporite salts and sulfates. Mining is performed by pumping aqueous solutions into underground wells that dissolve the whole solid minerals into a liquid. This liquid containing the minerals is then pumped out of the ground and the mineral can then be recovered by various techniques. The method has also been applied to the extraction of uranium ores hosted in porous sandstone, where the uranium salts are flushed into solution. Solution mining is more economical, safer and more environmentally friendly than conventional underground mining.

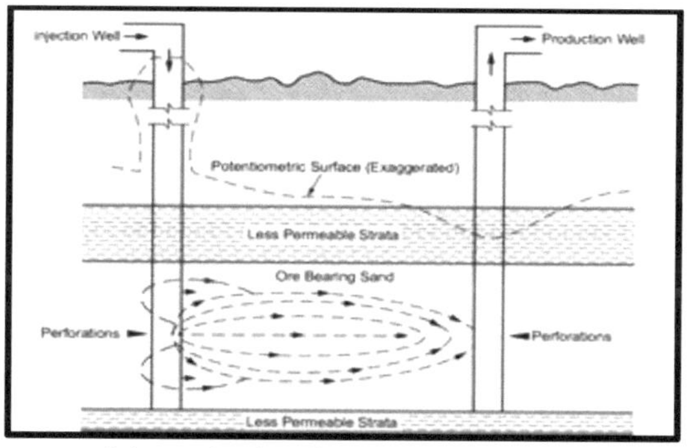


Fig. 7.11: (a) In situ leaching of ore in sand bed

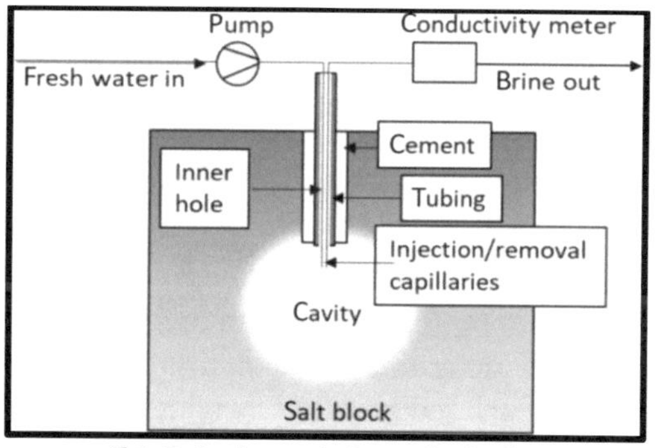


(b) Solution mining of salt deposit

## 7.8 Underground Mining

Underground or subsurface mining is the extraction of ores from deposits that occur beneath the surface that cannot be mined by open-pit methods due to technical or economic reasons. Such deposits are buried and can only be reached by digging tunnels or shafts into the Earth's crust (Fig. 7.12). The main objective of underground mining is to extract ore from the subsurface as safely and economically as possible while minimizing the rock waste that is generated. Underground mining is most applicable to ore deposits that occur deep below the surface, usually where the overburden is thick or the mineral occurs as veins in hard rock. Apart from the characteristics of the ore deposit, other technical, economic and environmental factors may determine the choice of underground mining.

Once a decision has been made to go underground, the specific mining method selected depends on several factors including the size, shape, and orientation of the ore body, the grade of mineralization, the strength of the rock materials, and the depths involved. For example, if the ore is very high grade or carries a high price, then a higher cost method can be used. In order to minimize the mixing of ore and waste, highly selective extraction methods are available, but if ore and waste can be separated easily after mining, then a less-selective bulk mining method may be chosen. The dip of the ore body is particularly important in method selection and ore transportation. If the dip is greater than about 50°, then systems using gravity to move the ore can be considered. If the dip is less than about 25°, then systems using rubber-tired equipment for ore transport can be considered. The strength of the rock mass and ore are also a factor in deciding if mine workings will require artificial support.

For minerals that occur deep below the surface - where the overburden is thick or the mineral occurs as veins in hard rock - underground mining methods extract the valued material most efficiently and cost effectively. Sometimes, ore deposits that exist at or near the surface may extend to greater deposits which may necessitate a transition from surface mining to underground mining.

Fig. 7.12: Underground mining in South Africa

## 7.8.1 Orebody Properties Influencing Underground Mining Method

Underground mining is a more complicated process than surface mining, and several factors, particularly the orebody characteristics are considered in the choice of an appropriate underground mining method. The orebody properties of interest are:

1. Geometric configuration of an orebody which indicates the relative dimensions and shape is a major factor in determining an appropriate mining method. For example, coal seams or stratiform deposits of sedimentary origin; veins, lenses, and lodes of hydrothermal emplacement or metamorphic processes; regular shapes of porphyry copper orebodies are distinct geometric factors that determine the method selection.
2. Disposition and orientation are concerned with depth, dip, conformation, etc. For example, mining in a heavily faulted environment may require a capacity for flexibility and selectivity in stoping due to sharp changes in ore distribution.
3. Size of orebody. Large, geometrically regular deposits may be suitable for mining using a mechanized, mass-mining method such as block caving. Small deposits of the same ore type may require selective mining.
4. Geomechanical setting - the response of a rock mass to a particular mining method depends on the geotechnical properties: strength, deformation characteristic and nature of discontinuities (joints, faults, and shear zones) and pre-mining state of stress.
5. Orebody value and spatial distribution of value are important parameters that define selective mining and boundaries of production. Average grade defines the size and value of the deposit while marginal grade of ore may be excluded from the production. Devising a mining strategy for recovering higher-grade domains is required where grade varies.

## 7.8.2 Underground Mine Design -Geological/Geotechnical Considerations

The main purpose of mine design is to ensure that the ore is accessed and extracted efficiently and cost effectively. The design and development of underground mines, just like in surface mining, usually require geological and geotechnical information on the distribution and characteristics of the ore body and the surrounding wall rocks. This is a very important consideration in determining where and how to access the orebody and the size of openings that can be constructed to establish stopes for the ore extraction. The geotechnical consideration in underground mine design relates to the strength of the ore and wall rocks to ensure the overall stability of the complete mine structure.

The geological characteristics of relevance include the thickness and orientation of the mineralization, the ore and rock strength, the distribution of mineralization within the ore body, and the depth of mineralization and surface conditions. The geotechnical features of importance are the strength of the rock mass and the nature and size of discontinuities such as faults, folds and fractures that can affect the stability of mine workings. The hydrogeological conditions, particularly groundwater flow and seepages within mine workings must be controlled to prevent water from entering the shaft and other mine workings. The need for mine support (rock support) must be determined based on the geological characteristics and strength of the mine rocks and ores. Depending on the geomechanical properties of the rock mass, underground mines may require no support at all, or a little amount of support, or complete support. Obviously, rock support is needed in the following situations: (a) where weak or unconsolidated parts of rocks are present in the roof and walls of mines: (b) where there are some geologic structures (e.g. shear zone) that weaken rocks; (c) where there are voids and spaces in the rocks; (d) where clay beds or weathered rocks occur; (e) where there is groundwater seepage. The availability of geological information and appropriate remedy implementation are critical to underground mining.

## 7.8.3 Mine Development and Operations

Prior to the production of ore, a certain capital investment in mine development work is required. In open-pit mines this consists of building access roads and stripping the overlying waste material in order to expose the ore and establish the initial bench geometries. For an underground mine the development work is considerably more complicated because it requires providing access to the orebody at various levels by constructing vertical, sub-vertical and horizontal structures. Some of the development components of an underground mine are illustrated in Figure 7.13.

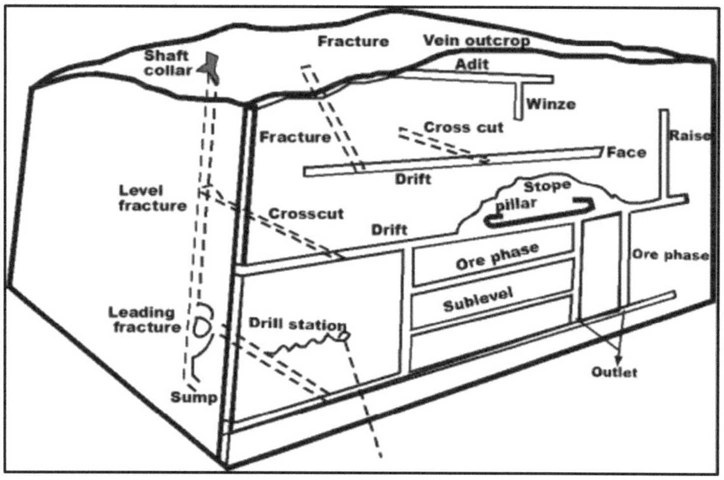


7.13: 3D Sketch map of mine openings in underground mine

### 7.8.3.1 Vertical Access Ways

Access to an underground mine is provided from the outside in three ways, by a shaft, ramp (decline) and adit. The ***shaft*** is excavated, or sunk, from the surface downward to a depth somewhat below the deepest planned mining horizon. At regular intervals along the shaft, horizontal openings called drifts are driven toward the ore body. Each of these major working horizons is called a ***level***. Most shafts are divided into a number of compartments each with a different purpose, such as one compartment for moving people, a second for transporting ore to the surface, and other compartments for ventilation and electrical wiring. Other subsidiary shafts may originate inside the mine. Another means of providing access to a mine is through a ***ramp*** *or decline* which is an inclined tunnel usually sunk at a low slope angle (<20° dip) with a portal to the surface. Ramp access is the common selection in shallow ore bodies, especially where the mineralization is near horizontal. A ramp from the surface can facilitate machine movements and transport of people and materials. Mines can both have shafts and decline access ways. Another access way is an ***adit*** - a horizontal or inclined tunnel excavated into the side of a hill or mountain to provide access to an ore body.

Ore that is mined on the different levels is dumped into a vertical or near-vertical opening called ***ore pass*** or *ore chute*, through which it falls by gravity to the lowest level in the mine. Other alternative techniques for ore transport are conveyor belts and truck haulage. Vertical or subvertical connections between levels generally are driven from a lower level upward are called raises with diameters of 2 to 5 meters and lengths up to several hundred meters, drilled by powerful raise-boring machines. The openings so created may be used as ore passes, waste passes, or ventilation openings. An underground vertical opening driven from an upper level downward to a lower level is called a *winze*.

### 7.8.3.2 Horizontal Openings

After accessing the ore body through a shaft, underground workings are continued on successive horizontal openings called ***levels.*** These are simply horizontal tunnels developed from stations cut into the side of the shaft with a size and shape depending on their use e.g. haulage, ventilation, or exploration. Level development provides haulage ways and means **of** access to the working places. ***Drifts*** are horizontal or nearly horizontal underground openings developed on the underground levels along the strike of ore body from which ore extraction may begin. A drift running parallel to the ore body and lying in the footwall is called a footwall drift (located at the lower ore-waste contact, at the 'foot' of the ore body), and drifts driven from the footwall across the ore body are called ***crosscuts*** which are horizontal tunnels intersecting the ore body across its strike. The crosscuts are usually developed to connect the drifts with the area in which stoping occurs. Thus, the main drifts and crosscuts serve as a base from which to drive auxiliary drifts, crosscuts, raises, and winzes for the purpose of exploring and blocking out the ore bodies and preparing them **for** economical extraction by stoping.

Extraction of ore takes place from the stopes which **are** the underground excavation made by removing ore from surrounding rock. Stoping is usually initiated by blasting the orebody and excavating the broken pieces into the slot. Within a mine there are different types of stopes; *production stopes* where ore is being extracted and *development stopes* where preparations are beginning. Stoping techniques are only practiced in underground mining when the rocks on the hanging wall or roof of a mine are strong enough to withstand the drilling, blasting, and removal of ore without artificial support. *Overhand* stoping is the process of removing the ore from above the level, while *underhand* stoping is removing the ore from below. The ***face*** of a stope is the surface where the mining work is advancing.

### 7.8.3.3 Underground Mine Infrastructure

Underground mines are complex structures with varied infrastructure and a variety of underground workings. Geologists must be conversant with the distinctive terminology that is used to describe these features. The infrastructure of underground mines is more complex than that of the open-pit mi, and can be classified into three groups: (a) access openings; (b) haulage openings and (c) development or production openings (see Fig. 7.14).

***Shaft*** is a vertical or sub-vertical excavation sunk adjacent to an ore body in which elevators are used to transport people and ore in and out of the mine.

***Skip*** is the portion of the shaft used for ore transport.

***Ramp*** or Decline is an inclined tunnel sunk at a low slope angle (<20° dip) with a portal to the surface.

***Adit*** is a horizontal or inclined tunnel excavated into the side of a hill or mountain to provide access to an ore body. The surface entrance to an adit is called a *portal.*

***Manway*** is underground opening that is intended for personnel access and communication

***Level*** is a system of horizontal underground workings connected to the shaft. A level forms the basis for excavation of the ore above or below.

***Sublevel*** is a system of horizontal underground workings located between the main development level for more effective drill and blast control.

***Raise:*** Underground opening driven upward from one level to a higher level or to the surface; a raise may be either vertical or inclined.

***Winze*** is a vertical or inclined underground opening driven downward from one level to another level or from the surface to a level

***Drift*** is a horizontal or nearly horizontal underground opening developed on the underground levels along the strike of ore body.

***Crosscut*** is a horizontal underground tunnel intersecting the ore body across its strike. The crosscuts are usually developed to connect the drives with the area in which stoping occurs

***Chute*** is a loading arrangement that utilizes gravity to move material from a higher level to a lower level.

***Cone*** is a funnel-shaped excavation located at the top of a raise used to collect rock from the area above.

***Stope*** is the underground excavation made by removing ore from surrounding rock.

***Face*** is the surface where the mining work is advancing.

***Back***: Roof or overhead surface of an underground excavation

***Slot*** is a vertical or inclined ore section excavated to open up for further stoping.

***Pillar*** is a block of ore or barren rock left intact in the mined out stope or between two stopes to act as a means of support.

***Footwall*** is wall or rock under the deposit

***Hanging wall*** is wall or rock lying above the deposit

***Draw Point*** is a place from which ore is extracted from the stope and loaded for haulage by trucks or conveyor belt.

***Ore Pass*** is a steeply dipping underground opening used for passing ore from one level to another level.

***Wall Rock*** is the wall in which an ore deposit is enclosed.

***Waste*** is the barren rock or rock of too low a grade to be mined economically.

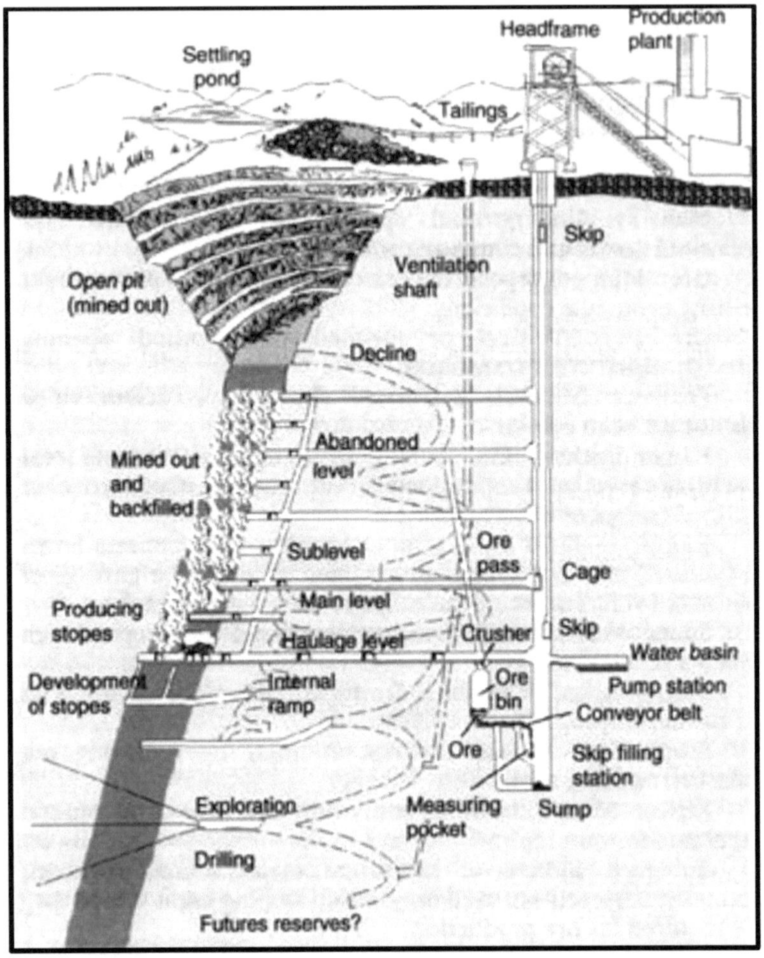

Fig. 7.14 Major Infrastructure of an Underground Mine

## 7.8.4 Underground Mining Methods

Underground mining methods are complex and varied because some were developed to satisfy the needs of different shapes of orebodies and the geotechnical characteristics of host rocks. The classification of underground mining methods is complex because of the unique ways in which the orebodies occur and the mining method had to be adapted to fit prevalent conditions. However, some parameters have been developed, both geological and mining related, to help classify underground mining methods including the type of access shafts used, the extraction method, morphology of the ore body, and the extent of support required. Mining in hard and soft rocks may require different techniques, due to the fact that hard rocks are more stable under unconfined mining conditions than soft rocks. Also, underground mining can involve *bulk* or *selective* methods to reflect whether the ore is recovered with or without dilution respectively. Bulk methods are used to extract low-grade and large deposits while selective methods which are more extensive are used in mining high grade metal deposits and narrow veins.

Reflecting the importance of ground support, underground mining methods can be classified into three groups on the basis of the extent of support required: *unsupported, artificially supported*, and *naturally supported (caving)* (Fig. 7. 15)

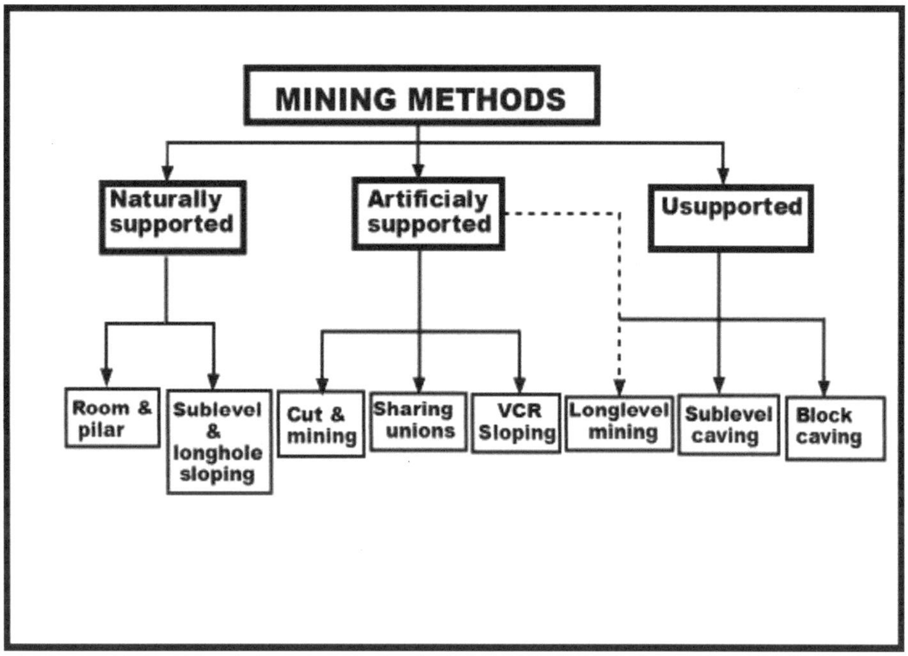


Fig. 7.15: Classification of Underground Mining Methods

*Unsupported* methods are essentially naturally or self-supported and require no major artificial system of support to carry superincumbent loads, instead relying on the natural competence of the walls of the openings and pillars. Examples of these methods are room-and-pillar mining, shrinkage stoping and sublevel stoping. *Supported* methods require substantial amounts of artificial support to maintain stability of its mine openings, as well as the need for systematic ground control throughout the mine. The common artificial supports include backfilling/stowing, timbers/stulls, cribs/packs, and hydraulic/frictional props. Examples of mining methods are cut-and-fill stoping, sharing unions and square set stoping. And stull stoping. Naturally supported methods include longwall mining, block caving and sublevel caving.

From the geological viewpoint, the morphology or mode of occurrence of ore bodies can be used to classify mining methods into three groups: flat-lying deposits, steeply dipping deposits and massive deposits. From the mining engineering point of view, mining methods can be classified into three groups based on the extent of ground support (Table 7.2). The geological grouping will be used in describing the mining methods as follows:

1. Flat Lying Deposits
2. Steeply Dipping Deposits
3. Massive Deposits

## 7.8.5 Mining of Flat-Lying Deposits

A large proportion of ore deposits that are mined occur in consolidated rocks as tabular flat-lying deposits mostly of sedimentary origin. Unsupported methods of mining are used in such ore deposits that have highly competent ore and wall rocks. Such deposits are mined by means of either of two basic techniques, longwall or room-and-pillar, depending on the thickness, uniformity, and depth of the ore-bearing layer or seam, the strength of the overlying layers, and whether surface disturbance is permitted.

Table 7.2: Classification of Underground Mining Methods

| *GEOLOGICAL CLASSIFICATION* | | |
|---|---|---|
| **Flat Lying Deposits** | **Steeply Dipping Deposits** | **Massive Deposits** |
| Room and pillar mining | Open-stope mining | Block and panel caving |
| Longwall mining | Sublevel stoping | Sublevel caving |
| Shortwall mining | Shrinkage stoping | *********** |
| *************** | Cut and fill stoping | *********** |
| | | |
| *MINING ENGINEERING CLASSIFICATION* | | |
| **Self-Supported Methods** | **Artificially Supported Methods** | **Unsupported Methods (Caving)** |
| Open-stope mining | Cut and fill stoping | Sublevel caving |
| Room and pillar mining | Square set and fill stoping | Block and panel caving |
| Sublevel stoping | ********************* | Longwall mining |
| Shrinkage stoping | ********************* | Shortwall mining |

### 7.8.5.1 Room and Pillar Method

Room and pillar is the most common method of subsurface mining, and is best suited for mining of flat, bedded or slightly inclined (stratiform) deposits. In this system a series of parallel drifts are driven, with connections made between these drifts at regular intervals. Ore is removed from these drifts to form rooms while leaving pillars in place to provide 'natural' support to the hanging wall or roof of the room but in some mines, after mining has reached the deposit's boundary, some or all of the pillars may be removed. The goal is to recover the maximum amount of ore, and leave the smallest possible pillars. The roof must remain intact, and rock bolts are often installed to reinforce rock strata (Fig. 7.15a).

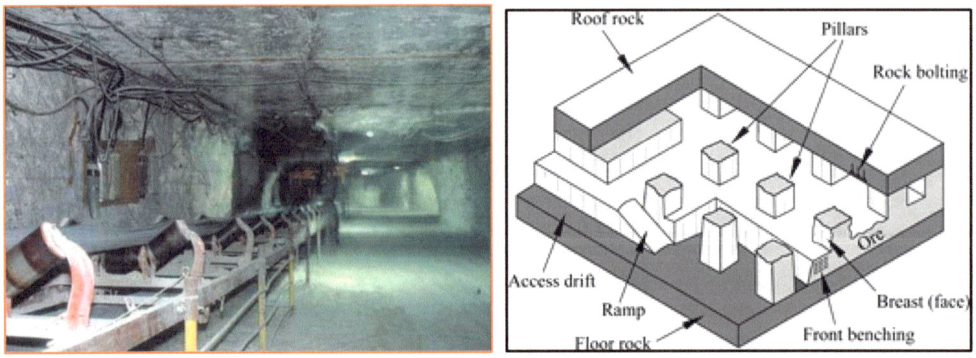

Fig. 7.15: (a) Room and pillar mine, Kansas (b) Block diagram of room and pillar method

Pillars may be spaced out at regular intervals, and designed with circular or square cross sections or shaped as elongated walls separating the rooms (Fig. 7.15 b). The flat ore body and wide open spaces permit production from various stopes in the room. After the ore is blasted and the material loaded in the room where it was extracted, it is transported to a central gathering point to be taken out to the mine. It is a very highly mechanized, flexible, relatively inexpensive, and selective mining system leaving waste material on pillars.

### 7.8.5.2 Longwall and Shortwall Mining

Longwall mining is used mainly to extract ore from thin-bedded deposits of uniform thickness and large horizontal extent (Fig.7.16a). Typical deposits of interest are sedimentary coal seams, potash layers, or fossil placers in pebble conglomerates, such as those in South African gold mines. Longwall mining can occur in both hard and soft rock as the working area along the mining face can be artificially supported where the hanging wall tends to collapse.

Longwall mining is a bulk mining technique in which the orebody is divided into panels or blocks, and a cutting machine which moves back and forth is used in grinding a long straight ore surface, followed by mechanized removal and transportation by chain conveyor that carries the mineral to the haulage drift (Fig. 7.16b). Longwall mining is most suitable for narrow, gently-dipping or flat-lying ore bodies of uniform thickness and large horizontal extent in soft sedimentary rocks, and is a common method for extracting coal, trona, and potash from seams of various thick- ness. Mining using this method does not require blasting, except in very hard rocks, such as the gold- and platinum-bearing reefs of South Africa, where the seam is removed by drilling and blasting. Long wall mining machines are about 150 to 300 m long and 2 to 6 m high.

Shortwall mining is very similar to longwall mining but with shorter face lengths, ranging between 40 and 90m, with the aim of controlling the caving nature of the overlying upper strata, the load on support and the overall operation of the supports applied at the face. The shortwall system provides advantages in flexibility of operations, cost savings and increased productivity.

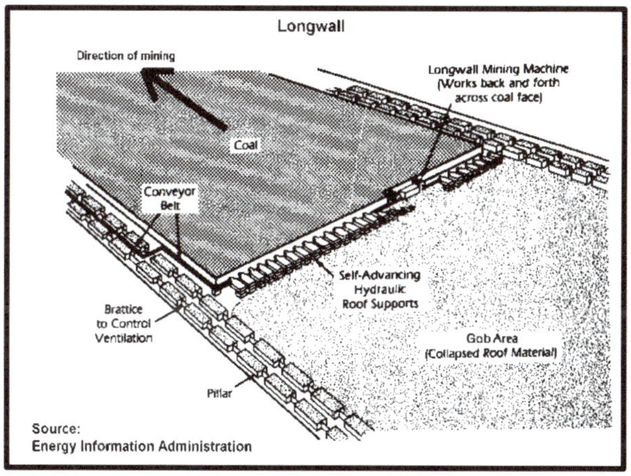

Fig. 7.16: (a) Longwall mining of coal (b) Longwall grinding of coal seam

## 7.8.6 Mining of Steeply-Dipping Deposits (Veins)

Several hydrothermal fissure-filling deposits such as veins, breccia pipes and structurally deformed deposits may appear in vertical or steeply dipping orientations. Their dimensions vary considerably, ranging from small gold-quartz veins less than a meter to massive bodies hundreds of meters wide. Mining is often selective (selective mining) which means miners prefer to extract only the valuable ore leaving the waste rock in the hanging wall and the footwall intact. Manual labor is often used for the smaller veins while in some thicker deposits, machine operations may be used within the ore body walls. Selective mining methods that are most applicable to these deposits are open stope mining, cut and fill stoping, shrinkage stoping, and sublevel stoping.

### 7.8.6.1 Narrow Vein Mining

This mining technique is used for very narrow ore bodies, as small as a half a meter wide (Fig. 7.16c). It is a very selective method in which waste rock is left in the hanging wall and the foot wall. A drift is driven along the bottom of the ore body, and this is eventually enlarged into the shape of a trough where stoping begins. From this point the ore is blasted removed from the stope as soon as it is mined. In a wide vein, a standard load haul dump (LHD) loader can operate inside the drift. "Slim-size" machines including drill rigs, jumbos, and 2 $m^3$ bucket LHDs, are available for working in drifts as narrow as 2 m. However, in such narrow veins the use of machines produces waste which dilutes the ore. The alternative is to use a manual technique to extract only the higher grade material in the vein... The excavation is left as a permanent void, with or without artificial support; after mining.

### 7.8.6.2 Sublevel Open Stoping

Sublevel stoping is an unsupported mining method in which ore is blasted from different levels of elevation in the ore body, and broken ore is passed through a chute (ore pass) from higher to lower levels, and removed from the lowest level at the bottom of the mine (Fig. 7.17). Ore is fragmented in the stope using ring-drilled or long parallel blast holes. This method is used for mining steeply dipping ore bodies with strong or competent ore and wall rocks. The ore body is vertically divided into levels, and between two levels, the stopes of convenient size are formed, hence the name sublevel stoping. Leaving a crown pillar at the top of the stope safeguards the level above, while lower level is utilized as haulage level to collect the mineralization from the stopes. The level developments, commonly in the footwall, range from 30 to 150 m, based on the vertical extent of the ore body and the number of production openings that can be extracted in each level. *Longhole stoping* is similar to sublevel stoping but uses longer (~100 m) and larger diameter blastholes, thus requiring less drilling than sublevel stoping. Greater drilling accuracy is required and non-planar irregularities in the orebody shape cannot be recovered. .Development of the infrastructure for both these stoping methods is time-consuming, costly, and complex.

Fig. 7.16 (c): Classification of underground mining methods (b) Narrow vein 2.5 km long 1.5 m wide

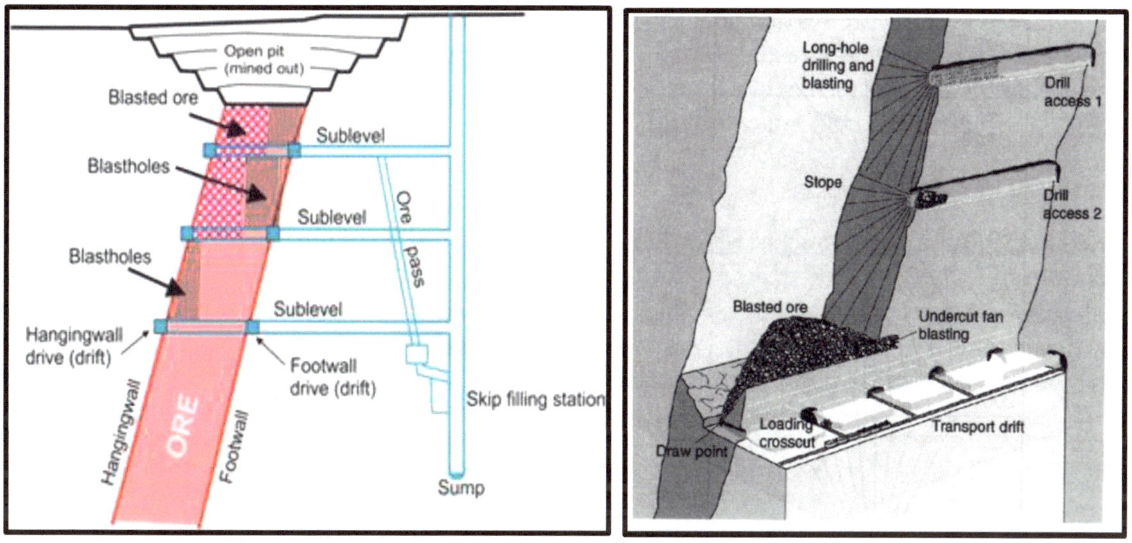


Fig. 7.17: (a) Sublevels in underground mine (b) Diagram of Sublevel Stoping

### 7.8.6.3 Cut and Fill Mining

Cut and fill is a mining method in which an opened stope from which ore has been removed is back-filled with waste rock (Fig. 7.18a). This technique can be adapted to many different ore body shapes and ground conditions. Together with room-and-pillar mining, it is the most flexible of underground methods. Cut and fill is a highly selective method that is considered most suitable for steeply dipping high grade deposits found in weak host rock. In this method, mining begins at the bottom of the ore body and advances upwards. During the extraction process, the back of the excavated stope is temporarily supported with rock bolts before it is finally backfilled to form the floor of the next level of development and excavation support. Mining progression is achieved between stopes by the construction of raises driven upward through the ore body.

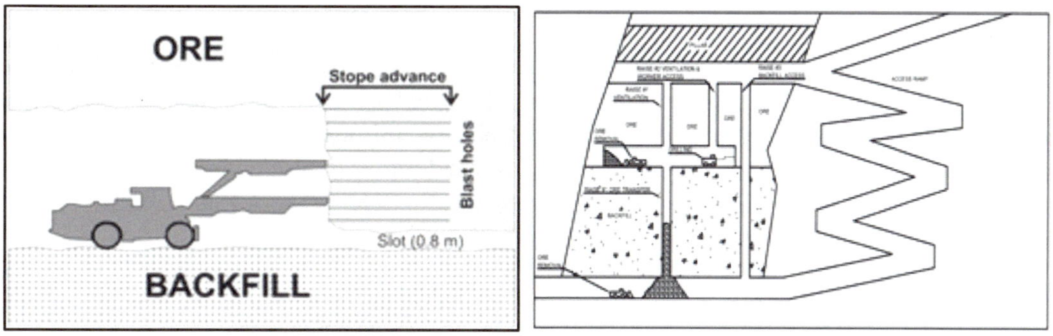


Fig. 7.18: (a) Cut and Fill mining (b) Cut and Fill stoping

### 7.8.6.4 Shrinkage Stoping

Shrinkage stoping is a method in which the ore is mined in horizontal slices, starting from the bottom of the ore body and advancing upwards within a stope. The hanging wall and footwall are supported by broken ore left in the mined-out stope (Fig. 7.19a). The broken ore also serves as a working floor or platform for additional blasting as the mined stope retreats. Shrinkage stoping is a highly selective method used in the mining of narrow, steeply dipping ore bodies (veins) with well-defined ore/waste boundaries, and where the hanging wall and footwall are very competent enough to support themselves during the extraction process. Shrinkage stoping is cost-effective and flexible, but labor- intensive, low-productivity method that requires a long lead time for total extraction of a zone. Shrinkage stoping has been replaced by mechanized methods, although it is still in use in small mines around the world.

### 7.8.6.5 Vertical Crater Retreat Stoping

This method involves open stoping in which large-diameter vertical holes are drilled bottom-up into the ore body, bottom-up mining method that involves vertically drilling large-diameter holes into the orebody from drives developed between two levels, and then blasting horizontal slices of the orebody into an undercut. The blasting creates downward facing craters from which the broken ore from the stope at a lower level. The stope is then backfilled to provide wall support for the blasting of successive stopes (Fig. 7.19b).

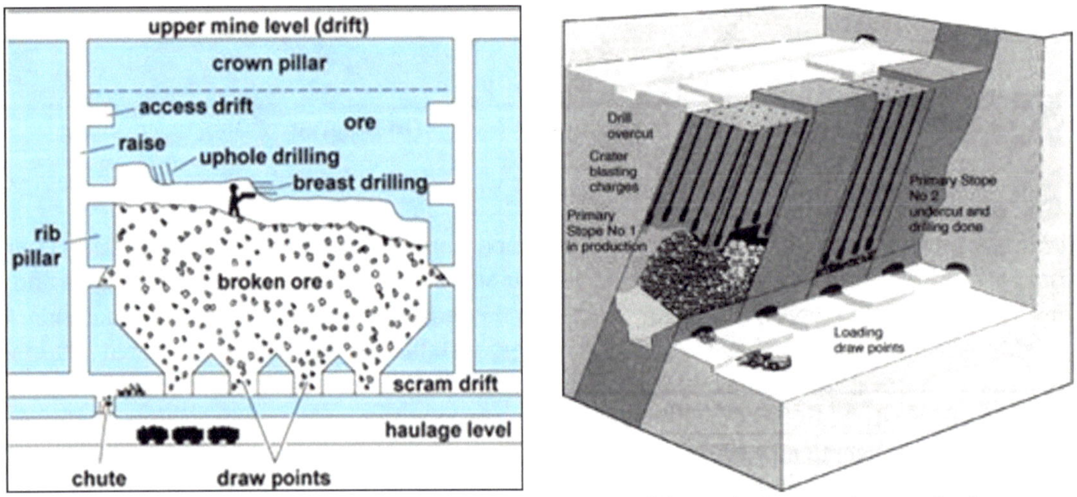


Fig. 7.19: (a) Shrinkage Stoping (b) Vertical Cratering Method

## 7.8.7 Mining of Massive Ore Deposits

Massive ore deposits consist of thick and continuous masses of ore minerals such as sulfides and oxides (e.g. chromite) that may be of substantial thickness and lateral extent. Also they can be steeply dipping or subhorizontal lenticular massive orebodies. The most applicable mining methods for these massive deposits are block caving and sublevel caving. These methods are used under the following conditions: (1) large ore bodies of steep dip, (2) massive ore bodies of large vertical extension, (3) rock that will cave and break into manageable fragments, and (4) surface that permits subsidence.

### 7.8.7.1 Block Caving

Block caving is a mining method in which gravity is combined with internal rock stresses to fracture and break the rock mass into pieces that can be handled by miners (Fig. 7.20a). The ore body is divided into "blocks" or large sections of several thousand square meters, and the "caving" of the rock mass is induced by undercutting the block by blasting which destroys its ability to support the overlying rock. Consequently, gravity forces act to fracture the block while continued pressure breaks the rock into smaller pieces to pass the draw points where the ore is handled by dump trucks.

Block caving is a bulk mining technique used very successfully on large-scale mines exploiting massive orebodies or pipes, and large low-grade or disseminated ore bodies which are at too great a depth to be exploited through open-pit mining. After the block collapses, the orebody is then essentially rubble making it easy to extract the minerals using gravity driven processes. The ore can flow out through a draw-point by using a funnel-shaped structure into which the blasted ore can be collected on belt conveyor systems or mine carts until they are transferred to the surface for processing. Block caving is regarded as one of the most productive and lowest cost underground mining methods and as such is often applied to low grade ore bodies.

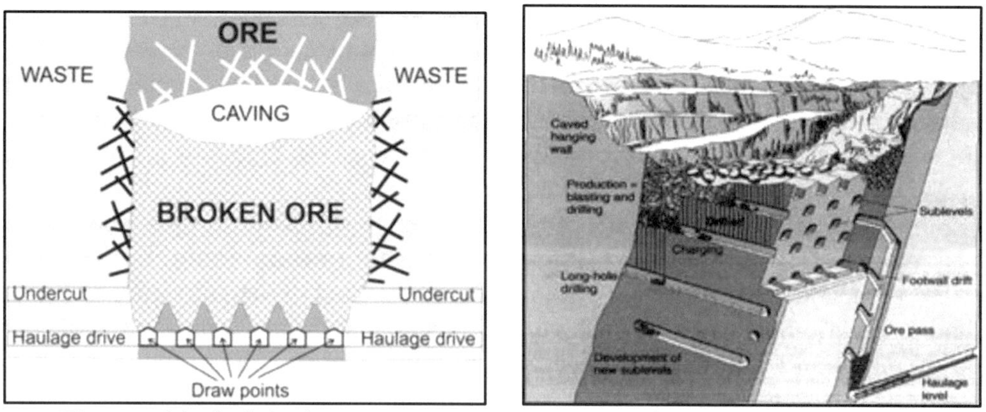

Fig. 7.20 (a) Block Caving (b) Sublevel Caving

### 7.8.7.2 Sublevel Caving

Sublevel caving is a mining method in which ore is extracted from sublevels spaced at regular intervals throughout the entire ore body (7.20b). Mining usually starts at the top of the orebody and develops downwards. A series of long blast holes are drilled in ring patterns and blasted from each sublevel in such a way to generate controlled caving when the roof fractures and collapses following the cave. The broken ore is mucked out after each blast. This mining method is suitable for extraction of ore from massive, steeply-dipping orebodies with considerable strike length, and usually has a high amount of dilution and low recoveries. The method is also used when open-pit mines transition into underground mining.

Caving requires a rock mass where both ore body and host-rock fracture easily. As the mining extracts ore and rock without backfilling, the unsupported hanging wall carries on caving into the open stopes. The development work required to prepare sublevel caving stopes is relatively extensive and involves driving multiple openings and blastholes into the orebody to prepare sublevels. Ore is usually fragmented from using an array of blastholes. Sublevel caving is a sophisticated method, capable of supporting high production volumes, and flexible due to the availability of multiple working places at all times. It requires strict control as dilution is unavoidable and must be kept within planned, accepted limits.

## 7.8.8 Artisanal and Small-Scale Mining (ASM)

Artisanal and small scale mining (ASM) is a crude form of mining involving the extraction of minerals from hand-dug pits and holes in the ground by poor and often illiterate people. This mining method involves both surface and underground techniques using low technology, and is common in parts of sub-Saharan Africa, South America and Southeast Asia. Officially, ASM is defined by the United Nations agencies as the "formal or informal mining operations with predominantly simplified forms of exploration, extraction, processing tools, such as diggers, pixes, shovels and pans. ASM uses low technology and can include "men and women, youth and children working on an individual basis as well as those working in family groups, in partnerships, or as members of cooperatives or other types of legal associations and enterprises involving hundreds or even thousands of miners." In some countries, a distinction is made between *artisanal* and *small-scale* mining, whereby 'artisanal' typically refers to pure manual mining while 'small-scale' may have fixed installations or use mechanized equipment. However, the diversity of ASM operations is vast and generalizations are easily contradicted.

ASM is an important socio-economic sector for the rural poor in many developing nations, many of whom have few other options for supporting their families. It is widespread in approximately 90 countries in the global south from developing countries in Africa, Asia, Oceania, and Central and South America with an estimated 40.5 million men, women and children directly engaged in ASM. Twenty percent (20%) of the global gold supply is produced by the ASM sector, as well as 80% of the global gemstone and 20% of global diamond supply, and 25% of global tin production. More than 150 million people - workers and their families - depend directly or indirectly on ASM for their livelihood. 70 - 80% of small-scale miners are informal, and approximately 30% are women, although this ranges in certain countries and communities from 5% to 80%.

Obtaining detailed information about ASM mine production, revenue and taxation is very challenging because most of ASM activity is informal or illegal in nature. From the geological viewpoint, most of the ASM production is mostly from shallow placer deposits in river alluvium and residual deposits in the weathered regolith. Such deposits such as gold, diamonds, tin and gemstones with relatively high value are the most common ore minerals. Revenue generated from the sale of the extracted minerals are mostly not reported to government authorities, and a substantial amount of smuggling is involved in the trade. The higher the value and portability of the material, the more likely it is to be traded illegally and the harder it is to quantify production. Miners may be migratory, work may be seasonal, and mines may be short lived, all of which lead to erratic production and challenges in quantifying the scale and value of the sector. At the same time, more professional small-scale miners may be leading rural entrepreneurs, employers and exporters. Despite being a major source of mineral production and an engine of economic activity, especially in rural areas, the true value of ASM is rarely captured and the sector typically fails to generate official revenues for the state. Countries in Africa with the largest numbers of ASM operators are Nigeria (gold, tin and gemstones), the Democratic Republic of Congo (gold) where civil strife and child labor are rampart; Ghana (gold), South Africa (gold and diamonds), Central African Republic (diamonds) and Burkina Faso (gold).

In Nigeria, artisanal mining has long been practiced particularly for the extraction of gold and precious gemstones. Approximately 95% of mining is carried out by ASM who extract minerals of high value such as gold, gemstones, cassiterite, tantalite, as well as other minerals that are preserved in alluvial and eluvial deposits. It is estimated that about one million people in Nigeria are involved in artisanal mining They are concentrated in the gold mining areas of Zamfara, Kebbi, Kaduna, Niger and Oyo State, the tin mining areas of Plateau and Bauchi States and the gemstone tantalite mining areas of Nasarawa, Kaduna, Kogi and Oyo State. Artisanal mining is mostly informal - that is, illegal mining that operates outside existing laws and regulations, and thereby generates little or no revenue for the Federal Government. Small scale commercial entities in mining are growing in Nigeria, particularly in the mining and processing of gold and tin-tantalite placers and the extraction of minerals like copper, lead-zinc, manganese, barite, coal, and some gemstones from weathered or hard rocks. These activities may require fixed and/or mobile equipment, use of explosives for blasting and hazardous chemicals for mineral processing. According to Nigeria's mining regulations, small-scale mining leases give the right to miners or companies to exploit minerals in areas of 5 acres to 3 square miles, while artisanal miners are expected to organize into cooperatives, be licensed and registered as a corporate entity to be able to apply for a mining lease and receive any technical assistance from the Ministry of Mines and Steel Development (MMSD. The requirements have created a major obstacle for many artisanal miners who are mostly poor and illiterate or semi-illiterate former farmers or herdsmen. Consequently, they have no easier choice but to operate illegally, sometimes with the connivance or cooperation of corrupt government officials.

The ASM mining sub-sector operates under hazardous, labor-intensive, highly disorganized and illegal conditions which have resulted in low productivity, lack of safety (from bandits), exposure to hazardous mining conditions (pits and holes), exposure to poisonous metals (lead) and chemicals (mercury), poor health care and lack of job security (employment may be seasonal). For the future development of mining in Nigeria, there is a need to formalize artisanal mining and organize the miners into cooperatives which will generate a lot of benefits including increased productivity, improved performance, better safety and more revenue for the Federal Government. According to the International Institute of Sustainable Development, there are six critical factors in formalizing ASM: (1) developing conducive and comprehensive legal framework; (2) providing access to geoscientific information; (3) ensuring access to capital through bank loans and government support funds; (4) providing access to equipment that are locally designed and affordable; (5) develop more capacity building through training on best mining practices; and (6) enabling dialogue between ASM stakeholders and government.

## 7.9 Geological Mapping in Mines

A mine geologist is required, as part of his/her main responsibilities, to provide geological information and services in support of mine development and production activities. Geological maps in mines are probably one of the major sources of information about the distribution of rock types, delineating geological structures and accurately locating geological contacts. This activity is important for understanding favourable geological features controlling distribution of the economic minerals which is essential for accurate exploitation of the deposits. Even with the best organized ore grade control system of drilling, detailed sampling and assaying in a mine, the boundaries of ore can only be adequately defined with the assistance of a detailed geological map which shows the spatial distribution of the geological features. Moreover, it is only by the preparation of detailed and up-to-date geological plans and sections of the mine can a drilling program be effectively planned to locate extensions to ore beyond the confines of the mine openings, particularly in underground mines. Of great importance is the role of geological mapping in providing mine support, such as in ore grade control, selective mining, and reduction of dilution of ore by waste and mixing of ore types. Services also include geotechnical mapping of faults and fractures that could represent potential slope failure zones and mitigating hydrogeological problems such as water seepage into mines

### 7.9.1 Approaches to Mine Mapping

Geologic mapping should be designed as the most effective method of recording and organizing geologic observations in a mine, and geologic maps are used as a means of presenting the observations as well as constructing geologic hypotheses on the ore deposit and associated rocks. Mine geological mapping takes place in both open-pit and underground mines usually with somewhat different parameters and perspectives. Mapping therefore provides essential data for geological interpretation to assist operating mine engineers in conducting periodic reevaluations involving production optimization modeling and associated risk management. Mapping is also an essential part of geotechnical modeling and planning; for example, of pit and underground mine wall stability, which affects the behavior of rock masses being mined, mine installations and equipment, and safety of miners.

## 7.9.2 Traditional and Digital Mapping

A geological map is a graphical presentation of geological observations and interpretations on a horizontal plane. A geological section is identical in nature to a map except that data are recorded and interpreted on a vertical rather than a horizontal surface. Maps and sections are essential tools in visualizing spatial, three-dimensional, geological relationships. Traditional geological mapping involves the use of simple tools such as the geological compass, and measuring tape. Observations including sketches, measurements and narratives are recorded in field notebooks and base maps. In the later part of the 20th century when inexpensive digital cameras became readily available, they became a useful additional tool for geological mapping in a busy working environment. While field observations are still very much in use and valid, digital photographs now frequently supplement sketches which saves time for the geologists to complete work in a short time frame. In addition, a new instrumentation such as the global positioning system (GPS) has in recent years enhanced measurement accuracy and made locations as precise as possible. In the 21st century, starting from 2000, geologists have seen digital technologies and computers transform geological mapping in general and mine mapping in particular. The whole procedure of the mine mapping had become an interactive process between direct observations in the field and desktop mapping using georeferenced digital photos plotted on the relevant mine faces, such as open pit benches or underground workings. Increasingly, field narratives are written and organized on a notebook computer or a personal digital assistant (PDA).

*Digital mapping* is a modern process of *geo-mapping* in which georeferenced digital photos and geological features that are directly observed and recorded in the field or mine (e.g. strike and dip of the geological contacts, faults, shears, lineation, cleavage, axial plans of the folds and other structures), are combined and interpreted together to produce a digital geological map of the mine which can be displayed in real-time on a computer or personal digital assistant (PDA). The primary function of this emerging technology is to produce spatially referenced geologic maps that can be utilized and continuously updated as mine production progresses and new geological information is obtained. Moreover, such maps can be constructed in a digital 3D geological model. Additionally, computers can now perform some tasks that were difficult to accomplish in the field, for example, handwriting or voice recognition and annotating photographs on the spot

Recent trends in digital geological mapping involve the conversion of geological information into databases that could be utilized in generating different types of interactive geological, geotechnical and ore distribution maps in mines. Such efforts however use database design or "data model" that is sufficiently robust to manage complex geologic concepts such as three dimensional (spatial) and temporal relations among geological map units, faults, and other features entered into the system. In addition, mapping sheets are currently being replaced by electronic data capturing devices, such as electronic mapping tablet Digital capturing of the data allows direct uploading of information into computers and therefore expedites construction of the digital 3D models. However, despite the advantages of this digital approach, it is more standardized and less flexible than traditional mapping on paper sheets which is the format in low technology countries.

### 7.9.2.1 Scale of Mine Maps

Scale of the mine mapping varies approximately from 1:500 to 1:2000 depending on mining preferences, reserves definition parameters and grade control practices. The common approach is to record the map observations at a scale larger than the final geological map, which often is represented as a simplified geologic map of the entire mine. There may be certain parts of the mine that are of special interest in mine production requiring more detailed mapping, usually on a larger scale of up to 1:200 but rarely at 1:100.

(Note that the smaller the scale ratio, the larger the scale). For example, there may be a series or swarm of gold-bearing quartz veins within a well-defined mineralized zone that need to be shown on the mine map. Different scales for the vein swarm on the maps of 1:200 and 1:2000 scales can be chosen. The maps of 1:200 scale allow to document and plot each vein composing the veins swarm, whereas on the smaller scale maps, such as 1:2000, the individual veins may not be distinguishable and the veins swarm will be shown as one zone of mineralized veining.

## 7.9.3 Mapping of Open Pit Mines

Open pit mines are exposed at the surface, and therefore the geological features can easily be observed on a *horizontal plane* at the bench (floor) and pit floor of the mine, and in the *vertical plane* along the bench faces (Fig. 7.21). In open pit mines, observations are best made and recorded at the pit faces (vertical plane) and in particular along the bench toes where geological features are exposed on the face above the broken rocks and on the bench floor (horizontal plane)

Observed geological features are accurately recorded on the base maps or field sheets (e.g. graph paper) using global positioning systems (GPS) instruments or surveyed control points, which are marked by paint on the pit faces or on the broken rock at the toe and their exact location is marked on the base maps. The mapping sheets should also display the pit infrastructure, in particular toes and crests of the benches (Fig. 7.21), which are used as background data to help in the accurate plot of observed geological features on the base map. A scale of 1500 is generally adequate for pit mapping but, in areas of complex geological features, mapping at more detailed scale such as 1:250 may be applicable.

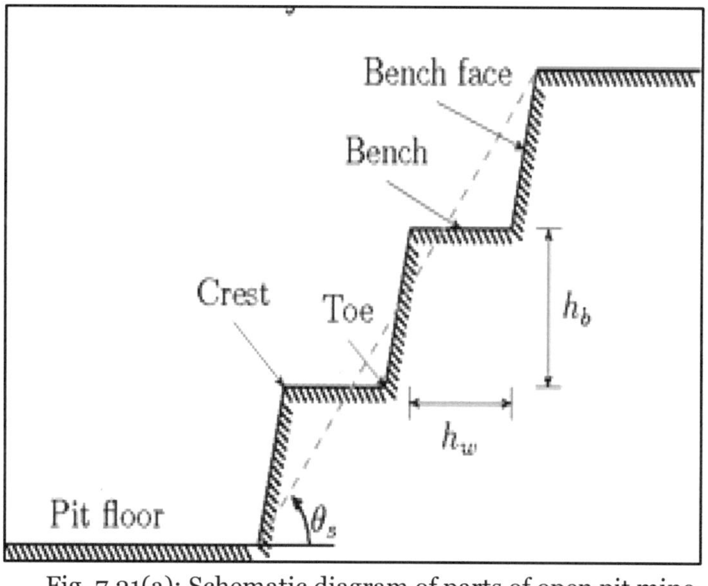


Fig. 7.21(a): Schematic diagram of parts of open pit mine

### 7.9.3.1 Mapping Procedures

The actual mapping procedure in most mines depends on the use of base maps or mapping base which is usually a survey plan of the mine opening prepared by the mine surveyor and supplemented by accurately established survey points from which distances can be taped. The application of the standard "compass and tape" procedure allows the geologist to accurately mark locations and direction (strike) of geological features observed in mine faces. If there is no measuring tape available, then pacing distances between established survey points can still allow a rough map to be constructed, assuming that the normal pace length is known. In some cases, particularly where extreme complexity is encountered, plane tabling can be used. In open-pit mines, most available rock surfaces are vertical or near-vertical; observations are thus best recorded onto sections and afterwards transferred to the standard level plans, a composite open-cut plan or mine sections.

### *Mapping of Pit Face (Vertical Section)*

Face mapping is the process of mapping the exposed bench faces of a pit at various bench levels throughout the open pit (Fig. 7.21b). The vertical face of a pit usually contains the highest density of information that can be directly observed, although fallen rubble may cover some parts. The geological information can be projected into the horizontal plane to produce a surface map (Fig. 7.21b). Mapping the exposed faces in the pit should be done on field sheets (such as graph paper). The position of the mapped face relative to the mine coordinates is established by measuring a distance and bearing to some known surveyed point or points in the mine. The map is compiled along the face and toe of the mining advance, from any outcrop on the pit floor, and from assay data. Extra geological information can be obtained by digging trenches or scrapes on the pit floor to expose geological features. A linear strip map of the base or 'toe' of the face is prepared at the same time with plotted structural measurements and the position and strike of all significant contacts or structures observed (see Fig. 7.21b). Figure. 7.22 is a face map for one bench level of the plan shown in Fig, 7, 21b.

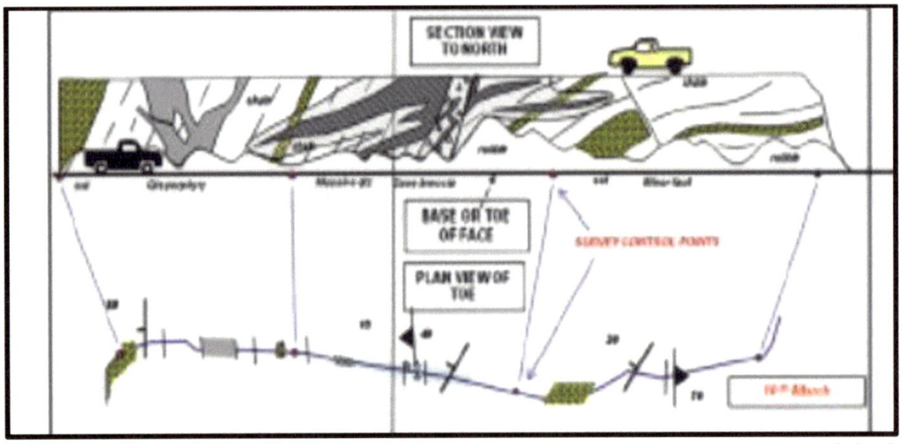


Fig.7.21 (b): Mapping the active mining face of an open cut mine. A linear strip map of the base or toe of the face (After Majoribanks, 2010).

Faces mapped on different levels or benches are usually combined as composite face maps to create a complete vertical mine section through the ore body from the pit floor to the top of the pit. Such vertical sections based on near-continuous observation, can readily be combined with drillhole data, to produce an accurate and powerful tool for resolving details of geology and mineralization. During the geological mapping of the pits it is a good practice to take a photo of the entire wall first and to sketch the main geological features prior to commencing a detailed mapping of the faces. The mapped faces should contain the marked and surveyed reference points for accurately locating observed features. A map can be produced from such photographs. Fig. 7.23 is a composite map and geological interpretation of a standard mine vertical section.

Fig. 7.22: Geological interpretation of a mining level in an open cut mine. The map is compiled from "toe" maps of the mining advance (Fig. 7.21),

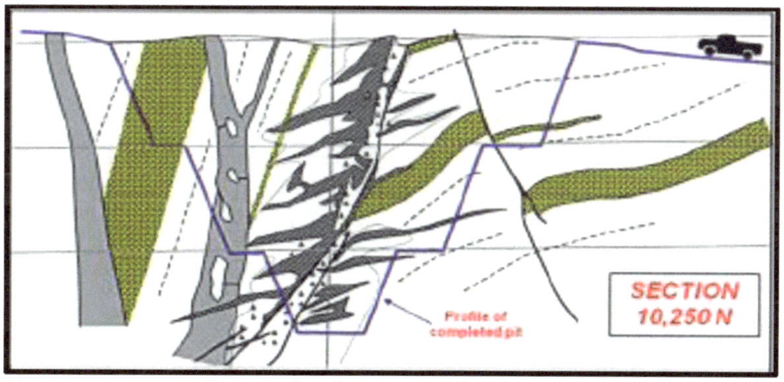

Fig. 7.23: Geological interpretation of a standard mine section. The section is compiled from face mapping on each mine level (After Majoribanks, 2010).

***Mapping of Pit Floor (Horizontal Plane)***

Maps of the pit floor can be produced as a composite of bench floor maps. The toe method allows the geologist to map the open pit levels as a record of successively retreating toe positions (Fig. 7.22). The map in this case represents a single horizontal plane and all contacts are shown with their true strike direction. The plan view of the toe of the face is now used with plan views from previous face positions to construct a geological map of the floor of the pit. The map can be further improved by adding the data collected during grade control sampling of the blast holes. Blast holes are usually drilled at intervals of 5–10 m between them and accurately located by the mine surveyors, which provides a unique opportunity of recording detailed information on rock types and their grades. Combination of toe mapping and blast holes documentation allows to produce the detailed geological maps of the mine levels (Fig.7.23). Figure 7.24 is an illustration of the importance of digital photographs in the mapping of surface mines. With adequate knowledge of the rock types, a useful geological map of the open pit walls in a vertical section can be generated.

## 7.9.4 Mapping of Underground Mines

Geological mapping in underground mines is more complex and challenging than in surface mines because of the narrow openings that limit viewing and observations underground. Many of these openings are "4 by 4", that is, about 4 m wide and 4 m high. Moreover, mapping in underground mines represents observations made at various levels of the mine which the geologist needs to correlate and assemble into a single mine map. Because of the great detail that can be observed and need to be recorded, underground mapping is usually undertaken at very large scales, generally from 1:100 to 1:10. Rock exposures can generally be observed on the roof (backs) and walls of the opening, and also on advancing faces.

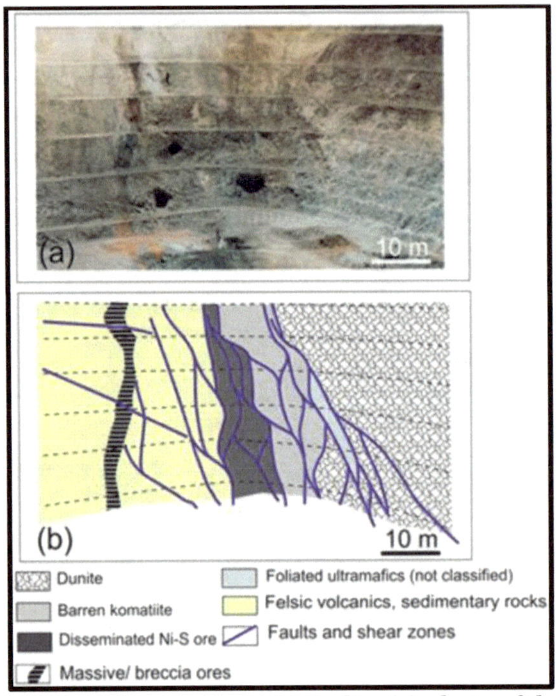


Fig. 7.24: Perseverance nickel-sulfide open pit, Australia: (a) photo of the pit wall; (b) Geological interpretation of the data (After Abzalov, 2016)

Therefore, the surfaces that are mapped within underground workings are: (a) *faces* freshly excavated on the stopes, (b) *side walls* exposed along the drive, and (c) *backs* (roof) of the opening (Fig. 7.25). Depending on the geometry and geological complexity of the orebody, a decision is usually made about which of these surfaces (face, walls and back) should be mapped in detail. Four maps may be involved here, and sometimes, in order to understand the complex geometry of an orebody, it may be necessary to conduct detailed mapping of all the four exposed surfaces. For example, in a case of narrow, gold bearing, quartz reefs that are offset by a series of faults, it may require detailed mapping of exposed surfaces to help understand the complexity and improve selective mining of the ore. Figure 7.26 is an example of .maps showing the roof (backs), both walls and advancing face of a complex mineralized zone. This is sometimes called "exploded box" mapping. The separate maps made for each advance can then be compiled on to level plans, longitudinal sections and standard mine cross sections.

#### 7.9.4.1 Roof and Wall Mapping

In most mines, the preferred approach is the mapping of the roofs of the drifts, supplemented occasionally by mapping of the walls of selected crosscuts when they are developed across the strike of the ore body. These crosscuts can provide more structural details of the mineralization. As a matter of fact, in most underground open stopes, virtually the only exposure available may be the roof of the opening, which means a map of the backs would be the obvious and indeed the only available choice. Flat-lying or gently-dipping structures such as bedding, veins, etc., are best observed on vertical surfaces, and would be appropriate in mapping of the walls and/or face of the opening.

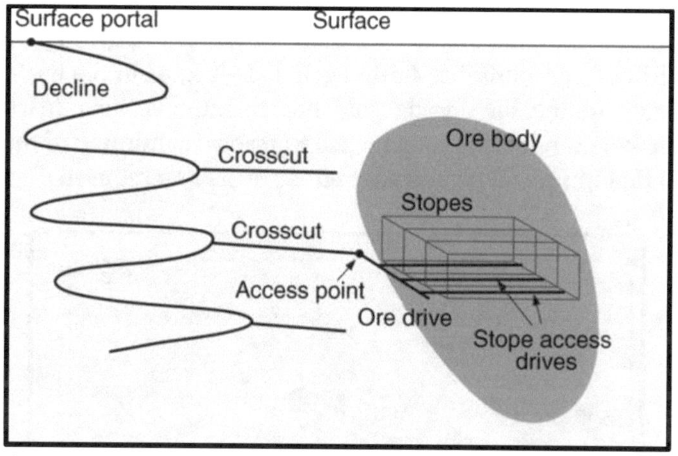


Fig. 7.25: Schematic diagram of underground mine openings

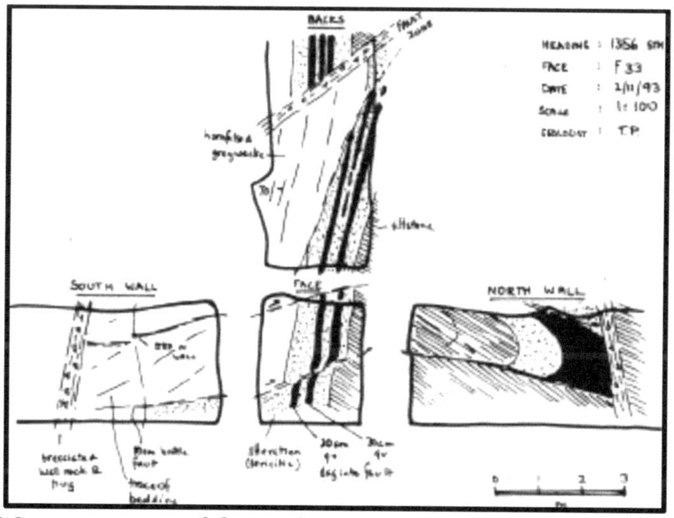


Fig. 7.26 Separate maps of the roof, both walls and advancing face of a mine

### 7.9.4.2 Face Mapping

In ore drives that run along the dominant strike of an ore body, the advancing face usually gives the most effective cross-section view of the mineralized structures. Consequently, geological mapping of the faces of ore drifts are undertaken regularly as mining progresses through the ore body. The faces are dynamic features, and get destroyed as mining advances, and mapping which is often done quickly not to delay mine production becomes a permanent record along with digital photographs. Face mapping is particularly important when drifts are developed along narrow steeply dipping ore bodies, in which case the mapping of the faces provides vital geological information to help the mining engineers and miners to accurately plan the mine production, design the mining stopes, and to control quality of the excavated ore (grade control). Figure 7.27 is a face map on a face sheet that shows the vein (face) at the base of a raise from the 120 sub-level. Here the high-grade footwall leader vein is well-exposed with visible gold.

### 7.9.4.3 Mapping Procedures

Procedures for underground geological mapping are very similar to those of surface mapping. It is normally conducted on a map plan or section of the opening provided by the mine surveyor. Mapping of the roof is usually the preferred format if the drift follows the strike of the ore body. Figure 7.28 is an example of a mapped section along a drift of a granodiorite-hosted gold lode located in the Eastern Cordillera, Peru. Map shows discontinuous nature of the high-grade gold quartz-Sulphide vein (early high-grade vein). The routine introduction of backs and raise mapping helped to resolve continuity of the high-grade vein and led to optimized stope design that improved high-grade vein recovery and reduced dilution.

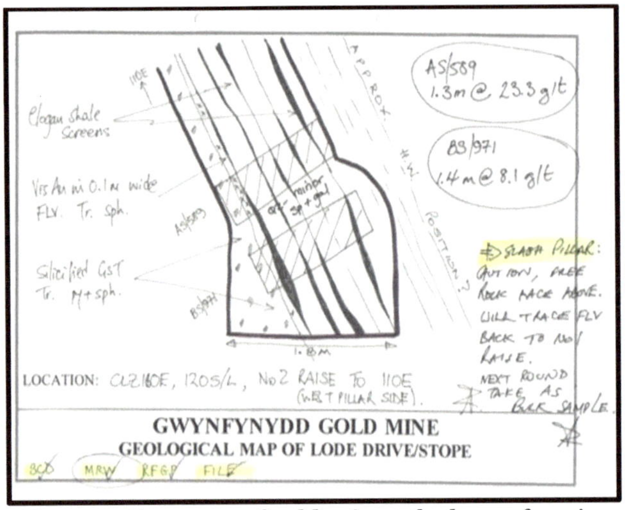


Fig. 7.27: Face map of gold vein at the base of a raise

In underground mine maps geological features are recorded on the special mapping sheets usually made of water resistant paper and containing the printed surveyed grid lines (Fig. 7.28). The sheets prepared for mapping should also include the outlines of the underground workings copied from the detailed engineering maps (Fig. 7.29). Equipment used by geologists for underground mapping include compass, geological hammer/pick, digital camera and hand lens and measuring tapes, usually one 20 or 30 m long and another small pocket tape, preferably 5 m long and made of metal for measuring vertical dimensions. The geologist also carries two or three cans of spray paint of different colors to mark contacts or interesting features, as well as a powerful electric torch. The torch is very important as a source of light in addition to the cap lamps for observing geological features in the roofs of the mine openings which are sometimes at a height of 4m (12 ft) to 6 m (18 ft). These mines are often dark and rocks may appear different in color when observed under inadequate lighting. The standard procedure in mapping an advancing face is; after each blast, miners remove the ore then bar down the roof and walls to make them safe, and the geologist hoses down the rock, measures the face advance with a tape, then maps face and walls at 1:25 scale. A portable, rechargeable flood light illuminates the scene. Although compasses work well underground but the nature of underground rock exposure makes it difficult to measure structures directly by laying a compass on the structure. The best way to measure the strike of a planar structure is to take a bearing between the points where the structure is intersected by the walls of the mine opening. The intersection of a planar structure with a vertical mine surface is only a true dip if the structure is at right angles to the wall. In the general case, intersections seen on mine walls are apparent dips. To obtain a true dip, it is necessary to sight the compass along the strike bearing.

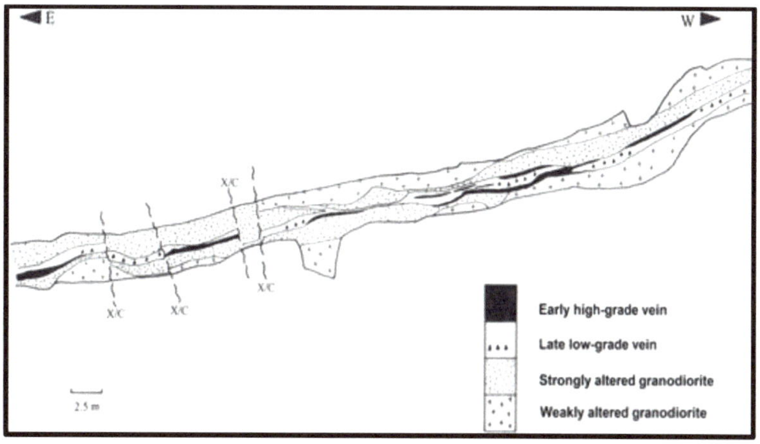

Fig. 7.28: Mapped section of a granodiorite-hosted gold lode Peru

### 7.9.4.4 Digital Mapping

Procedures of using digital cameras for geological mapping in underground mines are similar to their application for geological mapping of the open pits (Fig. 3.7. Main difference is that GPS techniques are not applicable for georeferencing the photos made underground. The photos can be accurately located only by determining location of the painted reference marks relative to surveying stations. Therefore, it is important that reference marks are clearly visible on a photo and their location has been determined before the mapped face was blasted away, shotcreted or became inaccessible. Mapping in underground mines can be significantly improved by using the georeferenced digital photos. Geological contacts can be digitized directly from these images.

Fig. 7.29 Sample of mapping sheet prepared for geological mapping Outlines of the mine workings and ore lenses are shown.

# 7.10 Geotechnical Support and Geotechnical Mapping

In the absence of a geotechnical or geological engineer in a mine, it may become the responsibility of a mine geologist with some knowledge of *engineering geology* to provide geotechnical information and support on the stability of rocks in mine openings. Geotechnical properties of importance are the rock mass strength and nature of rock discontinuities that can produce rock failures. Also, geotechnical data is needed to assist the blasting personnel (blasters) to develop a blast pattern design based on the rock mass properties of the wall rocks, and to optimize blasting in producing ideal fragmentation profiles required by the processing plant. The strength and related stability of the rock masses are important for all types of mining excavation, and are assessed using rock mass properties and classification systems throughout a mine. A *geotechnical model* of a mine in 3D is often constructed based on the available data, and provides layers of information on the geology, structures, rock mass geomechanics and hydrogeology which are continuously updated as mining progresses.

## 7.10.1 Rock Mass Properties

*Rock mass* is defined as an assemblage of rock material separated by rock discontinuities, such as joints, bedding planes, fractures and faults. The behavior of intact rock mass is influenced by the conditions the rock mass is subjected to, mainly the in-situ stress and groundwater pressure and flow. *Rock mass properties* are governed by the properties of intact rock materials and of the discontinuities in the rock. These parameters are determined by strength measurements - compressive strength, tensile strength and elastic modulus - in the laboratory, and observations of the orientation and spatial distribution of discontinuities in the rock mass (Table 7.3). Such geotechnical data can be obtained from already existing documentation on drill cores and from mine rock faces. Geotechnical data are widely used in the design of the roof support of the underground workings and parameters of the rock bolting and placement of concrete support (shotcreting).

Table 7.3: Prime parameters governing rock mass properties

| Joint Parameters | Material Parameters | Boundary Conditions |
|---|---|---|
| Number of joint sets Orientation<br>Spacing Aperture<br>Surface roughness<br>Weathering and alteration | Compressive strength<br>Modulus of elasticity | Groundwater pressure and flow<br>In situ stress |

Apart from the strength of an intact rock mass, the existence of rock joints and other discontinuities "weakens" the rock, and plays an important role in governing the behavior and properties of the rock mass under stress. Therefore, a rock mass that contains several joints, can be treated as a jointed rock mass, and is no longer fully intact. The most important parameters of joints are shown in Table 7.3. To facilitate design of mine workings, rock mass has been classified into groups or classes based on defined relationships, and assigning a unique description (or number) to it on the basis of similar properties/characteristics such that the behavior of the rock mass can be predicted.

#### 7.10.1.1 Rock Mass Classification Schemes

*Rock mass classification* systems allow the user to follow a guideline and place the rock in an appropriate class which serves as a way of communicating the estimated rock mass characteristics that can be useful in early mine design effort, but should not be taken as an alternative to detailed engineering design procedures. There are several rock mass classification schemes used in rock engineering for assisting in designing underground structures, but the most common are (1) Rock Load Factor (RLF), (2) Geological Strength Index (GSI), (3) Rock Mass Quality (Q), and (4) Rock Mass Rating (RMR). *Rock load factor* classifies rock mass into nine classes, ranging from hard and intact rock to blocky, and to squeezing rock, based on the estimated ability to carry load. The RLF is defined for each rock class and accordingly the appropriate support intensity is recommended (Table 7.4). This classification is qualitative and can be used as a rough estimate of strength in designing roof support in underground workings, including the type and strength of artificial support that may be needed. The *geological strength index* is a system of quantitative rock-mass characterization that has been developed in engineering rock mechanics using the geological character of rock material, together with the visual assessment of the mass it forms, as a direct input to the selection of parameters relevant for the prediction of rock-mass strength and deformability. The rock mass quality (Q) and the rock mass rating (RMR) use a similar quantitative approach in calculating index numbers by assigning values to six basic geomechanical parameters: - compressive strength, drill core quality, discontinuity spacing, and condition of discontinuity surfaces, groundwater conditions and orientation of discontinuities. The Q-value is the product of the ratio of parameters while RMR is the sum of parameters. All these parameters for characterizing rock mass properties are useful in the design and operation of underground mines.

## 7.10.2 Rock Properties and Blasting Performance

Mine development and production generally involve the fragmentation of rocks and ores so as to separate the valuable material from the waste rock for further processing. This is achieved through drilling and blasting, which is the process of breaking bulk rock masses into loose forms, using explosives. *Fracture blasting* is the appropriate designation for the type of blasting in mines. The process of blasting is carried out by well-trained personnel, and includes the following stages: (1) creation of blast holes by drilling, and cleaning by using suitable tools. (2) placing a charge of explosives in the holes (3) filling of the remaining portion with clay and other inert material .(4) inserting a fuse that is kept projecting 15-20 cm above the rock surface, and (5.) firing the free end of fuse by using a suitable detonator.

The positions and depths of the holes (and the amount of explosive each hole receives) are determined by a carefully constructed pattern, which, together with the correct timing of the individual explosions, will guarantee that the explosion generates the desired effect. A good blast design and proper execution of it, are very much essential for a successful mining operation. Usually, the main aim of blast design is to use explosive energy to break rock masses into sizes and shapes that can facilitate subsequent operations (e. g., excavation, loading, hauling, crushing, and milling). The design of the blast and the numerous in-situ characteristics of the rocks ultimately determine the mechanisms of the breakage and control the fragmentation.

Mine geologists often provide the needed geological and geotechnical information that is needed for blasting design to achieve the acceptable fragmentation outcome and target productions. The physical characteristics of the rocks are more important than the characteristics of the explosives used in blasting. The blasting team (or blasters) should have adequate knowledge of the different types of rock to be blasted and the rock properties that are described in terms of the strength, composition and structural characteristics. The most important characteristics of the rock influencing blasting outcome include: (a) rock composition (b) tensile and compressive strength; (c) bulk density; (d) sonic-wave velocity (acoustic velocity); and (e) structural features.

Table 7.4 Rock class and rock load factor classification

| Rock Class | Definition | Rock Load Factor Hp (feet) (B and $H_t$ in feet) | Remark |
|---|---|---|---|
| I. Hard and intact | Hard and intact rock contains no joints and fractures. After excavation the rock may have popping and spalling at the excavated face. | 0 | Light lining required only if spalling or popping occurs. |
| II. Hard stratified and schistose | Hard rock consists of thick strata and layers. Interface between strata is cemented. Popping and spalling at excavated face is common. | 0 to 0.5 B | Light support for protection against spalling. Load may change between layers. |
| III. Massive, moderately jointed | Massive rock contains widely spaced joints and fractures. Block size is large. Joints are interlocked. Vertical walls do not require support. Spalling may occur. | 0 to 0.25 B | Light support for protection against spalling. |
| IV. Moderately blocky and seamy | Rock contains moderately spaced joints. Rock is not chemically weathered and altered. Joints are not well interlocked and have small apertures. Vertical walls do not require support. Spalling may occur. | 0.25 B to 0.35 (B + $H_t$) | No side pressure. |
| V. Very blocky and seamy | Rock is not chemically weathered, and contains closely spaced joints. Joints have large apertures and appear separated. Vertical walls need support. | (0.35 to 1.1) (B + $H_t$) | Little or no side pressure. |
| VI. Completely crushed but chemically intact | Rock is not chemically weathered, and highly fractured with small fragments. The fragments are loose and not interlocked. Excavation face in this material needs considerable support. | 1.1 (B + $H_t$) | Considerable side pressure. Softening effects by water at tunnel base. Use circular ribs or support rib lower end. |
| VII. Squeezing rock at moderate depth | Rock slowly advances into the tunnel without perceptible increase in volume. Moderate depth is considered as 150 ~ 1000 m. | (1.1 to 2.1) (B + $H_t$) | Heavy side pressure. Invert struts required. Circular ribs recommended. |
| VIII. Squeezing rock at great depth | Rock slowly advances into the tunnel without perceptible increase in volume. Great depth is considered as more than 1000 m. | (2.1 to 4.5) (B + $H_t$) | |
| IX. Swelling rock | Rock volume expands (and advances into the tunnel) due to swelling of clay minerals in the rock at the presence of moisture. | up to 250 feet, irrespective of B and $H_t$ | Circular ribs required. In extreme cases use yielding support. |

*Rock composition* is definitely a major factor in rock blasting. Rocks that are hard and compact such as some igneous rocks (e.g. diorite, dolerite) and non-foliated metamorphic rocks (e.g. quartzite and marble) are more difficult to blast than some bedded sedimentary and foliated metamorphic rocks. A finer-grained texture generally leads to higher fracture strengths. The strength of a rock is defined as the maximum load it can bear at the point of failure, which is when the rock fractures and can no longer support a load. Rock mass strength depends on the strength of intact rock and the strength of rock discontinuities. In general, compared to intact rock, a rock mass has reduced tensile strength. Most types of rock have a compressive strength which is 8 to 10 times greater than the tensile strength. These properties are important factors in rock blasting (Table) 7.4). Surface weathering tends to reduce the strength of rocks, consequently, the same rock at depth may be of greater strength near the surface. Stronger, more competent rock requires an explosive with high energy output, higher density, and generally higher cost.

*Bulk density* or specific gravity of rocks is a function of the minerals present in the rock, as well the ratio of solid mineral grains to open pores such as small voids and micro-fractures. In general, the greater the rock density, the higher the explosive energy needed to break the rock.

*Sonic wave velocity* is the speed of sound in the rock or the velocity of a compression wave traveling in it. The more competent the rock (with less weathering and greater density), the higher the speed of a compression wave traveling in it and the greater the explosive energy needed to break the rock.

*Structural features* are the most important determinant of blasting performance. Both large-scale structures such as bedding planes, schistosity, foliation, and faults, and orientation and cohesion of the small-scale structures e.g. joints and fractures are relevant to geomechanical behavior of rocks. All the geological and geotechnical properties of rocks and the blast design systems if well executed make production in mines efficient and on schedule. Table 7.5 is a summary of the geotechnical properties of major rock types.

Table 7.5: Geotechnical Properties of Rocks

| | Compressive Strength (Kg/cm3) | Tensile Strength (Kg/cm3) | Density (Kg/cm3) | Sonic-Wave Speed (m/s) |
|---|---|---|---|---|
| Granite | 2000 - 3600 | 100 - 300 | 2.3 – 2.8 | 4000 - 6000 |
| Diabase | 2800 - 4000 | 150 - 300 | 2.8 – 3.1 | 6000 - 7000 |
| Gneiss | 2500 - 3500 | 100 - 200 | 2.5 – 2.6 | 4000 - 6000 |
| Marble | 1500 - 1900 | 150 - 250 | 2.8 – 3.0 | 6000 - 7000 |
| Limestone | 1300 - 2000 | 170 - 300 | 2.4 – 2.7 | 3000 - 4500 |
| Sandstone | 3000 | 300 | 1.8 – 2.0 | 1500 – 2000 |
| Clay/Mudstone | - | 300 - 1300 | 2.5 – 2.7 | 4000 - 5000 |

## 7.10.3 Groundwater Seepage/Flow in Mines (Mine Hydrogeology)

Groundwater seepage and flow in mines are a few of the problems encountered during extraction of minerals, both in the surface and underground. Mine geologists with knowledge of *hydrogeology* are often the mine personnel assigned the tasks of identifying and mitigating the impacts. Groundwater seepage is most commonly associated with infiltration of groundwater through discontinuities in rocks, whereas groundwater flow occurs mostly when mine excavation cuts through water-bearing layers (aquifers), usually sedimentary rocks. The groundwater emanation from seepages is miniscule compared to flow from breached aquifers. Moreover, because underground mines in particular are dug at great depths, there is often a high propensity to encounter groundwater flow. The presence of groundwater in mines has considerable adverse impacts on mine production, ground control and safety. Consequently, many countries have strict regulations regarding control of groundwater flow in mines for reasons of safety and groundwater quality protection (contaminated mine waters can pollute groundwater).

In surface mining operations, groundwater seepage and flow can occur from fissures and aquifer incisions, particularly in coal mines. The most important problem by the ingress of groundwater is the stability of pit slopes which is caused primarily by pore water pressures, and probably hydrodynamic shock due to frequent blasting which reduce the shear strength, and seepage pressures, which promotes movement of water in tension cracks. Pooling of water in open pits may increase the cost of pit drainage, drilling and blasting, tire wear and equipment maintenance. Flow of water into the pit can lead to erosion and failure of the slopes.

In underground mines, the situation with groundwater flow is more dire where operations below the water table can cause several operational, economic and safety implications. Mineralized and unmineralized discrete fractures, joint sets, and faults may control groundwater flow in and near the underground workings. These are the sources of seepages. The ability of fractures to transmit water depends on factors such as the degree of alteration or weathering of the rock, the presence or absence of fault gouge, and the connectivity of the fractures. The type of host rock, whether they are igneous, intrusive, metamorphic, volcanic or sedimentary, controls the density, orientation, and connectivity of fractures. Where sedimentary rocks are host rocks to mineralization, an incision into aquifers can lead to discharge of thousands of gallons per minute which if not controlled, could lead to mine flooding. In addition, groundwater in mines can get contaminated and constitute environmental threats to surface and groundwater resources.

#### 7.10.3.1 Groundwater Control Methods

Groundwater control methods are of three main types (a) prevention, (b) mitigation, and (c) monitoring. Prevention strategies include all actions that limit the flow of groundwater into mines, and mitigation actions are ways of dewatering or pumping of water after entering into the mine. The most common method of preventing seepage is grouting the openings with concrete or other effective sealants. Mine dewatering is the most effective method of water control in both surface and underground mines. The most relevant concerns with mine dewatering are related to acid mine drainage and the dispersal of the contaminated water so as to avoid surface and groundwater pollution. Sometimes, it may be necessary to capture and treat the water before discharge into surface waters or for other beneficial end use such as wetland irrigation,

Dewatering methods for underground mines may range from simple collection of the water seeping into the mine from fractures or roof bolts to complex drilling of dewatering wells from the ground surface, drilling of drainage boreholes from inside the mine and running of drainage galleries. For dewatering open pit mines the following are used: filter wells, disposal wells, vacuum drainage, horizontal drains, sealing walls, cut off walls and guard wells. Dewatering by pumping from vertical wells can be very expensive for deep mines. Environmental monitoring of water levels in piezometers installed in aquifer surrounding mines can be helpful in predicting water level changes and potential groundwater inflow.

### 7.10.4 Geotechnical Mapping

Geological mapping provides a perfect opportunity for the collection of geotechnical information by a trained geologist who is able to observe and record rock geomechanical properties from faces and/or side walls of mines. Such relevant data may include rock type, major and minor structures, and water seepage. Geotechnical mapping of rock surfaces can provide additional data on the nature of discontinuities, in particular, length of the discontinuities which is finite and can be measured on the rock faces. Length is measured along strike of discontinuity and in the down dip direction. This information will support the definition of rock mass quality and structural regimes, which can help in stope design (e.g. dilution control) and ground control. A visual geotechnical assessment can be made across the face and along the side walls back to the previous face position, as well as the stability of the roof, although it may already be bolted and meshed for support.

Rock strength of mine rocks can be qualitatively estimated during underground mapping using simple field equipment, such as geological hammer, pocket knife and rock scratcher. Rock hardness is a reflection of strength. Rock defects or discontinuities which are natural breaks in rocks are of considerable concern. They are of two types: major discontinuities such as folds and faults, and minor discontinuities such as fractures, joints, and slickensides. Discontinuities can be open or sealed, and represent different geological phenomena, such as cleavage, bedding planes, joints, stress release fractures, faults, brittle shear zones, geological contacts, dykes, veins and many others. Geological nature of the defects can be well documented during geological mapping.

Discontinuities in rocks affect rock stability and are often terminated or tied to other defects, such as faults, shear zones and joints or fading in the rocks. The type of termination of each end of the defect should be recorded on the map sheet. Spacing between discontinuities is another parameter which is difficult to measure in a drill core but is relatively easy to deduce from rock faces. It is measured as the perpendicular distance between two defects of the same set. Another important geotechnical parameter that is preferably documented in the mine workings is water seepage which can create problems for slope stability and blasting and excavation in stope faces.

### 7.10.5 Mine Safety Issues

Mines are one of the most dangerous environments for geologists to carry out their professional responsibilities. There are several hazards to be faced particularly in underground mines, including the ever present danger of geological hazards from unsupported surfaces, of being run over by heavy mine machinery or of falling into holes or over steep rock faces. In many mining operations, close access to the face is not permitted, or is allowed for just a few minutes so as to prevent delay in excavation. This often leads to the application of 'remote' mapping and grab sampling during development. There are many safety considerations to be had when working underground. To maximize safety, the mine geologist should conduct safety monitoring and keep the following rules at the forefront of his mind at all times:

1. Be adequately trained and familiar with the mine safety rules and procedures
2. Ensure that a responsible person knows where and when the geologist is working in the mine.
3. Avoid unsafe ground and ensure new mine advances are barred down, secured or supported.
4. Keep clear of heavy machinery and always wear safety gear including hard hats, safety glasses, steel-capped boots, and brightly colored reflective clothing.
5. Conduct regular monitoring of mine faces and walls, and groundwater level and seepage.

## 7.11 Geological Sampling in Mines

Geological sampling is the process of taking a small portion of a rock/ore mass that accurately represents the whole mass. Sampling of ores and wall rocks for geochemical analysis or grade control and petrographic studies is a responsibility of a mine geologist. Such samples can be obtained from drill cores, rock chips or blasted (broken) ore or rock, known as "muck". Sampling in mining differs from sampling during exploration in that, unlike drill cores, ore bodies are well exposed in the walls of the underground drives and open pit benches which allows for systematic sampling of all the exposed surfaces (roof, walls, benches and faces). Collecting samples of ores from rock surfaces and broken ore can be challenging because of the need to be unbiased and representative which can be a delicate and difficult task for geologists. Important decisions on what ore to extract or reduce dilution is usually based on the results of sampling.

### 7.11.1 Sampling of Mine Faces

Mineralization can be exposed on the front faces of the underground drives, along the side walls and on the backs and floors. All these faces are suitable for sampling and the choice depends on orientation of the mine workings relative to the ore body and accessibility of the faces. Face sampling is the cutting of pieces of ore and rock from exposed faces of ore and waste in the backs, walls, and floors of development openings and stopes. Face samples may be taken: in the following ways (a) by cutting grooves or channels of uniform width and depth across the face or sections of the face or (b) by picking off small pieces all over the face, more or less at random. The channel method is the more systematic, and accurate while the random "pick" sampling may not be as accurate, because of possible bias.

Sampling of the roof of mine openings distributed along the strike of an ore body during geological mapping of the underground drives can facilitate geological interpretations and estimation of the ore reserves. Sampling of side walls is easier for implementation, and commonly used for sampling cross cuts and raises. Sampling of broken ore (muck) is a common task undertaken by geologists. It occurs when mine geologists have to sample piles of broken ore as well as sampling of the ore stockpiles, and conveyor belts, tram cars and draw points of the underground stopes

## 7.11.2 Sampling Techniques

Sampling techniques are varied, but those methods applicable in mining are grab or muck sampling, channel sampling and chip sampling from percussion drilling.

### 7.11.2.1 Channel Sampling

Rocks exposed in the walls of mine workings are usually sampled by cutting narrow channels on the surface through the entire ore body and extending to the wall rocks. The resulting chips, fragments and dust are collected as samples. Channels are made by chisel and hammer or diamond saw. Cutting by diamond saw is a superior technique which produces samples of significantly better quality than conventional sampling by chisel and hammer. Location of the samples on the ore or rock faces should be chosen in such a manner that allows obtaining the most representative samples for the given face. As shown in Fig. 7.30, orientation of sample points should be perpendicular to the orebody. Channel sampling is the most acceptable method in mine sampling.

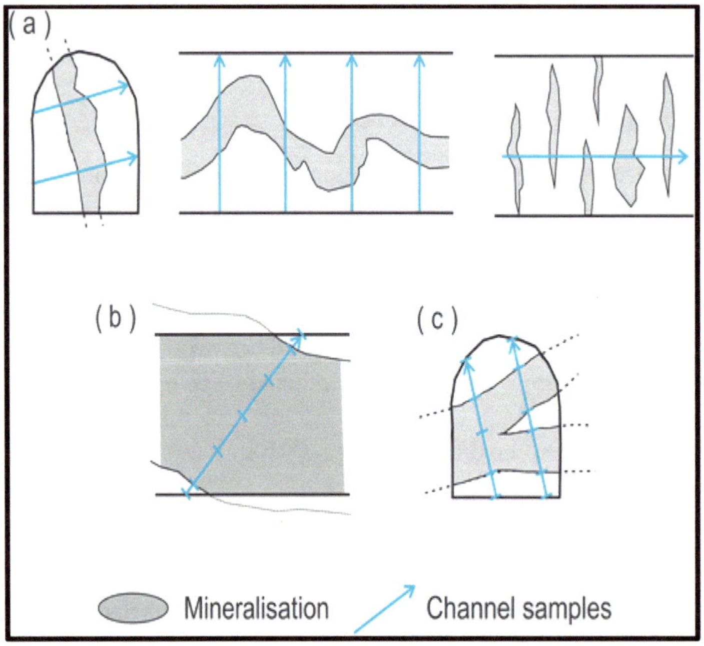


Fig, 7.30: General principles of the channel sampling:
(a) Orientation of the channels at right angle to mineralisation;
(b) Assuring that channel intersects upper and lower contacts
(c) Samples taken to geological contacts.

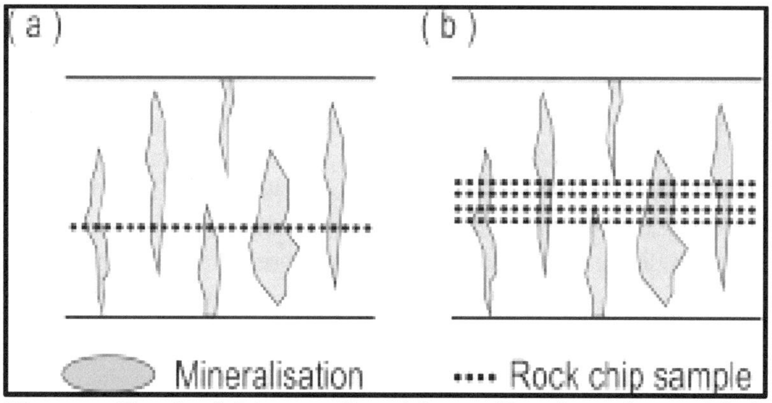

Fig. 7.31: Sketch of chip sampling: (a) sampling along the straight line; (b) Sampling along the rectangular grid distributed as a narrow band

#### 7.11.2.2 Chip Sampling

Chip sampling is another technique that is commonly used in sampling rock faces in mines. The process involves a systematic collection of rock chips which are taken at regular intervals along a straight line or rectangular grid. Points where rock chips are to be collected can be marked on the rock face before sampling, which ensures a representative sample free of potential bias due to subjective choices of sampling spots. The method is fast and inexpensive, but samples are less representative than channel samples. It is usually used for sampling uniformly distributed or disseminated mineralisation where rock chip sampling approach produces representative samples. Rock chips are collected by geological pick and chisel by hand held mechanical devices such as air-pick, hand held core drill, hand-held percussion drill or electric drills equipped with a drill cuttings catcher.

#### 7.11.2.3 Muck or Grab Sampling

Muck or grab sampling involves the collection of a large sample (or series of smaller samples that are later combined) from the broken ore or muck pile at a face or from tram cars or trucks transporting the ore from these points. Samples ranging in size from 1kg to 5kg are generally collected either by hand or shovel by a geologist, field technician or miner. This approach is commonly used for ore grade control at the underground operations in particular at the mines which use block cave or sub-level cave mining methods. Sometimes, muck sampling is used to reconcile the mined grade at the ore source to the predicted grade and/or predict the mill feed grade.

Muck sampling is known to be problematic because piles of the broken ore consist of fragments varying in the very wide range, from large boulders to the fine debris, and samplers tend to oversample the fines, and/or pick out high-grade fragments; surface sampling of piles does not test material within the pile. Non-scientific protocols are commonly used to define a sample mass and number of samples per blast or per truck, etc. However, grab samples from trucks or muck piles provide highly variable results, but in some cases have been shown to be effective in high "nugget-effect" environments (high variability of grades in gold mines) when samples in excess of 15 kg are collected. Owing to the difficulty of collecting representative samples, this sampling technique is used only when no other sampling means are applicable.

#### 7.11.2.4 Bulk Sampling

Bulk sampling is the process of taking very large samples, as part of the general procedure for the exploration and evaluation of a mineral deposit. In mines, bulk samples are sometimes collected to evaluate ore grades where there is high variability in concentrations obtained by other types of sampling, e.g. in gold mines where this is known as the "nugget effect". Bulk samples may also be collected for metallurgical testing. Bulk samples range in size from 1 to 100 tons depending on the size of the ore shoot being sampled. Often the bulk sample may be a composite of material from development drifts and raises. Ideally, drift rounds or other portions should be selected from geological mapping and prior sampling, handled separately, and should be stored temporarily on a pad for sampling if not directly processed.

Compared with other sample types, bulk samples are restricted in scope and only evaluate the grade of a particular zone of mineralization; and they may not be representative of the whole deposit. Before bulk sampling commences, the nature of the mineralization and geological and grade continuity should be established.

## 7.12 Grade Control in Mines

*Grade control* is a delicate process of accurately separating *ore* from *waste rock*, ensuring that the extracted ore that gets to the mill for processing is of optimum grade by minimizing dilution. For mining operations, grade control maximizes value and increases profits. One of the most important roles mine geologists can perform in mining operations is to supervise the implementation of grade control. Mine geologists know a lot about the distribution of ores and their contacts with host rocks. The foundation of any successful grade control program depends on high quality sampling supported by geological knowledge and a suitable data management and modelling system that can be used for data processing and analysis.

In open-pit mines, grade control involves the careful examination of blast-hole cuttings produced by down-the-hole hammer drills to ensure the visual classification of bench reserves into three categories of ore is accurate: ore, *low grade ore* and *waste material*, or into various metallurgical types. In some cases, grade control may also require the sampling and assay of the blasthole cuttings or muck samples from truck or shovel loads to ensure that a particular ore-rock is assigned to the correct appropriate stockpile or waste dump.

In underground operations, grade control may involve both the mapping and sampling of stope faces, tram car loads or draw-point muck piles, and broken rock at a recently blasted face. Sometimes, these may be supplemented by the sampling of DTH or jackhammer cuttings or diamond drill cores. These samples are sent to the laboratory for assay. It is the job of the mine geologist to ensure that mining is closely following the mineralized zone and that overbreak during ore blasting or stoping is kept to a minimum to avoid dilution. The main goal of grade control is thus to ensure that material being fed to the mill is of economic grade and that large fluctuations in grade are minimized by blending ores from different benches, or parts thereof, or from different stopes.

For a mining operation, minimizing misclassification of ore and waste during grade control is of critical importance because it affects profits in mining and processing of ores It is essential for profitable mineral processing that the ore feed to the mill is of optimum economic grade and that large fluctuations in grade are minimized and kept as close as possible to what was incorporated in the original design specification of the mill and concentrator.. The process of ore grade control also includes data collection, data integration and interpretation, local resource/reserve estimation, input into final stope design, supervision of development and mining, and stockpile management, leading ultimately to *reconciliation*.

The term 'reconciliation' is a process that allows determination of the ability of a mining operation to produce the tonnage, grade and contained metal that were estimated in the ore reserves documentation. Regular reconciliations are recommended during mining operations, between the estimated stope grades, the grades indicated from stope/truck sampling and those reported by the mill. It is essential that this is undertaken so that necessary modifications can be made to sampling practice or to the methods or parameters used to calculate grade, tonnage or contained metal.

The successful implementation of all grade control programs devolves partly on the mine geologist's understanding of the ore body, led by clear and accurate geological mapping and representative sampling protocol to guide appropriate resource estimation and mining. It is the responsibility of the mine geologist to ensure that mining is closely following the mineralized zone and that overbreak and dilution with waste rock are kept to a minimum. Sometimes, the challenge facing mining companies is the ability to establish sampling processes and techniques that allow the appropriate estimation of the geological grades of the mined out ore. The mine geologist has an important responsibility to develop an acceptable sampling and assay plan.

## 7.12.1 Grade Control Methods

There are generally two approaches provided by geologists in support of grade control in mines: (a) accurate geological mapping, and (b) representative and unbiased sampling and accurate assays. Data management and interpretation can be handled by other mine professionals.

### 7.12.1.1 Grade Control Mapping

Geological mapping of the ore zones particularly in underground mines provides important information in defining the boundaries of mineralized zones which has a positive impact on grade control in support of efficient and cost-effective mine operations. It is particularly applicable where orebody geology and geometry are variable and mine openings need to be planned accordingly. The primary objective is to identify and accurately define and locate the ore zone or at least the section likely to contain the ore with respect to the minimum stoping width. The secondary objective is to determine the attitude and position of relevant structures that control ore shoots particularly in high value minerals such as gold and platinum. Both surface and underground geological mapping has a positive impact on resource estimation and grade control. Grade control mapping supports dynamic mine development, particularly where geology is variable and mine openings may need to respond accordingly. Mapping contributes to the wider geological knowledge of the deposit and is part of the learning process about an evolving ore body. Details on the geological mapping of mines were described in a preceding section of the chapter.

Grade control mapping often takes place over a short period of time because of tight production schedules. Systematic digital photography can very much help to speed up data capture where shotcrete is applied for roof support almost immediately after mucking, and mapping time is very limited. However, an outline field sketch is still required. Moreover, face or roof mapping can be made more time efficient through the use of field tablet computers where mapping is done directly onto the tablet. This results in an immediate digital format that can be easily transferred to other software for editing and/or visualization.

### 7.12.1.2 Grade Control Sampling

Sampling of the ore zone is the more common and traditional approach in grade control. This process takes place prior to blasting or excavation, and assaying takes place to determine ore grades. Misclassification of ore grades results in lost revenues and the need for representative sampling procedures is increasingly important in all mining industries. To achieve accuracy of results, most mines establish grade control programs. The need is for high-quality grade control samples that are representative, unbiased, safe and operationally timely (bot to impact production schedule). Sampling methods that are used depend on the morphology of the orebody and the mining method. Preferred methods include chip sampling of percussion drill and reverse circulation drill cuttings, drill core sampling, channel sampling, grab or muck sampling, and rarely bulk sampling. These methods have been described in a preceding section of this chapter. To achieve quality samples, collection techniques employed should minimize errors related to sample delimitation, extraction and preparation.

Grade control staff should be supervised by the mine geologist, well-trained and motivated, and mining staff should understand the need for grade control. Implementing best practices and minimization of sampling errors require suitably skilled individuals to train and mentor staff, with ongoing Quality Assurance/Quality Control (QA/QC) monitoring and review will allow protocols and staff to be updated as required.

### *In Open Pit Mines*

In open-pit operations, grade control usually involves the sampling of blast-hole cuttings produced by percussion drilling or reverse circulation drilling using down-the-hole hammers followed by the classification of bench reserves into three categories of: ore, low grade or waste material. The preference to these drilling methods is made because of the limited time available for drilling and necessity to collect large amounts of representative samples. The constraints imposed by the mine logistics dictate the needs of choosing the low cost drilling techniques characterized by a high drilling rates, which can be achieved without compromising the samples quality. Auger drilling can also be used for grade control if the ground is soft which permits using the auger drilling. In some cases, grade control may also involve the sampling of truck or shovel loads to ensure that rock is assigned to the correct stockpile or waste dump.

Blast holes for grade control in open-pits are drilled as regularly distributed patterns fully covering the exploited ore body and their host rocks in the open pit bench Grids can be square, rectangular or parallelogram. Spacing between blastholes ranges from 3m to 10 m and their depth approximately matches the heights of the mining benches, about 6-15 m. Holes are usually 6 in to 12 in wide. A good spatial coverage of the mining area by the blastholes and their tight spacing make them a very convenient media for grade control. High density of the grade control data points allows for the most accurate location of the ore body contacts and demarcate waste by using the cut-off grade. The defined contacts are then marked on the open pit benches by the mine geologists, usually using the marking tapes; Average grade of the ore prepared for mining is estimated by averaging the grades of all blasthole samples collected from the delineated ore body polygons.

The common practice of sampling blastholes consists of collecting crushed rock fragments from the blasthole cones. Samples can be collected by shovel, scoop, hand-held auger drill or special devices allowing to split samples during drilling. General principles of collecting representative samples from the blast whole cone (pile) are as follows. Samples are better collected by a special instrument, such as hand held auger drill, al-lowing easy penetration through the entire pile. To ensure a good representative sample for the entire pile it should be collected by several increments which are distributed around the pile.

### *In Underground Mines*

In underground operations, grade control usually involves the mapping and sampling of stope faces, sampling of tram car loads or draw-point muck piles, broken rock at a recently blasted face, jackhammer cuttings or diamond drill cores. It is the geologist's job to ensure that mining is closely following the mineralized zone and that overbreak during stoping is kept to a minimum. Grade control procedures at underground mines differ depending on the mining method. At the mines where ore is selectively mined using cut-and-fill or shrinkage stoping methods the grade control is largely made by channel or chip sampling of the ore drive faces In mines where there is bulk mining underground operations, the grade control is usually made by additional infill drilling, using diamond core or, percussion techniques. Percussion drilling is commonly made by Jumbo-rigs which can be located on the footwall drives. The drilling has to be coupled with mapping and sampling of mineralisation exposed in the underground workings, including ore drives faces and backs and the side walls of the cross-cuts. If the procedure is systematically applied with a due rigor it provides a good quality information with the level of the details sufficient for accurate design of the mining stopes. High density spacing of grade control samples allows for accurate estimates of ore grade in the stope. In some instances, the grade control samples are collected from the draw points of the underground stopes. This is a very common procedure where mining is by block caving.

## 7.13 Review Questions

1. Describe the role and responsibilities of a mining geologist, and compare with those of a mining engineer. How do they work together in mining operations?
2. List the stages of the mining cycle, and describe the main ore deposit types in relation mining.
3. Define stripping and dilution. List the geological and environmental factors that are important in the selection of a mining method?
4. Draw a labeled diagram of an open-pit mine and label the major features. What type of orebodies are suitable for surface mining.
5. Attempt a geological and mining engineering classification of underground mining methods. What are the major access ways of underground mines?
6. Write short notes on the following mining methods describing the type of deposit they are most suitable for (a) strip mining (b) hydraulic mining © solution mining (d) longwall mining.
7. Define grade control and reconciliation, and briefly describe two methods of geological grade control in underground mines.
8. What is blasting? Explain its role in mineral extraction and list the geological and geotechnical rock properties that affect blasting
9. Describe the main geotechnical properties of mine rocks and explain how geotechnical data are collected and used in mine operations
10. Geological mapping in underground mines is complex. Give reasons why. Describe how to conduct mapping of (a) Roof and wall (b) Mine face

## References


Abzalov, M. (2016) *Applied Mining Geology*, Springer Publishers, 438pp.

Abzalov MZ, Menzel B, Wlasenko M, Phillips J (2010) Optimisation of the grade control procedures at the Yandi iron-ore mine, Western Australia: geostatistical approach. Appl. Earth Sci 119(3):132–142

Abzalov MZ (2008) Quality control of assay data: review of procedures for measuring and monitoring precision and accuracy. Exp Min Geol J 17(3–4): 131–144

Brady BHG, Brown ET (2004) Rock mechanics for underground mining. Kluwer Academic Publishing, New York, p 628

Carter PG (2011) Selection process for hard-rock mining. In: Darling P (Ed) SME mining engineering handbook, rd edn. Society for Mining, Metallurgy, and Exploration, Inc, Englewood, pp 357–376

Clark, GB, Mero, JL and Hustrulid, WA (2017) "Mining". *Encyclopedia Britannica* https://www.britannica.com/technology/mining.

Davis JG (1992) Grade control for Australian open pit old mines. In: Annels AE (Ed) Case histories and methods in mineral resource evaluation, v. 63. Geological Society Special Publication, London, pp 219–232

Dominy, SC, Platten4, IM. Xie, Y and Minnitt, RCA (2010) Underground grade control protocol design: case study from the Liphichi gold project, Larecaja, Bolivia. Applied Earth Science (Trans. Inst. Min. Metall. B, v. 119, no. 4, 205-219

Dominy, S. C. (2010). Grade control sampling methods in underground gold mine grade control, Proc. Sampling Conf., 7–20, Melbourne,

Dunbar, WS (2012) Basic Mining and Mineral Processing. Americas School of Mines Presentation. 75pp.

Ericsson M (2007) Trends in underground mining. In: Smith M (Ed) Mining methods in underground mining. Atlas Copco Rock Drills AB, Örebro, pp 3–5

Hartman H.L. ed. (1992). *Mining Engineering Handbook, 2nd Ed.* 2170 pp., Society for Mining, Metallurgy, and Exploration, Inc. (Littleton, USA).

Hamrin H (1982) Choosing an underground mining method. In: Hustrulid WA (Ed) Underground mining methods handbook. AIME, New York, pp 88–112

Hamrin H (2001) Underground mining methods and applications. In: Hustrulid WA, Bullock RL (Eds) Underground mining methods: engineering fundamentals and international case studies. Society for Min

Macdonald E.H. (1983). *Alluvial Mining,* 508 pp., Chapman and Hall (London, UK).

Peters WC (1987) Exploration and mining geology, 2nd edn. Wiley, New York, p 70

Ramani, RV (2012) Surface Mining Technology: Progress and Prospects. Procedia Engineering 46, 9 – 21

Revuelta, MB (2018) *Mineral Resources – From Exploration to Sustainability Assessmen*t. Springer Verlag, 638 pp.

Rossi, M. E. (1999). "Improving the Estimates of the Recoverable Reserves", in Mining Engineering, Volume 51, Society for Mining, Metallurgy, and Exploration (SME), pp. 50-54.

Thomas L.J. (1973). *An Introduction to Mining*, pp.436, Hicks Smith & Sons. (Sydney, Australia).

# Chapter 8
# Environmental Geology in Mining Operations

## 8.1 Introduction

**Environmental geology** is the application of geological knowledge in the study, interpretation and solution of environmental problems. Mining operations both on the surface and subsurface are major sources of environmental problems that require identification and remedial solutions; including mitigation of geological hazards in mines, evaluating and mitigating impacts of mining on the landscape, soils, surface water, groundwater, and the air surrounding mines. Geologists employed in mines are often tasked with the responsibilities that may include environmental protection, conducting investigations, environmental monitoring, and providing oversight of restoration and remediation efforts. Sometimes these tasks may be contracted to outside consultants.

In most advanced countries and some developing countries, most mining operations are subject to environmental regulations that protect the environment from degradation or pollution related to mining activities. Mining companies are often required to apply for approval of operations through obtaining 'permits' from several levels of government – local, state and federal – before starting a new mine project. Environmental impact assessments are often required which may delay mining projects for years. Multiple governmental agencies may be involved at each of these levels, such as agencies in environmental protection, water resources, forestry, land management etc. There are often several federal and state environmental laws that affect mining in areas of water, air, and soil protection. Mining practices should include measures such as reducing water and energy consumption, minimizing land disturbance and waste production, preventing soil, water, and air pollution at mine sites, and conducting successful mine closure and reclamation activities. Mining companies are therefore required to have mining personnel to oversee environmental issues and ensure compliance with mining regulations. The most likely professional staff knowledgeable in environmental science is the mine geologist, (or sometimes the mine engineer) who is asked to handle such responsibilities.

## 8.2 Responsibilities of Mine Geologist in Environmental Protection

The main role of a mine geologist in environmental protection is to ensure that mineral extraction is undertaken with minimal geological hazards and contamination of the environment within and proximate to mining sites. A mine geologist with some training in environmental science should be aware of and capable of taking some actions including, preventing geohazards, monitoring of mine workings, implementing mitigation actions, and supervising remediation and land reclamation efforts. Some of the specific responsibilities include but not limited to the following:

- Identify, mitigate or monitor potential geological hazards such as rock slides, roof collapse, water seepages, lethal gas accumulation (coal mines) etc.
- Conduct or coordinate geological field investigations related to soil and groundwater pollution, and groundwater quality monitoring in and around mining sites.
- Participate in environmental remediation studies and/or interact with consultants to negotiate remedial work and oversee all aspects of remediation.
- Conduct or supervise sampling and analysis of soils, mine waste, surface water and groundwater.
- Ensure compliance with environmental laws and regulations relating to mining activities including obtaining required permits or exemptions.
- Evaluate special mining projects (e.g., acid leaching), salt solution mining wells, LPG storage areas, surface mining and underground storage tanks to detect and prevent sources of surface and subsurface pollution.
- Participate in the management of mine waste to minimize water and air pollution.
- Conduct environment-related investigations and prepare relevant reports.

Environmental problems associated with mining are listed as follows:

1. Mining geological hazards
2. Land disturbance and soil contamination
3. Surface water diversion and pollution
4. Groundwater pollution
5. Air pollution
6. Ecosystem damage

## 8.3 Mining Geological Hazards

A mining geologic hazard refers to a naturally occurring source of danger or potential danger as a result of the extraction of minerals in a mine that can cause harm to life or damage to a mine's infrastructure, utilities, land, or the environment. Most of mine geohazards are related to sudden movement of mine rocks and soils, gas outbursts, and seepage or flood of groundwater that cause landslides (soil or rock slides, rock fall), slope failures, roof collapse, subsidence, rock bursts etc., A *rock burst* is a sudden and potentially violent dislodging of a rock mass from the walls of a mine, tunnel or quarry.

Accidents caused by geohazards can lead to loss of lives and expensive machinery along with substantial ore loss and loss in productivity. Excluding the effects of natural hazards such as earthquakes or volcanic activity, geologic hazards in mines are typically caused by the rapid or even instantaneous release of accumulated energy that produces sudden movement and related processes such as flooding due to groundwater seepage along cracks created by the movement. A mine geologist has an important role to play in identifying zones of potential structural weakness in mine rocks such as discontinuities, cracks, fractures, seepage, or noticeable changes in the characteristics or orientation of discontinuities. The following geohazards are of considerable importance;

- Slope failures and landslides in open-pit mines
- Rock failures and movement in underground mines
- Mine Subsidence and ground failures
- Groundwater bursts and flooding (inundation)
- Gas outbursts in coal mines

### 8.3.1 Slope Failures

Slope failure or collapse in open-pit mines and quarries is a product of slope instability which is a critical hazard that can be of any scale, small or large, and directly affect people, equipment, and production processes. Landslides, rock falls and rock slides driven by gravitational forces are the most common hazards. The geotechnical properties of rock mass and slope angle design are important in preventing slope failures. The overall slope of the pit walls and that of the benches are designed based on the strength and geotechnical properties of the rock mass. Maintaining slope stability is often a challenging aspect of large open-pit mines due to complexity of geological structures and heterogeneous characteristics of mine rocks. In open-pit mining, there is a general tendency to increase the slope angle as an attempt to decrease the stripping ratio, which in turn can produce a higher return on investment. However, an increase of the slope angle decreases the stability of the slope, and it could lead to safety implications and higher operating costs due to slope failure and work stoppages. Consequently, the analysis and monitoring of slope stability are of vital importance for preventing disasters that occur because of such instabilities.

Three types of slope failures occur in open pit mines: *bench failure, interramp failure* and *overall slope failure* (Fig. 8.1). Bench failures which produce rock fall or rock slide from the crest or face of benches have less serious consequences other than mostly related to cleanup costs. Damage to equipment and injuries to personnel are unlikely provided that the failure does not occur when the slope is under construction]. The consequences of inter-ramp failures are more significant than the bench failures because injuries to personnel and damage to equipment are likely. The most severe case of failure is the overall slope failure. Injuries and fatalities to personnel and damage to equipment are highly likely to be serious if failure occurs in working hours. The economic impacts may be irreversible because failure can lead to ore dilution and consequently, a decrease in the ore's economic value.

Fig. 8.1: (a) Three types of slope failure (b) Massive slope failure in open pit mine

Disasters associated with slope failures in surface mines can be prevented by ensuring that the initial slope design is based on sound geological and geotechnical data and modelling of the ore and wall rocks. The implementation of the design should also involve minimizing unnecessary damage to slopes during blasting, excavation control, groundwater and surface water control, as well as installation of ground support and reinforcement where necessary. The role of the mine geologist becomes relevant in the monitoring of slope stability because, slope monitoring is crucial in predicting and preventing slope failures, and, when failure is imminent, it helps in mitigating the effects of a slope failure by taking appropriate actions to minimize damage. Surface monitoring can be easily implemented by routine visual surveys and geotechnical mapping of the pit and bench walls for fractures, cracks and other discontinuities that may indicate potential weakness. Direct measurement techniques can also be used to evaluate the geometry of rock discontinuities using instruments such as crack width meters, tilt meters, and other similar devices.

### 8.3.2 Rock Failures in Underground Mines

Underground mining in tunnels, drifts, crosscuts and stopes takes place in a more confined space than surface mining which makes geohazards more dangerous. Failure of rocks in the side walls, roof and face of mine openings can occur as roof collapse, rock slides, rock falls, rock bursts and ultimate ground failure. Such failures can be attributed to several reasons, including strength failure or weakness in rock properties, increased compressive and horizontal stress, and sudden movement along discontinuities such as cracks, fractures, slickensides, and even faults. These failures occur in two modes: *ductile* and *brittle* failure. Ductile failure takes place as a relatively slow process that consumes sufficient energy to prevent its acceleration, thereby allowing time to remove any hazard to miners and equipment although the economic and environmental consequences (inundation, sinkholes, etc.) may still be catastrophic. In contrast, brittle failure, such as rock bursts, involves a sudden release of energy which can be hazardous and impact work areas leading to fatal accidents and major mine disasters. That is, the failure process will be sufficiently slow to give miners time to retreat.

*Roof falls* or roof collapse in underground mines is one of the most common and destructive geological hazards particularly in coal mines where the rocks are heterogeneous sedimentary rocks with inadequate artificial support (Fig. 8.2). Many of the roof instabilities are caused by sudden change in geologic structures, weak rocks, the presence of fault zones or major discontinuities, and sometimes, mining induced damages. Other natural causative factors include high horizontal and compressive stress and influx of groundwater or gas outbursts.

Mine geologists can be tasked with the responsibility of geotechnical monitoring of mine workings to evaluate the stress state and delinquent properties of mine rocks in order to get a better understanding of the rock mass deformation mechanisms. Although brittle failure may be sudden, there are often signs of structural weakness that develop over time, such as widening of discontinuities and water seepage. A proactive monitoring plan can be implemented involving comprehensive examination of drilling logs; diligent evaluation of roof, rib and stope; accurate mapping of discontinuities, rock falls, and geologic structures; and installation of measuring devices such as extensometers and other microseismic and remote monitoring systems.

Fig. 8.2 (a) Collapsed underground mine

(b) Ground failure in coal mine

## 8.3.3 Mine Subsidence

Mine subsidence can be described as the depression or settlement of the ground surface as a result of readjustments of mine overburden overlying voids or open spaces created during or after the mining process. Surface depressions vary from potholes to large troughs and complete collapsed holes (Fig. 8.3). When valuable minerals are extracted using self-supported mining methods such as "room and pillar" and longwall mining of coal and solution mining of salt deposits, voids are left behind that may collapse resulting in the shifting or sinking of the ground surface. The void created by the underground extraction causes significant changes in the magnitude and orientation of the in-situ stress field and results in deformations both in mine and surrounding rocks, sometimes affecting surface structures. A very different type of subsidence is produced by the extraction of irregularly shaped metalliferous ore bodies by block caving or by leaching of disseminated deposits (in situ leaching) where the extracted width is small compared to the depth, a prominent collapse structure, as much as tens to hundreds of meters in depth, may develop rapidly.

Moreover, there are instances where underground mines are overlain by a considerable thickness of consolidated rock, subvertical fractures can propagate upward toward the surface, resulting in downward settling of the strata. Alternatively, shallow mines overlain by a thinner rock overburden may collapse, causing overlying soil and unconsolidated sediment to sink into the resulting void. Both processes may result in a surface depression that worsens over time and can lead to injury, loss of life and damage of mine infrastructures and buildings. Sinkholes occur in areas overlying underground mines which are relatively close to the ground surface (Fig. 8.3b). This type of subsidence is fairly localized in extent and is usually recognized by an abrupt depression evident at the ground surface as overburden material collapses into the mine void. Sinkhole subsidence is perhaps the most common type of mine subsidence and has been responsible for extensive damage to many engineered structures, including roads, buildings, and water supply infrastructure.

Visual inspections and evaluation of rock strength and structural deformation are helpful in identifying potential zones of weakness which can lead to mitigation to avoid potential mine subsidence. Ground control measures including reinforced support, partial extraction etc., can be implemented in active mining sites particularly in coal fields to prevent or minimize the effects of mine subsidence.

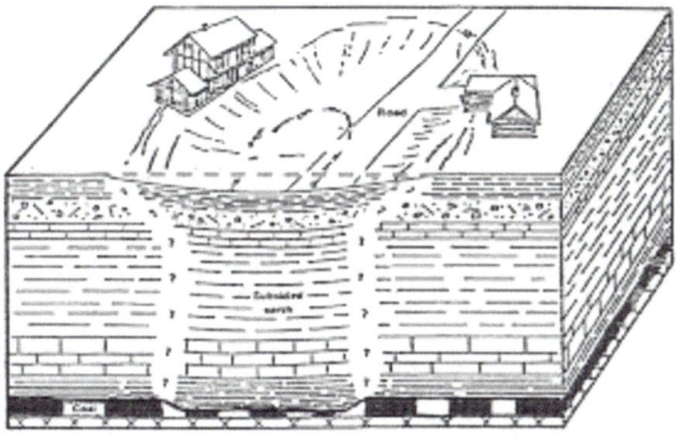

Fig. 8.3: (a) Depression from mine subsidence

(b) Sinkhole in grass field above old coal mine

## 8.3.4 Groundwater Bursts and Flooding

In deep underground mines, particularly coal mines, due to the complicated geological conditions and structural failures along faults, mining disasters have occurred as a result of sudden outbursts of groundwater into mine tunnels resulting in sudden flooding that trapped miners, and produced loss of life and damage to mine infrastructure and equipment. Groundwater hazard in mines can be attributed to several reasons, including sudden movement along rock discontinues, collapse of water-bearing layers and water accumulation in abandoned mine workings. Although groundwater seepage is common in mines, the sudden water outbursts are very hazardous. Water incursion can also originate from dams or water bodies within or around mines.

For example, one of the worst underground mine accidents in the world was the so-called "Mufulira disaster", recorded in Zambia in 1970 when 89 miners died due to flooding. The accident took place in the morning of 25th September 1970, when half of the mine was flooded because mud and water from the slime dam seeped through cracks in an old stope, causing a section of the overhanging wall to give way. Thus, the mud and water rushed into the eastern section of the mine and flooded all shafts below 500 m. Water outburst can also be caused by human error that produces serious disasters as well. The Knox Mine disaster was a mining accident on January 22, 1959, at the River Slope Mine in Jenkins Township, Pennsylvania. The disaster occurred when workers were ordered to dig illegally under a river bed of the Susquehanna River without proper safety precautions, creating a hole in the riverbed which caused the river to flood into the many interconnected mine galleries. Twelve miners were killed. Plugging the hole in the riverbed took three days, and mitigation efforts created several new islands that altered the western-side flow of the river around the newly created islands.

Mine geologists are capable of visual observation of water seepage along cracks in underground mines and their relationship to fractures and faults. Monitoring of groundwater levels and groundwater flow in abandoned tunnels can also be very helpful in risk evaluation and mitigation.

## 8.3.5 Gas Outbursts and Explosion in Coal Mines

Gas outbursts are the sudden and violent release of gas in an underground coal mine as a result of a complex function of geology, stress regime, and gas pressure and content. Coal deposits are known to generate various gases from organic decay, and the accumulation of excess gases such as carbon dioxide and methane. Gas outbursts are a major geohazard in coal mines, and whenever they occur, can be very serious disasters, possibly resulting in fatalities and damage to mine infrastructure and mining equipment. For instance, an outburst at Tahmoor Colliery in New South Wales, Australia in June 1985 involved the ejection of 350 tons of coal and rock and over 3000 cubic meters of gas, resulting in massive destruction and one fatality.

The gases involved in mine outbursts are generally methane and carbon dioxide, but there are also cases of nitrogen gas being the predominant outburst fuel. These gases when ejected not only cause injury to workers and damage to equipment around the working face, but also cause the mine atmosphere to become toxic for the workers to breathe and can create an explosive gas combination and start fires. Mixtures of methane in the air can also become explosive at concentrations of 5% to 15%, with 9.5% being its most dangerous concentration, as its ability to react with oxygen within the mine allows for the massive production of heat that will induce an explosion (Fig. 8.4).

The conditions that promote gas outbursts are varied but include some of the following; failure along geological structures (e.g. faults, mylonite zones), increased gas content, gas composition (predominant $CO_2$), high compressive strength of coal and stress regime at the coal face. The ultimate method for preventing an outburst event during mining is to predict potential zones of outbursts and avoid them or at least reduce their outburst potential. As the outburst mechanism is a complex relationship between geologic structure, mining-induced stresses, and gas content and pressure, the removal or mitigation of one or more of these elements can possibly reduce outburst potential.

Fig. 8.4: (a) Mine explosion in coal mine

(b) Mine fire caused by gas explosion

Mine geologists have a role to play in the adequate monitoring of gas levels and visual inspections to identify structural discontinuities in wall rocks, such observations can help prevent or mitigate gas outbursts. Geologic structures such as faults or sheared zones can be avoided to a certain extent. Stresses can be reduced by changes in mining rates, methods, and geometry. Gas contents and pressures can be reduced by drainage. Increased ventilation in mines using giant fans to circulate the air that will dilute methane gas levels and provide adequate ventilation for workers. Most of the methods practiced throughout the world involve relieving stress concentrations and/or in situ gas pressure.

## 8.4 Environmental Impacts of Mining

Mining is the human activity that, perhaps, has the greatest negative impact on the environment, and human health and safety. Mining usually involves excavation of the land and rocks, both on the surface and underground. However, underground mining is a less environmentally harmful method for mineral extraction than surface mining because it occurs below the surface with limited exposure to surface water and atmosphere. The ores and other minerals that are exploited are composed of elements (metals and non-metals) that could become toxic to the environment if they ultimately accumulate in the soil or leach into surface water and groundwater that are used by humans for drinking and other uses.

Waste rock and tailings generated by mining are sometimes disposed in a manner that is inappropriate and dangerous to human health and safety. Mine waste can pollute air and drinking water, harm wildlife and habitat, and permanently scar natural landscapes. Modern mines as well as abandoned mines are responsible for significant environmental damage throughout the mining countries in the world, particularly in China, Canada, Australia, western United States and South Africa. The impacts of mining on the environment will be briefly described in the following areas of concern:

- Land disturbances
- Safety hazards
- Waste rock disposal
- Soil contamination
- Surface water pollution
- Groundwater pollution
- Air pollution
- Ecosystem damage

## 8.4.1 Land Disturbance

The extraction of minerals, particularly in surface mining, usually involves the digging of open pits, and other mine workings. The excavated or dug-out materials and wastes are dumped in heaps over the surrounding lands, sometimes in a haphazard manner, thereby creating widespread disturbance of the landscape. Underground mining causes less damage, and generally produces relatively small waste rock disposal areas ranging from a few to tens of acres in size. Such areas are usually located near the entrance of the underground workings, not far from the shafts and adits. In addition to the poor visual and aesthetic effects, mining activities have exposed the lands to degradation and erosion. In some places, deep gullies have developed from loose mine waste and abandoned pits that continue to accelerate and cut deeper into the surrounding areas thereby threatening nearby human habitat and ecosystem (Fig. 8.5a). In coal mining areas, the land disturbance can be very profound (Fig. 8.5b). Strip mining of coal severely alters the landscape while mountaintop removal obviously destroys the natural landscape which accelerates erosion and makes reclamation near impossible.

Fig. 8.5: (a) Disturbed mine landscape (b) Coal mining landscape in Appalachia, USA

In many developed countries, mining companies are required to comply with environmental regulations that stipulate land reclamation or restoration, ensuring that the areas mined are returned close to their original state or adapted to other beneficial end uses. In contrast, most developing countries do not have environmental codes, or they fail to enforce such codes so that there are no plans for land rehabilitation after mining. Moreover, the presence of uncontrolled illegal mining makes the situation worse. Many of these illegal miners do not have any professional training or expertise and therefore arbitrarily dig the ground in search of the minerals, leaving the landscape riddled with deep gullies and abandoned pits which have made gully erosion a recurrent problem in tropical environments with heavy annual precipitation. This devastating effect of mining on the landscape has seriously affected mining regions such as the Jos Plateau in Central Nigeria where tin mining has occurred for many decades.

## 8.4.2 Safety Hazard of Abandoned Mines

Abandoned mines are common in the mining landscape of many countries. Many of these structures contain dilapidated frames, open shafts, and water-filled pits (Fig. 8.6). Abandoned mines and uncontrolled mining both at the surface or underground, creates issues of public safety. Pits, holes and shafts, if uncovered after mine closure, can become death traps. Innocent people and children have been killed by falling into abandoned mine pits and underground holes while playing around or hunting... Illegal miners are known to have died when they fall into pits while digging them. Some also lose their lives while searching for mineral stones or fragments inside pits and the pits suddenly collapse, closing in on them.

Another safety hazard around some underground coal mines are sinkholes or collapsed ground due to subsidence above underground coal mines. Because an unexpected collapse can occur without warning, such areas create safety concerns and usually can be identified and avoided.

Fig. 8.6: (a) Abandoned mine

Fig. 8.5 (b) Abandoned shaft

## 8.4.3 Mine Waste Disposal

Mining generates a lot of waste materials, more so in surface mines (Fig. 8.6a), that are of no economic value and need to be disposed appropriately. These wastes produced during extraction of valuable minerals include soils, rock, tailings and mineral matter usually piled up near the mine site. However, waste rock and tailings usually contain metals and/or sulfide minerals in concentrations considered too low to be profitably extracted, but may be at levels that become toxic to the environment and human health. As the amount of waste rock in open pit mines is commonly two to three times the amount of ore produced, tremendous volumes of waste rock are removed from the pits and deposited in areas nearby. Underground mines also yield mine waste in smaller quantities that are piled near the mine access ways. In addition, waste piles from processing, such as tailings impoundments, leach piles, and slag piles (products of smelting) can be very large, and further contribute to exposed toxic wastes (Fig. 8.7a). Mine waste storages are sometimes contained in dangerous structures such as embankments that could give way and lead to accidental deaths and damage to the environment and local infrastructure.

In most developed countries, mine waste dumps may be covered with soil and re-vegetated after mine closure, but in developing countries where there are no regulations or regulations are poorly enforced, these waste rock dumps become not only an eyesore but a problem to the environment. Sometimes they are located near cities or mining shanty towns, such as in South Africa (Fig. 8.6a).Erosion of the waste dumps and mine tailings, as well as hillside slopes usually result in siltation that block drainages of rivers, streams and creeks producing devastating effects and enhanced risk of flooding of surrounding areas. In some areas, ecosystems have suffered extensive damage particularly in wilderness areas. Agricultural lands and grazing areas close to mining sites have also been impacted by waste dumps and often suffer loss of productive land. Leaching of metals from mine waste can lead to surface and groundwater contamination.

## 8.4.4 Soil Contamination

Soils surrounding metal mining sites are likely to be contaminated with heavy metals such as Pb, Zn, Cu, Cd, As, Ni and Se. These contaminants in soils may be products of direct transfer of mine waste and tailings or from deposits of particulates that settle from mine dust that may be airborne. Wind, precipitation, surface water and dust actions are agents that can further transport or transfer these elements to great distances far away from the mining areas. Heavy metals such as Pb (lead), Cd (cadmium), As (arsenic) and

Se (selenium) are toxic if ingested or inhaled or absorbed through the skin over a long period of time. Such human interactions can have long term effects on health and cause diseases such as cancer, or lead to birth defects in unborn children. The area surrounding mines can also be contaminated by the various chemicals that are used in the processing of minerals, such as mercury that is used in gold extraction. There were reports of lead or mercury poisoning of children in the gold mining areas of Anka in Zamfara State, northwestern Nigeria. The toxic metals originated from the processing of gold-bearing ores in residential compounds where children play and ingested the metals in soil.

Fig. 8.7: (a) Mine dump and Contaminated Soil (b) Acid mine drainage in surface water

## 8.4.5 Surface Water Pollution

Surface water in streams and rivers near mining sites are not only disturbed but could also be contaminated by discharge of mine waste through drainage and precipitation. Sometimes the river courses may be silted or diverted by mine waste. *Acid mine drainage* which is the outflow of acidic water from mining areas, including coal fields, is the primary source of water pollution from mining. This corrosive water with low pH is derived from the leaching of ore material from the mine (mine water), waste rock or tailings (Fig. 8.7b). The yellowish-brown color of the acid water is caused by the oxidation of pyrite (FeS2), a common but valueless sulfide mineral, as well as other sulfides of arsenic, lead and mercury associated with metallic ores. When such sulfides are exposed to water and air, sulfuric acid is created, which can seep from mines and waste rock piles into streams, rivers, and groundwater. Acid mine drainage can sometimes last for decades and even centuries after mining operations cease. Problems associated with mine drainage include contaminated surface water and drinking water, disrupted growth and reproduction of aquatic plants and animals, and the corroding effects of the acid on parts of infrastructures such as bridges. For surface water contamination, the best approach to protection is to prevent the incursion of mine discharge, because it is difficult to stem the flow as soon as pollution enters the drainage system.

## 8.4.6 Groundwater Pollution

Groundwater is a precious natural that is used for drinking and industrial purposes, and a source of recharge to surface water. Depending on the mining site's local hydrology, mining activities may affect groundwater quantity by lowering the water table elevation due to pumping, which in turn may impact nearby lake levels and base flow in streams. Additionally, groundwater quality may be affected by the handling, storage, and disposal of mining wastes and the mine excavation, particularly if it is underground. Other sources of pollution include the wastewater discharge; and the storage and handling of chemicals and fuel. Acid runoff from the mines is one of the most toxic sources of dissolved heavy metals such as copper, lead, mercury that may enter into the groundwater system in shallow aquifers. The rate and degree by which acid-mine drainage proceeds can be increased by the action of certain bacteria.

Groundwater quality and quantity should be monitored by the company and verified by the government regulators throughout the mine's operation and well after reclamation to reveal any mining induced impacts to groundwater. Early detection of potential groundwater impacts should lead to intervention before the actual violation of specific environmental protection requirements.

## 8.4.7 Air Pollution

Mining usually involves the use of heavy equipment for drilling, blasting, excavation and transportation of soil, rock and ore particularly in open pit mines. Such activities can generate a lot of noise, heavy metal particles and dust that become air pollutants. Limestone quarries used in the production of raw material for cement manufacture in some factories in Nigeria, such as Ewekoro and Sagamu in Ogun State have generated so much dust pollution that it has negatively impacted the quality of life in communities surrounding the quarries. Everywhere near these quarries is enveloped by dust that has caused respiratory diseases and turned these communities into blighted areas or "brownfields"

Historical smelting operations are known to be associated with the release of sulfur dioxide gas and particulate emissions. The sulfur dioxide gas reacts with atmospheric water vapor to produce acid rain that has a corrosive effect on buildings and other urban infrastructure and a source of human health concerns. In other cases, the emissions from older metal smelters may have affected human health, and accounted for elevated lead (Pb) in the blood of residents living near lead-zinc smelters. However in recent years, smelting operations, combined with environmental controls, are implemented to prevent potential environmental and health issues related to emissions.

In the confined spaces of underground mines, different kinds of aerosols can be produced by mining activities that are harmful to human health. Coal and coal waste generally contain trace impurities of toxic metals such as arsenic, lead, mercury, nickel, vanadium, beryllium, cadmium, which can be hazardous if released into the atmosphere by substantial coal burning.

Fires and gas outbursts in coal mines are additional hazards that can impact human health and are capable of causing the death of miners in addition to environmental impact. Presence of methane is probably the most characteristic source for fires although flammable and combustible liquids are often stored underground in most mines and pose a special fire hazard. Also, it is well established that coal mining is associated with chronic health problems among miners due to inhalation of coal dust causing respiratory diseases such as black lung disease, which causes permanent scarring of the lung tissues. Studies have shown that long-term exposure to radon gas and its decay products from uranium mining and milling increases the risk of developing lung cancer.

Fig. 8.8: (a) Dust from surface coal mining (b) Radioactive dust from uranium mining

### 8.4.8 Ecosystem Damage

Ecosystem is a community of living organisms that live and interact with each other in a specific environment. .Ecosystems of various types (e.g. marine, terrestrial etc.) function because of the continuing exchange between the biotic (living) and abiotic (nonliving) components. Mines are highly damaging to the ecosystems surrounding them both from the physical and chemical viewpoints related to land degradation and release of toxic metals to the environment. For example, the removal of vegetation as a result of mining operations, particularly in tropical vegetated areas leads to the destruction of plant life. With reduced plant biomass, less carbon is being processed via photosynthesis, which leads to less oxygen production, less standing biomass and reduced transfer and cycling of nutrients. Soil contamination and acidification are toxic and can limit the amount of plant life and the absence of vegetation cover accelerates erosion and environmental degradation.

Mining destroys animal habitats and ecosystems. Pits that mines create could have been home to some animals. Also, the activity that surrounds the mine, including people movement, explosions, road construction, transportation of the goods, the sounds made, etc. are harmful to the ecosystems and will change the way the animals have to live, because they will have to find a new way to cope with the mine and live around it. Many of the toxins and tailings that are discharged from the mines can disrupt and poison the environment where the animals live, and their health. Mining can completely destroy ecosystems by adding or eliminating substances animals need for sustenance. Mining also affects human activities by reducing land available for farming in rural areas and reduced productivity. Mine dust and contaminated drinking water create hazards to human health.

## 8.5 Environmental Monitoring

Environmental monitoring is the systematic measurement of key environmental indicators over a period of time to help characterize and assess the quality of the environment. Environmental monitoring is used to evaluate the potential or ongoing environmental impacts in which mining activities carry a risk of harmful effects on the natural environment and surrounding commonly. Therefore, environmental monitoring is considered critical to the protection of human health and the environment. Mining companies undertake environmental monitoring using mine personnel as well as install monitoring devices to collect all the relevant data needed for continuous evaluation of the mining environment. All monitoring strategies and programs have reasons and justifications which are often designed to establish the current status of an environment or to establish trends in environmental parameters. In all cases, the results of monitoring will be reviewed, analyzed statistically, and sometimes published. Environmental parameters that are monitored include; (1) Geohazards. (2) Air quality. (3) Water quality, and (4) Soils.

### 8.5.1 Geohazard Monitoring

Most of the mine geological hazards are attributable to natural forces or geological factors that produce sudden movements which may result in landslides, rock slides, rock falls, rock burst, roof collapse, groundwater burst and gas outbursts. These geohazards can lead to major disasters involving loss of life and damage to mine infrastructure and equipment. While natural events like earthquakes cannot be easily predicted, other natural hazards can be identified by regular geological and geotechnical mapping or monitoring of mine walls and openings. In addition, several direct measuring devices can be installed and monitored manually or remotely to provide information on changes in the behavior of cracks, faults and other discontinuities. Various mitigation measures proposed by geomechanical engineers can be implemented.

## 8.5.2 Air Quality Monitoring

Air quality monitoring is very important in mining sites because of its relevance to the health and safety of mine workers. Different types of aerosols containing toxic heavy metals and gases are generated in surface and underground mines. Monitoring the dust concentration from a mining operation can help ensure the protection of the mine employees and the surrounding communities. In developed countries, there are regulations and standards that are established for compliance monitoring of dust particles (particulate matter) and some gases such as sulfur dioxide, nitrous oxide and methane in coal mines and radon in uranium mines. Air quality monitoring instruments can be installed throughout the interior of underground mines, in addition to being deployed around the perimeter of the site, ensuring the particulate matter is not at dangerous levels when it leaves the property and makes its way into the environment and adjoining neighborhoods if present. In open pit mines and quarries that are exposed directly to the atmosphere, monitoring of dust and other particulates are of top priority

## 8.5.3 Mine Waste Monitoring

Waste materials derived from mining, such as waste rock and tailings from ore processing are usually stored in waste disposal facilities not too far from the mine premises. Normally, they should be located in areas that minimize environmental impacts and long-term environmental liabilities. Monitoring of such waste facilities is required so that the long-term chemical behavior of the mineral waste can be evaluated to minimize the impact on the surrounding environment and human health. Also, mine waste storages are dangerous structures in terms of associated accidental deaths and continuing environmental hazards.

Physical monitoring of miner waste disposal facilities commonly includes regular visual inspections of surface structures and facilities; regular visual inspections for signs of excessive surface erosion and shallow or deep-seated failure on the outer slopes of waste repositories, and (monitoring of water levels and pore pressure within embankments and the waste dump. Groundwater monitoring and sampling from monitoring wells may provide information on the migration of toxic elements into the subsurface.

## 8.5.4 Water Quality Monitoring

Surface water, and sometimes groundwater near mining sites need to be monitored for acidity and metal content to evaluate the potential impact of mine drainage on water quality. Groundwater in particular can be adversely impacted by plumes of contaminated water emanating from mine waste dumps. Waste disposal facilities are regularly monitored to evaluate the geochemical behavior of waste components. Geochemical monitoring of surface water can be implemented by (a) periodic sampling of runoff water and water discharging from waste disposal facilities in order to monitor flow volumes, solute concentrations, and the solute mass that is being released from the waste (Fig. 8.9a), (b) periodic sampling of down-gradient monitoring wells and surface water bodies to ensure that seepage from the waste is not adversely impacting receiving environment water quality. It is important to continue such monitoring for several years after the closure of the mining facilities.

Protecting groundwater is a top priority of environmental protection programs in many countries that depend on groundwater as a major source of drinking water. Groundwater monitoring at mining sites often has some or all of the following primary purposes: (1) identify baseline (pre-mining) groundwater quality, water table elevations, and flow patterns; (2) measure water table levels and flow patterns during mining to quantify effects from mine pumping and dewatering; (3) discover if spills or accidental discharges have contaminated groundwater; (4) determine the extent of any contaminant migration from project facilities. Several monitoring wells or piezometers are usually installed within the mining areas and down gradient from such facilities (Fig. 8.9b). Monitoring results are usually included in annual reports to government regulators.

Fig. 8.9: (a) Surface water sampling

(b) Groundwater monitoring

## 8.6 Environmental Mitigation and Remediation

Environmental mitigation refers to all the activities or actions taken with the intention to avoid, reduce, remedy or offset known negative impacts on the environment from human activities such as mining. Mitigation, therefore minimizes adverse environmental impacts. The major environmental impacts that require mitigation are physical disturbance and erosion of landscapes, mine waste disposal, soil and water quality, and air quality. These negative impacts are addressed by the following mitigation measures:

a. reclaiming areas of physical disturbance to prevent erosion
b. stabilizing soils containing metals or chemicals
c. preventing and/or treating water contamination, and
d. controlling air emissions.

### 8.6.1 Mine Reclamation

Mine reclamation is the process of restoring land that has been impacted by mining to a natural state or economically usable state. This is applicable mostly to areas impacted by surface mining. Such beneficial end uses may include conversion to open space, wildlife habitat, agriculture, or residential and commercial development. Erosion control measures may involve modifying slopes and other surfaces and planting vegetation as part of the process stabilizes the soil material and prevents erosion and surface water infiltration. The final step is usually to completely replace the topsoil and revegetate with suitable plant species that can be used for recreational activities. Although the process of mine reclamation usually occurs when mining is completed, the planning of mine reclamation may be concurrent with mining operations. In the United States, mine reclamation is a regular part of modern mining practices, and a reclamation plan is often included in the feasibility study. Modern mine reclamation has been very successful in countries like Canada and the USA in minimizing and mitigating the environmental impacts of surface mining. Examples of reclamation projects include converting a mining pit to orange grove; a gravel extraction area to a riparian wildlife habitat, and reclaimed land for grazing and production of crops.

### 8.6.2 Soil Treatment

Soils that are contaminated with heavy metals and elevated acidity can be treated to reduce their toxicity, or contained to minimize their impact on surrounding areas and reduce human exposure. A simple way to reduce acidity is to add lime or other materials that will neutralize acidity, and cover with top soil. Chemical methods are also used to stabilize metals in soils, making them less mobile and biologically available. There are successful attempts to reduce or eliminate the oxidation of pyrite that yields sulfuric acid in contaminated soils by applying bactericide to stop the bacterial growth that promotes the oxidation. Another approach that is commonly used in dealing with contaminated soil and some mine waste is to move them to specially designed repositories. This approach is expensive and controversial because the volume and toxicity of the soil is not reduced by this process, but the soil is just relocated elsewhere. *Geoliners* which are very low permeability synthetic membrane or *bioliners* such as low permeability and compacted manure, have been used as barriers at the base of waste piles to reduce seepage of toxic elements.

## 8.6.3 Water Treatment

The most common treatment for acidic and metal-bearing waters is the addition of a neutralizing material, such as lime, to reduce the acidity. However, where large quantities of contaminated groundwater are involved, large-scale treatment or "remediation" is required. There are two approaches: active and passive. Active (compared to passive) treatment processes involving the removal of dissolved metals by precipitation can be implemented for the remediation of contaminated drinking water sources by "pump and treat" method. Such a treatment process usually involves the construction of a treatment facility which may be required to operate for several decades. The ongoing maintenance that such a plant requires makes this treatment technique very expensive. Using "passive" methods where the contaminated water is not physically treated have been applied in wetland systems to treat metal-bearing water. Such efforts have been successful where the volumes and acidity of the water are not too great. (A wetland is an area covered or flooded with water throughout the year). Also, passive wetland systems have the added advantage of creating desirable wildlife habitat. Other methods used in water treatment include the creation of in-situ treatment zones where reactive materials or electric currents are placed in the subsurface so that water passing through them would be treated. Some of these processes are quite complicated and require the use of advanced technology for remote control of the operations.

## 8.6.4 Controlling Air Quality

The control of gaseous emissions and aerosol in mines is a major regulatory requirement to protect human health of the workers and surrounding communities. Emissions from smelters, especially sulfur dioxide and particulate materials, have historically presented significant environmental problems. Modern smelting technology has met this challenge by drastically reducing the amount of emissions. New smelters as they come online are getting very clean by installing devices that capture more than 99% of emitted sulfur. Particulate emissions, particularly PM10 (particulates of 10microns in diameter or less) have been drastically reduced below regulatory levels.

Coal mines generate gaseous emissions some of which are toxic or flammable. More recently, attention has been focused on gas emissions from coal mining as contributions to the production of greenhouse gases such as methane ($CH_4$) that cause climate change. Available statistics indicate that methane from coal mines account for about 10% of total U.S. methane emissions, and about 1% of total U.S. greenhouse gas emissions. In coal mines, emissions occur when air is pumped into underground coal mines for ventilation, thereby releasing methane into the atmosphere. This reduces the risk of explosions in mines but significantly increases greenhouse gas levels. Gas emission from ventilation accounts for over 60% of coal methane. Methane is about 23 times more potent than carbon dioxide in terms of its contribution to global warming. Consequently, the capture and use of ventilation air methane (VAM), broadly referred to as VAM abatement has been a priority for the coal industry for some time. New technologies are being developed to capture and use VAM for energy production and other uses. Currently, it seems the mining industry is in full control of most gas emissions that are detrimental to human health and increased global warming.

Concentrations of radon in uranium mines and mills are strictly controlled and must be monitored in air to protect workers. Controls include sophisticated detection and ventilation systems that effectively protect uranium workers

## 8.7 Review Questions

1. Briefly discuss the relationship between mining and the environment, and explain the role of governmental agencies in environmental protection.
2. Briefly describe the role and responsibilities of the mine geologist in protecting the surface environment surrounding mines.
3. Describe mine geological hazards and what action can be taken to prevent or mitigate disasters.
4. Write short notes on the following and explain their causes: (a)Rock falls (b)Groundwater busts (c) Gas explosion (d) Mine roof failure (e) Sinkholes
5. What is environmental monitoring? Explain why it is important in mines.
6. Briefly describe the impacts of mining on surface water and groundwater, highlighting the major sources and pathway of mine pollutants.
7. Discuss the roles of government agencies in the protection of the environment from mine pollution.
8. Define environmental mitigation. Briefly describe the mitigation actions that are applicable to soil contamination and air landscape disturbance problems in mines.
9. What is groundwater remediation and why is it required? Explain the active and passive approaches to contaminated water treatment.
10. Write short notes on the following: (a) Geohazards in mines (b) Mine waste management, and (c) Environmental sampling

## References

Bell FG, and Donnelly LJ (2006) Mining and its impact on the environment. Taylor and Francis, Oxford, 47pp

Call, R.D. and J.P. Savely (1990) Open Pit Rock Mechanics. Surface Mining, 2nd edition. Society for Mining, Metallurgy and Exploration, Inc., pp. 860-882. Kennedy B.A. (ed).

Chamley H (2003) Geosciences, environment and man. In: Chamley H (ed) Developments in earth and environmental sciences, 1. Elsevier, Amsterdam, 527 p

Donnelly LJ (2006) A review of coal mining-induced fault reactivation in Great Britain. Q J Eng. Geol. Hydrogeol. 39:5–50

Donnelly L. (2018) Mining Hazards. In: Bobrowsky P.T., Marker B. (eds) Encyclopedia of Engineering Geology. Encyclopedia of Earth Sciences Series. Springer, Cham.

Gamble J.C., Gray R.E. (1988) Mine subsidence control. In: General Geology. Encyclopedia of Earth Science. Springer, Boston, MA. https://doi.org/10.1007/0-387-30844-X_80

Hodge RA (2011) Mining and sustainability. In: Darling P (ed) SME mining engineering handbook, 3rd edn. Society for Mining, Metallurgy, and Exploration, Inc, Englewood, pp 1665–1688

Hudson, T.L, Fox, F.D., and Plumlee, G.S. (1999). Metal Mining and the Environment, p. 7,20-27,31-35,38-39, 41-46. Published by the American Geosciences Institute

Lee, FT and Abel Jr, JF (1993) Subsidence from Underground Mining: Environmental Analysis and Planning Considerations Geol Surv Circular 876, 29pp

Plumlee, GS (1999) The environmental geology of mineral deposits. Series title, Reviews in Economic Geology. Volume, 6A. 46pp

Molinda, G. M. (2003) Geologic hazards and roof stability in coal mines. NIOSH Information Circular 9466, 30pp.

Revuelta, MB (2018) *Mineral Resources – From Exploration to Sustainability Assessmen*t. Springer Verlag, 638 pp.

Shepard, Jo, L. Rixon, KO and Griffiths. L (1981) Outbursts and Geological Structures in Coal Mines: A Review. Int. J. Rock Mech., Min. Sci. & Geomech. Abst, v. 18, 1981, pp. 267-283.

Verbug R (2011) Mitigating acid rock drainage. In: Darling, P (ed) SME mining engineering handbook, 3rd edn. Society for Mining, Metallurgy, and Exploration, Inc, Englewood, pp 1721–1732

Wu, Q., Wang, M (2006). Characterization of water bursting and discharge into underground mines with multilayered groundwater flow systems in the North China coal basin. Hydrogeol J **14,** 882–893.

# Glossary of Common Terms in Economic and Mining Geology

**Acid mine drainage** - The run-off water from mines that is acidic due to presence of sulfides oxidized to sulfuric acid.

**Adit - A** horizontal or sub-horizontal passage driven into a mine from a hill or mountain side to extract ore.

**Alluvial** – Accumulation of sediments in a river bed.

**Assay-** The geochemical analysis of ores and rocks for their metal content which is expressed either in percentages (%), parts per million (ppm) or ounces per ton (oz/t) for precious metal**...**

**Backfill** - Waste material used to fill the void created by mining an orebody.

**Banded Iron Formation** – A sedimentary deposit consisting of alternating bands of iron minerals and beds of silica common in the Precambrian.

**Beneficiation** - The treatment of mined material, making it more concentrated or richer.

**Blasthole** - A drill hole in a mine that is filled with explosives to blast loose a quantity of rock.

**Block Caving** - An inexpensive method of mining in which large blocks of ore are undercut, causing the ore to break or cave under its own weight.

**Borehole** - Any deep or long drill-hole, usually associated with a diamond drill.

**Brownfield exploration** - Exploration work that's done close to or at existing mines.

**Chip sample** - A method of sampling a rock exposure whereby a regular series of small chips of rock is broken off along a line across the face.

**Chute** - An opening usually constructed of timber and equipped with a gate, through which ore is drawn from a stope into mine cars.

**Concentrate** – what is left after the raw ore has been ground down and the waste has been removed.

**Core logging** - A process of recording geological observations of drill core either on paper or computer disc.

**Core or Core sample** - A cylinder sample generally 1-5" in diameter drilled out of an area to determine the geologic and chemical characteristic of the rock or ore.

**Crosscut** - A passageway driven between the entry and its parallel air course or air courses for ventilation purposes.

**Cut-off grade** -The minimum metal concentration or grade of ore that is considered minable at a profit. This may change with time and changing economic considerations.

**Cut-and-fill** - A method of stoping in which ore is removed in slices, or lifts, and then the excavation is filled with rock or other waste material (backfill), **Decline -T**his is a tunnel driven from the surface towards the base of the mine to allow vehicles and equipment to go in and out of the mine.

**Deposit** - This is an amount of ore or mineral that makes exploitation worthwhile.

**Diamond drill** – A drill used to collect core samples from rock.

**Disseminated ore - T**he scattering of ore mineral uniformly throughout host rock.

**Dragline** - A large excavation machine used in surface mining to remove overburden (layers of rock and soil) covering a coal seam.

**Dredge** – A common piece of mining equipment, used to sucks up dirt and gravel from within a stream bed by the use of water pressure.

**Drift** - A horizontal passage underground. A drift follows the vein, as distinguished from a crosscut that intersects it, or a level or gallery, which may do either.

**Drift mine** – An underground coal mine in which the entry or access is above water level and generally on the slope of a hill, driven horizontally into a coal seam.

**Drilling** - The process of installing a cylindrical hole in the ground or in the rocks primarily to investigate the subsurface and or collect samples as drill cores

**Economic geology -** The science **of** mineral deposits, and a study of their characteristics, distribution, mode of occurrence and origin.

**Eluvial deposit** – Mineral deposit or soil derived from in situ weathering of rocks or gravitational movement or accumulation.

**Endowment** – The full economic potential of an area due to be mined.

**Environmental pollution** – The release of toxic substances into the environment.

**Epigenetic** – Mineral deposit precipitated after the host rock had formed.

**Epithermal** - Hydrothermal mineral deposits that formed at depths of less than 6km and temperatures of 100 to 200°C.

**Explosive** - Any rapidly combustive or expanding substance. The energy released during this rapid combustion or expansion can be used to break rock

**Extraction** - The process of mining and removal of coal or ore from a mine.

**Face** - The end of a drift, crosscut or stope in which mining work is taking place.

**Face drilling** - A method used in conventional mining to drill a hole in the coal bed for inserting explosive charges. The holes are drilled into the rock and packed with explosives.

**Feasibility study- A**n assessment of the economic viability of an ore deposit taking into consideration all aspects relevant to mining.

**Footwall** - The area below where ore is present in a mine.

**Gangue** – The worthless material surrounds or is mixed with the mineral in ore.

**Glory hole** - An open pit from which ore is extracted, especially where broken ore is passed to underground workings before being hoisted.

**Graben** – Elongated depression of the earth's crust lying between two faults.

**Grade** – The amount of useful mineral contained in a piece of ore.

**Greenfield exploration** –The search for minerals in an area without previous information or mining.

**Hanging wall** - The rock or rocks stratigraphically and/or structurally above an ore deposit or the rock on the upper side of an ore deposit.

**Haulage** - The horizontal transport of ore, coal, supplies, and waste. The vertical transport of ore is called hoisting

**Hydrothermal mineral deposit** - Deposit that formed from hot aqueous solutions.

**Hoist** - The machine used for raising and lowering the cage or other conveyance in a shaft Horst – Uplifted and elongated block of the earth's crust lying between two faults.

**Host rock** - The rock surrounding a mineral deposit.

**Hypabyssal** – Intrusive igneous rock emplaced at shallow depths within the crust.

**Hypogene** - Primary minerals in an ore deposit that formed at great depth and high temperature from ascending hydrothermal solutions of igneous origin.

**Hypothermal-** Hydrothermal ore deposits formed at depths >10km and temperatures between 300 and 500 degrees centigrade.

**Incline** - Any entry to a mine that is not vertical (shaft) or horizontal (adit).

**Industrial mineral - A** non-metallic mineral that can be used in construction and manufacturing industries.

**Kimberlite- A** variety of peridotite; the most common host rock of diamonds

**Level** - An excavation or passageway driven in the coal, establishing a base from which other workings begin

**Lode** - A series of associated but closely separated veins that are rich in ore.

**Longwall mining** - A form of underground coal mining where a long wall of coal is mined in a single slice.

**Manhole** - A safety hole constructed in the side of a gangway, tunnel, or slope in which miners can be safe from passing locomotives and cars.

**Mesothermal** - Hydrothermal ore deposits formed at depths of 5 to 10 km and temperatures of 200 - 300 degrees centigrade.

**Metallic mineral - A** mineral that contains a useful metal.

**Metallogeny** - The study of the origin (genesis) of ore deposits, particularly in relation to tectonic processes or orogeny. **Metallogenic epoch** - A period of geological time favorable for the deposition of mineral deposits characterized by specific deposit types.

**Metallogenic province**- An area of the earth's crust characterized by a particular assemblage of mineral deposit types.

**Metallurgy- The** process of separation and purification of metals in ores.

**Miarolitic -** A term applied to small irregular cavities in igneous rocks, especially granites, into which small crystals of the rock-forming minerals protrude;

**Microthermometry** – Temperature of small fluid inclusions in ore minerals.

**Mineral deposit** - An accumulation of useful minerals in a form that has economic potential.

**Mineral economics** - An academic discipline that focuses on economics and policy issues related to the production and supply of mineral commodities.

**Mineral exploration** - The systematic search for, discovery and evaluation of economic minerals.

**Mineral occurrence**- A locality or wherever a mineral of potential economic interest is found, without reference to quantity or quality

**Mineral showing** - A mineral occurrence that is so small and only of geological interest but not necessarily of economic interest.

**Mineral processing** - The separation of ore minerals from gangue after mining.

**Mineral prospect** - Any promising mineral occurrence.

**Mineral prospecting** - Searching for minerals by visual observation in the field.

**Mineral reserves** - The portion of indicated a mineral resource that is economically feasible to extract.

**Mineral resources - The** useful substances in the earth's crust that can be of economic benefit.

**Mineral resource potential** - The probability that minerals will occur in an area.

**Mineral zoning** – A systematic variation in the composition of a mineral or systematic variation in the distribution of metals or minerals within a deposit, district or region.

**Mine** - An opening or excavation constructed either above the land surface or underground from which ore minerals are extracted.

**Mine development- The** operations involved in preparing a mine for ore extraction, which include tunneling, sinking, cross-cutting, drifting and raising.

**Mine workings - The** entire system of openings in a mine for the purpose of mineral extraction.

**Miner** - One who is engaged in the business or occupation of extracting ore, coal, precious substances, or other natural materials from the earth's crust.

**Mining - The** process of extracting valuable minerals from an ore body.

**Mining engineer** - A person qualified by education, training, and experience in mining engineering.

**Mining geology** is a branch of economic geology that applies geological knowledge and techniques in the extraction of mineral deposit.

**Native elements -T**hose elements that occur in nature in uncombined form.

**Open pit** - A mine that is entirely on the surface. Also referred to as open-cut or open-cast mine

**Open pit mining-** The process of mining an ore body from the surface in progressively deeper steps.

**Opencast mining** – A method of mining rock or minerals by removing them from an open pit situated close to the surface above ground.

**Ore** - A mixture of ore minerals and gangue from which at least one of the metals can be extracted at a profit. **Ore body- A**n extensive or continuous mass of ore and gangue distinctly distinguishable by form and character from the enclosing host rocks.

**Ore deposit** - A mineral deposit that has been tested and found to be of sufficient size, grade, and accessibility to be mined at a profit.

**Ore grade- A** measure of the concentration of a mineral or metal contained in an ore body.

**Ore mineral - A** mineral from which one or more metallic elements can be extracted.

**Ore reserves** - The calculated tonnage and grade of mineralization which can be extracted profitably.

**Ore shoot** - The portion, or length, of a vein or other structure that carries sufficient valuable minerals to be extracted profitably.

**Orogeny -** A period of tectonic activity or mountain building characterized by the folding of a portion of the earth's crust.

**Oxide minerals - M**inerals that contain one or more metals that are combined with oxygen.

**Oxidized ore** - The alteration of metallic minerals by weathering and action of surface waters that converts them wholly or partially into secondary minerals; oxides, carbonates or sulfates.

**Outcrop** - An exposure of rock or mineral deposit that can be seen on surface that is, not covered by soil or water.

**Overburden** - Layers of soil and rock covering an orebody or a coal seam. Overburden is removed prior to surface mining and replaced after the coal is taken from the seam.

**Pillar** - A column or body of coal left unmined to support the roof.

**Placer deposits - C**oncentrations of ores in gravels usually transported by water.

**Porphyry copper** - A deposit of disseminated copper minerals in or around a large body of intrusive rock.

**Protore** - Initial non-economic concentration of metallic minerals that may be economic if enriched by geological processes e.g. weathering.

**Quarry** - Large deep pit from which stone or other materials are extracted

**Raise** - A secondary or tertiary inclined opening, vertical or near-vertical opening driven upward form a level to connect with the level above, or to explore the ground for a limited distance above one level.

**Ramp** - A secondary or tertiary inclined opening, driven to connect levels, usually driven in a downward direction, and used for haulage.

**Reclamation** – The restoration of land and environmental values to a surface mine site after the ore is extracted.

**Reef** - Term commonly used in Australia and South Africa to describe a mineralized zone that occurs in a rock layer'

**Room-and-pillar mining** - A method of mining flat-lying ore deposits in which the mined-out area, or rooms, are separated by pillars of approximately the same size

**Run-of-mine** - Raw material as it exists in the mine; average grade or quality.

**Salting** - Act of introducing metals or minerals intentionally or by accident into a deposit or samples, resulting in false assays which may defraud the public or investors.

**Shaft** - A vertical passageway to an underground mine for moving miners, equipment, supplies and broken ore and waste rock.

**Seam** - A stratum or bed of coal.

**Stratabound deposits** - Mineral deposits that lie within sedimentary strata and of sedimentary origin

**Stratiform deposits-** Mineral deposits laid out like stratified rock layers.

**Sedimentary exhalative deposits (SEDEX)** - Sediment hosted deposits formed by the discharge of hydrothermal fluids into the sea floor.

**Shaft** - A vertical excavation connecting surface and mine workings.

**Shaft mine** – An underground mine in which the main entry or access is by means of a vertical shaft.

**Shaft mining** - Using shafts driven vertically from the top down into the ground to reach and mine the orebody

**Shrinkage stoping** - A stoping method which uses part of the broken ore as a working platform and as support for the walls of the stope.

**Shortwall** – An underground mining method in which small areas are worked (15 to 150 feet) by a continuous miner in conjunction with the use of hydraulic roof supports.

**Sludge** - Rock cuttings from a diamond drill hole, sometimes used for assaying.

**Stope** - An excavation in a mine from which ore is, or has been, extracted.

**Slope mining** - type of underground mining that uses diagonally sloping access shafts to reach the orebodies particularly where drilling vertical shafts may be difficult.

**Solution mining** – Type of mining where minerals are dissolved into solution and extracted

**Stockwork** - A large number of closely spaced, planar and irregular veinlets filled with economic and/or gangue minerals.

**Stoping -** The act of extracting ore from a face or breaking down a stope and excavating it with a pick.

**Stripping ratio** – the amount of waste rock mined relative to the of ore rock mined.

**Sump** - The bottom of a shaft, or any other place in a mine, that is used as a collecting point for drainage water.

**Strip** - To remove the overburden or waste rock overlying an orebody in preparation for mining by open pit methods.

**Strip mine** - An open-pit mine, usually a coal mine, operated by removing overburden, excavating the coal seam, and then returning the overburden

**Stripping ratio** - The ratio of tons removed as waste relative to the number of tons of ore removed from an open-pit mine.

**Sublevel** - A level or working horizon in a mine between main working levels.

**Sulfide minerals** - Minerals that contain one or more metals that are combined with sulfur. Example is chalcopyrite, a copper sulfide.

**Supergene** - Mineralization and enrichment caused by secondary processes or leaching by descending surface waters.

**Surface mine** – A mine in which the coal lies near the surface and can be extracted by removing the covering layers of rock and soil.

**Syngenetic** – Mineral deposit formed at the same time as the host rock.

**Tailings** - Material rejected from a mill after most of the recoverable valuable minerals have been extracted.

**Tenor** is the percentage of the ore mineral that is actual metal to be extracted. Often used as synonymous with ore grade.

**Tonnage** - The quantity of metal or useful mineral in a deposit.

**Trench** - A long, narrow excavation dug through overburden or blasted out of rock, to expose a vein or ore body during mineral exploration.

**Tunnel** - A horizontal, or near-horizontal, underground passage, entry, or haulageway, that is open to the surface at both ends. A tunnel (as opposed to an adit) must pass completely through a hill or mount

**Underground mine** – Also known as a "deep" mine. Usually located several hundred feet below the earth's surface, an underground mine's coal is removed mechanically and transferred by shuttle car or conveyor to the surface.

**Underground station** - An enlargement of an entry, drift, or level at a shaft at which cages stop to receive and discharge cars, personnel, and material.

**Vein** - A fissure, fault or crack in a rock filled by minerals that have travelled upwards from some deep source.

**Vug** - A small cavity in a rock, frequently lined with well-formed crystals. Amethyst commonly forms in these cavities.

**Waste** - That rock or mineral which must be removed from a mine which has no value.

**Wall rock** - Rock units on either side of an orebody.

**Winze** - Secondary or tertiary vertical or near-vertical opening sunk from a point inside a mine for the purpose of connecting with a lower level or of exploring the ground for a limited depth below a level.

**Working face** is any place in a mine where material is extracted during a mining cycle.

**Workings** - The entire system of mine excavations.

# About the Author

Moses A. Olade is a retired Professor of Economic Geology who taught for more than two decades at several higher institutions including the University of Ibadan, Nigeria, the University of Phoenix, Arizona, Federal University of Technology, Akure and the Achievers University. He received his B.Sc. (Hons.) degree in Geology from the University of Ibadan, M.Sc. in Economic Geology from the University of Alberta, Edmonton, and a Ph.D. in Applied Geochemistry from the University of British Columbia where he worked on porphyry copper deposits. He served his postdoctoral research fellowship at the Geological Survey of Canada, Ottawa. His professional experience included mineral prospecting and exploration for uranium and base metals in northwest Canada and California and employed for over a decade in environmental program management in Arizona, USA. He has visited several underground mines in Montana and Idaho, and open-pit mines in Califoenia, Arizona and Nigeria. His research papers on mineral deposits and applied geochemistry are widely published, and he is the author of the recent book Mineral Deposits and Exploration Potential of Nigeria. He was a member of several professional organizations and a Mining and Geosciences Society (FNMGS).

Made in the USA
Monee, IL
05 November 2026

67805992R00247